Algorithms for Intelligent Systems

This book series publishes research on the analysis and development of algorithms for intelligent systems with their applications to various real world problems. It covers research related to autonomous agents, multi-agent systems, behavioral modeling, reinforcement learning, game theory, mechanism design, machine learning, meta-heuristic search, optimization, planning and scheduling, artificial neural networks, evolutionary computation, swarm intelligence and other algorithms for intelligent systems.

The book series includes recent advancements, modification and applications of the artificial neural networks, evolutionary computation, swarm intelligence, artificial immune systems, fuzzy system, autonomous and multi agent systems, machine learning and other intelligent systems related areas. The material will be beneficial for the graduate students, post-graduate students as well as the researchers who want a broader view of advances in algorithms for intelligent systems. The contents will also be useful to the researchers from other fields who have no knowledge of the power of intelligent systems, e.g. the researchers in the field of bioinformatics, biochemists, mechanical and chemical engineers, economists, musicians and medical practitioners.

The series publishes monographs, edited volumes, advanced textbooks and selected proceedings.

Indexed by zbMATH.

All books published in the series are submitted for consideration in Web of Science.

João Manuel R. S. Tavares · Souvik Pal ·
Vassilis C. Gerogiannis · Bui Thanh Hung
Editors

Proceedings of Second International Conference on Intelligent System

ICIS 2023

Springer

Editors
João Manuel R. S. Tavares
Faculdade de Engenharia da
Universidade do Porto (FEUP)
Porto, Portugal

Vassilis C. Gerogiannis
Department of Digital Systems
University of Thessaly
Larissa, Greece

Souvik Pal
Department of Computer Science
and Engineering
Sister Nivedita University (Techno India
Group)
Kolkata, India

Bui Thanh Hung
Data Science Laboratory, Faculty
of Information Technology
Industrial University of Ho Chi Minh City
Ho Chi Minh City, Vietnam

ISSN 2524-7565 ISSN 2524-7573 (electronic)
Algorithms for Intelligent Systems
ISBN 978-981-99-9012-2 ISBN 978-981-99-8976-8 (eBook)
https://doi.org/10.1007/978-981-99-8976-8

This Springer imprint is published by the registered company Springer Nature Singapore Pte Ltd.
The registered company address is: 152 Beach Road, #21-01/04 Gateway East, Singapore 189721, Singapore

Organizing Committee and Key Members

Conference Committee Members

General Chairs

Dr. João Manuel R. S. Tavares, Faculdade de Engenharia da Universidade do Porto, Portugal
Dr. Huynh Trung Hieu, Professor, Dean of FIT, IUH University, Vietnam
Dr. Prasant Kumar Pattnaik, School of Computer Engineering, KIIT Deemed to be University, India

Honorary Chairs

Dr. Phan Hong Hai, President of IUH University, Vietnam
Pr. Dam Sao Mai, Vice President of IUH University, Vietnam
Pr. Le Van Tan, Vice President of IUH University, Vietnam
Dr. Nguyen Xuan Hong, Vice President of IUH University, Vietnam
Pr. Trinh Ngoc Nam, Head of Office of Science Management and International Affairs, IUH University, Vietnam
Dr. Pham Tran Bich Thuan, Deputy Head of Office of Science Management and International Affairs, IUH University, Vietnam
Dr. Mai Thang Long, Dean of FET, IUH University, Vietnam

Conference Chairs

Dr. Bui Thanh Hung, IUH University, Vietnam
Dr. Souvik Pal, Department of Computer Science and Engineering, Sister Nivedita University (Techno India Group), Kolkata, India
Dr. Raghvendra Kumar, Department of Computer Science and Engineering, GIET University, India

Program Chairs

Dr. Brajesh Kumar Kaushik, Department of Electronics and Communication Engineering, Indian Institute of Technology Roorkee, India
Dr. Le Hoang Son, VNU Information Technology Institute, Vietnam National University, Vietnam
Dr. Sanjeevikumar Padmanaban, Department of Energy Technology, Aalborg University, Esbjerg, Denmark
Ahmed J. Obaid, Faculty of Computer Science and Mathematics, University of Kufa, Iraq

Publicity Chairs

Ha Huy Cuong Nguyen, The University of Danang, Da Nang, Vietnam
Subrata Chowdhury, Sreenivasa Institute of Technology and Management Studies, India

Technical Program Committee Members

Dr. Moe Moe Myint, Department of Computer Engineering and Information Technology, Myanmar
Dr. Noor Zaman Jhanjhi, Taylor's University, Malaysia
Dr. Manju Khari, Ambedkar Institute of Advanced Communication Technology and Research, India
Dr. Mohammad Israr, Sur University College, Oman
Dr. Bharat Bhushan, HMR Institute of Technology and Management, India
Dr. Victor Hugo C. de Albuquerque, University of Fortaleza, Brazil
Dr. D. Jude Hemanth, Karunya University, India
Dr. Kusum Yadav, University of Hail Kingdom of Saudi Arabia, Kingdom of Saudi Arabia

Dr. Dac Nhuong Le, Faculty of IT, Haiphong University, Haiphong, Vietnam
Dr. Anchit Bijalwan, Arba Minch University, Ethiopia
Dr. Allan R. Lao, Lorma Colleges, Philippines
Dr. Ishaani Priyadarshini, University of Delaware, USA
Dr. Kenan Kiliç, Niğde Ömer Halisdemir University, Engineering Faculty, Turkey
Dr. Abhishek Kumar, Banaras Hindu University, India
Dr. David Taniar, Monash University, Australia
Dr. Saw Mon Yee Aung, Lecturer, Technological University (Toungoo), Myanmar
Dr. Magesh Chandramouli, Purdue University, USA
Dr. Surjeet Dalal, SRM University, Delhi-NCR, Sonipat, Haryana, India
Dr. Ganeshsree Selvachandran, UCSI University, Kuala Lumpur, Malaysia
Dr. Binh Thai Pham, University of Transport Technology, Vietnam
Dr. Srinivas Sethi, IGIT Sarang, India
Dr. Alok Ranjan Prusty, MSDE, India
Dr. Hoang Viet Long, Ton Duc Thang University, Ho Chi Minh City, Vietnam
Dr. Manisha Bharti, National Institute of Technology (NIT), Delhi, India
Dr. Sanjaya Kumar Panda, National Institute of Technology, Warangal, India

Organizing Committee Members

Dang Thi Phuc, IUH University, Vietnam
Le Nhat Duy, IUH University, Vietnam
Bui Thanh Hung, IUH University, Vietnam
Huynh Tuong Nguyen, IUH University, Vietnam
Pham Thi Thiet, IUH University, Vietnam
Nguyen Thi Hanh, IUH University, Vietnam
Ta Duy Cong Chien, IUH University, Vietnam
Nguyen Chi Kien, IUH University, Vietnam
Le Trong Ngoc, IUH University, Vietnam
Dang Quang Vinh, IUH University, Vietnam
Nguyen Huu Tinh, IUH University, Vietnam
Ho Dac Quan, IUH University, Vietnam
Ton Long Phuoc, IUH University, Vietnam
Luu Giang Nam, IUH University, Vietnam

Invited Speaker(s)

Prof. Ts. Dr. Rosziati Ibrahim, Faculty of Computer Science and Information Technology, Universiti Tun Hussein Onn Malaysia (UTHM), Malaysia
Prof. Masoud Mohammadian, Information Systems/Computer Science, University of Canberra, Canberra, Australian Capital Territory, Australia
Prof. Mohd Helmy Bin Abd Wahab, Universiti Tun Hussein Onn Malaysia (UTHM), Malaysia

Preface and Acknowledgements

The primary objective of this proceedings volume is to facilitate communication between leading academic scientists, researchers, and research scholars working in the fields of intelligent ecosystems, data sciences, and mathematics. The International Conference on Information Systems 2023 (ICIS 2023) is an event planned to unite researchers, educators, and students from all branches of engineering and technology. The International Conference on Information Systems (ICIS) 2023 is your chance to network with the world's top scholars, hear about cutting-edge research being conducted all across the globe, and observe emerging patterns in the intersection of science and technology. Attendees and presenters at the conference will have the opportunity to collaborate with academic institutions from all around the world to advance scientific knowledge and advance technological progress. The purpose of this conference is to facilitate the application of both basic and applied research in institutional and industry settings.

ICIS 2023 proceedings are comprised of the 50 best papers presented at the conferences after being vetted by both the conference committee and experts from across the world. The speakers have delivered their PowerPoint presentations either online or in person. To better educate the future generation of leaders through education, professionals in the industry have convened from all over the world, including India, Malaysia, Vietnam, Iraq, Germany, Australia, and UAE. This gathering brought together experts from a wide range of research traditions and countries. Conferences in the academic world put a premium on the expertise of their writers and speakers. We are grateful that the authors chose to present their findings at this conference in light of the ongoing global pandemic.

We express profound gratitude to the divine entity for its unwavering presence in our lives, providing support during both favorable and unfavorable circumstances, as well as bestowing upon us the means to empower ourselves. From the initiation of the Call for Papers until the completion of the chapter finalization process, the collaborative efforts of all team members were evident, indicating the presence of a cohesive and robust team. The conference editors and organizers express their sincere gratitude to all members of Springer, with special recognition given to

Mr. Aninda Bose, for his invaluable recommendations and for providing the opportunity to successfully complete the conference proceedings. We would like to express our gratitude for the assistance provided by Prof. William Achauer and Prof. Anil Chandy. We express our gratitude to Mr. Ramamoorthy Rajangam, employed as a project coordinator at Springer, for his valuable assistance. We express our gratitude for the reviewers who provided their support and remained committed to the objective of obtaining high-quality chapters, representing diverse geographical locations worldwide.

Finally, we would like to extend our best wishes to all participants for their upcoming presentations and engagement in social networking activities. The success of this conference is contingent upon your unwavering support. It is anticipated that individuals who attended the conference had satisfaction from both the technical program and the virtual presence of speakers and delegates. We anticipate that you will experience a fruitful and enjoyable period during your attendance at ICIS 2023.

Porto, Portugal — João Manuel R. S. Tavares
Kolkata, India — Souvik Pal
Larissa, Greece — Vassilis C. Gerogiannis
Ho Chi Minh City, Vietnam — Bui Thanh Hung

Contents

About the Editors

Professor João Manuel R. S. Tavares graduated in Mechanical Engineering at the Universidade do Porto, Portugal, in 1992. He also earned his M.Sc. degree and Ph.D. degree in Electrical and Computer Engineering from the Universidade do Porto in 1995 and 2001, and attained his Habilitation in Mechanical Engineering in 2015 from the same University. He is a senior researcher at the Instituto de Ciência e Inovação em Engenharia Mecânica e Engenharia Industrial (INEGI) and a Full Professor at the Department of Mechanical Engineering (DEMec) of the Faculdade de Engenharia da Universidade do Porto (FEUP). He has been the Head of DEMec since June 2023. João Tavares is co-editor of more than 90 books, co-author of more than 55 book chapters, 650 articles in international and national journals and conferences, and 3 international and 3 national patents. He has been a committee member of several international and national journals and conferences, is co-founder and co-editor of the book series "Lecture Notes in Computational Vision and Biomechanics" published by Springer, founder and Editor-in-Chief of the journal "Computer Methods in Biomechanics and Biomedical Engineering: Imaging & Visualization" published by Taylor & Francis, Editor-in-Chief of the journal "Computer Methods in Biomechanics and Biomedical Engineering" published by Taylor & Francis, and co-founder and co-chair of the international conference series: International Symposium on Computational Modeling of Objects Presented in Images, ECCOMAS Thematic Conference on Computational Vision and Medical Image Processing, International Conference on Computational and Experimental Biomedical Sciences and International Conference on Biodental Engineering. Additionally, he has been (co-) supervisor of several MSc and PhD theses and supervisor of several post-doc projects and has participated in many scientific projects both as a researcher and as scientific coordinator. His main research areas include computational vision, medical imaging, biomechanics, biomedical engineering, and new product development. (More information can be found at www.fe.up.pt/~tavares)

Dr. Souvik Pal is an Associate Professor in the Department of Computer Science and Engineering at Sister Nivedita University (Techno India Group), Kolkata, India. Prior to that, he was associated with Global Institute of Management and Technology;

Brainware University, Kolkata; JIS College of Engineering, Nadia; Elitte College of Engineering, Kolkata; and Nalanda Institute of Technology, Bhubaneswar, India. Dr. Pal received his MTech, and PhD degrees in the field of Computer Science and Engineering from KIIT University, Bhubaneswar, India. He has more than a decade of academic experience. He is author or co-editor of more than 20 books from reputed publishers, including Elsevier, Springer, CRC Press, and Wiley, and he holds three patents. He is serving as a Series Editor for "Advances in Learning Analytics for Intelligent Cloud-IoT Systems", published by Scrivener-Wiley Publishing (Scopus-indexed); "Internet of Things: Data-Centric Intelligent Computing, Informatics, and Communication", published CRC Press, Taylor & Francis Group, USA; "Conference Proceedings Series on Intelligent Systems, Data Engineering, and Optimization", published CRC Press, Taylor & Francis Group, USA; Dr. Pal has published a number of research papers in Scopus/SCI/SCIE Journals and conferences. He is the organizing chair of RICE 2019, Vietnam; RICE 2020 Vietnam; ICICIT 2019, Tunisia. He has been invited as a keynote speaker at ICICCT 2019, Turkey, and ICTIDS 2019, 2021 Malaysia. He has also served as Proceedings Editor of ICICCT 2019, 2020; ICMMCS 2020, 2021; ICWSNUCA 2021, India. His professional activities include roles as Associate Editor, Guest Editor, and Editorial Board member for more than 100+ international journals and conferences of high repute and impact. His research area includes cloud computing, big data, internet of things, wireless sensor network, and data analytics. He is a member of many professional organizations, including MIEEE; MCSI; MCSTA/ACM, USA; MIAENG, Hong Kong; MIRED, USA; MACEEE, New Delhi; MIACSIT, Singapore; and MAASCIT, USA.

Vassilis C. Gerogiannis is a full-time Professor and Head in the Department of Digital Systems at the University of Thessaly in Greece. His academic subject is "Analysis & Design of Systems & Projects with Emphasis on Decision Making". He is also adjunct Professor at the Hellenic Open University (Greece). He holds a Diploma in Computer Engineering (1992) and a PhD in Software Engineering (2001) from the University of Patras in Greece. His research interests include fuzzy decision making, software engineering, requirements engineering, systems analysis and design, software project management and machine learning. From 1992 until present, he has been participated in several R&D projects. He has been author/co-author of over 150 papers which have been published in international journals/conference proceedings and cited in a plethora of citations. He has been co-author/editor of three scientific books, member of the editorial board, guest editor and reviewer in international journals. He has served as conference chair/program chair and invited speaker in several international conferences. He has received the "best paper award" in five international conferences. He has served as member of the management board of the Hellenic National Academic Recognition Information Centre (NARIC) and member of the Council for Research & Innovation in Thessaly region of Greece. At present, he is a member of the management committee of the Entrepreneurship & Innovation Research Institute of the Research Center IASON at the University of Thessaly, member of the management committee of the Technical Chamber of Engineers in Central and Western Greece, member of the central assembly of the Technical

Chamber of Engineers in Greece, member of the scientific committee of Electronics Engineers in the Technical Chamber of Greece and also member of the ACM.

Dr. Bui Thanh Hung received his M.S. degree and Ph.D. degree from Japan Advanced Institute of Science and Technology (JAIST) in 2010 and in 2013. He was the Director of Data Analysis Artificial Intelligence Laboratory–DAAI Lab, Thu Dau Mot university (2018–2021); Director of Artificial Intelligence Laboratory–AI Lab, Faculty of Information Technology, Ton Duc Thang University (2021–2022). Now he works for the Data Science Laboratory, Data Science Department, Faculty of Information Technology, Industrial University of Ho Chi Minh city. He has completed 2 projects, published 3 books, 28 journals, 26 book chapters, 45 International conference papers and 17 domestic conference papers. He had four best paper awards of FICTA 2018, RICE 2020, ICAIAA 2021 and CIEMA 2022. He was selected as Excellent Scientific Researcher of Thu Dau Mot University in 2019, 2020, 2021; and got CIEMA Research Excellence Awards 2022. He is a reviewer of many reputed journals and conferences. He is also a keynote speaker of many domestic and international conferences. His main research interests are Natural Language Processing, Machine Learning, Machine Translation, Text Processing, Data Analytics, Computer Vision and Artificial Intelligence.

Chapter 1
Deep Convolutional Neural Networks for Brain Tumor Image Segmentation

Masood Ahamed Shaik and J. Jabez

1 Introduction

Brain tumors get even more prevalent as people get older [1]. The concentration of this paper would be gliomas which are prevalent in many cases of these tumors. The World Health Organization divides these gliomas into categories called grades based on their location, cell type, and tumor magnitude. Low-grade gliomas are classified as Classes I and II, whereas high-grade gliomas are categorized as Classes III and IV [2]. Various imaging methods are used to allowing doctors to reliably eliminate the tumor inside the maximum effective range. These methods involve many popular scans like the Positron-Emission Computed Tomography scans and Computed Tomography scans.

Magnetic Resonance Imaging, on the other hand, has a wealth of anatomical details about the mass. It also has substantial contrast of the soft mass. MRI is frequently employed to detect gliomas preoperatively, intraoperatively, and postoperatively in the medical evaluation of gliomas [2].

Glioma is a tumor with a necrotic nucleus, a tumor activity margin, and edema tissue. As seen in Fig. 1, various MRI scans would assist in the imaging of diverse tumor masses [3]. The MRI scans mentioned in Fig. 1 are among the four modal sequences currently used in MRI imaging of gliomas. Glioma tissues are represented differently by various sequences [4]. The FLAIR sequence in general is helpful to look at edema masses, while the T1-weighted image is helpful to look at the active components of the tumor nucleus [5].

M. A. Shaik (✉) · J. Jabez
Sathyabama Institute of Science and Technology, Chennai, India
e-mail: masooda100@gmail.com

J. M. R. S. Tavares et al. (eds.), *Proceedings of Second International Conference on Intelligent System*, Algorithms for Intelligent Systems,
https://doi.org/10.1007/978-981-99-8976-8_1

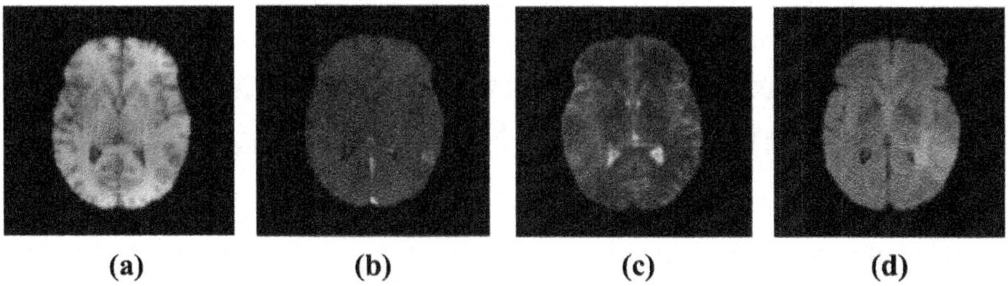

Fig. 1 **a** T1-weighted MRI, **b** postcontrast T1-weighted MRI, **c** T2-weighted MRI, and **d** FLAIR MRI

The illumination of the particular mass in question is key in the process of manually segmenting a tumor. This is readily influenced by the characteristics of the person performing the segmentation and the nature of the image. Hence, the method is vulnerable to redundant and imprecise segmentations [6]. As a result, in medical care, a highly automated glioma segmentation system [7] with high precision is needed. Nevertheless, the following are the issues that have been identified during the analysis of the methods employed for automated glioma segmentation:

(1) The difference between pixel strengths of the region with a lesion compared to the adjacent natural tissue is also used to distinguish glioma in images. A grayscale offset field would smoothen the strength differential between neighboring tumor tissues [8]. This emanates in fuzzy edges for the tumorous mass.
(2) A glioma can vary in scale or form or location. This makes the creation of an algorithm for segmentation very laborious. A machine learning algorithm [9] can effectively be the answer to medical image classification, regression, and aggregation. It can be used in primarily training the model parameters using different features of medical images. Later, it can employ the learned model and simulate derived features. Simultaneously, machine learning's deep learning technologies can robotically modify model parameters employing forward propagation along with back-regulation algorithms. This is done after extracting high-dimensional features from data [10]. This allows the output of the model for similar functions to be drastically improve.

Methods for segmenting brain tumors can be loosely classified into three categories: standard image algorithms [11–16], methods based on machine learning [17–20], and methods which have a basis of deep learning [21–24]. Deep learning seems to have been the tool of choice for complicated tasks in past years due to its high precision. As per the suggestion of [20] in the domain of image processing, the convolutional neural networks have made considerable advances. As a result, the convolutional neural network-based segmentation approach is commonly used in segmentations of many organs like lung, glioma, and liver [21]. A lot of researchers started to use convolutional neural networks for glioma segmentation in deep learning. Segmentation of cancer of the brain which is rooted using dual-path convolutional neural network is proposed in Ref. [16]. A couple of convolutional neural networks was qualified for the segmentation of low- and high-grade gliomas in Ref. [17]. For glioma segmentation, Ref. [18] has a suggestion. It suggested a two-channel three-dimensional convolutional neural network.

The focus of this paper is on a glioma segmentation approach focused on deep learning. The goal is to innately and correctly perform glioma segmentation using a deep learning algorithm from an MRI. The suggestion of this article is DCNN-fusion-SVM deep classifier for glioma segmentation.

Following are the key research topics:

(1) The creation of a new depth classifier is recommended. A deep CNN along with an integrated SVM algorithm makes up the classifier. The test images and the projected labels from CNN were fed to the integrated SVM classifier. To improve the accuracy of the data, we intensified the classification procedure and repeated the above-mentioned couple stages to provide the skeleton for the next CNN-SVM.
(2) Standard approach involves training an appropriate classifier. This is done using the training set. This is followed by a verification test on the set. The approach proposed in this analysis differs significantly from conventional methods. The method that is currently proposed consists of three steps. The first step is preprocessing. This is followed by feature extraction and lastly followed by CNN and SVM training. The subsequent step is validating the results of segmentation and producing them. Order of the CNN-SVM cascade classifier is iteratively deepen in the final step.
(3) Evaluate the proposed model using publicly available and self-created datasets [25]. The dominance of the suggested model is seen across different assessments. This is as compared to the results of just CNN and SVM segmentation.

2 Related Survey

2.1 *Deep Learning-Based Brain Tumor Segmentation Algorithm Method [26]*

Machine learning-based glioma segmentation approaches have tasks that involve selecting image features. The effect is restriction in the generalization potential as this category of method is dependent on artificial features.

Deep learning-based glioma segmentation system [8] will extract image features and segment the glioma region automatically using a neural network model. As a result, the above method's flaws of strict prior restrictions and manual control are eliminated [27].

Flow of a deep learning based glioma segmentation algorithm is shown in Fig. 2. The procedure is described in the following: Initially, collect the patient's MRI which will act as the input to the algorithm; then, partition this to the training set.

Next, verify and test it. Around this stage, the separated data must be preprocessed due to factors like noise and irregular intensities of the initial MRI.

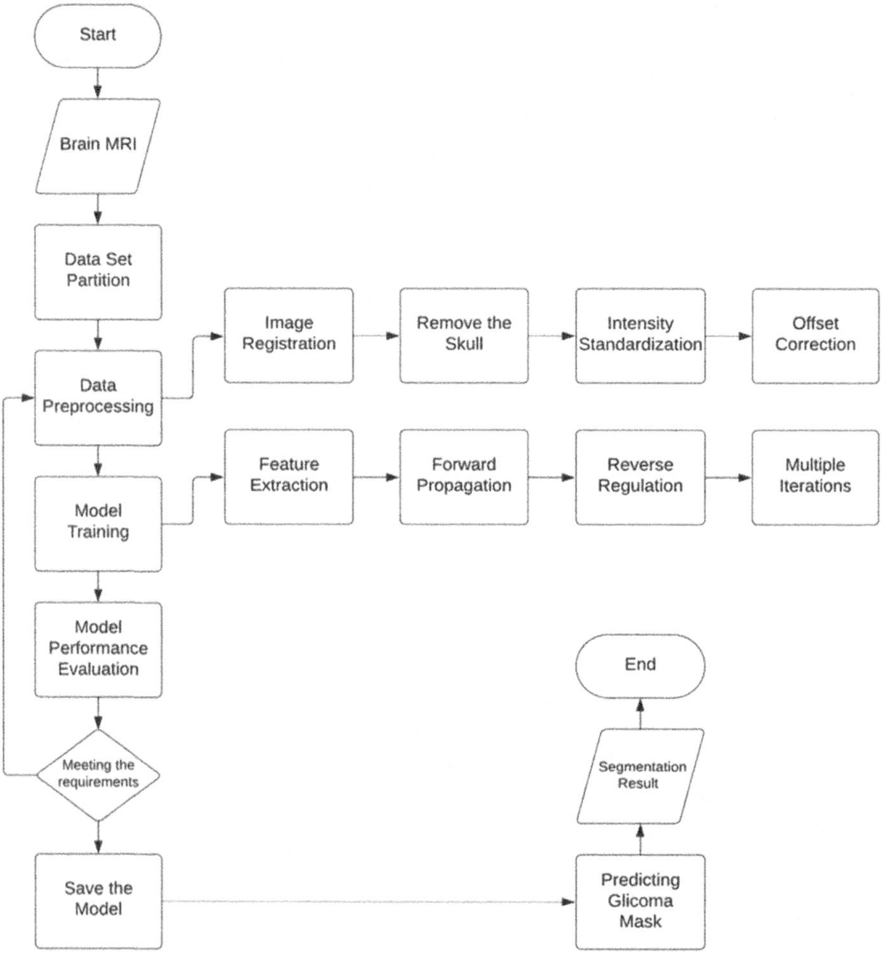

Fig. 2 Flowchart for an algorithm for glioma segmentation which is based on deep learning

Image registration, skull replacement, intensity standardization, and offset correction are all popular glioma image preprocessing techniques. After that, train the deep learning algorithm with the preprocessed input results. Synchronously, the glioma multi-region mask acts as the label which quantifies the loss value. This allows the parameters of the model to be reversed modified in several iterations for optimum model efficiency. When each iteration ends, various assessment metrics are employed to assess the results of the model. Models which satisfy the criteria of the indicators end up chosen. In the end, to achieve the final glioma segmentation results, the test dataset is segmented using this highly evaluated model.

2.2 A Method for Generating Deep Brain Tumor Features

CNNs are very popular deep learning practical models. Human brain nerves are processed and this is the source of their groundbreaking insights. The original CNN is the perceptron model, which was proposed in 1980. It is a well-known model. But, it has significant flaws which are widely known in the domain of machine learning. One of the major issues is XOR problems, which it is unable to solve satisfactorily. The LeNet model, consists of several convolutional layers, each of this is a fully connected model. Each uses the backpropagation algorithm for training. Displacement invariance is an artificial neural network proposed in Ref. [21] and it has investigated the parallel structure of the convolutional neural network. Nevertheless, experimental data and hardware are factors that hinder these models. As a result, it is unsuitable for more difficult tasks like object detection and scene classification. AlexNet model which was suggested by Krizhevsky et al. to address certain issues in the training phase of CNNs like overfitting. This model suggests local convolution and Relu technology to address the problem. The overfitting problem is quite well resolved.

A convolution layer, a down sampling layer, and a fully connected layer make up the hidden layer, which includes several neurons in each plane. The convolution layer appears in alternate with the downsampling layer. These both along with the fully connected layer can have several layers. Figure 3 shows the LeNet CNN network structure.

The learned convolution kernel convolves the feature maps' output from the previous layer, and the resulting partial derivatives act as the input to the activation function. These in conjunction make the output feature maps in the convolution layer. The downsampling layer is used to pick representative features for feature selection. A neural network layer responsible for mapping two-dimensional distributed features onto feature vectors is the fully connected layer. This is done in order to improve the classification. Output layer is usually just a simple classification layer, which is typically classified using logistic regression. Regularly used activation functions are shown below (Relu and Softplus functions):

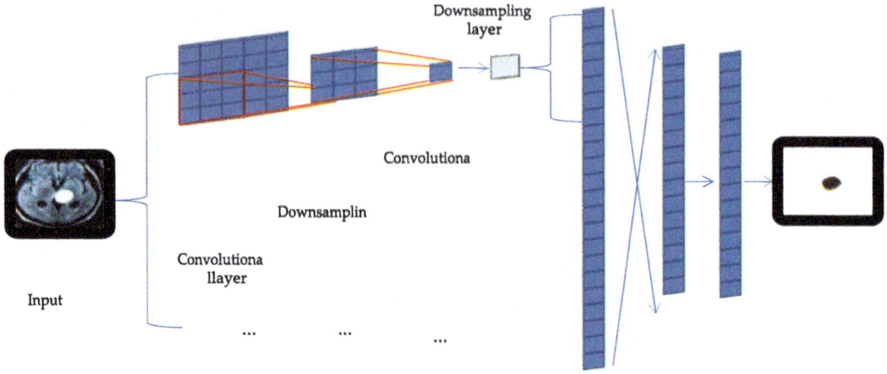

Fig. 3 LeNet CNN structure

(1) Relu function: $f(x) = \max(0, x)$.
(2) Softplus function: $f(x) = \log(1 + e^x)$.

Compared to that of a neurocognitive machine, the framework of a convolutional neural network model is less complicated and effortless to extend. Downsampling layer alternates with the convolutional layer in the neurocognitive system to shape the function of feature extraction and abstraction, whereas the convolutional layer and the downsampling layer alternate in the convolutional neural network, and their roles are very comparable. Several nonlinear functions are replaced by the excitation function of the neurocognitive machine. Also, the pooling procedure is often simplified by the convolution operation for the purposes of feature extraction. Figure 4 depicts the flow of the CNN algorithm.

2.3 Dataset of the Brain Tumor

In 2012, a challenge called the BraTS Challenge took place. It presented a dataset of MRIs of the brain that included low-grade as well as high-grade gliomas. MRIs from many people have affected with gliomas.

A combination of twenty segmentation algorithms later labeled manually by several specialists is called a ground truth [28]. The dataset mentioned above also included multi-region glioma segmentation ground truth respectively for all the cases.

Competition includes a public glioma dataset also.

After BraTS17, the glioma dataset given has been noteworthily divergent to the ones which were given prior to 2016. The difference is being:

Datasets provided by BraTS14 and BraTS16 include images pre and postoperation of the glioma. Ambiguous glioma segmentation criteria in these dataset, which lack the parameters to be valid segmentation criteria, were the result of this.

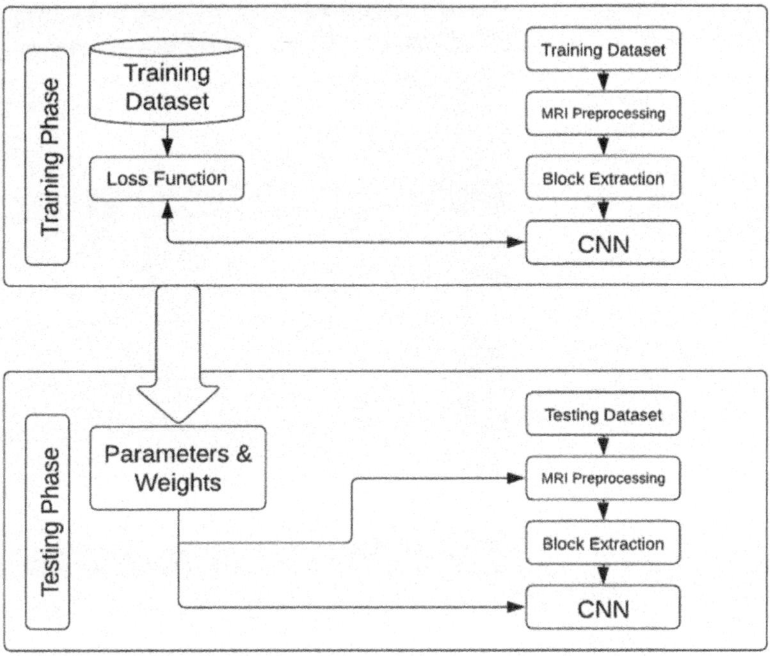

Fig. 4 Flowchart of a convolutional neural network

As a result, after 2017, BraTS14 and BraTS16 types were discontinued in the games. The TCIA glioma dataset which consists of 262 high-grade and 199 low-grade glioma patient images has been added to the BraTS18 dataset, which is built on the dataset of BraTS17. This TCIA dataset, which includes the MRIs of 543 glioma patients along with their respective ground truths, is the most used glioma segmentation dataset at the moment. Table 1 shows the specifics of the datasets in the BraTS rivalry datasets over time.

Table 1 BraTS dataset from 2020 to 2014

Year	Sample count of various sets			
	Test	Validation	Training	Aggregate
2020	191	74	285	543
2019	146	69	285	412
2018	191	67	205	465
2017	110	46	205	384
2016	156	N/A	274	456
2015	125	N/A	274	387
2014	25	10	40	65

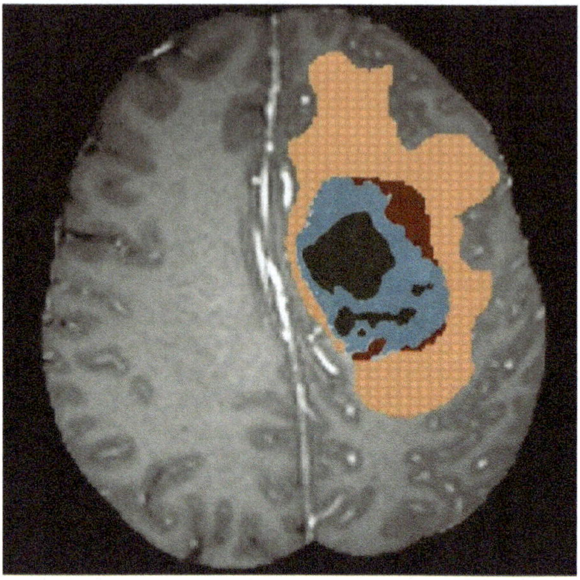

Fig. 5 Division of the tumor area of the glioma

Gliomas are classified into four tumor areas, as seen in Fig. 5, namely necrotic core (NCR), enhanced tumor core (ET), nonenhanced tumor core (NET), and edema around the tumor (ED). ED, NET, and NCR can be classified as legitimate glioma tumor tissues. Primary aim of the tumor core enhancement is to make it easier to examine the tumor core.

2.4 *Evaluating the Methods*

Table 2 shows the most common approaches for analyzing each model's output in the domain of image segmentation.

There are other assessment metrics, such as Hausdorff Li and positive predictive value, in addition to the ones mentioned above. DSC and sensitivity are the most extensively used.

3 Methodology of DCNN-F-SVM Model

Research here uses a convolutional neural network fusion SVM-based brain tumor segmentation model. The model flowchart is shown in Fig. 6.

Suggested model can be partitioned into the following:

Table 2 Expressions for indices used

Indices	Description
Specificity	It is the dividend of true negative and the sum of true negative and false positive
Sensitivity	It is the dividend of true positive and the sum of true positive and false negative
Dice similarity coefficient (DSC)	It is the dividend of twice the true positive and the aggregate sum of twice true positive with false positive and false negative
False negative (FN)	Indicates the wrong segregation of pixels of tumor area as the background
True negative (TN)	Correct segregation of background area pixels
False positive (FP)	Indicates the wrong segregation of pixels of background area as the tumor region
True positive (TP)	Correct segregation of glioma area pixels with the manual markings of a doctor

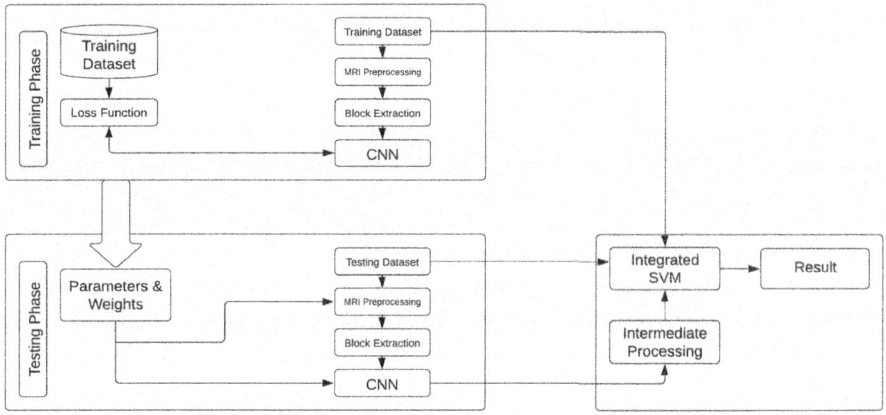

Fig. 6 Flowchart of the introduced model

(1) Preprocessing the data. This is followed by feature extraction. The last part involves training CNN and SVM.
(2) Testing the results followed by the production of final segmentation data. It can be broken down into three stages [29]:
    (a) Mapping onto tumor label domain from the gray image domain to is procured by training CNN and integrated SVM.
    (b) The labeled CNN output along with test image is fed to the integrated SVM classifier.

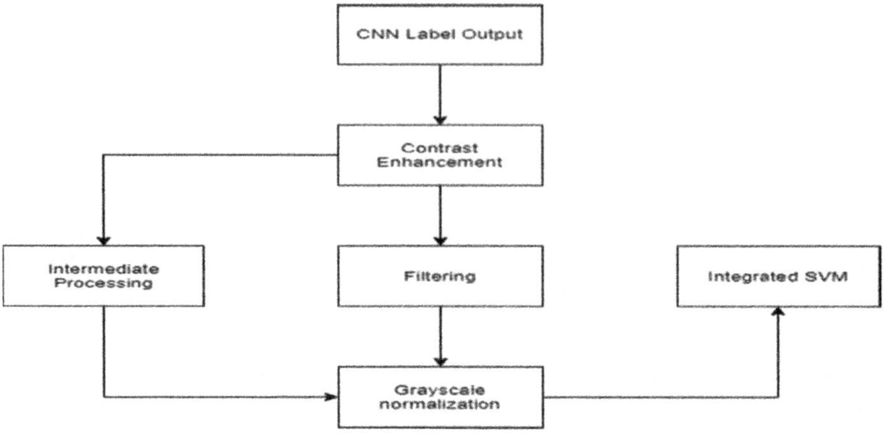

Fig. 7 Figure showing the intermediary processing step in the suggested model

(c) A reiterative step to link the CNN and the integrated SVM classifier increasing the count of layers. An intermediary stage of processing is applied to choose the best optimal feature. This can be seen in Fig. 7.

Each pixel is characterized by grayscale, mean, and median. These are the same properties that are used in the training of the CNN. The result of which is a mapping from the input labels to the features. The obtained mapping is nonlinear in nature. The features along with the map produced in the previous step are used to train the SVM classifier in the stage of testing. The preprocessed input image is subjected to a classification procedure which is iterative in nature.

Initially, CNN categorizes pixels belonging to the region which can be considered as key, resulting in a rather presegmentation that is then submitted to the integrated SVM classifier. Then, using presegmentation, a Region of Interest (ROI) will be formed. On this (Region of interest), classification based on integrated SVM will be done in addition to presegmentation. The integrated SVM then investigates the CNN—convolutional neural network—output's immediate neighborhood.

4 Experiment on Simulation

Directives related to the experiment: A public dataset and the self-made dataset were used in this analysis as experimental datasets. DCNN-F-SVM, SVM, and CNN are the reference models. The window size is set to 5, the value of σ to 0.1, and $C = 1000$ in the experimental parameters. The BraTS18 dataset was used as the public dataset. Twenty-six MRI scans of patients make up the self-created dataset. The DSC, sensitivity, and specificity evaluation indexes were used in the experiment. The hardware used for the experiment ran on an Ubuntu 14.04 operating system with an

Table 3 Evaluation indexes of various models

Model	Sensitivity	DSC	Specificity
DCNN-F-SVM	0.9108	0.8956	0.9980
CNN	0.8874	0.8554	0.9960
SVM	0.8304	0.8266	0.9843

AMD A8-5600K processor. The RAM supporting the architecture was of 16 GB with 479 MB of video memory. Python is the language chosen for the experiment, which was set in a PyCharm environment. TensorFlow was the deep learning algorithm library and the library chosen for image algorithm was OpenCV.

Experiment with a public dataset: Estimation of the test set is possible for the model after the training is complete, to achieve the glioma segmentation outcome obtained by the model segmentation. Evaluation index pairs are shown in Table 3 for each of the models. These results are for the BraTS18 dataset. This test set is divided by three-fold cross-validation. The results of Table 3 clearly reveal that the suggested model outperforms the combination of SVM and CNN at tumor segmentation. When compared to CNN, the proposed algorithm improved by 4.7%, 2.6%, and 0.2% on the three indicators of DSC, sensitivity, and specificity, respectively. Coming to the comparison to Fusion Support Vector Machine, the algorithm that is proposed improved by 8.3%, 9.6%, and 1.3% respectively.

Experiment on the self-made dataset: MRI scans of 26 people were obtained in this section, and segmentation of brain tumors was performed using all the three models in this section. Results of this experiment are given. Table 4 demonstrates the DCNN-F-SVM segmentation outcomes for 26 patients, respectively.

The general distribution of values of sensitivity, DSC, and specificity is about 0.92, 0.89, and 0.97. The up and down floating errors are nearly 0.13, 0.16 and 0.09, respectively.

Table 5 shows the results for the three models. The proposed model has improved in all the aspects. Thus, we can conclude that the proposed model is better than the other two models of CNN and SVM.

Table 4 Evaluated patient data by DCNN-F-SVM

Patient no.	Sensitivity	DSC	Specificity
1.	0.9218	0.8921	0.9661
2.	0.9061	0.8865	0.9366
3.	0.9191	0.9089	0.9700
4.	0.9012	0.8780	0.9586
5.	0.9287	0.9024	0.9793
6.	0.9096	0.8996	0.9403
7.	0.9194	0.9054	0.9741
8.	0.9227	0.9028	0.9694
9.	0.9108	0.8925	0.9709
10.	0.9287	0.9124	0.9804
11.	0.9296	0.9183	0.9883
12.	0.9108	0.8787	0.9603
13.	0.9166	0.8823	0.9691
14.	0.9220	0.8954	0.9783
15.	0.9183	0.8894	0.9667
16.	0.9102	0.8874	0.9676
17.	0.9084	0.8780	0.9583
18.	0.9101	0.9018	0.9784
19.	0.9121	0.9021	0.9750
20.	0.9114	0.8883	0.9598
21.	0.9203	0.8961	0.9694
22.	0.9285	0.9002	0.9743
23.	0.9256	0.9100	0.9796
24.	0.9113	0.8761	0.9596
25.	0.9086	0.8687	0.9467
26.	0.9303	0.8994	0.9795

Table 5 Comparison between evaluation indexes between the various models

Model	Sensitivity	DSC	Specificity
DCNN-F-SVM	0.9234	0.9008	0.9887
CNN	0.9150	0.8867	0.9655
SVM	0.9001	0.8703	0.9584

5 Conclusion

The assessment of brain disorders necessitates a precise and error-free diagnosis. Any diagnoses' error can result in irreversible harm. The increase in the incidence of brain tumors in people with brain disorders is soaring, and this towering has been observed regularly. An effect of this has been increased in the demand of professionals for this area. A reliable and coherent method must be proposed as soon as possible to meet the growing need for segmenting brain tumor images. A depth classifier is proposed in this paper with the aim to eradicate manual arbitration based on the context to increase segmentation precision and accomplish automated segmentation. A series of DCNN along with an integrated SVM primarily makes the classifier. Proposed model is split into three phases. To create a map to tumor marker space from image space, a deep CNN is made to undergo training in the first phase. Projected labels from the CNN preparation, as well as the test images, are fed into the integrated SVM in the second phase. To train a deep classifier for the last phase, a deep CNN is connected in series to an integrated Fusion SVM.

The proposed model's supremacy and effectiveness were confirmed by simulation. The model, however, also does come with flaws, like long computation times. The next research topic will be aimed at simplifying this model along with trying to reduce factor of time and thus to making this system more efficient.

References

1. Muir, C.S., Storm, H.H., Polednak, A.: Brain and other nervous system tumours. Cancer Surv. **19**(20), 369–392 (1994)
2. Louis, D.N., Ohgaki, H., Wiestler, O.D., et al.: The 2007 WHO classification of tumours of the central nervous system. Acta Neuropathol. **114**(5), 547 (2007)
3. Bauer, S., Wiest, R., Nolte, L.P., Reyes, M.: A survey of MRI-based medical image analysis for brain tumor studies. Phys. Med. Biol. **58**(13), R97–R129 (2013)
4. Gillies, R.J., Kinahan, P.E., Hricak, H.: Radiomics: images are more than pictures, they are data. Radiology **278**(2), 563–577 (2016)
5. Wang, S., Summers, R.M.: Machine learning and radiology. Med. Image Anal. **16**(5), 933–951 (2012)
6. Stadlbauer, A., Moser, E., Gruber, S., et al.: Improved delineation of brain tumors: an automated method for segmentation based on pathologic changes of ^{1}H-MRSI metabolites in gliomas. NeuroImage **23**(2), 454–461 (2004)
7. Qian, P., Friel, H., Traughber, M.S., et al.: Transforming UTE-mDixon MR abdomen-pelvis images into CT by jointly leveraging prior knowledge and partial supervision. IEEE/ACM Trans. Comput. Biol. Bioinform. (2020)
8. Jiang, Y., Zhao, K., Xia, K., et al.: A novel distributed multitask fuzzy clustering algorithm for automatic MR brain image segmentation. J. Med. Syst. **43**(5), 118:1–118:9 (2019)
9. Qian, P., Xu, K., Wang, T., et al.: Estimating CT from MR abdominal images using novel generative adversarial networks. J. Grid Comput. **18**, 211–226 (2020)
10. Xia, K., Zhong, X., Zhang, L., Wang, J.: Optimization of diagnosis and treatment of chronic diseases based on association analysis under the background of regional integration. J. Med. Syst. **43**(3), 46:1–46:8 (2019)
11. Deng, W., Xiao, W., Deng, H., Liu, J.: MRI brain tumor segmentation with region growing method based on the gradients and variances along and inside of the boundary curve. In: 2010 3rd International Conference on Biomedical Engineering and Informatics, vol. 1, pp. 393–396. Yantai, China (2010)
12. Jayadevappa, D., Kumar, S.S., Murty, D.S.: A hybrid segmentation model based on watershed and gradient vector flow for the detection of brain tumor. Int. J. Signal Process. Image Process. Pattern Recognit. **2**(3), 29–42 (2019)

13. Prastawa, M., Bullitt, E., Ho, S., Gerig, G.: A brain tumor segmentation framework based on outlier detection. Med. Image Anal. **8**(3), 275–283 (2019)
14. Gooya, A., Pohl, K.M., Bilello, M., et al.: GLISTR: glioma image segmentation and registration. IEEE Trans. Med. Imaging **31**(10), 1941–1954 (2012)
15. Kwon, D., Shinohara, R.T., Akbari, H., Davatzikos, C.: Combining generative models for multifocal glioma segmentation and registration. In: International Conference on Medical Image Computing and Computer-Assisted Intervention, pp. 763–770. Springer, Cham (2014)
16. Khotanlou, H., Colliot, O., Atif, J., Bloch, I.: 3D brain tumor segmentation in MRI using fuzzy classification, symmetry analysis and spatially constrained deformable models. Fuzzy Sets Syst. **160**(10), 1457–1473 (2009)
17. Bauer, S., Nolte, L.P., Reyes, M.: Fully automatic segmentation of brain tumor images using support vector machine classification in combination with hierarchical conditional random field regularization. In: International Conference on Medical Image Computing and Computer-Assisted Intervention, pp. 354–361. Springer, Berlin (2011)
18. Geremia, E., Menze, B.H., Ayache, N.: Spatial decision forests for glioma segmentation in multi-channel MR images. In: MICCAI Challenge on Multimodal Brain Tumor Segmentation. Springer, Germany (2012)
19. Wu, W., Li, D., Du, J.: Department of Epidemiology and Health Statistics, School of Public Health, China (2020)
20. Kadkhodaei, M., et al.: Automatic segmentation of multimodal brain tumor images based on classification of super-voxels. In: IEEE 38th Annual International Conference of the Engineering in Medicine and Biology Society (EMBC), pp. 5945–5948 (2020)
21. Jafari, M.H., et al.: Skin lesion segmentation in clinical images using deep learning. In: 2016 23rd International Conference on Pattern Recognition (ICPR). IEEE (2016)
22. Nasr-Esfahani, M., et al.: Left ventricle segmentation in cardiac MR images using fully convolutional network. In: 40th Annual International Conference of the IEEE Engineering in Medicine and Biology Society (EMBC) (2021)
23. Rafiei, S., et al.: Liver segmentation in CT images using three dimensional to two dimensional fully convolutional network. In: 2018 25th IEEE International Conference on Image Processing (ICIP). IEEE (2018)
24. Havaei, M., et al.: Brain tumor segmentation with deep neural networks. Med. Image Anal. **35**, 18–31 (2022)
25. Qian, P., Chen, Y., Kuo, J.-W., et al.: MDixon-based synthetic CT generation for PET attenuation correction on abdomen and pelvis jointly using transfer fuzzy clustering and active learning-based classification. IEEE Trans. Med. Imaging **39**(4), 819–832 (2020)
26. Qian, P., Xi, C., Xu, M., et al.: SSC-EKE: semi-supervised classification with extensive knowledge exploitation. Inf. Sci. **422**, 51–76 (2018)
27. Jiang, Y., Deng, Z., Chung, F.-L., et al.: Recognition of epileptic EEG signals using a novel multiview TSK fuzzy system. IEEE Trans. Fuzzy Syst. **25**(1), 3–20 (2017)
28. Jiang, Y., Wu, D., Deng, Z., et al.: Seizure classification from EEG signals using transfer learning, semi-supervised learning and TSK fuzzy system. IEEE Trans. Neural Syst. Rehabil. Eng. **25**(12), 2270–2284 (2020)
29. Qian, P., Zhou, J., Jiang, Y., et al.: Multi-view maximum entropy clustering by jointly leveraging inter-view collaborations and intra-view-weighted attributes. IEEE Access **6**, 28594–28610 (2018)

Chapter 2
A Comparison and Evaluation of Handwritten Digit Recognition (HDR) Algorithms

Shashwat Singh, Riya Agrawal, Swapnil Singh, Ayush Ranjan, Sujoy Datta, and Monideepa Roy

1 Introduction

Handwritten digit recognition using machine learning algorithms has been an area of intense research in recent years. Convolutional neural networking (CNN) is a widely used technique in deep learning, which has been applied to various applications, such as face recognition, object detection, spam detection, and image categorization [1, 2]. Handwritten digit recognition has numerous practical applications in professional and commercial contexts, as well as in our daily lives. It has the potential to assist individuals who are visually impaired or blind, making it a vital technology [3, 4]. However, the infinite variation in writing styles poses a challenge for recognition systems, leading to misclassifications due to a lack of distinction, unclear image text, discontinuous text strokes, unwanted objects, deformation, and disoriented patterns.

S. Singh · R. Agrawal · S. Singh · A. Ranjan · S. Datta (✉) · M. Roy
School of Computer Engineering, KIIT, Bhubaneswar, India
e-mail: sdattafcs@kiit.ac.in

S. Singh
e-mail: 1905204@kiit.ac.in

R. Agrawal
e-mail: 1905765@kiit.ac.in

S. Singh
e-mail: 1905287@kiit.ac.in

A. Ranjan
e-mail: 1905838@kiit.ac.in

M. Roy
e-mail: monideepafcs@kiit.ac.in

J. M. R. S. Tavares et al. (eds.), *Proceedings of Second International Conference on Intelligent System*, Algorithms for Intelligent Systems,
https://doi.org/10.1007/978-981-99-8976-8_2

1.1 Objective of the Paper

The primary objective of this paper is to develop an accurate digit recognition system that produces the simplest output possible. This will be accomplished by utilizing a comprehensive database that contains a range of training experiences. The system will be designed and developed in Java, which will be capable of recognizing handwritten digits drawn on the screen in a similar manner to the training data. Extensive research has been conducted to address the challenges associated with digit recognition, and numerous algorithms have been proposed. In order to develop an accurate and efficient handwritten digit recognition system, a wide range of techniques will be utilized. These include guide matching, neural networks, supreme rejection classification, and model-based detection. These techniques have been extensively researched and have shown promising results in recognizing handwritten digits. By combining these techniques and utilizing the comprehensive training database, the developed system will be capable of accurately recognizing a variety of handwritten digits. This approach will help address the challenges posed by infinite variation in writing styles and improve the accuracy of the recognition system.

2 Related Work

2.1 Early Scanners

The initial impetus behind written text classification was the implementation of digit classification for the purpose of communicating mail. The early communication readers created by Jacob Rabinow featured scanning instrumentation as well as hardwired logic to recognize mono-spaced fonts [5–7]. Allum et al. presented an improved and new scanning method that can handle variations in text formatting. The method encodes data onto a Universal Product Code (UPC) directly on the letter.

2.2 The Digital Age

In 1974, Ray Kurzweil created the first optical character recognition (OCR) software that could recognize any font. This was a significant achievement in OCR technology because previous software could only recognize specific fonts [8]. The OCR software utilized matrix methodology, specifically pattern matching. It compared the bitmaps of example characters with the scanned characters to determine the closest match. However, the software was limited in its sensitivity to variations in handwriting and individual writing styles [9–11].

OCR software improved upon the use of templates by adopting feature extraction techniques. Instead of relying on predefined templates, the software would analyze each character and search for features such as projections, histograms, zoning, and geometric moments. These features were used to recognize and classify characters, allowing for greater flexibility and accuracy in OCR technology.

3 The Proposed Method

The aim of the projected system is to develop a system of improved facilities. The projected system can overcome a few limitations of the existing system. The system provides proper accuracy and reduces manual [12–14].

- Accuracy
- High speed
- Multiplied performance
- Reduced error.

The projected model contains the following four stages to classify and observe the digits [15, 16].

3.1 Preprocessing

HDR also includes another component known as preprocessing. Examining the limits will be a lot less difficult if there are certain guidelines to follow, such as allocating a box for each digit in the number. Filtering, normalization, and standardization are three common techniques used in preprocessing to eliminate noise and improve the quality of the data. The operation of converting a grayscale image into binary format is known as binarization.

3.2 Feature Extraction

There are numerous primary classifications of feature extraction methods, and each of these categories is associated with a different error rate. The errors that are produced by each kind of algorithm do not overlap, so combining all these algorithms leads to a perfect recognition rate. This also helps us to reject the ambiguous digit and improve the recognition rate of misclassified digits.

3.3 Classification and Recognition

During the phase of classification and recognition, the extracted feature vectors are presented as a single input value to each classifier. It is possible for the CNN convolution layer, and sub-sampling layer as a result, to have multiple separate layers. In addition, the layer that was used for down-sampling is now known as the pooling layer. The image is then cut up into numerous little parts, and a value is determined for each of those segments. After then, the values that have been calculated are reorganized in order to produce an entirely new picture [17–19].

3.4 Training and Testing

Finally, to gauge a model, the check dataset was used. Work may be less difficult as the result of each module is supposed to handle a particular subproblem. It is expected that by training each module separately the machine learning model will work with high efficiency and accuracy. Moreover, the modular approach gives you the flexibility and ability to improve the model over time so that it can be added or deleted as required.

A convolutional neural network (CNN) typically consists of multiple layers that perform different operations on the input data (Fig. 1). The three main types of layer used in a CNN are convolutional layer, pooling layer, and fully connected layer [20–22].

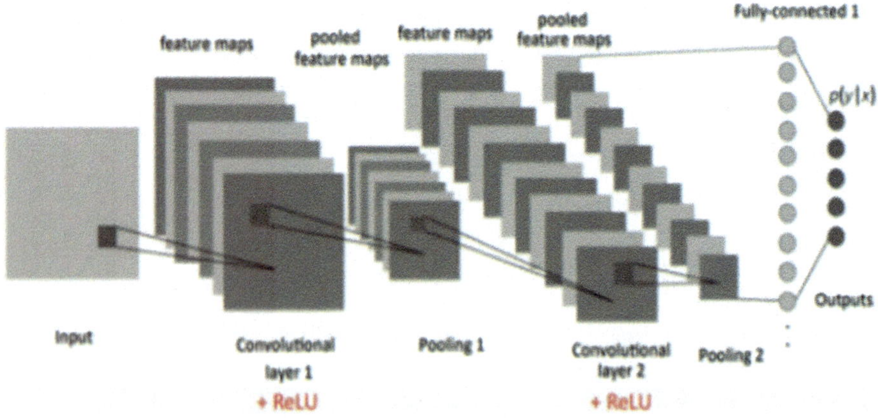

Fig. 1 Example CNN with two convolutional layers, two pooling layers, and a fully connected layer which decides the final classification of the image into one of several categories

Convolutional Layer: This layer performs a convolutional operation on the input image. It uses a set of learnable filets or kernels to extract features from the input image.

Pooling Layer: This layer reduces the spatial size of the output and is used to down-sample the feature maps and reduce the computational cost of the network.

Fully Connected Layer: This layer connects all the neurons in the previous layer to the neurons in the next layer. It is used to classify the image into one of several classes.

4 Steps in the Algorithm

A number of different processes [23–25] are performed for generating the final model by applying CNN (Fig. 2), which are as follows.

4.1 Loading the Dataset

We know that the images are pre-aligned and they all have the same square size of 28 $\times$ 28 pixels, and the images are grayscale. Therefore we have loaded the images and reshaped the data arrays to have a single color channel. We have used load dataset function which implements these behaviors and can be used to load the dataset.

4.2 Preparing the Pixel Data

The pixel values for each of the images in the dataset are unsigned integers within the range between black and white or 0–255. However some amount of pixel scaling is needed. A nice point to start will be to normalize the pixel values of grayscale images, e.g., they will be rescaled to the range (0, 1). This consists of first converting the data type from unsigned integers to floats, and to divide the pixel values by the maximum value. Here the integer values have been converted to float and then normalize to the range of 0–1.

4.3 Creating TensorFlow Model

The model is made up of two primary parts: the feature extraction forepart, which is made up of convolutional and pooling layers and the classifier posterior, which has produced a prediction. Both of these parts are located at the beginning of the model.

Fig. 2 CNN model

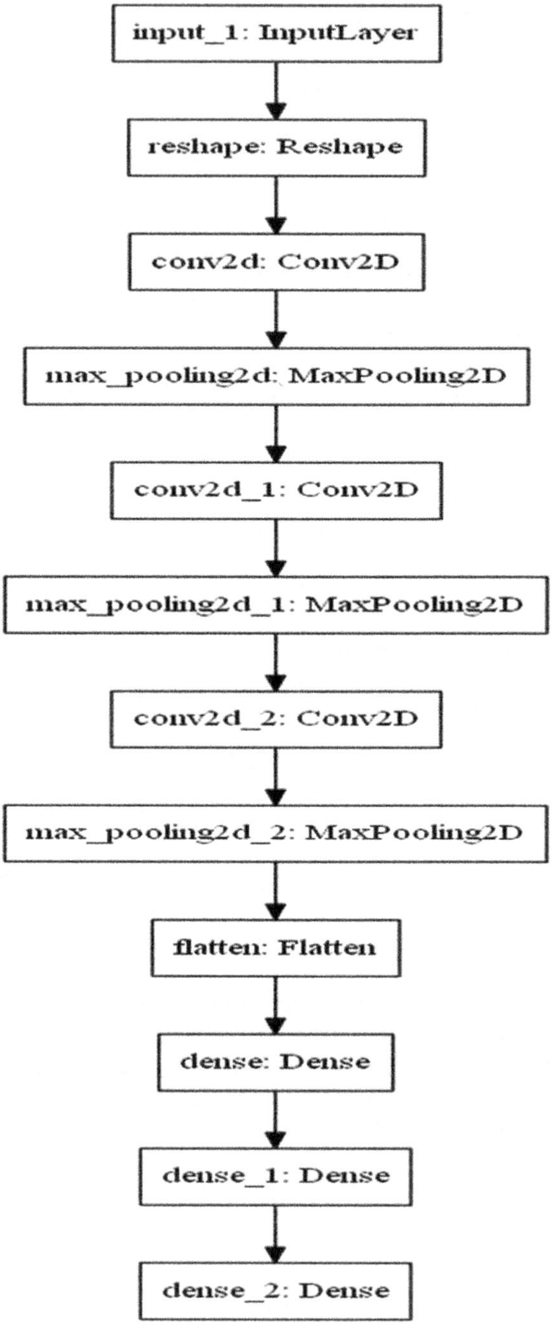

We commonly use one convolutional layer for the convolutional front end. This layer has a modest filter size (3, 3) and a small number of filters (32), followed by a soap pooling layer.

4.4 Evaluating the Model

The Keras evaluate() function was used for the evaluation of the model. Each of the test sets included a total of around 12,000 examples, which is a size that is equivalent to the actual size of the test set that was used for this issue. Any model we evaluated had the same train and test datasets in each fold, which allowed for an end-to-end comparison of models to be performed. The training dataset was shuffled before it was split, and the sample shuffling was performed each time. This was done to ensure that any model we evaluated had the same data. An evaluation of the model was carried out using five-fold cross-validation [3]. The value $k = 5$ was decided upon to act as a starting point for both of the continually recurring evaluations. To train the baseline model, we used a group size of 32 by default and trained it for a total of ten training epochs. The test set for each fold was utilized for the evaluation of the model during each epoch of the training run, so that learning curves could be generated later, so that at the end of the run, the performance of the model can be estimated.

4.5 Visualizing the Training Process

The visualization of the training process was done by plotting a graph with accuracy versus epoch number in order to check the difference between the trained model and the validation model.

4.6 Predicting the Result

The evaluated model was saved at the last and afterward it was used for the prediction on new images. The assumptions for the model are that new images are grayscale, the images have been aligned so that one image contains one centered handwritten digit, and that the size of the image is square with the size 28×28 pixels.

The workflow diagram is shown in Fig. 3.

A comparison between the existing model and the proposed model is shown in Table 1.

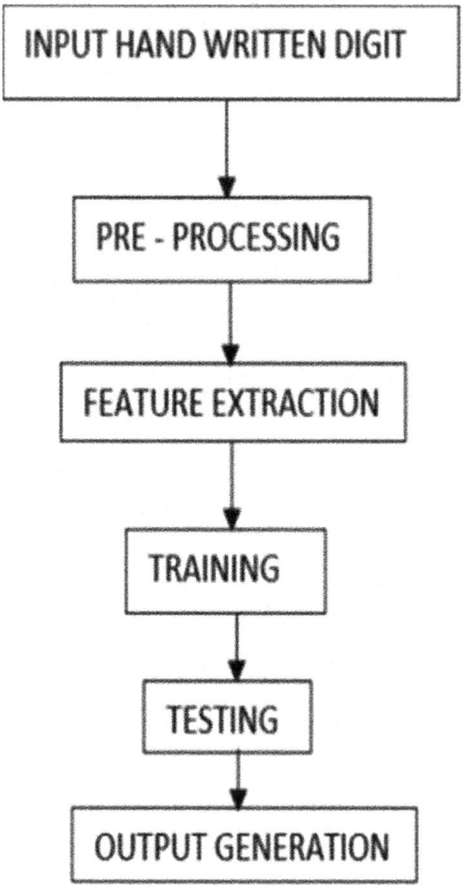

Fig. 3 Workflow diagram

Table 1 Comparison table between the existing model and the proposed model

No hidden layer activation function = SIGMOID		Hidden layer activation function = SIGMOID	
Epochs	Accuracy (%)	Epochs	Accuracy (%)
1	87.52	1	92.43
2	91.54	2	96.46
3	92.05	3	97.42
4	92.37	4	97.95
5	92.57	5	98.41
6	92.80	6	98.76
7	92.83	7	98.96

5 Results and Data Visualization

There is a nice intersection between front-end development and machine learning, but we frequently find it daunting to obtain knowledge in both fields and get them to play together. This work is going to be a practical one that is supported by some developments in engineering science. Every day, people all around the world search for novel approaches in the field of engineering science to transform the restrictions imposed by humans into those imposed by machines in order to collect an increasing number of precise and essential pieces of data. The approach taken in computer science and machine learning has no back-slopping that goes in the wrong direction; rather, it only uses forward momentum. This project can only be a very simple blueprint for implementing these ideas. Neural networks are the subject of this project, which focuses on one of the most cutting-edge educational approaches now available. There are many other means by which one can accomplish the goal of reaching the desired output; but, in the field of machine learning, neural networks give the means by which the computer learns the steps necessary to reach the goal. This work has been successful, thanks to the notions of applied mathematics modeling, the pc vision, and machine learning libraries, all of which have a significant amount of research pertaining to these concepts. The graphical representation of accuracy versus epoch values is shown in Fig. 4. Figure 5 shows the character 0 stored at position 1 in the MNIST dataset. Figure 6 shows the heat map without the hidden layer and using SIGMOID function as activation function; keeping epochs value as 7, the accuracy is coming out to be 92.83% as shown in Fig. 7. Figure 8 shows the heat map with ReLu as the hidden layer and Fig. 9 shows that the accuracy is 98.96% with hidden layer and SIGMOID activation function and keeping epoch values as 7 [8, 22].

Fig. 4 Graph of accuracy versus epoch values

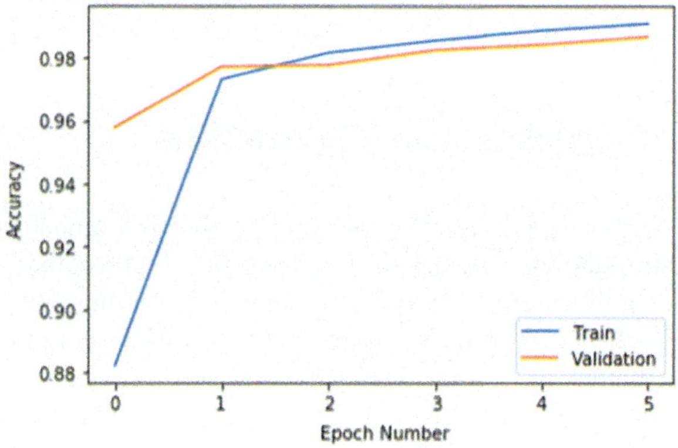

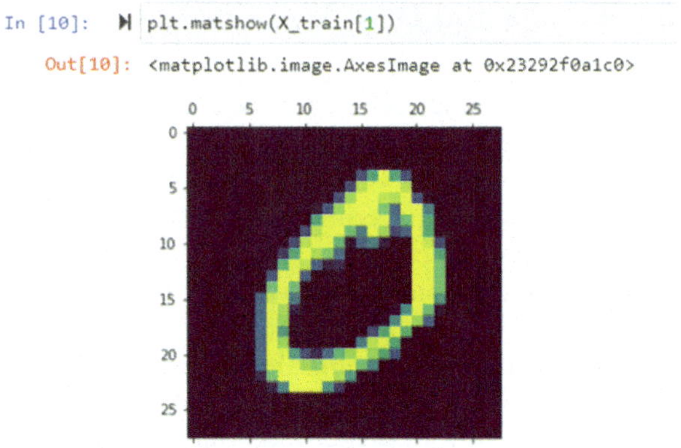

Fig. 5 0 stored at position 1 in MNIST dataset

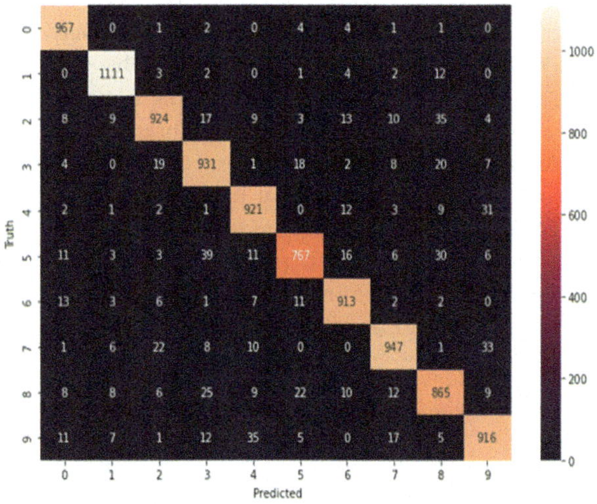

Fig. 6 Heat map without hidden layer

6 Conclusion and Future Work

Convolutional neural networks (CNNs) are trained using real-time data, which both simplifies the model by lowering the total number of variables and improves its overall accuracy. In order to achieve the highest possible level of precision in our work, we decided to make use of CNN in conjunction with a number of other libraries, including Keras, Matplotlib, CV2, and TensorFlow. When compared with other machine learning algorithms such as the random forest classifier, the convolutional neural network, linear regression, K-nearest neighbors, and the support vector machine, the accuracy of CNN comes out on top with a score of 99.01%. We can improve our training by using techniques like jittering and normalizing data. Due to time and budget constraints, we only had 20 training examples for each word. To enhance our character segmentation model, we can use a more exhaustive coding

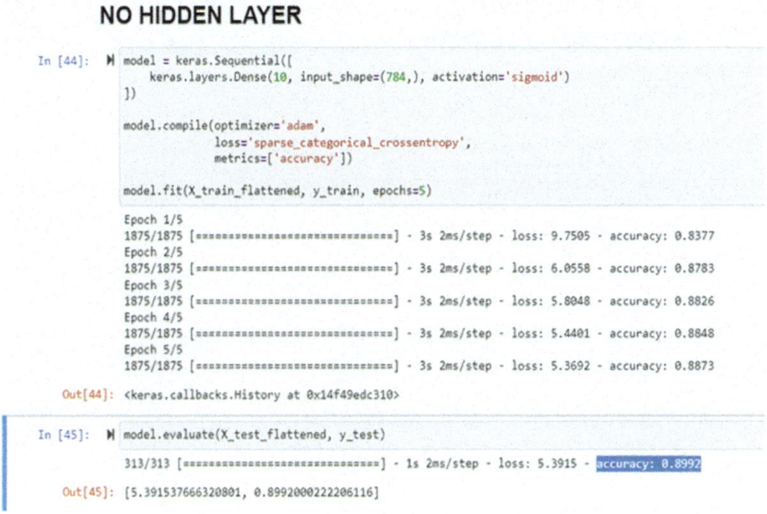
NO HIDDEN LAYER

```
In [44]: model = keras.Sequential([
             keras.layers.Dense(10, input_shape=(784,), activation='sigmoid')
         ])

         model.compile(optimizer='adam',
                       loss='sparse_categorical_crossentropy',
                       metrics=['accuracy'])

         model.fit(X_train_flattened, y_train, epochs=5)

         Epoch 1/5
         1875/1875 [==============================] - 3s 2ms/step - loss: 9.7505 - accuracy: 0.8377
         Epoch 2/5
         1875/1875 [==============================] - 3s 2ms/step - loss: 6.0558 - accuracy: 0.8783
         Epoch 3/5
         1875/1875 [==============================] - 3s 2ms/step - loss: 5.8048 - accuracy: 0.8826
         Epoch 4/5
         1875/1875 [==============================] - 3s 2ms/step - loss: 5.4401 - accuracy: 0.8848
         Epoch 5/5
         1875/1875 [==============================] - 3s 2ms/step - loss: 5.3692 - accuracy: 0.8873

Out[44]: <keras.callbacks.History at 0x14f49edc310>

In [45]: model.evaluate(X_test_flattened, y_test)

         313/313 [==============================] - 1s 2ms/step - loss: 5.3915 - accuracy: 0.8992

Out[45]: [5.391537666320801, 0.8992000222206116]
```

Fig. 7 Accuracy with epoch values [without hidden layer]

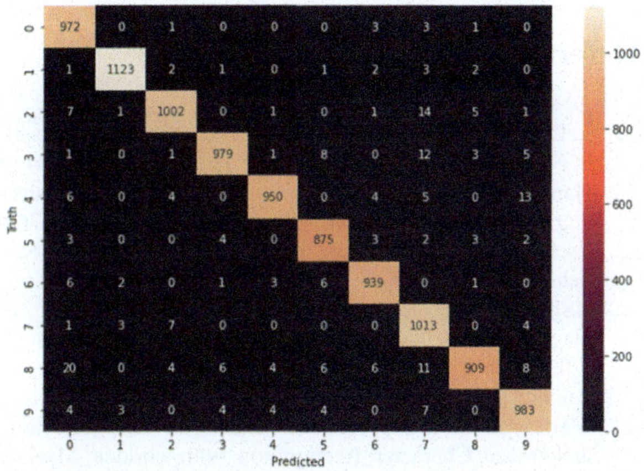

Fig. 8 Heat map with hidden layer added

algorithm like beam search. This algorithm considers penalty/benefit scores and individual probabilities to find the most likely candidate paths. By using a language-based model, we can correct our model if it suggests an unlikely candidate word.

ADDING HIDDEN LAYER

```
In [27]:  model = keras.Sequential([
              keras.layers.Dense(100, input_shape=(784,), activation='relu'),
              keras.layers.Dense(10, activation='sigmoid')
          ])

          model.compile(optimizer='adam',
                        loss='sparse_categorical_crossentropy',
                        metrics=['accuracy'])

          model.fit(X_train_flattened, y_train, epochs=5)

          Epoch 1/5
          1875/1875 [==============================] - 4s 2ms/step - loss: 0.2706 - accuracy: 0.9223
          Epoch 2/5
          1875/1875 [==============================] - 4s 2ms/step - loss: 0.1237 - accuracy: 0.9638
          Epoch 3/5
          1875/1875 [==============================] - 4s 2ms/step - loss: 0.0869 - accuracy: 0.9740
          Epoch 4/5
          1875/1875 [==============================] - 4s 2ms/step - loss: 0.0669 - accuracy: 0.9795
          Epoch 5/5
          1875/1875 [==============================] - 4s 2ms/step - loss: 0.0538 - accuracy: 0.9837

Out[27]:  <keras.callbacks.History at 0x14f48e4f6d0>

In [28]:  model.evaluate(X_test_flattened,y_test)

          313/313 [==============================] - 1s 1ms/step - loss: 0.0875 - accuracy: 0.9745

Out[28]:  [0.08753662556409836, 0.9745000004768372]
```

Fig. 9 Accuracy with epoch values [hidden layer]

References

1. LeCun, Y., et al.: Handwritten digit recognition with a backpropagation network. In: Advances in Neural Information Processing Systems, pp. 396–404 (1990)
2. Han, X., Li, Y.L.: The application of convolution neural networks in handwritten numeral recognition. Int. J. Database Theory Appl. **8**(3), 367–376 (2015)
3. Tschopp, F.: Efficient convolutional neural networks for pixelwise classification on heterogeneous hardware systems. In: IEEE 13th International Symposium on Biomedical Imaging (ISBI) (2016)
4. Nagy, G.: Reflections of an ancient document processor. Pattern Recognit. Lett. **166**, 76–79 (2023)
5. Krevat, E., Cuzzillo, E.: Improving off-line handwritten character recognition with hidden Markov models. https://www.researchgate.net/publication/229034911_Improving_Offline_Handwritten_Character_Recognition_with_Hidden_Markov_Models (2011)
6. Shima, Y., Nakashima, Y., Yasuda, M.: Pattern augmentation for handwritten digit classification based on combination of pre-trained CNN and SVM. In: 6th International Conference on Informatics, Electronics and Vision (ICIEV) and 7th International Symposium on Computational Medical and Health Technology (ISCMHT) (2017)
7. Teow, M.Y.W.: Understanding convolutional neural networks using a minimal model for handwritten digit recognition. In: 2017 IEEE 2nd International Conference on Automatic Control and Intelligent System (12CACIS 2017), Kota Kinabalu, Sabah, Malaysia, 21 Oct 2017, pp. 167–172
8. Siva Ajay, T.: Handwritten digit recognition using convolutional neural networks. Int. Res. J. Eng. Technol. (IRJET) **04**(07), 2971–2976 (2017)
9. Cireşan, D.C., Meier, U., Gambardella, L.M., Schmidhuber, J.: Deep, big, simple neural nets for handwritten digit recognition, Mar 2010, pp. 1–14. arXiv
10. Deng, L.: The MNIST database of handwritten digit images for machine learning research. IEEE Signal Process. Mag. 141–142 (2012)

11. Le Cun, Y., Jackel, L.D., Boser, B., Denker, J.S., Graf, H.P., Guyon, I., Henderson, D., Howard, R.E., Hubbard, W.: Handwritten digit recognition: applications of neural network chips and automatic learning. NATO ASI Ser. F Comput. Syst. Sci. **68**, 41–46 (1990)
12. Deng, L.: The MNIST Database of Handwritten Digits Images for Machine Learning Research. MIT Press (2012)
13. Singh, V., Lal, S.P.: Digits recognition using single layer neural network with principal component analysis. In: 2014 Asia-Pacific World Congress on Computer Science and Engineering (APWC on CSE), 4–5 Nov 2014. IEEE
14. Larasati, R., Lam, H.K.: Handwritten digits recognition using ensemble neural networks and ensemble decision tree. In: 2017 International Conference on Smart Cities, Automation & Intelligent Computing Systems (ICON-SONICS), 8–10 Nov 2017
15. Bottou, L., Cortes, C.: Comparison of classifier methods a case study in handwritten digit recognition. In: Proceedings of the 12th IAPR International Conference on Pattern Recognition. Vol. 2 Conference B; Image Processing, 06 Aug 2002. IEEE (1994)
16. Bharati, V.C., Veningston, K., Venkateswara Rao, P.V.: Query-based word spotting in handwritten documents using HMM. In: Data Engineering and Communication Technology. Advances in Intelligent Systems and Computing. Springer Nature Singapore Ltd. (2020)
17. Sudhakar, R., Venkateswara Rao, P.V.: Video super resolution using non-linear regression and deep learning. Imaging Sci. J. (2019). https://doi.org/10.1080/13682199.2019.1652445
18. Zhan, H., Lyu, S., Lu, Y.: Handwritten digit string recognition using convolutional neural network. In: 24th International Conference on Pattern Recognition (ICPR), Aug 2018, pp. 3729–3734
19. Uchida, S., Ide, S., Iwana, B.K., Zhu, A.: A further step to perfect accuracy by training CNN with larger data. In: 15th International Conference on Frontiers in Handwriting Recognition, pp. 405–410 (2016)
20. Giusti, A., Cireşan, D.C., Masci, J., Gambardella, L.M., Schmidhuber, J.: Fast image scanning with deep max-pooling convolutional neural networks. In: The 17th IEEE Conference on Image Processing, pp. 4034–4038 (2010)
21. Golovko, V., Egor, M., Brich, A., Sachenko, A.: A shallow convolutional neural network for accurate handwritten digits classification. In: 13th International Conference, PRIP, Minsk, Belarus, Oct 2016, pp. 77–85
22. Uchida, S., Ide, S., Iwana, B.K., Zhu, A.: A further step to perfect accuracy by training CNN with larger data. In: 15th International Conference on Frontiers in Handwriting Recognition, pp. 405–410 (2016)
23. Wu, M., Zhang, Z.: Handwritten digit classification using the MNIST data set. In: Course Project CSE802: Pattern Classification & Analysis (2010)
24. Mizukami, Y, Sato, T., Tanaka, K.: Handwritten digit recognition by hierarchical displacement extraction with gradual prototype elimination. In: Proceedings 15th International Conference on Pattern Recognition (ICPR), pp. 339–342 (2000)
25. Calderon, A., Roa, S., Victorino, J.: Handwritten digit recognition using convolutional neural networks and Gabor filters. In: Proceedings of the International Congress on Computational Intelligence (CIIC), pp. 1–8 (2003)

Chapter 3
An Analysis of Machine Learning Algorithms for AQI Prediction

Naresh Sharma and Rohit Sharma

1 Introduction

A severe issue that has an impact on both health of the humans and the environment is air pollution. The Air Quality Index (AQI) is a measure of air quality that indicates the degree of air pollution. It is an essential tool for monitoring air quality and informing the public about potential health hazards. AQI is calculated depending on various pollutants such as particulate matter ($PM_{2.5}$ and PM_{10}), nitrogen dioxide (NO_2), sulfur dioxide (SO_2), carbon monoxide (CO), and ozone (O_3). Higher values of the AQI, which ranges from 0 to 500, indicate poorer air quality. Accurately predicting AQI levels can help individuals and policymakers to make informed decisions regarding air quality management. The conventional methods based on statistics and probability are exceedingly intricate and ineffective. It has been demonstrated that the ML-based AQI prediction models are more dependable and consistent. Modern technologies and sensors have made data collection for environmental analysis more precise and easier. Only machine learning algorithms can effectively handle the rigorous analysis required to make accurate predictions from such vast environmental data [1]. The purpose of this study is to evaluate how well various machine learning models predict AQI results.

N. Sharma
Department of Computer Science and Engineering, SRM Institute of Science and Technology, Delhi NCR Campus, Ghaziabad, Uttar Pradesh, India

R. Sharma (✉)
Department of Electronics and Communication Engineering, SRM Institute of Science and Technology, Delhi NCR Campus, Ghaziabad, Uttar Pradesh, India
e-mail: rohitapece@gmail.com

J. M. R. S. Tavares et al. (eds.), *Proceedings of Second International Conference on Intelligent System*, Algorithms for Intelligent Systems,
https://doi.org/10.1007/978-981-99-8976-8_3

An Air Quality Index (AQI) prediction system is a tool or model that uses historical and real-time data to forecast the future air quality levels in a given area. It employs various techniques, such as machine learning, statistical analysis, and atmospheric modeling, to estimate the AQI values for different pollutants like particulate matter ($PM_{2.5}$ and PM_{10}), ozone (O_3), nitrogen dioxide (NO_2), sulfur dioxide (SO_2), and carbon monoxide (CO).

The purpose of an AQI prediction system is to provide timely and accurate information about the expected air quality conditions in advance. This allows individuals, communities, and policymakers to take proactive measures to mitigate the impact of poor air quality on public health, the environment, and various activities. By forecasting AQI values, the system enables people to make informed decisions such as adjusting outdoor activities, planning commuting routes, implementing pollution control strategies, and issuing public health advisories.

The functioning of an AQI prediction system involves several steps:

1. Data collection: The system gathers data from different sources as ground-based monitoring systems, satellite imagery, weather data, and historical air quality records. This data provides information on pollutant concentrations, meteorological conditions, and other relevant factors [2].
2. Data preprocessing: The collected data is processed and cleaned to remove outliers, missing values, and inconsistencies. It may involve data interpolation, smoothing, and transformation techniques to make it suitable for analysis.
3. Feature selection: For the creation of the model, relevant features or variables, such as temperature, humidity, wind speed, and pollutant emissions, are identified and chosen [3].
4. Model development: Various techniques, such as regression models, time-series analysis, or machine learning algorithms, are employed to develop a predictive model. The model is trained using historical data and incorporates relevant variables to forecast future AQI values.
5. Validation and evaluation: The developed model is validated using independent datasets and evaluated based on performance metrics like accuracy, root mean square error (RMSE), and correlation coefficients. This step ensures the reliability and accuracy of the model predictions.
6. Deployment and visualization: The AQI prediction system is deployed in a user-friendly interface, such as a website or mobile application, where users can access the predicted AQI values. The system may also provide visual representations, such as maps or graphs, to make the information easily understandable and accessible [4].

It is important to note that AQI prediction systems are based on mathematical models and historical data, and they have inherent limitations and uncertainties. The accuracy of predictions can vary depending on the quality and availability of data, the complexity of atmospheric conditions, and other factors. Continuous monitoring and refinement of the system are necessary to improve its performance and provide more reliable predictions [5].

Table 1 Air Quality Index (AQI) values for major metro cities in India (from CPCB.NIC)

Delhi	Mumbai	Kolkata	Chennai	Bengaluru	Hyderabad
The average AQI value over the past year was around 169, with occasional spikes reaching over 300	The average AQI value over the past year was around 90, with occasional spikes reaching over 150	The average AQI value over the past year was around 61, with occasional spikes reaching over 200	The average AQI value over the past year was around 54, with occasional spikes reaching over 150	The average AQI value over the past year was around 38, with occasional spikes reaching over 150	The average AQI value over the past year was around 56, with occasional spikes reaching over 150

Here are the latest Air Quality Index (AQI) values for major metro cities in India, as of cutoff of June 2023 (Table 1).

It is important to note that AQI values can fluctuate significantly depending on the time of year, weather patterns, and other factors.

1.1 Reasons for Poor AQI in India

The quality of air, we breathe, is measured by Air Quality Index (AQI), with higher values indicating poorer air quality [6]. India has been struggling with high levels of air pollution, particularly in urban areas, which leads to a high AQI. Here are some of the main reasons for bad AQI in India:

- Industrial emissions: In urban areas of India, industrial emissions are a significant cause of air pollution. These emissions consist of various harmful substances, such as particulate matter (PM), sulfur dioxide (SO_2), and nitrogen oxides (NO_x), that can negatively impact human health.
- Vehicular emissions: In recent years, the quantity of vehicles on Indian roads has notably surged, resulting in an increase in vehicular emissions. These emissions contain various pollutants such as CO, NO_x, and PMs, which significantly contribute to air pollution.
- Agricultural burning: Agricultural burning is a common practice in India, particularly during the winter months, when farmers burn crop residues to prepare their fields for the next crop. The burning of crop residues releases large amounts of PM, which can have a significant impact on air quality.
- Construction activities: Construction activities are another significant source of air pollution in India, particularly in urban areas. Construction activities generate a range of pollutants, including PM, which can have a significant impact on air quality.

- Natural sources: Natural sources such as dust, pollen, and forest fires also contribute to poor air quality in India. Dust from construction sites, unpaved roads, and other sources can lead to high levels of particulate matter in the air.
- Climate and geography: India's climate and geography can also contribute to poor air quality. For example, during the winter months, temperature inversions can trap pollutants close to the ground, leading to high levels of pollution. In addition, the geography of the Indo-Gangetic plain, which covers a large part of northern India, can contribute to poor air quality by trapping pollutants in the region.

Overall, the sources of air pollution in India are numerous and varied, and addressing the issue requires a multi-faceted approach that involves government action, public education, and individual action.

1.2 Ways for Improving the AQI

Improving the Air Quality Index (AQI) in India requires a multi-faceted approach, involving government action, public education, and individual actions. Here are some ways to improve AQI in India:

- **Reduce industrial and vehicular emissions**: Government and industry can work together to reduce industrial emissions by introducing stricter regulations and investing in clean technologies. In addition, promoting the use of public transport and electric vehicles can help to reduce vehicular emissions.
- **Encourage the use of renewable energy**: Encouragement of the use of renewable energy sources, such as wind and solar energy, can help to lessen dependency on fossil fuels and, consequently, emissions.
- **Control agricultural burning**: Alternatives to crop burning such as the use of mulching machines and other methods can help to reduce the amount of particulate matter released into the atmosphere.
- **Increase green cover**: The cultivation of greenery such as trees and other plants in urban settings can aid in the assimilation of pollutants and enhance the overall air quality.
- **Promote awareness and education**: Raising awareness about the effects of air pollution on both environment and human health can be an effective way to educate people and encourage them to take steps toward reducing emissions on an individual level.
- **Implement strict enforcement of environmental regulations**: Enforcement of existing environmental regulations should be strict, and the consequences of violating these regulations should be severe enough to deter polluters from continuing to pollute.
- **Encourage public transportation**: Government can promote the use of public transportation and promote cycling and walking as alternatives to reduce vehicular traffic on roads.

- **Improving waste management practices**: Poor waste management can also lead to air pollution. Implementing better waste management practices, such as segregation of waste at the source, composting, and recycling can help to reduce the amount of waste that is burned in the open air.

Overall, improving AQI in India requires a sustained and coordinated effort by individuals, industries, and the government. By adopting these measures and promoting clean and sustainable practices, we can work toward a cleaner and healthier environment for all.

2 Methods

The suggested method intends to compute the Air Quality Index (AQI) for various pollutants, with a special emphasis on fine material ($PM_{2.5}$), which is a serious issue for human health when levels in the air are high [7]. $PM_{2.5}$ refers to microscopic particles in the air that, when present in high concentrations, can decrease visibility and make the air hazy.

To assess the air quality level in a certain city, the system calculates the AQI of all contaminants using the AQI equations and employs gradient descent and Box-Plot analysis. Using current AQI values, the algorithm may also anticipate AQI values for the following years.

The suggested method, by delivering correct AQI values for various contaminants, might be a helpful resource for policymakers and people managing air quality in a specific city [8]. The capacity of the system to forecast future AQI readings can also help in making educated judgments about air quality management. Overall, the suggested approach has the potential to improve air quality monitoring and management, which is critical for public health and well-being.

2.1 Data Source

We collected AQI data from several air quality monitoring stations in different cities. Data was collected over a period of six years (2017–2022). The dataset consists of hourly AQI values and weather parameters such as temperature, humidity, wind direction, and speed of the wind. We used Python programming language and Scikit-learn library to implement different machine learning models.

2.2 Preprocessing Data

Data preprocessing is a critical stage in any data analysis assignment that involves cleaning, converting, and organizing data before it can be analyzed. Outliers and missing data are two prevalent errors in data extraction and monitoring applications. The process of data preprocessing involves several operations, such as filling in missing data and removing or modifying outlier data, to ensure the accuracy and quality of the data [1]. For data preparation in this investigation, we utilized Python's Pandas package. The steps taken were as follows:

- Import the required libraries, which are Pandas, NumPy, Matplotlib, and Seaborn.
- Load the information: The Pandas read csv function was used to load the dataset, which read the data from a CSV file and built a Pandas Data Frame object.
- Examine the data types: The dtypes function was used to determine the data types of the dataset's columns. We discovered that certain columns had the incorrect data type; therefore, we had to convert them.
- Convert data types: Using a for loop, we ran over the columns in the dataset and used the Pandas to numeric function to convert the data types of the columns to numeric. The errors option was set to 'coerce,' which means that all non-numeric values were replaced with NaN.
- Handle missing information: The isna and sum functions were used to verify the dataset for missing data. We utilized Pandas' dropna function to remove any rows with missing data.
- Handle outliers: We checked the frequency of the target variable, $PM_{2.5}$, using the value counts function. We then removed any rows with $PM_{2.5}$ equal to 0.0 since they were considered outliers.
- Split the input variables and output variables as follows: To segregate the input and output variables from the preprocessed data, we utilized Pandas' iloc function.
- Exploratory data analysis: To display the link between the input and the output variable, we utilized Seaborn's pairplot and correlation matrix.

3 Machine Learning Models

We used the following machine learning models to predict the AQI values:

- KNN.
- ANN.
- XGBoost.
- Decision tree.
- Random forest.

We created training and testing sets from the dataset. Eighty percent of the data is in the training set, and the final twenty percent is in the testing set. The training set was used to develop the machine learning models, while the testing set was used to evaluate their performance. As evaluation metrics, we employed the mean squared error (MSE) and the coefficient of determination (R^2).

3.1 K-Nearest Neighbors (KNNs)

The k-nearest neighbors' (KNNs) algorithm is a kind of non-parametric supervised learning classifier. It makes use of nearby data points to categorizes or forecast how a single data point will group itself [9].

From the dataset, 70% of data is used as training dataset, which is used to train the model, while the remaining 30% data is used as testing dataset, which is used to test the model. The KNN algorithm is first fit with $k = 1$ (where k is the number of neighbors considered for prediction), and the coefficient of determination (R^2) score is calculated for both the training and test sets. While the model performs flawlessly on the training set ($R^2 = 1.0$), it performs poorly on the test set ($R^2 = -0.049$). The model is further assessed using cross-validation, and the result is a negative mean squared error (MSE) score of − 0.26.

To increase the performance of the model, hyperparameter tuning is done to discover the ideal value of k. The negative MSE score for each value is determined using fivefold cross-validation after the KNN model has been trained for k values ranging from 1 to 39.

The results show that $k = 7$ gives the best negative MSE score of − 0.27. After fitting the KNN model with $k = 7$, the cross-validation score—which is an improvement over the previous score—is discovered to be 0.26. The mean absolute error, mean squared error, and root mean squared error are additional metrics calculated for the model. 46.56, 3984.08, and 63.12 are the outcomes, respectively.

Finally, the accuracy of the model is calculated using the R^2 score. The model's accuracy was found to be 39.75%, which shows that it is not functioning well. Overall, the KNN model used to predict air pollution performs poorly, and to increase prediction accuracy, further feature engineering or the usage of more complex models may be needed.

3.2 Artificial Neural Network (ANN)

The model's hyperparameters are an important component of machine learning because they have a direct impact on the model's performance [8]. In this case, the hyperparameters used are the number of epochs, learning rate, batch size, and optimizer.

The total number of iterations the model undergoes over the entire training dataset depends on the number of epochs. Better accuracy may result from using more epochs, but overfitting may happen as well. The optimizer changes the weights during training in steps of varying sizes, according to the learning rate. The model may overshoot the ideal weights if the learning rate is large, whereas a low learning rate may result in a sluggish convergence of the model.

During the training process, the batch size determines the number of training samples used per iteration. Using larger batch sizes can result in shorter training times, while smaller batches may promote better generalization. Additionally, the optimizer used in this model is Adam, a widely used optimization technique that adjusts the learning rate based on the gradient of the loss function.

Accuracy, precision, recall, and F1-score are used to assess the model's performance. Precision measures the percentage of genuine positives among the anticipated positives, whereas accuracy reflects the percentage of correctly classified occurrences. The F1-score is the harmonic mean of precision and recall. Recall quantifies the percentage of genuine positives among the real positives.

Overall, the performance of the model is impressive, achieving an accuracy of over 90% on the test dataset. The precision and recall scores are likewise excellent, demonstrating that the model is able to recognize both positive and negative events with high accuracy. The F1-score is a good indicator of the overall performance of the model, and in this case, it is high, indicating that the model is performing well in both precision and recall.

3.3 XGBoost

An XGBoost regressor, a form of boosting method that can handle a combination of continuous and categorical data, is the model that was employed. On a dataset with features including temperature, humidity, wind speed, and particulate matter 2.5 ($PM_{2.5}$) concentration, the model was trained and evaluated. The default hyperparameters were used for training the XGBoost regressor, which included 100 decision trees, a maximum depth of 6, a learning rate of 0.3, and subsampling of 1.0. The model attained a coefficient of determination (R^2) of 0.997 on the training set and 0.344 on the test set, indicating overfitting. The model's performance on unknown data was evaluated using cross-validation, and the average R^2 score across the three folds was 0.223, indicating that the model's predictive capacity was not very strong.

Using random search and a fivefold cross-validation, hyperparameter adjustment was done to enhance the model's performance [1]. The number of estimators (decision trees), learning rate, maximum depth, subsampling rate, minimum child weight, and goal function were among the hyperparameters modified. In comparison to the default hyperparameters, the model's performance has improved after adjusting.

3.4 Decision Tree

The DecisionTreeRegressor() function from the Scikit-learn library is then used to train the decision tree regression model using the training data, with the Friedman mean squared error as the criterion for splitting nodes [10]. Utilizing the coefficient of determination R^2, which determines how much of the variance of the target variable can be anticipated from the independent variables, the model's performance is assessed on both the training set and the test set. In contrast to the test set, where the R^2 value is − 0.119, which indicates subpar performance, the training set's R^2 score is 1.0, indicating perfect fit.

Cross-validation is carried out using the cross val score() function with a threefold cross-validation in order to further assess the model's performance. The mean score from the cross-validation is − 0.287, which is even worse than the R^2 score on the test set. The distribution is not centered around zero when the difference between the actual and predicted $PM_{2.5}$ values is plotted using a histogram, which suggests that the model has poor predictive ability.

To enhance the performance of the decision tree model, the GridSearchCV() function is utilized for hyperparameter tuning. This includes adjusting the maximum depth, minimum sample size required to split a node, the minimum weighted fraction of samples required to split a node, maximum number of features considered when splitting a node, maximum number of leaf nodes in the tree, and splitter. The GridSearchCV() function performs a grid search across all possible hyperparameter combinations and uses fivefold cross-validation to evaluate each combination's performance.

The model's accuracy of 37.25% indicates inadequate performance. Therefore, further feature engineering or using more complex models may be necessary to increase prediction accuracy for air pollution prediction using the decision tree model.

3.5 Random Forest

To predict a continuous numerical value, a machine learning approach known as a random forest regressor employs many decision trees [11]. Random forest is a supervised machine learning technique, which is frequently used for classification and regression problems. The average for regression and the classification vote, along with a variety of samples, it builds decision trees [12]. To produce a more precise and reliable prediction, it combines the predictions of each decision tree. The model is then trained using the training set, and the coefficient of determination R^2 is determined on both the training and testing sets to evaluate the model's performance. The training set's R^2 score is 0.91, which indicates a good fit, while the testing set's R^2 score is 0.39, which indicates a moderate fit. The mean score from cross-validation, which is used to obtain a more accurate assessment of the model's performance, is 0.35, which is lower than the mean score from the testing set.

Table 2 Performance of ML models

Model	MSE	Coefficient of determination (R^2 in %)
KNN	3984	39.75
ANN	6346	79.6
XGBoost	119.15	98.20
Decision tree	4149.65	37.25
Random forest	2212	66.54

The next step in the code involves hyperparameter tuning using RandomizedSearchCV, which aims to find the optimal hyperparameters that can minimize the mean squared error.

It is determined that 1200 estimators, a minimum samples leaf of 5, a maximum feature setting of sqrt, and a maximum depth of 20 are the best hyperparameters. The model's best score, derived via cross-validation, is − 4383.84. The mean absolute error(), mean squared error(), and sqrt() functions of Scikit-learn are used to calculate the performance metrics. The trained model with the optimized hyperparameters is used to make predictions on the testing set. The mean absolute error is 34.42, mean squared error is 2212.37, and the root mean squared error is 47.04.

4 Results

We compared the performance of different machine learning models in predicting AQI values. The results are shown in Table 2.

The results show that XGBoost has the lowest MSE and the highest R^2 value, indicating that it is the best model for predicting AQI values.

5 Discussion

The outcomes show that highly accurate predictions of AQI values can be made using machine learning algorithms. When it comes to prediction accuracy, the XGBoost model performs better than the competition. This is so that the intricate interactions between the characteristics and the target variable may be handled by the XGBoost model.

6 Conclusion

In this research, we evaluated how well several machine learning models predicted AQI values. The outcomes demonstrate that the XGBoost model performs the best in terms of AQI value prediction. The results of this study can help with policy and control of air quality. To increase the accuracy of AQI prediction, future study can be done to include other elements including traffic data, industrial pollutants, and land-use data.

7 Future Works

Future research can focus on developing machine learning models for AQI prediction that incorporate real-time data from air quality monitoring stations. Additionally, research can investigate the use of ML algorithms for AQI prediction in different geographical locations and under other environmental conditions. Finally, research can focus on developing machine learning models that can predict the potential health effects of exposure to air pollutants.

7.1 Future Scope of AQI Prediction in India

There are ample opportunities for employing machine learning (ML) to forecast Air Quality Index (AQI) in India. Here are some potential areas of future work for AQI prediction using machine learning:

- **Data collection and analysis**: Collecting and analyzing data on various factors that affect air quality such as weather, emissions, and land-use patterns can help to build more accurate ML models for predicting AQI. This includes exploring the use of satellite data, sensor networks, and other sources to improve data collection.
- **Feature engineering**: The selection of important features or input variables that have a significant impact on AQI, and their extraction from different datasets, can be further optimized for building effective ML models for AQI prediction.
- **Model selection and tuning**: Researchers can explore different types of ML models, such as deep learning models, decision trees, and ensemble models, and tune their hyperparameters to achieve better accuracy in AQI prediction.
- **Spatial and temporal modeling**: Modeling the spatial and temporal variations in air quality can help in developing targeted interventions for areas with poor air quality.
- **Integration with IoT**: Integration of IoT-based sensing devices and ML models can provide real-time, high-resolution information on air quality in specific regions.

- **Enhancing the interpretability of models**: Developing models that are interpretable and can explain how specific factors or variables contribute to AQI predictions can help in making better decisions and guiding actions for policymakers and other stakeholders.

References

1. Kumar, K., Pande, B.P.: Air pollution prediction with machine learning: a case study of Indian cities. Int. J. Environ. Sci. Technol. (2022)
2. Soundari, A., Jeslin, J.: Indian air quality prediction and analysis using machine learning (2019)
3. Corrigendum to "Impacts of the natural gas infrastructure and consumption on fine particulate matter concentration in China's prefectural cities: a new perspective from spatial dynamic panel models". J. Clean. Prod. (2019)
4. Kalbande, S.R., Gaikwad, S.T.: IoT based air quality monitoring system using Arduino. In: 2019 International Conference on Communication, Computing and Electronics Systems (ICCCES), pp. 1045–1049 (2019)
5. Yang, X., Zhang, Y., Li, X., Li, C., Li, L.: Air quality forecasting using machine learning algorithms: a case study in Beijing. J. Environ. Manage. **232**, 858–865 (2019)
6. Han, S., Hong, S., Kim, S., Lee, S.: A novel hybrid approach for air quality forecasting using machine learning techniques. Environ. Pollut. **252**, 1681–1690 (2019)
7. Ghude, S.D., Beig, G., Fadnavis, S., Polade, S.D., van der A, R.J., Patil, S.D.: Spatial decision support system for air quality management in India. Environ. Model. Assess. **21**(6), 703–716 (2016)
8. Usha, N.H., Srinivas, S.: Air quality index prediction using artificial neural networks and multiple linear regression. Int. J. Eng. Technol. **7**(3.23), 59–62 (2018)
9. Kothandaraman, D., Praveena, N.: Intelligent forecasting of air quality and pollution prediction using machine learning (2001)
10. Mousavi, S.M., Pirsaheb, M.: Forecasting of air quality index using machine learning algorithms (support vector regression, random forest, and decision tree) in Tehran, Iran. Sustain. Cities Soc. **47**, 101474 (2019)
11. Devata, R., Ashish, R.: IoT based air quality prediction using SVM and random forest (2022)
12. Gupta, N.S., Mohta, Y., Heda, K., Armaan, R., Valarmathi, B., Arulkumaran, G.: Prediction of air quality index using machine learning techniques: a comparative analysis. J. Environ. Public Health **2023**, 26 pages. Article ID 4916267 (2023)

Chapter 4
Application of the Harmonic Runge–Kutta with Forward–Backward Technique by Parallelism

Mahmood D. Jasim and Rabeea M. H. Darghoth

1 Introduction

The interest in developing the Runge–Kutta method for solving differential equations has lately increased by applying other means, for instance, the harmonic mean, the Heronian mean, and the contra-harmonic mean, see Jasim [1–3].

There have been a lot of studies in the last decades that dealt with this method using the harmonic mean; see Sanugi and Evans [4]; but here we present a new explicit process consisting of two steps to calculate the value in the next position.

In 2021, Jasim [1] submitted new Runge–Kutta methods that was sufficient for parallel computers by using contra-harmonic mean. Here we display the derivation of a new method, where we linked the forward backward technique with harmonic mean in order to obtain our new method, which is appropriate for parallel processors [5].

2 Derivation of a New Harmonic Mean Method

Sanugi and Evans [2] introduced the Harmonic Mean, which took the form

$$m_{i+1} - m_i = \frac{\hbar}{i}\left(\sum_{j=1}^{i} \frac{2\mathcal{T}_j \mathcal{T}_{j+1}}{\mathcal{T}_j + \mathcal{T}_{j+1}}\right), \quad \mathcal{T}_j = \varphi\left(m_i + \hbar \sum_{k=1}^{j-1} s_{jk}\mathcal{T}_k\right), \tag{1}$$

M. D. Jasim (✉) · R. M. H. Darghoth
Department of Mathematics, College of Basic Education, University of Mosul, Mosul, Iraq
e-mail: mahoodalani@uomosul.edu.iq

R. M. H. Darghoth
e-mail: r.darghoth@uomosul.edu.iq

J. M. R. S. Tavares et al. (eds.), *Proceedings of Second International Conference on Intelligent System*, Algorithms for Intelligent Systems,
https://doi.org/10.1007/978-981-99-8976-8_4

where $j = 1, 2, \ldots, i$, $\hbar$ step size, $\varphi = \varphi(m)$.

Now, our new method calculates m_{i+1}^{p} which depends on m_{i-1}^{c} while m_i^{c} depends on m_{i-1}^{c} and m_i^{p}, which has a form:

$$m_{i+1}^{\mathrm{p}} - m_{i-1}^{\mathrm{c}} = 2\hbar\left(\frac{2\mathcal{T}_1\mathcal{T}_2}{\mathcal{T}_1 + \mathcal{T}_2}\right). \tag{2}$$

Here $\mathcal{T}_1 = \varphi(\mathcal{X}_{i-1}, m_{i-1}^{\mathrm{c}})$, $\mathcal{T}_2 = \varphi(\mathcal{X}_{i-1} + 2\hbar, m_{i-1}^{\mathrm{c}} + 2\hbar\mathcal{T}_1)$,

$$m_i^{\mathrm{c}} - m_{i-1}^{\mathrm{c}} = \hbar\left(\frac{2\mathcal{S}_1\mathcal{S}_2}{\mathcal{S}_1 + \mathcal{S}_2}\right). \tag{3}$$

Here $\mathcal{S}_1 = \varphi(\mathcal{X}_i, m_i^{\mathrm{p}})$, $\mathcal{S}_2 = \varphi(\mathcal{X}_i - \hbar, m_i^{\mathrm{p}} - \hbar\mathcal{S}_1)$.

To drive the first part of the new method, we can rewrite (1) as

$$m_{i+1}^{\mathrm{p}} - m_{i-1}^{\mathrm{c}} = 2\hbar\left(\frac{\gamma\mathcal{T}_1\mathcal{T}_2}{\alpha\mathcal{T}_1 + \beta\mathcal{T}_2}\right). \tag{4}$$

Here $\mathcal{T}_1 = \varphi(m_{i-1}^{\mathrm{c}})$, $\mathcal{T}_2 = \varphi(m_{i-1}^{\mathrm{c}} + \delta\hbar\mathcal{T}_1)$.

Expansion of $\mathcal{T}_1$ and $\mathcal{T}_2$, first $\mathcal{T}_1 = \varphi(m_{i-1}^{\mathrm{c}}) = \varphi$, also

$$\mathcal{T}_2 = \varphi + \delta\hbar\mathcal{T}_1\varphi_{m_{i-1}^{\mathrm{c}}} + \mathrm{O}(\hbar^2) = \varphi + \delta\hbar\varphi\varphi_{m_{i-1}^{\mathrm{c}}} + \mathrm{O}(\hbar^2)$$

for denominator multiply α with $\mathcal{T}_1$ adding to multiplying β with $\mathcal{T}_2$ as the following:

$$\mathcal{T}_1\mathcal{T}_2 = \varphi * (\varphi + \delta\hbar\varphi\varphi_{m_{i-1}^{\mathrm{c}}}) + \mathrm{O}(\hbar^2).$$

Then, we get

$$\frac{\gamma\mathcal{T}_1\mathcal{T}_2}{\alpha\mathcal{T}_1 + \beta\mathcal{T}_2} = \frac{\gamma\varphi^2 + \gamma\delta\hbar\varphi^2\varphi_{m_{i-1}^{\mathrm{c}}} + \mathrm{O}(\hbar^2)}{(\alpha + \beta)\varphi + \beta\delta\hbar\varphi\varphi_{m_{i-1}^{\mathrm{c}}} + \mathrm{O}(\hbar^2)}. \tag{5}$$

Substituting (5) into (4) to get

$$m_{i+1}^{\mathrm{p}} - m_{i-1}^{\mathrm{c}} = 2\hbar\left(\frac{\gamma\varphi^2 + \gamma\delta\hbar\varphi^2\varphi_{m_{i-1}^{\mathrm{c}}} + \mathrm{O}(\hbar^2)}{(\alpha + \beta)\varphi + \beta\delta\hbar\varphi\varphi_{m_{i-1}^{\mathrm{c}}} + \mathrm{O}(\hbar^2)}\right). \tag{6}$$

By comparing (6) with Taylor series [3] which takes the form

$$m_{i+1} - m_{i-1} = 2\hbar\varphi + \hbar^2\varphi\varphi_{m_{i-1}} + \mathrm{O}(\hbar^3). \tag{7}$$

From Eq. (6), we cannot allow direct replacement, because it shares in the division of two chins. Therefore, it will rewrite the Eq. (6) as the following

$$m_{i+1}^{p} = m_{i-1}^{c} + \frac{\text{Numerator}}{\text{Denominator}}. \tag{8}$$

Numerator $= \gamma\varphi^2\hbar + \gamma\delta\varphi^2\hbar^2\varphi_{m_{i-1}^{c}} + O(\hbar^3)$,
Denominator $= (\alpha+\beta)\varphi + \beta\delta\hbar\varphi\varphi_{m_{i-1}^{c}} + O(\hbar^2)$.
Taylor series takes the form:

$$\text{Taylor} = m_{i-1} + 2\hbar\varphi + \hbar^2\varphi\varphi_{m_{i-1}} + O(\hbar^3). \tag{9}$$

From (8) and (9), it obtains

$$\text{Error} = \text{Taylor} - \frac{\text{Numerator}}{\text{Denominator}}.$$

Thus

$$\text{Denominator. Error} = \text{Denominator. Taylor} - \text{Numerator}. \tag{10}$$

Compare both sides of the equation to the same exponent of $O(\hbar^n)$. Then, we have $\alpha+\beta=\gamma$, $\alpha+\beta+2\delta(\beta-\gamma)=0$, choosing $\alpha=\beta=1, \Rightarrow \delta=1, \gamma=2$. So, we get the derivation of (2) above.

To drive the second part of the new method, we can rewrite (1) as

$$m_i^{c} - m_{i-1}^{c} = \hbar\left(\frac{\gamma S_1 S_2}{\alpha S_1 + \beta S_2}\right). \tag{11}$$

From (11), it can say $S_1 = \varphi$ and $S_2 = \varphi - \delta\hbar\varphi\varphi_{m_i^{p}} + O(\hbar^2)$, multiplying S_1S_2 with γ to have $\gamma S_1 S_2 = \gamma\varphi^2 - \gamma\delta\hbar\varphi^2\varphi_{m_i^{p}} + O(\hbar^2)$.

Then, $\alpha S_1 + \beta S_2 = (\alpha+\beta)\varphi - \beta\delta\hbar\varphi\varphi_{m_i^{p}} + O(\hbar^2)$, so, it obtains

$$\frac{\gamma S_1 S_2}{\alpha S_1 + \beta S_2} = \frac{\gamma\varphi^2 - \gamma\delta\hbar\varphi^2\varphi_{m_i^{p}} + O(\hbar^2)}{(\alpha+\beta)\varphi - \beta\delta\hbar\varphi\varphi_{m_i^{p}} + O(\hbar^2)}. \tag{12}$$

By substituting Eq. (11) into (12),

$$m_{i-1}^{c} - m_i^{c} = -\hbar\left(\frac{\gamma\varphi^2 - \gamma\delta\hbar\varphi^2\varphi_{m_i^{p}} + O(\hbar^2)}{(\alpha+\beta)\varphi - \beta\delta\hbar\varphi\varphi_{m_i^{p}} + O(\hbar^2)}\right), \tag{13}$$

comparing Eq. (13) with Taylor series of the form,

$$m_{i-1} - m_i = -\hbar\varphi + \frac{\hbar^2}{2}\varphi\varphi_{m_i} + O(\hbar^2). \tag{14}$$

Therefore, it gets, $\alpha+\beta=\gamma$, $\frac{(\alpha+\beta)}{2} + \delta(\beta-\gamma) = 0$.

Choosing $\alpha = \beta = 1,\ \gamma = 2,\ \delta = 1$.

So, we get the derivation of (3) above. Finally, Eqs. (2) and (3) together represent our new method.

3 The Analysis of Stability

We use the test equation $m' = \lambda m$, where $\lambda = \partial\varphi/\partial m$ constant [4, 6, 7]. We will apply Eq. (2) to the test equation in order to calculate the absolute stability interval of this method.

$$\mathcal{T}_1 = \varphi\left(\mathcal{X}_{i-1}, m_{i-1}^{c}\right) = \lambda m_{i-1}^{c}, \tag{15}$$

$$\mathcal{T}_2 = \varphi\left(x_{i-1} + 2\hbar,\, m_{i-1}^{c} + 2\hbar\mathcal{T}_1\right) = \lambda\left(m_{i-1}^{c} + 2\hbar\lambda m_{i-1}^{c}\right). \tag{16}$$

Substitute Eqs. (15) and (16) in (2),

$$m_{i+1}^{p} - m_{i-1}^{c} = 2\hbar\left(\frac{2\left(\lambda m_{i-1}^{c}\right)\left(\lambda m_{i-1}^{c} + 2\hbar\lambda^2 m_{i-1}^{c}\right)}{\left(\lambda m_{i-1}^{c}\right) + \left(\lambda m_{i-1}^{c} + 2\hbar\lambda^2 m_{i-1}^{c}\right)}\right).$$

Therefore

$$m_{i+1}^{p} = m_{i-1}^{c} + 2\hbar\lambda m_{i-1}^{c}\left(\frac{1 + 2\hbar\lambda}{1 + \hbar\lambda}\right). \tag{17}$$

Dividing (17) by m_{i-1}^{c} and putting $\mathcal{W} = \hbar\lambda$, we get,

$$d = \frac{m_{i+1}^{p}}{m_{i-1}^{c}} = 1 + 2\mathcal{W}\left(\frac{1 + 2\mathcal{W}}{1 + \mathcal{W}}\right). \tag{18}$$

The Eq. (18) identifies the condition of absolute stability if $|d| < 1$, where $\mathcal{W} \in (-0.5, 0)$, so this interval is the stability interval of the our new method.

4 Test Example

We test our new method on the initial value problem, $m' = m$, $m(0) = 1$, with $\hbar = 0.05$, where result appeared in Table 1.

The new method applied to example when $m' = m$, $m(0) = 1$, when $\hbar = 0.05$.

Table 1 Comparison of results

$\mathcal{X}$	Exact sol. $m(\mathcal{X})$	Forward sol. m^{p}	Backward sol. m^{c}	$\lvert m - m^{\mathrm{p}}\rvert$	$\lvert m - m^{\mathrm{c}}\rvert$
0	1	1	1	0	0
0.05	1.05127109637	1.10476190476	1.04871794871	0.053490808385	0.002553147658
0.10	1.10517091807	1.15858363858	1.10253968253	0.053412720507	0.002631235535
0.15	1.16183424272	1.21804383975	1.15898350082	0.056209597029	0.002850741898
0.20	1.22140275816	1.28040081996	1.21832409815	0.058998061804	0.003078660008
0.25	1.28402541668	1.34595805129	1.28070259963	0.061932634603	0.003322817051
0.30	1.34985880757	1.41487144340	1.34627491495	0.065012635832	0.003583892620
0.35	1.41906754859	1.48731323937	1.41520454937	0.068245690786	0.003862999214
0.40	1.49182469764	1.56346407359	1.48766339950	0.071639375957	0.004161298139
0.45	1.56831218549	1.64351385087	1.56383216206	0.075201665388	0.004480023428
0.50	1.64872127070	1.72766219808	1.64390078556	0.078940927387	0.004820485133
0.55	1.73325301786	1.81611896310	1.72806894393	0.082865945234	0.005184073932
0.60	1.82211880039	1.90910473806	1.81654653444	0.086985937670	0.005572265945
0.65	1.91554082901	2.00685140948	1.90955420117	0.091310580467	0.005986627843
0.70	2.01375270747	2.10960273653	2.00732388522	0.095850029061	0.006428822247
0.75	2.11700001661	2.21761495891	2.11009940315	0.100614942299	0.006900613456
0.80	2.22554092849	2.33115743586	2.21813705500	0.105616507375	0.007403873491
0.85	2.33964685192	2.45051331790	2.33170626341	0.110866465979	0.007940588511
0.90	2.45960311115	2.57598025291	2.45109024556	0.116377141758	0.008512865587
0.95	2.58570965931	2.70787112843	2.57658671942	0.122161469122	0.009122939886
1	2.71828182845	2.84651485194	2.70850864619	0.128233023481	0.009773182260

5 Discussion the Results of the New Method

As shown in Table 1, the first column represents the step length, the second column represents the exact solution, the third column represents the forward solution, the fourth column represents the backward solution, and the fifth and sixth columns represent the absolute differences between the exact and the forward and backward solutions in the order.

From the last column, it can notice that new method is more accurate and its results are closer to the exact compared to the forward one, which at the same time represents the traditional method [8].

Acknowledgements The others are grateful to College of Basic Education, Mosul University, for their provided facilities, which helped to improve the quality of this work.

References

1. Jasim, M.D.: A new method using the forward backward technique with contra harmonic mean formula. J. Phys. Conf. Ser. **1963**(1), 012018 (2021). IOP Publishing
2. Jasim, M.D.: Two new Runge–Kutta methods based on the root mean square formula for solving IVPs in ODEs. J. Interdiscip. Math. **24**(7), 1947–1953 (2021)
3. Jasim, M.D.: Implementation of the Heronion mean (HeM) in the predictor corrector mode. J. Interdiscip. Math. **24**(7), 1917–1923 (2021)
4. Sanugi, B.B., Evans, D.J.: A new fourth order Runge–Kutta formula based on the harmonic mean. Int. J. Comput. Math. **50**(1–2), 113–118 (1994)
5. Ahmed, H.: A review of hash function types and their applications. Wasit J. Comput. Math. Sci. **1**(3), 120–139 (2022)
6. Khalaf, B.M.S., Mahmood, D.A.-A.: Parallel Runge–Kutta methods for solving initial value problems in ordinary differential equations. Raf. J. Comp. Math's **11**(3), 64–83 (2000)
7. Lambert, J.D., Lambert, J.D.: Computational Methods in Ordinary Differential Equations, vol. 5. Wiley (1973)
8. Lambert, J.D.: Numerical Methods for Ordinary Differential Systems, vol. 146. Wiley, New York (1991)

Chapter 5
A Preconditioned Conjugate Gradient Multigrid Method for Multi-material Topology Optimization

Nam G. Luu and Thanh T. Banh

1 Introduction

Topology optimization (TO) is a computational methodology that focuses on optimizing the arrangement of materials within a specified design domain, taking into account prescribed loads, boundary conditions, and constraints, all with the overarching goal of maximizing the performance of the design. The field of TO has experienced remarkable growth since the pioneering work of Bendsøe and Kikuchi [1], attracting significant attention from researchers. In the context of implementing TO in MATLAB, researchers often turn to three influential papers. Firstly, Sigmund's work, detailed in Sigmund [2], provides a concise MATLAB-based TO code designed to minimize compliance. Secondly, Jensen et al., in their publication Andreassen et al. [3], introduced a modified version of this implementation that enhances the efficiency of TO within the MATLAB environment. Lastly, Tavakoli and Mohseni [4] tackled multi-material topology optimization (MTO) problems using the alternating active-phase algorithm.

In most topology optimization (TO) problems, a primary challenge lies in reducing computational costs and burdens, particularly when dealing with large-scale multi-material TO scenarios. The inefficiencies in terms of both cost and execution time in such large-scale problems have prompted numerous researchers to seek methods for cost and time reduction. Bruns et al. [5] explored various algorithms to mitigate the computationally expensive nature of TO. Liu et al. [6] introduced a novel approach employing the guide-weight method, presenting two formulations of TO

N. G. Luu (✉)
Faculty of Information Technology, Industrial University of Ho Chi Minh City, Ho Chi Minh City, Vietnam
e-mail: luugiangnam@iuh.edu.vn

T. T. Banh
Department of Architectural Engineering, Sejong University, Seoul, Republic of Korea

J. M. R. S. Tavares et al. (eds.), *Proceedings of Second International Conference on Intelligent System*, Algorithms for Intelligent Systems,
https://doi.org/10.1007/978-981-99-8976-8_5

using the SIMP method. Wang et al. [7] contributed a high efficiency isogeometric TO approach aimed at improving mesh scale reduction, accelerating solving processes, and reducing design variables. In the realm of multi-material problems, Zhang et al. [8] introduced multi-material topology optimization based on an MMC-framework, while Chin et al. [9] delved into large-scale MTO within a scalable framework using octree-based mesh adaptation.

This research introduces a novel approach employing a preconditioned conjugate gradient multigrid method with a V-cycle to alleviate the computational expenses inherent in multi-material topology optimization problems. Furthermore, it conducts a comparative analysis against the original MATLAB code as detailed in Tavakoli and Mohseni [4]. The central aim of this study is to underscore the efficacy of the multigrid technique in the domain of MTO, specifically emphasizing mean compliance-based approaches.

The structure of this paper is as follows: Sect. 2 presents the multigrid method and various numerical techniques for solving linear equations. Section 3 provides a concise overview of TO for multi-material and multi-phase scenarios. The central focus of this paper is Sect. 4, where we conduct a comparative analysis of the pCGMG V-cycle against several numerical methods. Finally, we draw conclusions from this study in Sect. 5.

2 Preconditioned Conjugate Gradient Multigrid

2.1 *Multigrid Method*

The multigrid method, employed in numerical analysis, is an algorithm designed to enhance the convergence of a fundamental iterative approach by globally refining the approximation of the fine grid solution (relaxation) and addressing the problem at a coarser level. To delve further into the concept, let's examine a linear problem denoted as $\mathbf{KU} = \mathbf{F}$, where $\mathbf{K}$ represents the stiffness matrix, $\mathbf{F}$ is a vector, and $\mathbf{U}$ is a vector solution within the domain Ω. A multigrid V-cycle method uses nested grids of increasing fineness: $\Omega_0 \subset \Omega_1 \subset \Omega_2 \subset \ldots \subset \Omega_M = \Omega$ where Ω is the finest mesh. With finite element spaces $V_0 \subset V_1 \subset V_2 \subset \ldots \subset V_M = V$, we also have a linear system corresponding on each level n: $\mathbf{K}_n\mathbf{U}_n = \mathbf{F}_n$ where $\mathbf{K}_n$ is the stiffness matrix obtained from discretization in the mesh Ω_n. To facilitate the transformation of linear systems across multiple grid levels, transfer operators are established to serve as linear mappings between finer and coarser grids. These operators take the form of a restriction operator, denoted as $\mathbf{R}_l : \mathbf{R}^{n_l} \to \mathbf{R}^{n_{l-1}}$, and a prolongation operator, denoted as $\mathbf{P}_l : \mathbf{R}^{n_{l-1}} \to \mathbf{R}^{n_l}$, with n_l representing the number of nodes within the grid Ω_l. Algorithm 1 describes the multigrid method with γ-cycle in general.

Algorithm 1 Multigrid γ-cycle $\mathrm{MG}_{\gamma}^{l}(U_l, F_l)$

Require: $\mathbf{K}_l, \mathbf{F}_l, \mathbf{R}_l, \mathbf{P}_l$

If $l = 1$ then $\mathbf{U}_l = \mathbf{K}_l^{-1}\mathbf{F}_l$

Else: Smoother ν_1 times $\mathbf{K}_l\mathbf{U}_l = \mathbf{F}_l$ with the initial guess $\mathbf{U}_l$

$\mathbf{F}_{l-1} \leftarrow \mathbf{R}_l(\mathbf{F}_l - \mathbf{K}_l\mathbf{U}_l)$. Set $\mathbf{U}_{l-1} = 0$

If $\Omega_l = \Omega$ then

While $\gamma \le N$ do: $\mathbf{U}_{l-1} \leftarrow \mathrm{MG}^{l-1}(\mathbf{U}_{l-1}, \mathbf{F}_{l-1})$, $\mathbf{U}_l \leftarrow \mathbf{U}_l + \mathbf{P}_l\mathbf{U}_{l-1}$

Smoother ν_2 times $\mathbf{K}_l\mathbf{U}_l = \mathbf{F}_l$

2.2 *Preconditioned Conjugate Gradient*

The conjugate gradient (CG) method is an algorithm employed for the numerical solution of specific linear equation systems characterized by a symmetric and positive-definite matrix, as introduced by Hestenes and Stiefel [10]. This method is geared toward minimizing the functional expression $T(u) = \|\mathbf{KU} - \mathbf{F}\|_{\mathbf{K}^{-1}}$. From the following expression, $\|\mathbf{U} - \mathbf{U}_k\|_{\mathbf{K}^{-1}} \le \|\mathbf{U} - \mathbf{U}_0\|_{\mathbf{K}^{-1}} \left(\frac{\sqrt{\kappa}-1}{\sqrt{\kappa}+1}\right)^k$, where κ is the condition number and k signify the iteration number. The convergence behavior is associated with the condition number κ. The process converges rapidly when κ is sufficiently small but greater than 1, and it slows down considerably when $\kappa \gg 1$. In scenarios where $\kappa \gg 1$, a preconditioner is employed to expedite the convergence of the linear system. A preconditioner is typically a matrix or operator $\mathbf{M}$ that satisfies the condition $\kappa\left(\mathbf{M}^{-1}\mathbf{K}\right) \ll \kappa(\mathbf{K})$ then the preconditioner $\mathbf{M}$ must ensure that $\kappa\left(\mathbf{M}^{-1}\mathbf{K}\right)$ is close to 1. The preconditioner CG is built as depicted in Algorithm 2.

Algorithm 2 Preconditioned conjugate gradient

Require: Stiffness matrix $\mathbf{A}$, load vector $\mathbf{b}$, preconditioner $\mathbf{M}$, initial value $\mathbf{x}_0$

Ensure: $\mathbf{r}_0 = \mathbf{b} - \mathbf{A}\mathbf{x}_0$ small enough. Set $i = 1$

While $\mathbf{r}_0 < \varepsilon$ do: $\mathbf{z}_i = \mathbf{M}^{-1}\mathbf{r}_i$

If $\mathbf{r}_i \neq 0$ then:

$\beta_{i-1} = \mathbf{z}_i^T\mathbf{r}_i/\mathbf{z}_{i-1}^T\mathbf{r}_{i-1}$, $(\beta_{-1} = 0)$

$\mathbf{p}_i = \mathbf{z}_i + \beta_{i-1}\mathbf{p}_{i-1}$, $(\mathbf{p}_0 = \mathbf{z}_0)$

$\alpha_i = \mathbf{z}_i^T\mathbf{r}_i/\mathbf{p}_i^T\mathbf{A}\mathbf{p}_i$

$\mathbf{x}_{i+1} = \mathbf{x}_i + \alpha_i\mathbf{p}_i$, $\mathbf{r}_{i+1} = \mathbf{r}_i - \alpha_i\mathbf{A}\mathbf{p}_i$

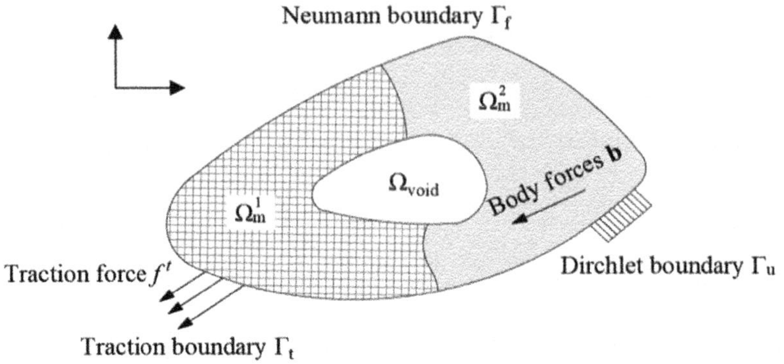

Fig. 1 Definitions of multi-material topology optimization problems

3 Multiple Materials Topology Optimization Model

Since the pioneering research Bendsøe and Kikuchi [1], many approaches have developed for solving topology optimization problems. Besides TO for single material, MTO is also an attractive problem due to composite structure issues and their structural effectiveness and has made outstanding periods. In this part, the main idea of multi-phase and multi-material is presented, including numerical interactions to save computational time by using the presented numerical methods in Sect. 2.

3.1 Multi-material Interpolation

We address the problem of topology optimization based on minimum compliance for multiple materials within a designated design domain denoted as Ω, which is discretized using quadrilateral isoparametric finite elements. Following the approach outlined by Bendsøe and Sigmund [11], the schematic design of the MTO problem involves dividing the domain into subdomains, as illustrated in Fig. 1. Within this framework, Ω_{void} represents void material regions, while Ω_{m}^1 and Ω_{m}^2 correspond to solid subdomains consisting of materials 1 and 2, respectively.

3.2 Minimum Compliance Topology Optimization Problem

This problem is considered in Tavakoli and Mohseni [4] as a first example. The relaxation for densities in [0, 1] is chosen by a very small nonzero value ε_i for this type of problems to avoid the singularity phenomenon. The general formulation of structural MTO problems can be presented as

$$\begin{aligned} \min_{\alpha} : & \ C(\alpha_i, \mathbf{U}) = \mathbf{U}\mathbf{K}^T\mathbf{U} \\ \text{s.t.} : & \quad \mathbf{K}(\alpha)\mathbf{U} = \mathbf{F} \\ & \quad \int_{\Omega} \alpha_i \mathrm{d}x \le V_i \\ & \quad 0 < \varepsilon_i \le \alpha_i \le 1 \end{aligned}, \tag{1}$$

where C is the compliance and the density vector α_i in the phase material ith. V_i is the per-material volume fraction constraint with values of i in range from 1 to n + 1 with unity $\sum_i V_i = 1$. Global stiffness matrix, global load vector, and global displacement vector are corresponding **K**, **F**, and **U**:

$$\mathbf{K}_{ij}^e = \int_{\Omega_e} \mathbf{B}_i^{uT} \overline{D} \mathbf{B}_i^u, \mathbf{F}_i = \{\mathbf{F}_i^u\}, \mathbf{U} = \{\mathbf{u}\}, \tag{2}$$

where $\overline{\mathbf{D}} = \sum_{k=1}^{n+1} \alpha_k^p D_k^0$ with D_k^0 is the material property matrix corresponding to the phase material including Poisson's ratio ν and nominal elastic modulus E_k^0. $\mathbf{B}_i^u$ is standard strain–displacement matrix as follows:

$$\mathbf{D}_k^0 = \frac{E_k^0}{1-\nu^2} \begin{bmatrix} 1 & \nu & 0 \\ \nu & 1 & 0 \\ 0 & 0 & \frac{1-\nu}{2} \end{bmatrix}, \mathbf{B}_i^u = \begin{bmatrix} N_{i,X_1} & 0 \\ 0 & N_{i,X_2} \\ N_{i,X_2} & N_{i,X_1} \end{bmatrix}. \tag{3}$$

4 Numerical Examples and Discussion

4.1 Time Consumption of pCGMG Comparing

In this section, a partial differential equation named Poisson's equation is considered. The statement of Poisson's equation is $\Delta u = f$ in domain Ω. Dirichlet boundary condition, which means u is zeros on $\partial\Omega$. After transferring this equation to weak formulation, finite element and Galerkin method are applied to create **K** and **F** for a linear system **KU** = **F** where **U** is a discrete solution of u. Multigrid V-cycle, Cholesky, Gauss–Seidel, and Jacobi method are dealt with to compare the time-running with five different grids $2^n \times 2^n$ using the same Poisson equation's parameters and maximum level of multigrid. It means direct solution method is utilized in the coarsest mesh 2 × 2: (1) 16 × 16 mesh, multigrid level 3; (2) 32 × 32 mesh, multigrid level 4; (3) 64 × 64 mesh, multigrid level 5; (4) 128 × 128 mesh, multigrid level 6; (5) 256 × 256 mesh, multigrid level 7.

Table 1 reveals that multigrid stands out as the most effective method for efficiently solving large-scale linear systems. In contrast, Cholesky, Gauss–Seidel, and Jacobi methods demonstrate satisfactory performance primarily for smaller-scale

Table 1 Time-running comparison of pCGMG, Cholesky, Gauss–Seidel, and Jacobi method in Poisson's equation

Method	Mesh				
	16 × 16	32 × 32	64 × 64	128 × 128	256 × 256
pCGMG	0.23 s	0.26 s	0.26 s	0.27 s	0.41 s
Cholesky	0.2 s	4.15 s	151.09 s	7880.85 s	No survey
Gauss–Seidel	0.027 s	0.03 s	0.25 s	3.66 s	131.05 s
Jacobi	0.017 s	0.083 s	0.66 s	10.96 s	523.43 s

problems. Multigrid's superiority lies in its mesh-independence, implying that the same number of pCGMG iterations is required regardless of the fine mesh resolution. This feature presents a substantial advantage over other preconditioning techniques. Consequently, the utilization of pCGMG proves optimal for minimizing both execution time and the number of iterations when conducting multi-material topology optimization.

4.2 Minimum Compliance Topology Optimization Results Using Multi-material and pCGMG Depending on Mesh Sizes

In this section, the performance of the proposed preconditioned conjugate gradient multigrid (pCGMG) method is illustrated. MTO of a square wall structure with three distinct materials is carried out. In order to investigate the time-running effect of pCGMG for multi-material topology optimization, several mesh sizes are applied: 64 × 64, 64 × 128, 128 × 128, and 128 × 256. Time-running results and converged curves of compliance are investigated to appeal the merit of the present numerical method. Design conditions of multi-materials of a square wall structure and multigrid parameters for all meshes are described as rf = 8, tolf = 0.05, tol = 1e−3, cgtol = 1e−6, cgmax = 1000, $p = 4$, $\mathbf{e} = [9, 3, 1, 1e-9]$, $\mathbf{v} = [0.16, 0.08, 0.08, 0.68]$. The parameter $p = 4$ illustrates the number of phases considered in the problem while elasticity **e**, volume fraction **v** are physical properties of all four phases. rf and tolf denote a filter radius and filter adjusting tolerance, respectively. tol is a tolerance that is the only solver's stopping criteria, because in this study maximum number of iterations is neglected. The other parameters cgtol and cgmax are tolerance and maximum number of iterations of preconditioned conjugate gradient, respectively.

4.3 Meshes 64 × 64 and 64 × 128

As can be seen in Table 2, pCGMG with $l = 3$ is the best choice with respect to time-running and the number of iterations in a mesh 64 × 64. In Fig. 2a, oscillations of objective function change in case of pCGMG with level 3 are smaller than that of accurate solutions and converged quickly to the tolerant after 210 times of iteration.

In mesh 64 × 128, from Table 2, pCGMG method with level 2 and 4 are outstanding in comparing with direct solution methods. However, the time-running of pCGMG with level 4 is shorter than that of level 2. The reason is that pCGMG with level 2 reaches where the cost of direct solution is not negligible in comparing

Table 2 Comparison of accurate solution and pCGMG in time-running and the number of iterations for minimum compliance TO problem with meshes 64 × 64 and 64 × 128

Mesh		Method					
		Accur. solution	$l = 2$	$l = 3$	$l = 4$	$l = 5$	$l = 6$
64 × 64	Iter.	230	232	210	217	235	216
	Time (s)	367.08	340.07	281.98	311.98	406.90	434.78
64 × 64	Iter.	328	263	274	262	326	321
	Time (s)	1079.46	861.7	729.7	722.01	987.98	1124.62

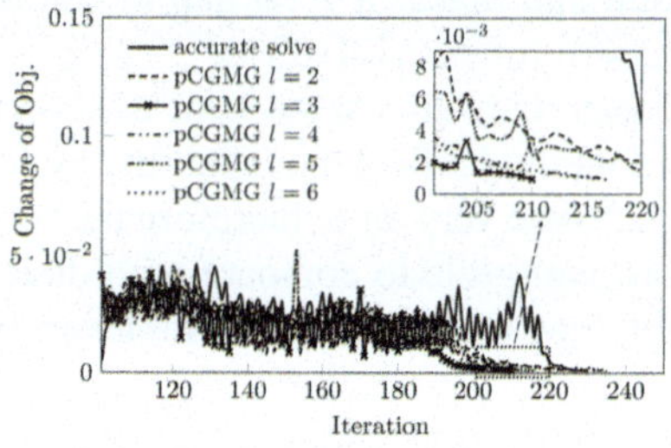

(a) Objective function - 64×64

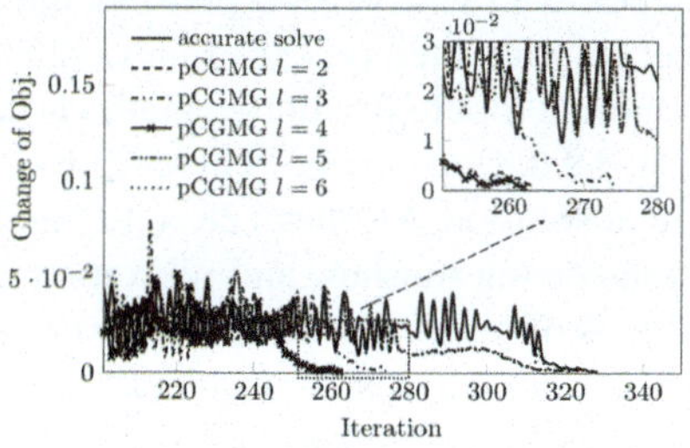

(b) Objective function - 64×128

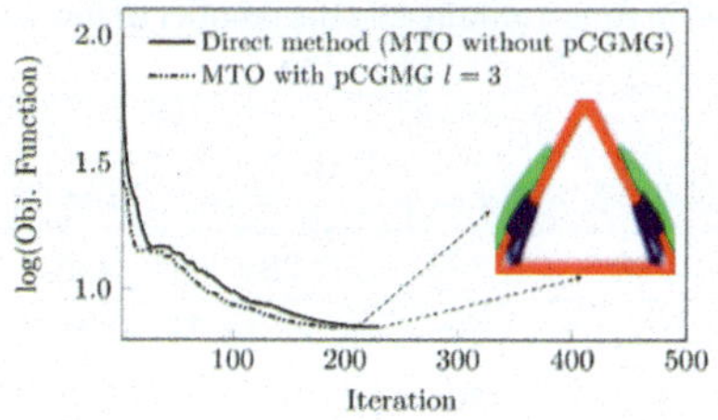

(c) Mesh 64×64. Log10(final compliance): 0.849298784(direct method) and 0.847712454(pCGMG l=3)

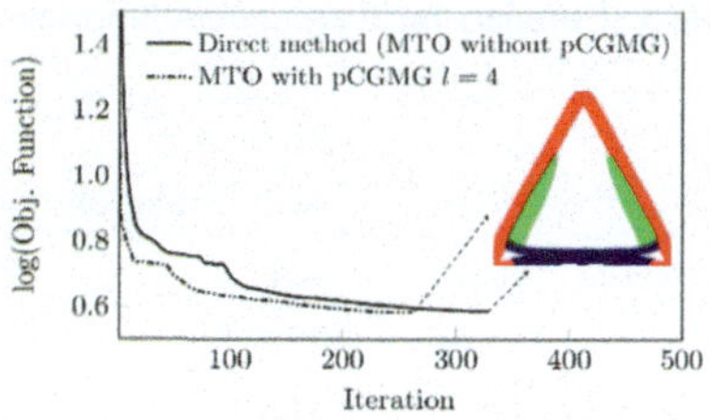

(d) Mesh 64×128. Log10(final compliance): 0.587278731(direct method) and 0.583904966(pCGMG l=4)

Fig. 2 Histories of objective function and the final optimal topologies with three material problem using pCGMG with the best choice of level in mesh 64 × 128 with the cost of one relaxation to sweep on the fine grid. With level 2 of pCGMG method in this case, a mesh 32 × 64 is not good for direct solutions after a mesh restriction step.

Figure 2b shows that objective function changes of pCGMG and direct solution method oscillate constantly until 250th iteration. After this iteration, pCGMG with $l = 4$ converges quickly and stops at 262nd iteration, while an original code using accurate solution keeps this oscillation and stops nearly 330th iteration. Figure 2b, c describes final optimal topologies as well as histories of compliance of the best choice with meshes 64 × 64 and 64 × 128, respectively. With meshes 64 × 64 and 64 × 128, the performance of pCGMG is obviously better than that of direct solution method. Level 3 or level 4 of pCGMG is considerable in these meshes.

4.4 Meshes 128 x 128 and 128 x 256

Meshes 128 × 128 and 128 × 256 have a large number of finite elements, since the number of nodes is 16,641 and 33,153, respectively. With these meshes, two-level is not good because the direct solution applied after one restriction step is not negligible in compared with the cost of one relaxation to sweep on the fine grid as shown in Table 3. pCGMG with level 3 produces the best result for a mesh 128 × 128 in both time-running and the number of iteration (270 times and 1524.1 s). In a mesh 128 × 256, pCGMG with level 5 presents the best result and makes a great gap to that of direct solution method (314 times, 4259.65 s vs. 412 times, 6120.83 s).

Optimal topologies of two methods in these meshes are almost the same as shown in Fig. 3c, d. Convergence curves of the proposed method reflect the reliability of the optimal solution. Meshes 128 × 128 and 128 × 256 are very large therefore the cost of transferring meshes, and relaxation sweeps is negligible to compare with direct solutions of original problems or two-level method. pCGMG methods with levels in a range from 3 to 5 are considerable in these cases.

Table 3 Comparison of accurate solution and pCGMG in time-running and the number of iterations for minimum compliance TO problem with meshes 128 × 128 and 128 × 256

Mesh		Method					
		Accur. solution	$l = 2$	$l = 3$	$l = 4$	$l = 5$	$l = 6$
128 × 128	Iter.	401	375	270	322	284	435
	Time (s)	2732.78	2780.47	1524.1	1969.86	2388.77	3217.66
128 × 256	Iter.	412	319	339	366	314	345
	Time (s)	6120.83	5453.93	5097.79	4903.15	4259.65	5197.33

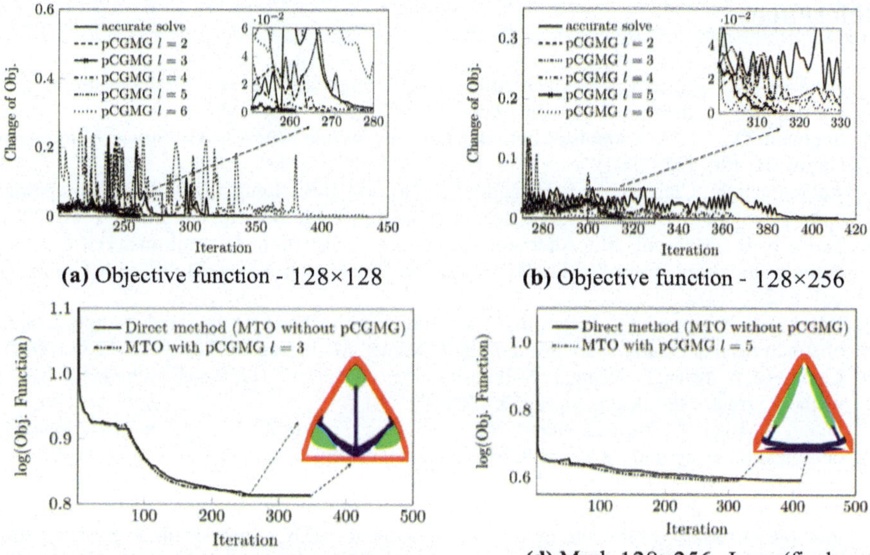

(a) Objective function - 128×128

(b) Objective function - 128×256

(c) Mesh 128×128 . Log_{10}(final compliance): 0.813341063(direct method) and 0.813403928(pCGMG l=3).

(d) Mesh 128×256 . Log_{10}(final compliance): 0.592475897(direct method) and 0.592475735(pCGMG l=5).

Fig. 3 Histories of objective function and the final optimal topologies with three material problem using pCGMG with the best choice of level in mesh 128 × 128

5 Conclusion and Future Works

This study proposes a pCGMG method for large-scaled MTO problems. The pCGMG method presents a great efficient computational approach not only on a large linear system raised by discretization of a PDE problem, but also on MTO in large scale. The effectiveness of pCGMG is to considerably reduce time-running and the number of iterations as well as stabilities of error. The main finding of this article is to evaluate an appropriate choice of level in pCGMG methods which is oriented on the size of mesh and efficient computation expense. In large-scale problems, higher levels may be considered because the cost of accurate solution in these levels is negligible in compared with the cost of transferring mesh. The effectiveness of pCGMG is to considerably reduce time-running and the number of iterations as well as stabilities of error. The positive results reported here clearly illustrate the suitability of pCGMG methods for applications running in large-scale MTO problems.

For the near future's work, this also encourages a new motivation based on high-performance parallel computing environments for enhancing results of crack topology optimization problems or functionally graded materials (FGMs) problems which usually take computational burdens.

References

1. Bendsøe, M., Kikuchi, N.: Generating optimal topologies in structural design using a homogenization method. J. Comput. Appl. Math. **71**(2), 97–224 (1998)
2. Sigmund, O.: A 99 line topology optimization code written in MATLAB. Struct. Multidiscip. Optim. **21**, 120–127 (2001)
3. Andreassen, E., Clausen, A., Schevenels, M., Lazarov, B.S., Sigmund, O.: Efficient topology optimization in MATLAB using 88 lines of code. Struct. Multidiscip. Optim. **43**, 1–16 (2011)
4. Tavakoli, R., Mohseni, M.: Alternating active-phase algorithm for multimaterial topology optimization problems: a 115-line MATLAB implementation. Struct. Multidiscip. Optim. **49**, 621–642 (2014)
5. Bruns, T.E., Sigmund, O., Tortorelli, D.A.: Numerical methods for the topology optimization of structures that exhibit snap-through. Int. J. Numer. Meth. Eng. **55**(10), 1215–1237 (2002)
6. Liu, X., Li, Z., Wang, L., Wang, J.: Solving topology optimization problems by the guide-weight method. Struct. Multidiscip. Optim. **6**, 136–150 (2011)
7. Wang, Y., Liao, Z., Ye, M., Zhang, Y., Li, W., Xia, Z.: An efficient isogeometric topology optimization using multilevel mesh, MGCG and local-update strategy. Adv. Eng. Softw. **139**, 102733 (2020)
8. Zhang, W., Song, J., Zhou, J., Du, Z., Zhu, Y., Sun, Z., Guo, X.: Topology optimization with multiple materials via moving morphable component (MMC) method. Int. J. Numer. Meth. Eng. **113**(11), 1653–1675 (2018)
9. Chin, T.W., Leader, M.K., Kennedy, G.J.: A scalable framework for large-scale 3D multimaterial topology optimization with octree-based mesh adaptation. Adv. Eng. Softw. **135**, 102682 (2019)
10. Hestenes, M.R., Stiefel, E.: Methods of conjugate gradients for solving linear systems. J. Res. Natl. Bur. Stand. **49**(6), 409–436 (1952)
11. Bendsøe, M.P., Sigmund, O.: Material interpolation schemes in topology optimization. Arch. Appl. Mech. **69**, 635–654 (1999)

Chapter 6
Generated Graph for Text Encryption Algorithm Based on BRHC Curve

Batool Hatem Akar AlKfari and Ruma Kareem K. Ajeena

1 Introduction

In this study, we study graph. $G(V, E)$ is a set of edges connecting a set of vertices named V to a set of edges named E. A path graph is a route from one vertex to another along which no vertex appears twice. When the cycle includes all of the graph's vertices, it is called a cycle graph. A graph is complete if there is an edge connecting every pair of vertices. There are two primary methods to representing graphs: Adjacency list and adjacency matrix are two terms for the same thing. The graph $G(V, E)$'s adjacency list representation is a matrix of V lists, one for every vertex in V. For each vertex v, the adjacency list of v contains all adjacent (connected by an edge) vertices. The adjacency-matrix graph representation $G(V, E)$ is comprised a $|V| \times |V|$ matrix gij. Any subgraph (a tree) of a graph whose edge weight is less than or equal to a threshold value is called a spanning tree. There are several algorithms exist to solve these problems, including the Kruskal algorithm and the prime algorithm [1–5]. Researchers that have employed graph theory with encryption methods include the following refrences [6–13].

B. H. A. AlKfari · R. K. K. Ajeena (✉)
Department of Mathematics, Education College for Pure Sciences, University of Babylon, Babylon, Iraq
e-mail: pure.ruma.k@uobabylon.edu.iq

B. H. A. AlKfari
e-mail: batool.hatem.pure291@student.uobabylon.edu.iq

J. M. R. S. Tavares et al. (eds.), *Proceedings of Second International Conference on Intelligent System*, Algorithms for Intelligent Systems,
https://doi.org/10.1007/978-981-99-8976-8_6

2 BRHC Curve

A new curve known as the BRHC curve is defined in this section. On this curve, the formulae for addition and doubling points are introduced. It has been demonstrated that an affine BRH curve is smooth. Thus, a discussion of the term will serve as the beginning point.

Definition 2.1 Suppose that field F with a char. $\neq 2$ and $\alpha, \beta \in F$. The BRHC curve over a field F is described by

$$\mathrm{BRH}_{\alpha,\beta} : x\left(\alpha y^2 - 1\right) = \beta y\left(x^2 - 1\right). \tag{1}$$

If and only if the condition $\beta\alpha\left(\beta^2 - \alpha\right) \neq 0$ exists, $\mathrm{BRH}_{\alpha,\beta}$ is a non-singular.

Definition 2.2 Suppose $P_1 = (x_1, y_1)$ and $P_2 = (x_2, y_2)$ on the BRHC. With $x_1x_2 \neq \pm 1$ and $\alpha y_1 y \neq \pm 1$, then the sum point $P_1 + P_2$ is determined by

$$x_3 = \frac{(x_1 + x_2)(\alpha y_1 y_2 + 1)}{(x_1 x_2 + 1)(1 - \alpha y_1 y_2)}$$

and

$$y_3 = \frac{(y_1 + y_2)(x_1 x_2 + 1)}{(1 - x_1 x_2)(\alpha y_1 y_2 + 1)}.$$

Definition 2.3 Let $P_1 = (x_1, y_1)$ is on the BRHC. With $x_1^2 \neq \pm 1$ and $\alpha y_1^2 \neq \pm 1$, the doubling point is determined by

$$x_2 = \frac{\alpha x_1\left(\alpha y_1^2 + 1\right)}{\left(x_1^2 + 1\right)\left(1 - \alpha y_1^2\right)}$$

and

$$y_2 = \frac{\alpha y_1\left(x_1^2 + 1\right)}{\left(1 - x_1^2\right)\left(1 + \alpha y_1^2\right)}.$$

3 Generated Graph and Its an Adjacent Matrix

This technique begins by representing the data as graph vertices, with each letter represented by a vertex and all neighboring letters represented by adjacent vertices in the data. Continue adding vertices until a cycle graph is formed. Each edge in this graph has a weight that signifies the distance between these two letters in the encoding table (the database used to encode all alphabetic characters). Then, every vertex in the graph will be connected with edges to form a complete graph, with the weight sequence for each new edge commencing with the final index in the encoding table. The adjacency matrix for the entire graph is constructed. Based on the entire graph, the spanning tree (ST) graph is then constructed. The (ST) graph is represented as a diagonal adjacency matrix preserving the character order of the data.

4 Text Encryption Algorithm Using BRHC Curve Based on Generated Graph

The proposed algorithm of encrypting the text can be presented as follows.

A Generation key process:

Input: a prime number P, $\delta = (x, y) \in$ BRHC with order h, e, $\phi \in (F_p)$.

1. Create a table to encode each alphabet using δ.

A	B	C	...	Y	Z
δ	2δ	3δ	...	25δ	26δ

2. Select public a matrix $\gamma \in GL_n(F_p)$.

Key creation

1. Alice chooses a privet key a.
2. She compute $A = \gamma^a (\text{mod } p) \in GL_n(F_p)$.

Encryption

1. Bob chooses a privet key b and compute $c_1 = \gamma^b (\text{mod } p)$.
2. He selects his message m.
3. Each character is converted to a vertex.
4. Connect every pair of related characters to create a cycle graph.
5. The weight of every edge, according to the table of encoding, corresponds to the separation between the two linked vertices.
6. Adding additional edges to complete the graph.
7. Add a special character to denote the beginning of the string (let A).
8. Compute ST and its adjacency matrix M_1.
9. Select the spreading tree.
10. Store the order of the vertices in the diagonal positions of the M_2 matrix.
11. Compute $c_2 = \gamma^b M_2$.

Bob send (c_1, c_2) to Alice.

Decryption

1. Alice compute $c_1{}^{-a}\ c_2$ to find M_2.
2. She knew that (0, 0) node is A. So by using encoding table, she computes the original text.

5 Study Case of the Text Encryption Algorithm Using BRH Curve Based on Generated Graph

Suppose $p = 103$ be a prime number and $\text{BRH}_{5,11}$: $x(5y^2 - 1) = 11y(x^2 - 1)(\text{mod } 103)$.

Choose $\delta = (101, 65)$ lies on $\text{BRH}_{7,11}$.

Then, a character encoding table is generated (Table 1).

And choose public matrix

$$Q = \begin{bmatrix} 2 & 3 & 5 & 6 & 3 \\ 7 & 0 & 1 & 1 & 16 \\ 9 & 7 & 2 & 2 & 11 \\ 12 & 11 & 2 & 1 & 7 \\ 13 & 9 & 8 & 2 & 1 \end{bmatrix}$$

Now, Alice performs the following steps:

1. She chooses her privet key $a = 2$.
2. Compute her public key Υ by

$$\Upsilon = \begin{bmatrix} 78 & 16 & 59 & 37 & 33 \\ 62 & 79 & 16 & 65 & 49 \\ 46 & 4 & 45 & 89 & 41 \\ 16 & 89 & 30 & 102 & 79 \\ 60 & 18 & 97 & 102 & 79 \end{bmatrix}.$$

Alice's keys are $a = 2$ and a matrix Υ.

Table 1 Table encoding

A	B	C	...	H	...	X	Y	Z
δ	2δ	3δ	....	8δ	...	24δ	25δ	26δ

Now, Bob does the following steps:

1. He chooses her privet key $b = 3$.
2. He computes $C_1 = Q^b$.

$$C_1 = \begin{bmatrix} 24 & 53 & 38 & 87 & 99 \\ 75 & 76 & 16 & 28 & 8 \\ 66 & 51 & 6 & 26 & 101 \\ 86 & 48 & 35 & 93 & 1 \\ 74 & 31 & 9 & 8 & 57 \end{bmatrix}.$$

3. He chooses his massage (BRHC).
4. He converts each character to vertex (Fig. 1).
5. Using the encoding table, the message is written in a form

$$M = \{(17, 24), (23, 51), (7, 90), (65, 43)\} \in \mathrm{BRH}_{5,11}.$$

   And compute the weight of each edge.
6. He continues to add edges to the graph until it is full (Fig. 2).

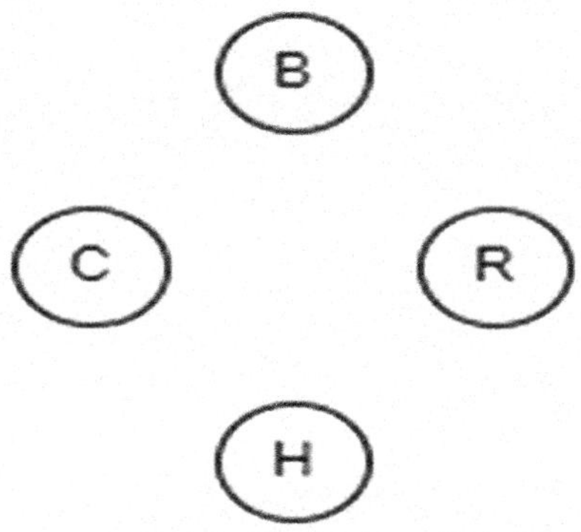

Fig. 1 Change each letter to a vertex

7. To signal to the first character, he inserted a special character in front of him (Fig. 3).
8. He finds the spanning tree (Fig. 4).
9. He represents as a matrix M_1.

$$M_1 = \begin{bmatrix} (0,0) & (19,62) & (0,0) & (0,0) & (0,0) \\ (19,62) & (0,0) & (6,27) & (0,0) & (0,0) \\ (0,0) & (6,27) & (0,0) & (87,39) & (0,0) \\ (0,0) & (0,0) & (87,39) & (0,0) & (58,56) \\ (0,0) & (0,0) & (0,0) & (58,56) & (0,0) \end{bmatrix}.$$

10. He stores the character order diagonally rather than (0, 0).

A	*B*	*R*	*H*	*C*
(0, 0)	(1, 1)	(2, 2)	(3, 3)	(4, 4)

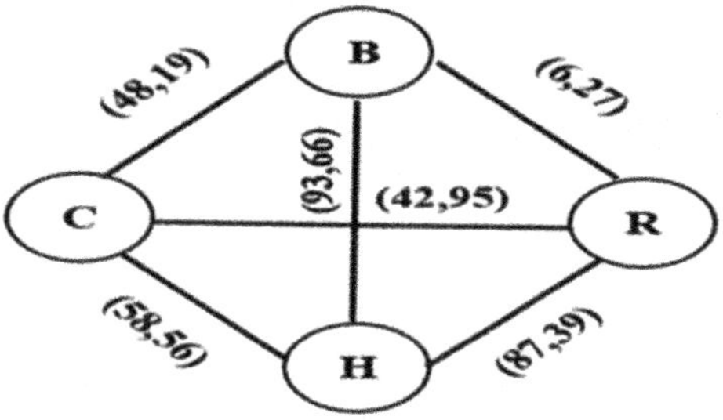

Fig. 2 Weighted graph contains plain text characters. Complete plain graph

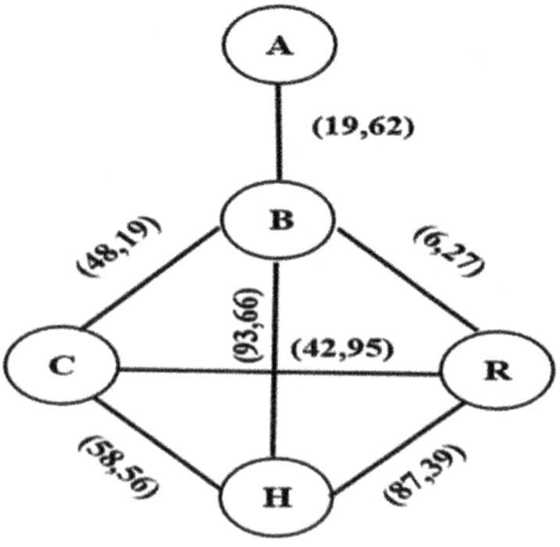

Fig. 3 Complete plain graph with a special character

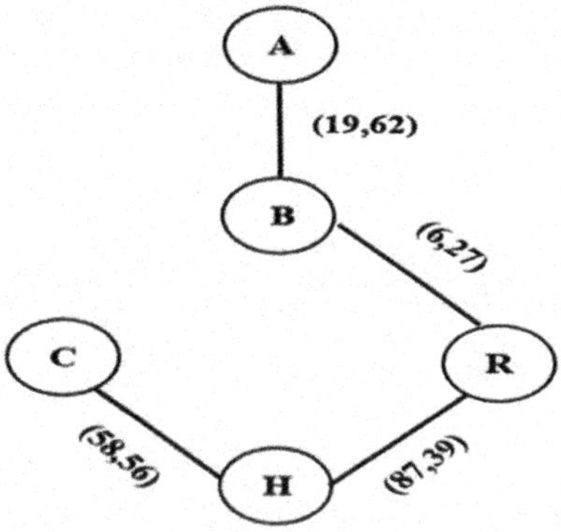

Fig. 4 Minimum spanning tree graph

$$M_2 = \begin{bmatrix} (0,0) & (19,62) & (0,0) & (0,0) & (0,0) \\ (19,62) & (1,1) & (6,27) & (0,0) & (0,0) \\ (0,0) & (6,27) & (2,2) & (87,39) & (0,0) \\ (0,0) & (0,0) & (87,39) & (3,3) & (58,56) \\ (0,0) & (0,0) & (0,0) & (58,56) & (4,4) \end{bmatrix}.$$

11. He computes $C_2 = Q^b M_2$.

$$C_2 = \begin{bmatrix} (75,55) & (86,8) & (76,1) & (81,8) & (40,20) \\ (5,38) & (57,90) & (79,88) & (58,82) & (64,89) \\ (36,47) & (54,33) & (90,43) & (76,48) & (83,12) \\ (73,81) & (40,101) & (26,101) & (14,45) & (53,50) \\ (61,31) & (24,20) & (62,2) & (36,59) & (48,9) \end{bmatrix}.$$

12. Bob sends (C_1, C_2) to Alice.

Decryption process

1. Alice computes $C_1{}^{-a}$.
2. Compute $C_1{}^{-a} C_2$

$$C_1^{-2} C_2 = \begin{bmatrix} (0,0) & (19,62) & (0,0) & (0,0) & (0,0) \\ (19,62) & (1,1) & (6,27) & (0,0) & (0,0) \\ (0,0) & (6,27) & (2,2) & (87,39) & (0,0) \\ (0,0) & (0,0) & (87,39) & (3,3) & (58,56) \\ (0,0) & (0,0) & (0,0) & (58,56) & (4,4) \end{bmatrix} = M_2.$$

3. She assumes that the node (0, 0) is A and uses an encoding table to compute its value.

$$(0,0) = \text{code } A = (101, 65).$$

$$
\begin{aligned}
\text{Node}\,(1,1) &= \text{code}\,A + w_1 \\
&= ((101, 65) + (19, 62))(\text{mod } 103) \\
&= (17, 24) = B. \\
\text{Node}\,(2,2) &= \text{code}\,B + w_2 \\
&= (23, 51) = R. \\
\text{Node}\,(3,3) &= \text{code}\,R + w_3 \\
&= (7, 90) = H. \\
\text{Node}\,(4,4) &= \text{code}\,H + w_4 \\
&= (65, 34) = C.
\end{aligned}
$$

So on until she got the original text (BRHC).

6 Conclusion

We provide a new cryptographic technique in this work, one that can be implemented with any algorithm already in use to encrypt data in transit utilizing an encoding table and the features of full graphs and spanning trees from graph theory. Using two keys, one for encryption and the other for decryption, would increase the complexity and security of public-key cryptography. Future advancements can be made to reduce the number of matrices required to encrypt and decrypt messages.

References

1. Corman, T.H., Leiserson, C.E., Rivest, R.L., Stein, C.: Introduction to Algorithms, 2nd edn. McGraw-Hill
2. Yamuna, M., Gogia, M., Sikka, A., Khan, M.J.H.: Encryption using graph theory and linear algebra. Int. J. Comput. Appl. (2012). ISSN: 2250-1797
3. Ustimenko, V.A.: On graph-based cryptography and symbolic computations. Serdica J. Comput. 131–156 (2007)
4. Paszkiewicz, A., et al.: Proposals of graph based ciphers, theory and implementations. ResearchGate
5. Lu, S., Ostrovsky, R., Manchala, D.: Visual cryptography on graphs. In: CiteSeerx, COCOON, p. 225 (2008)
6. Diffie, W., Hellman, M.: New directions in cryptography. IEEE Trans. Inf. Theory **22**(6), 644–654 (1976)
7. ElGamal, T.: A public key cryptosystem and a signature scheme based on discrete logarithms. IEEE Trans. Inf. Theory **31**(4), 469–472 (1985)
8. Al Etaiwi, W.M.: Encryption algorithm using graph theory. J. Sci. Res. Rep. 2519–2527 (2014)
9. Agarwal, S., Uniyal, A.S.: Prime weighted graph in cryptographic system for secure communication. Int. J. Pure Appl. Math. **105**(3), 325–338 (2015)
10. Shen, A., Tong, R., Deng, Y.: Application of classification models on credit card fraud detection. In: 2007 International Conference on Service Systems and Service Management, pp. 1–4 (2007)

11. Aljamaly, K., Ajeena, R.: Undirected complete graph to design new public key cryptosystem. J. Phys. Conf. Ser. **1897**(1) (2021). IOP Publishing
12. Jasim, T.H., Abed, S.A., Ghazi, A.: Concepts of bi-supra topological space via graph theory. Wasit J. Comput. Math. Sci. **1**(2) (2022)
13. Kalaivani, N., Rahman, K.F.U., Obaid, A.J.: GSβ-compactness of topological spaces with grills. In: Peng, S.L., Lin, C.K., Pal, S. (eds.) Proceedings of 2nd International Conference on Mathematical Modeling and Computational Science. Advances in Intelligent Systems and Computing, vol. 1422. Springer, Singapore (2022). https://doi.org/10.1007/978-981-19-0182-9_56

Chapter 7
Issues and Challenges of Digital Banking System

Sarita Kumari Singh, Prasant Kumar Pattnaik, and Sasmita Samanta

1 Introduction

Due to advances in computing power, Earth is becoming increasingly digitized, including the financial sector, enhancing its growth and competitiveness. This has prompted traditional financial institutions like banks to modernize through cutting-edge digital technologies. These innovations, spanning computer systems, digital communication, the Internet, and mobile phones, along with necessary databases and software apps, boost the speed, effectiveness, and security of banking services. Information technology has led to the integration of electronic, Internet, and digital banking, driven further by Internet expansion, broadband networks, social media, and new data processing systems, raising client expectations for convenience and customization.

The proliferation of smart phones and global connectivity has transformed the environment and technological landscape. Electronic payments, facilitated by innovations such as Big Data, widespread Internet access, and cloud computing, are predicted to significantly impact the economy. M-banking and online transactions are gaining traction, particularly post-COVID-19, due to their safety, speed, and convenience. Numerous AI companies are leveraging scientific research and machine learning to help financial institutions create advanced tools, aiming to enhance their operations.

S. K. Singh (✉) · P. K. Pattnaik · S. Samanta
KIIT Deemed to be University, Bhubaneswar, India
e-mail: sarita.singhfcm@kiit.ac.in

J. M. R. S. Tavares et al. (eds.), *Proceedings of Second International Conference on Intelligent System*, Algorithms for Intelligent Systems,
https://doi.org/10.1007/978-981-99-8976-8_7

The banking sector, a cornerstone of economies, drives economic reforms. The advent of digital banking has propelled financial inclusion by reducing costs compared to traditional banking, extending accessibility beyond physical bank branches. This inclusivity benefits large populations, including the financially marginalized, enabling them to access financial services conveniently and affordably. Financial inclusion also fosters sound financial management and future planning, promoting savings and wealth creation.

Recognizing the significance of digital banking in an economy's financial system, this review aims to understand digital banking-related issues in both developed and emerging countries. Drawing from existing literature, the characteristics of digital banking will be outlined to aid comprehension, followed by discussions on its importance and challenges in developing nations. The study's findings hold relevance for banking sector managers and the general public in effectively embracing digital banking.

Numerous studies have explored the application of Multiple Criteria Decision-Making (MCDM) techniques in digital banking. The Analytic Hierarchy Process (AHP) emerges as a prominent tool, lauded for its efficacy in assessing and prioritizing digital banking attributes based on customer preferences, thereby contributing significantly to heightened customer satisfaction and loyalty. Parallel to this, the Technique for Order Preference by Similarity to Ideal Solution (TOPSIS) is acknowledged for its adeptness in evaluating and ranking digital banking services in alignment with customer requirements. This approach facilitates banks in making informed choices regarding digital banking solutions, particularly in areas concerning security and user convenience. Additionally, the study touches upon other MCDM methodologies such as the Weighted Sum Model (WSM), Grey Relational Analysis (GRA), and the Electre Method. These methods are recognized for providing a holistic framework in the evaluation and selection of optimal digital banking solutions, considering a wide range of criteria including cost, reliability, and accessibility. The integration of MCDM techniques in digital banking yields various benefits, including improved customer satisfaction, increased loyalty, and enhanced operational efficiency and cost savings for banks.

2 Overview of the Digital Banking System

Digital banking has transformed financial management, offering convenience and accessibility. This system employs technology and online platforms to provide diverse banking services. It encompasses terms like electronic, Internet, and online banking [2]. Digital banking enhances traditional online and mobile services by integrating tech tools, social media, innovative payments, and customer-centric approaches. Customers can manage accounts, perform transactions, and access features through websites, apps, or other digital platforms. These systems often include personalized insights, budgeting tools, and integration with external services, empowering customers with resources to manage finances and achieve goals.

AI applications are reshaping the financial sector, reducing errors, strengthening customer relations, curbing fraud, and enhancing security. AI and ML streamline banking processes, resulting in a more efficient operation [21]. Digital banking encompasses banking services through electronic channels like ATMs, phones, and the Internet, digitalizing core and non-core operations, aiming to simplify transactions from any location.

3 Key Functionalities of Digital Banking

Digital banking systems enhance customer banking experiences by providing a range of functionalities. Users can access and manage accounts conveniently, eliminating physical visits. They can check balances, view transaction history, and update personal information. Funds transfer is simplified, allowing quick movement between accounts within the same bank or different financial institutions.

Customers can electronically pay bills, set recurring payments, and receive reminders for due dates. Mobile deposits enable cheque submission through app-captured images, eliminating the need for visits. Loan applications, document submission, and progress tracking are seamless. Card management enhances security by activating new cards, setting spending limits, and reporting losses. Personalized insights, budgeting tools, and guidance on savings and investments aid financial management.

Prompt customer support is available through chat, email, or phone. Robust security measures, including encryption and biometric authentication, protect data and transactions. Integration with external services ensures a smooth experience, while real-time alerts prevent fraud. Some systems offer investment and wealth management services [30].

Digital wallet integration enables secure mobile payments. Digital banking empowers users to manage finances efficiently, access insights, and receive support within a comprehensive platform.

4 Security and Privacy Concerns Related to the Digital Banking System

4.1 Privacy and Security Concerns

Digital banking raises significant privacy and security worries demanding attention to protect customer data. Key issues include data breaches, where cybercriminals gain unauthorized access, risking identity theft and financial harm. Fraudulent activities like phishing and malware attacks exploit trust, accessing sensitive info and causing financial damage. Weak authentication and security gaps might let unauthorized individuals compromise accounts, breaching confidentiality, and misusing funds [3]. Adhering to data protection laws such as GDPR is vital, as breaches result in penalties and harm reputation. Phishing and spoofing mimic real banking sites, tricking customers into revealing details for unauthorized access and thus leds to financial loss. Malware and hacking jeopardize data security, potentially leading to financial fraud. Insider threats from staff with data access can cause breaches without controls [34]. Tackling these issues demands strong security and regulatory adherence, securing data and building trust in digital banking.

4.2 Security Measures

To address these concerns, digital banking systems deploy a variety of security measures. These include strong authentication methods like multi-factor and biometric checks, data encryption for storage and transmission security, robust firewalls and intrusion detection systems, and proactive fraud detection. Regular audits are conducted to ensure compliance and assess vulnerabilities [25]. Additionally, two-factor authentication, network safeguards, and mobile security enhancements further bolster the defense. Educating customers about phishing risks enhances awareness. Integrating advanced biometrics, AI-driven fraud detection, and blockchain technology further enhance security [26]. Collaboration, real-time monitoring, and strict data policies contribute to a secure environment. Advocacy for stronger regulations plays a crucial role. These steps collectively ensure the security and evolution of digital banking in a dynamic landscape [3].

5 Challenges in Ensuring Service Availability in Digital Banking

5.1 Challenges

Digital banking platforms can experience various challenges that can disrupt their operations or affect their performance. Hardware or network failures, such as malfunctioning servers or connectivity issues, can lead to system downtime or degraded performance. Software glitches or bugs can also affect the stability and performance of the platform, resulting in slow response times, transaction errors, or system crashes [3]. Cyberattacks, hacking attempts, and data breaches pose a significant threat to digital banking platforms, potentially causing service disruptions, unauthorized account access, or compromise of sensitive financial information. Scheduled system upgrades, maintenance activities, or software deployments may require temporary service interruptions. Unexpected spikes in user traffic during peak periods or events can strain system resources and impact performance. Digital banking platforms often rely on third-party services like payment processors or data providers. Disruptions or performance issues with these services can affect platform functionality adversely [8].

5.2 *Strategies for Minimizing Such Disruptions*

Financial institutions can implement various measures within their digital banking platforms to mitigate customer impact. Key steps involve establishing redundant systems and backup infrastructure for high availability and fault tolerance. Regular testing and updates address software vulnerabilities, while robust cybersecurity measures protect against cyber threats and data breaches. Disaster recovery plans and drills ensure readiness for system failures [34]. Transparent communication channels inform customers about service disruptions and alternative access methods. Diverse customer support channels offer assistance during outages. Collectively, these measures minimize customer impact and ensure a reliable digital banking platform. For effective disaster recovery, financial institutions can adopt strategies including redundant infrastructure, backup processes, cloud solutions, real-time monitoring, incident response, testing, vendor management, employee training, and continuous improvement [9]. Redundant infrastructure enhances resilience, cloud solutions offer scalability and recovery, while real-time monitoring proactively handles disruptions. These efforts, along with collaboration with regulatory authorities, ensure compliance and adaptability [19].

6 Challenges in Customer Support and Communication in Digital Banking

Digital banking systems face challenges in providing customer support. Personalization can be lacking due to the absence of face-to-face assistance, necessitating personalized support channels. Timely responses are crucial, with chatbots and live chat addressing queries efficiently, but staffing and accuracy are also vital. Diverse accessible support options are important, along with seamless integration for consistency [5]. Overcoming language and cultural barriers requires multilingual support and training for agents. Empowering customers with self-service tools reduces dependency. Moreover, Continuous improvement via feedback and surveys identifies enhancement areas [12].

To enhance digital banking support, financial institutions can take specific steps. Personalization can be improved by using customer profiles and transaction history for tailored assistance. Omni-channel integration ensures a consistent experience. AI-powered chatbots with natural language processing manage inquiries, escalating complex issues when needed. Real-time communication features like chat notifications provide immediate help. Proactive channels update users about important information [19]. Multichannel feedback mechanisms gather insights, and continuous training empowers agents. These measures enhance customer support, satisfaction, and communication in digital banking systems.

7 Interoperability Challenges in Digital Banking System

Integrating digital banking systems with other financial platforms poses intricate interoperability challenges due to differences in technology infrastructure, data formats, and security standards, leading to compatibility issues [26]. A significant hurdle involves seamless data exchange between systems, essential for communication with payment processors, third-party services, and regulatory reporting. Discrepancies in data formats and communication protocols can cause disruptions and transaction delays. Compatibility concerns also extend to transaction processing across diverse systems, demanding standardized transaction formats, adherence to industry protocols like ISO 20022, and strong transaction reconciliation. Security and compliance pose additional interoperability complexities. Integrating external systems requires aligning security mechanisms, encryption standards, and authentication protocols. Mismatched security frameworks and regulations complicate sensitive data exchange. Addressing this entails secure data exchange protocols, industry security standards adherence (e.g., PCI DSS), and rigorous security assessments during integration [32]. To overcome these challenges, financial institutions can collaborate and develop industry standards. Engaging in initiatives like open banking promotes standardized APIs and data exchange. Building flexible systems for easy integration and implementing thorough testing and certification processes ensures compatibility with diverse financial systems. By addressing these hurdles, digital banking systems can achieve seamless integration, data exchange, and transaction processing, enhancing operational efficiency and customer experience.

8 Regulatory and Compliance Challenges in Digital Banking

Digital banking systems grapple with regulatory challenges arising from legal requirements, data protection mandates, and industry standards. Compliance spans anti-money laundering (AML), know your customer (KYC), consumer protection, and data privacy laws, compounded by Payment Card Industry Data Security Standard (PCI DSS) and General Data Protection Regulation (GDPR) complexities across regions. Adapting to evolving regulations requires vigilant monitoring, interpretation, and implementation. Effective compliance management, risk assessments, and regulatory communication also plays an important role [8]. Data protection is pivotal, with GDPR and data privacy laws demanding rigorous security, consent management, and secure practices. Regular audits, training, and vulnerability assessments reinforce safeguards. Cross-border operations further amplify challenges, necessitating alignment with varying regulatory frameworks and harmonizing practices. To address these, proactive compliance, robust frameworks, resource allocation, audits, and a compliance culture are vital. Technology, like regulatory compliance software, streamlines processes and tracks changes [30]. In conclusion, digital banking must adeptly navigate a multifaceted regulatory landscape. Staying updated, embracing data protection, and fostering proactive compliance engender trust and regulatory alignment.

9 Scalability and Performance Challenges in Digital Banking

Digital banking systems face scalability and performance challenges due to surging demand for services and an expanding customer base. As digital banking adoption grows, the system must handle higher traffic and transactions without sacrificing performance or user experience. Ensuring scalability for a growing customer base is a challenge. The platform must accommodate new users and their activities. Cloud computing, load balancing, and auto-scaling support this whereas, handling large transaction volumes poses another hurdle to this. Batch processing, parallel processing, and optimized databases enhance performance, whereas concurrency issues during peak periods also impact responsiveness. Techniques like rate limiting, session management, and caching can be very helpful. Optimizing code and queries, leveraging CDNs, and refining caching strategies further enhance performance. Financial institutions need to monitor and plan for the future. Load testing, capacity planning, and upgrades optimize performance. Advanced analytics and emerging tech like serverless or edge computing can also help. In conclusion, proactive strategies like scalable infrastructure, efficient transaction processing, and performance optimization ensure digital banking systems meet customer demands while offering a seamless experience.

10 Challenges in Integrating Emerging Technologies in Digital Banking

The integration of emerging technologies into digital banking presents financial institutions with both opportunities and challenges. While adopting technologies like AI, ML, blockchain, and biometric authentication offers significant benefits, it also entails intricate technical and operational complexities. The seamless integration of these technologies into existing systems requires meticulous consideration of compatibility, data integration, and system architecture. Technical challenges emerge while implementing AI and ML algorithms, demanding quality data, effective training, and robust privacy measures. Blockchain integration introduces its own set of hurdles, including scalability, interoperability, and compliance concerns. Selecting suitable platforms, designing smart contracts, and ensuring regulatory alignment become crucial [28]. Biometric authentication introduces challenges regarding security, user acceptance, and standardization. Prioritizing robust security measures, safeguarding user privacy, and driving standardization efforts are imperative to address these concerns. Operational challenges involve staff training and skill development to effectively utilize these technologies. Financial institutions must invest in comprehensive training programs for AI, ML, blockchain, and biometric technologies, while continuous monitoring ensures their successful integration [34]. By effectively addressing compatibility, data integration, and security concerns, alongside navigating technical and operational challenges, financial institutions can leverage the potential of emerging technologies to enhance customer experiences, bolster security, and drive efficiency in the realm of digital banking.

11 Analyzing and Enhancing User Experience

The user experience (UX) in digital banking is crucial for customer satisfaction, yet common usability issues can hinder effective system use. These include complex navigation, unclear interfaces, confusing processes, inconsistency, vague error messages, and inadequate feedback [5]. To improve UX, financial institutions should prioritize user research, simplify navigation, enhance interfaces, streamline processes, provide clear errors, and offer feedback [24]. Responsive design, personalization, integration with tools, support channels, accessibility, testing, and performance optimization are the key.

By refining UX based on feedback and trends, digital banking can create an engaging platform. User-centered design meets needs, while clear processes reduce friction. Navigation, branding, and design can enhance usability. Responsive design caters to devices, while intuitive interfaces ease use. Personalization empowers users, and performance ensures a smooth experience. Proactive notifications informs, and improvement based on feedback drives enhancements [30] as well as support and security bolster trust. Through these strategies, financial institutions can create a satisfying digital banking experience that meets evolving customer needs, enhances engagement, and fosters loyalty.

12 Role of Digital Banking in Development of Emerging Economy

Banks have invested significantly in technology to improve efficiency and customer experience, utilizing digital channels like ATMs, Internet banking, mobile banking, and kiosks to offer quality services and reduce costs [33]. Digitization aims to connect remote areas to the financial sector for economic growth, offering underserved populations new banking services through digital channels. Mobile apps also extend access to non-financial services. The COVID-19 pandemic has accelerated financial digitization. Contactless transactions have grown by 69%, with global consumers embracing digital payments. However, some regions, like India, still have low digital payment usage rates, urging the need for greater awareness. Developing nations see growth in branchless banking, especially in rural areas. Business Correspondent (BC) agents play a crucial role, providing financial access to the unbanked, boosting economic security [22].

Compared to government banks, private, multinational, and international commercial banks have achieved great technological adoption success. The impact of digital banking in other ASEAN nations is also substantial. The contributions are as follows:

- 24 × 7 ATMs accessibility.
- Enhanced customer service through the utilization of mobile phones.
- Increased turnaround time for payment processing compared to branch counter services.
- As a result of increased system automation and integration, banks have lower operational expenses, which boosts their overall earnings.
- Better financial services with lower fees, to the advantage of depositors.

Further digitalization of financial services can give consumers more tailored services and provide banks with capabilities in artificial intelligence and business analytics (Fig. 1).

A "word cloud" is a graphical depiction of the frequency of words. The larger a keyword appears in the resulting graphic, the more frequently it appears in the image generated. The word cloud shown in Fig. 1 is made by collecting data from the Scopus database, and it has been analyzed using R Studio.

Figure 1 shows that a huge amount of study has already been conducted on digital banking with its impact on financial inclusion. However, the figure shows that there has been less work done on digital banking with respect to developing countries, and even lesser work is being conducted in context of developed countries. Therefore, this study focuses on analyzing digital banking issues and challenges in developed and developing countries. Figure 2 shows an increasing trend of research in the field of

Fig. 1 Word cloud for digital banking

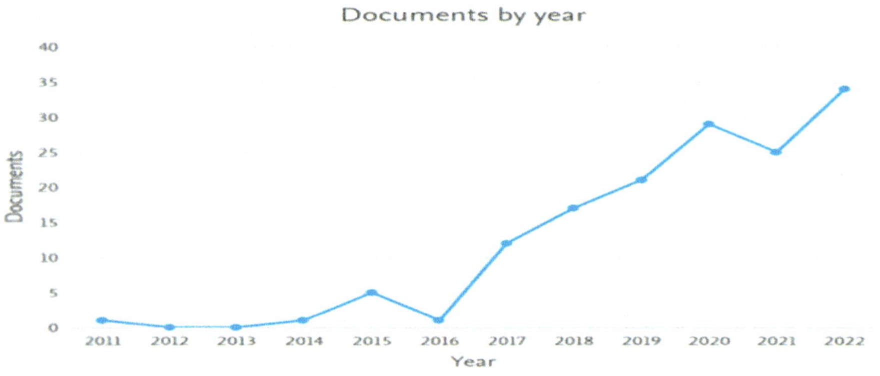

Fig. 2 Digital banking

digital banking. There is an increasing trend of research in the case of digital banking studies in developing countries, but there is a stagnant trend in case of developed countries.

13 Summary of Digital Banking System in Developing Countries

The review provides valuable insights into the adoption and impact of digital banking and fintech services in different developing countries. It collectively offer insights into the factors influencing digital banking adoption and the challenges and opportunities in different developing countries, contributing to the understanding of the global landscape of digital banking and fintech. Kaur analyzed India's adoption of digitalization and found that fintech had a significant impact, with people preferring fintech technologies over traditional banks during the COVID-19 pandemic [16]. In Vietnam, Pham discovered that despite the increasing use of digital banking services,

Vietnamese banks are expanding their physical branch networks, particularly to cater to individuals who lack access to traditional banking services [27]. Mufarih et al. studied digital banking users in Indonesia and found that perceived trust and risk were influential factors in user attitudes and intentions toward digital banking adoption [11]. Rathnaweera et al. identified perceived ease of use, platform, and intent to use as crucial factors for digital banking adoption in the Western Province of Sri Lanka [29]. Mastran highlighted the challenges faced by banks in Nepal in increasing e-banking adoption, including low customer response rates and concerns over security and privacy [23]. Alam et al. emphasized the need for e-banking in Bangladesh to bring non-banking individuals into the banking umbrella and highlighted the lack of infrastructure and awareness as barriers [1]. Cheng and Sharmayne discussed the appeal of digital banking applications in the Philippines, citing higher interest rates but the need for improvements in registration and bank services [6]. Gautam and Sharma highlighted the potential for fintech growth in Bhutan and the need for investments in fintech industries [10]. Kazmi and Hashim pointed out the challenges faced by Pakistan in the adoption of online banking, including issues of trust, security, and limited technology infrastructure [17]. Aziz et al. stressed the importance of branchless banking in Malaysia and encouraged its rapid deployment for increased acceptability and the advancement towards a cashless society [4].

From India's preference for fintech during the pandemic to Vietnam's expansion of physical branches for underserved individuals, and the factors influencing adoption in Indonesia, Sri Lanka, Nepal, Bangladesh, and Pakistan, these studies highlight the diverse challenges and opportunities. Additionally, the potential for fintech growth in Bhutan and the importance of branchless banking in Malaysia underscore the significance of digital banking. These findings inform policymakers and financial institutions in navigating the evolving landscape of digital banking and fintech in developing economies.

14 Summary of Digital Banking Systems in Developed Countries

Several studies from different developed countries have shed light on the adoption and impact of digital banking and fintech services. In Singapore, Gerrard and Barton Cunningham found that adopters of Internet banking favored its ease of use and compatibility, while innovativeness played a role in their adoption [11]. Lee highlighted the potential disruption virtual banks pose to traditional banks in Hong Kong, with customers dissatisfied with conventional banks being more inclined to enroll with virtual banks [20]. Choi discussed the catfishing effect of digital banks on the Korean banking industry, highlighting how incumbent banks improved their services to compete [7]. Iwashita explored the reasons for the slow change in Japanese banking despite fintech developments, attributing it to traditional practices and slow digitization [13]. Jünger and Mietzner emphasized the need for Germany to catch up in fintech adoption and identified trust, familiarity with technology, and financial literacy as influencing factors [15]. Rod et al. examined customer satisfaction with Internet banking in New Zealand, highlighting the importance of service quality [31]. Jhuremalani et al. found that the COVID-19 pandemic had a significant financial impact on elderly Australians, emphasizing the importance of human interaction during stressful times [14]. Khatoon et al. emphasized the role of e-banking service quality in customer satisfaction and purchase intentions in Qatar [18].

These studies collectively provide valuable insights into the factors influencing digital banking adoption, customer satisfaction, and the impact of fintech on traditional banking systems in different developed countries. Understanding these dynamics can inform policymakers, financial institutions, and technology providers in shaping the future of digital banking and fintech services.

15 Issues and Challenges of Digital Banking in Developed and Developing Economy

In addition to the comprehensive examination of the challenges and issues faced by digital banking systems, as discussed earlier in the paper, it is crucial to recognize that specific challenges emerge in both developing and developed economies due to a variety of factors. Through an extensive review, the following challenges have been explored and analyzed.

15.1 Developing Economy

Developing economies confront unique obstacles that impede the expansion of digital banking in the country. The important ones are discussed below.

A country requires a sufficient degree of infrastructure to adopt the technology and the necessary support for its growth and use. However, infrastructure in emerging nations is still in its infancy. Internet connectivity has yet to reach many people. Financial literacy is still poor in underdeveloped nations, making it difficult for banks to convince customers to use services beyond opening a savings account. Unlike other established economies such as Singapore and Indonesia, which have issued digital bank licenses for non-banking firms, legislation for digital banks is not clearly defined in most emerging nations. This necessitates that any startup wishing to build a digital bank join with an existing bank, which can drastically slow down the process.

Since the Internet is an open communication medium, the privacy and security of data carried over the network are vulnerable to theft, illegal access, and exploitation. As a result, many are wary of digital banking services, citing security, and identity theft worries. Cybercrime and hacking are significant, and even the most reputable businesses are susceptible. This has led to a loss in confidence in banks' digital activity.

Although the banking sector in emerging nations is extensively regulated in terms of capital adequacy standards, provisioning and asset recognition regulations, investments, bankruptcy norms, etc., applying these principles to digital banks is challenging. For example, although the Reserve Bank of India (RBI) is developing guidelines for specific activities (such as awarding NBFC status to peer-to-peer lending platforms), these regulatory standards have yet to be clearly defined and implemented promptly. Furthermore, many regulations, like KYC standards, necessitate interaction and face-to-face meetings between the banker and customer, which may not be achievable with a purely digital bank. The technological boom in the banking industry has produced new types of products/services and delivery methods. However, different legal standards, such as the definition of electronic signature and authorization, are required. At the same time, existing legal reports and licenses must be reconsidered.

15.2 Developed Economy

Although, in developed economies, independent digital banks are becoming famous. Banks and other financial organizations, especially digital financial service providers, are under increasing pressure to maintain client privacy proactively. In the World Retail Banking Report, 80% of bank executives highlighted cybersecurity and privacy as important concerns as digital banking develops.

16 Summary

Banks are expanding their e-banking services to match competition, harness technology, and optimize profits while minimizing costs. However, a major hurdle lies in addressing customers' low response rates and fortifying security measures. Customer reluctance to adopt e-banking stems from limited Internet usage, unfamiliarity with advanced tech, and concerns over data security and privacy. To fully benefit from e-banking, banks must invest resources in assuring customers of upgraded security measures, potentially through an online banking education program. Addressing cybersecurity issues is crucial for both emerging and developed economies. This involves updated education, awareness, and information sharing, as well as proactive monitoring and updating of security practices and resources. Developed nations have embraced independent digital banks, whereas developing countries lag behind. Nonetheless, customers occasionally require physical support, necessitating a sufficient number of branches for satisfactory service. Employing Multiple Criteria Decision-Making (MCDM) techniques in digital banking offers several advantages. Firstly, it empowers banks and customers to make well-informed decisions based on multiple factors, aligning with customer needs. Secondly, MCDM identifies areas for enhancement in digital banking services, enhancing customer satisfaction, and loyalty. Thirdly, it aids in optimizing these services, bolstering efficiency and cost savings. However, MCDM in digital banking comes with limitations. It demands considerable data for evaluating and selecting optimal solutions, potentially consuming time and resources. Furthermore, its reliance on subjective criteria may not accurately reflect customer preferences. Lastly, it might not account for changing customer needs, leading to suboptimal solutions. Despite these limitations, MCDM techniques enhance digital banking decision-making, amplifying service quality. Further research is essential to refine MCDM applications in digital banking and address these limitations effectively.

References

1. Alam, M.J., Jesmin, F., Faruk, M., Nur-Al-Ahad, M.: Development of E-banking in Bangladesh: a survey study. Financ. Mark. Inst. Risks **5**(2), 42–51 (2021). https://doi.org/10.21272/fmir.5(2).42-51.2021
2. Alkhowaiter, W.A.: Digital payment and banking adoption research in Gulf countries: a systematic literature review. Int. J. Inf. Manag. **53**, 102102 (2020). https://doi.org/10.1016/j.ijinfomgt.2020.102102
3. Asbaş, C., Tuzlukaya, Ş.: Cyberattack and cyberwarfare strategies for businesses. In: Conflict Management in Digital Business, pp. 303–328. Emerald Publishing Limited (2022)
4. Aziz, K.A., Jabar, M.A., Abdullah, S., Nor, R.N.H.: Challenges from the disastrous COVID-19 pandemic: exposure to opportunities for branchless banking in Malaysia. Bull. Electr. Eng. Inform. **11**(4), 2339–2347 (2022). https://doi.org/10.11591/eei.v11i4.3865
5. Chauhan, S., Akhtar, A., Gupta, A.: Customer experience in digital banking: a review and future research directions. Int. J. Qual. Serv. Sci. **14**(2), 311–348 (2022)
6. Cheng, L.C., Sharmayne, L.R.: Analysing digital banking reviews using text mining. In: Proceedings of the 2020 IEEE/ACM International Conference on Advances in Social Networks Analysis and Mining, ASONAM 2020, pp. 914–918 (2020). https://doi.org/10.1109/ASONAM49781.2020.9381429
7. Choi, Y.: Digital Banks: Lessons from Korea. World Bank Group, 2. http://hdl.handle.net/10986/34701 (2020)
8. Dimitrova, I., Öhman, P., Yazdanfar, D.: Barriers to bank customers' intention to fully adopt digital payment methods. Int. J. Qual. Serv. Sci. **14**(5), 16–36 (2022)
9. Egala, S.B., Boateng, D., Mensah, S.A.: To leave or retain? An interplay between quality digital banking services and customer satisfaction. Int. J. Bank Mark. **39**(7), 1420–1445 (2021)
10. Gautam, K.P., Sharma, D.: Fintech and digital transformation in the era of industry 5.0: where does Bhutan stand? Asian J. Econ. Financ. Manag. **7**(3), 21–27 (2022). https://globalpresshub.com/index.php/AJEFM/article/view/1587
11. Gerrard, P., Barton Cunningham, J.: The diffusion of Internet banking among Singapore consumers. Int. J. Bank Mark. **21**(1), 16–28 (2003). https://doi.org/10.1108/02652320310457776
12. Hussein, A.S., Sumiati, S., Hapsari, R., Abu Bakar, J.: Bank 4.0 experiential quality and customer loyalty: a serial mediating role of customer trust and engagement. TQM J. (2022)
13. Iwashita, N.: Why fintech is not changing Japanese banking. Asian Econ. Policy Rev. **17**(2), 297–310 (2022). https://doi.org/10.1111/aepr.12390
14. Jhuremalani, A., Potter, R., Lushington, K., Lowies, B.: COVID-19: financial well-being of older Australians in times of crisis. Australas. J. Ageing **41**(3), 440–447 (2022). https://doi.org/10.1111/ajag.13061

15. Jünger, M., Mietzner, M.: Banking goes digital: the adoption of FinTech services by German households. Finance Res. Lett. **34** (2020). https://doi.org/10.1016/j.frl.2019.08.008
16. Kaur, N.: FinTech Evolution to Revolution in India—From Minicorns to Soonicorns to Unicorns, pp. 1–6 (2021)
17. Kazmi, S.S.A., Hashim, M.: E-banking in Pakistan: issues and challenges. Int. J. Acad. Res. Bus. Soc. Sci. **5**(3), 50–54 (2015). https://doi.org/10.6007/ijarbss/v5-i3/1498
18. Khatoon, S., Zhengliang, X., Hussain, H.: The mediating effect of customer satisfaction on the relationship between electronic banking service quality and customer purchase intention: evidence from the Qatar banking sector. SAGE Open **10**(2) (2020). https://doi.org/10.1177/2158244020935887
19. Larsson, A., Viitaoja, Y.: Building customer loyalty in digital banking: a study of bank staff's perspectives on the challenges of digital CRM and loyalty. Int. J. Bank Mark. **35**(6), 858–877 (2017)
20. Lee, E.: Digital financial inclusion: observations and insights from Hong Kong's virtual banks. Law Contemp. Probl. **84**(1), 95–113 (2021)
21. Mahapatra, P., Singh, S.K.: Artificial intelligence and machine learning: discovering new ways of doing banking business. In: Artificial Intelligence and Machine Learning in Business Management, pp. 53–80. CRC Press (2021)
22. Maharjan, M.: System architecture for social security cash transfer through branchless banking in Nepal, pp. 60–64 (2015)
23. Mastran, S.: Opportunities and challenges of online banking in Nepal. The Batuk **7**(2), 37–51 (2021). https://doi.org/10.3126/batuk.v7i2.39498
24. Mbama, C.I., Ezepue, P., Alboul, L., Beer, M.: Digital banking, customer experience and financial performance: UK bank managers' perceptions. J. Res. Interact. Mark. **12**(4), 432–451 (2018)
25. Mir, R.A., Rameez, R., Tahir, N.: Measuring Internet banking service quality: an empirical evidence. TQM J. **35**(2), 492–518 (2022)
26. Pio, P.G., Sigahi, T., Rampasso, I.S., Satolo, E.G., Serafim, M.P., Quelhas, O.L., Filho, W.L., Anholon, R.: Complaint management: comparison between traditional and digital banks and the benefits of using management systems for improvement. Int. J. Product. Perform. Manag. (2023)
27. Pham, T.T.: What is the future of the bank branch in the midst of the 4.0 industrial revolution? Answers from Vietnam. https://doi.org/10.21511/bbs.17(2).2022.17 (2022)
28. Rahman, A., Abedin, M.J.: The Fourth Industrial Revolution and private commercial banks: the good, bad and ugly. Int. J. Organ. Anal. **29**(5), 1287–1301 (2021)
29. Rathnaweera, L., Karunasena, A.: Influencial factors of adopting digital banking by users in western province of Sri Lanka, pp. 132–137 (n.d.)
30. Rathnayake, N.D.N.B.: The era of the transition—from traditional to digital banking. In: Transformation for Sustainable Business and Management Practices: Exploring the Spectrum of Industry 5.0, pp. 41–55. Emerald Publishing Limited (2023)
31. Rod, M., Ashill, N.J., Shao, J., Carruthers, J.: An examination of the relationship between service quality dimensions, overall internet banking service quality and customer satisfaction: a New Zealand study. Mark. Intell. Plan. **27**(1), 103–126 (2009). https://doi.org/10.1108/02634500910928344
32. Santos, A.A.D., Ponchio, M.C.: Functional, psychological and emotional barriers and the resistance to the use of digital banking services. Innov. Manag. Rev. **18**(3), 331–348 (2021)
33. Sarel, D., Marmorstein, H.: Marketing online banking services: the voice of the customer. J. Financ. Serv. Mark. **8**, 106–118 (2003)
34. Tn, N., Shailendra Kulkarni, M.: Zero click attacks—a new cyber threat for the e-banking sector. J. Financ. Crime (2022)

Chapter 8
Impact of COVID-19 Pandemic on Assessment Process of Technical Education System-Analysis and Critical Review

Srabanti Chakraborty, Srinka Basu, and Indrajit Saha

1 Introduction

The deadly impact of novel corona virus began at the end of December 2019 in Wuhan city of China [1]. In late 2019, the effect of a new variant of the virus the coronavirus (SARS-CoV-2) spread across almost all the inhabited countries and became a dangerous pandemic that challenges the whole world [2]. As the post effect of COVID-19 pandemic, the whole world witnessed a big disruption of total education system along with a paradigm shift in the teaching–learning process. Along with the other sectors, the technical education system was also greatly affected by that pandemic that imposed far-reaching transformation and adjustments of the system. Due to rapid attack of COVID-19 pandemic, the total education system faced a big challenge due to sudden lockdown.

Therefore, in that pandemic situation, to compensate the hamper in regular education system, new approaches were required with effective and innovative technologies since the traditional face-to-face teaching–learning process was absolutely not possible during that pandemic situation. Hence to recover that situation and to ensure the continuity of teaching–learning process, most of the educational institutions chose the online-based teaching–learning process. As a tool of online education system,

S. Chakraborty (✉)
Elitte Institute of Engineering and Management, Kolkata, India
e-mail: srabanti2k@gmail.com

S. Basu
University of Kalyani, Kalyani, India

I. Saha
National Institute of Technical Teachers' Training and Research, Kolkata, India
e-mail: indrajit@nitttrkol.ac.in

J. M. R. S. Tavares et al. (eds.), *Proceedings of Second International Conference on Intelligent System*, Algorithms for Intelligent Systems,
https://doi.org/10.1007/978-981-99-8976-8_8

the application of learning management system (LMS) has been accepted as an efficient and effective alternative in respect of teaching, learning, and assessment [3]. The technical education system has always focused on live experiments and hand on practices. Thus, the pandemic effect has made a big impact on the learning and assessment process of technical education system that in turn affected the academic performance of the students. The evaluation process is a significant factor for any education system [1]. A fair assessment process encourages both the learner and teacher to continue with a successful teaching–learning system. Hence effects of the COVID-19 pandemic on assessment process plays an important role on students' career. During COVID-19 phase, to continue the learning process, the situation demanded a virtual mode of learning process. Therefore the learning methodology was flipped and the mode of education system changed from offline to online. As a result, in most of the cases, the online teaching–learning system followed the active learning system and that has a remarkable impact on students' assessment process. Our study tried to get answer to the question that what were the factors that affected the assessment process of technical education system during the time of COVID-19 pandemic. For that purpose, a questionnaire set were prepared for the technical students to assess the possible reasons behind the change in academic performance if any, and to know various obstructions they faced to continue the study and examination system during this pandemic time.

2 Literature Review

We have studied and reviewed various research papers, journals, conference papers, web sources, and documents as part of the analysis of the online education system and assessment process in technical education system. The online assessment processes are absolutely related to technologies and also dependent on modern and expensive licensed assessment software system along with huge number of computer systems and technology skilled members [4]. Over the last few decades, technology great impact on each and every sector of education system.

In [5], the author defined that the educational institutions want frequent occurrence of thorough assessment process of learning outcomes. Author [6] stated that assessment process in education system is surrounded by classroom assessment as well as external assessment which include the process of data collection and assessment regarding the student learning information. In [5], the researchers stressed and highlighted on assessment cycle where the assessment is used for betterment of the education system with a systematic and planned execution. As per their view, assessment and evaluation method in engineering education section is unsatisfactory. Authors focused in [7] on a feedback analysis report for exploring the students' views on assessment experiences which specified the summative learning system with a few formative learning characteristics. In [8] the authors demonstrated that the assessment process is related to the educational objectives of different factors like evaluation, guidance, analysis, judgment, selection, administration, grading, placement etc.

Author in [9] identified technical primary factors that have impact on the achievement of online assessment by applying the Delphi method. The online assessment processes are absolutely related to technologies and also dependent on modern and expensive licensed assessment software system along with huge number of computer systems and technology skilled members [10]. In [11], the author highlighted on two essential factors of online assessment—concept of academic dishonesty (AD) and trustworthy assessment (TA). The purpose of the study was to keep the honesty and excellence of online assessment by finding different forms of academic dishonesty and also the way to minimize AD and raise the TA. According to [12], authors stated that in current years, the application and implementation of online teaching–learning system and consequent online assessment process have increased due to the sudden COVID-19 attack and also for the development of ICT mode education. In this scenario, to handle this odd situation the online assessment system also expands with quiz, assignment, and other tools of evaluation by moving to online platform in Emergency Remote Teaching (ERT) period [13].

3 Data Collection Method and Participant Details

A Google form with questionnaire sets were designed and sent to students of different colleges from six universities of West Bengal, Odisha, and Agartala for getting feedback. Our survey process was conducted for 2 months.

A total of 293 technical students participated in our survey. Figure 1a is showing details of the participants from different universities. Figure 1b is presenting the male and female ratio details of participant and Fig. 1c is showing the age details of participants.

4 Result and Analysis

4.1 Different Parameters Distributed in Feedback

In our survey study, we have tried to distribute various parameters in feedback for getting the reviews namely:

- Background–demographic profile
- Family income and education.
- Role of online technologies in technical education system during COVID-19.
- Role of online technologies in the acquired subject knowledge of technical students during COVID-19.

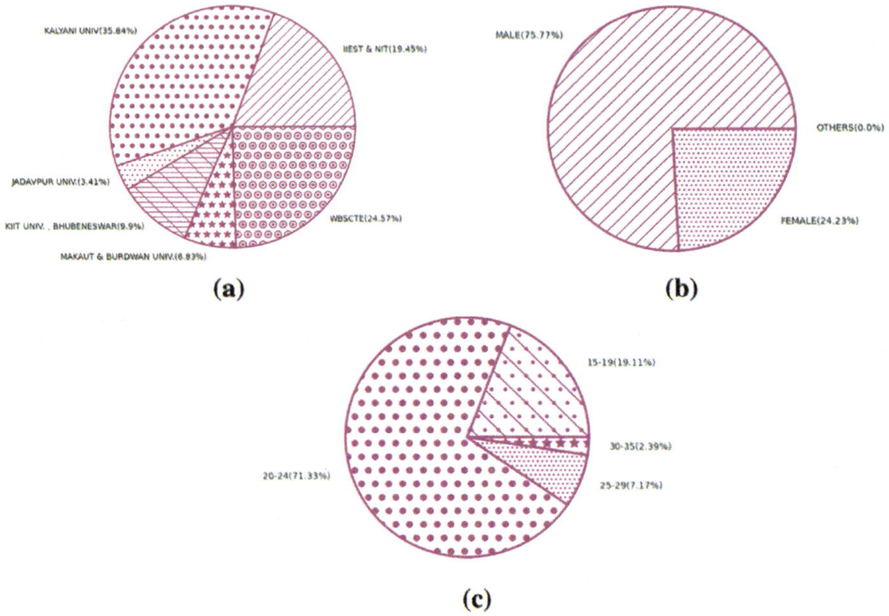

Fig. 1 Participant details. **a** Participants from different Universities. **b** Gender details of participants. **c** Age details of participants

- Role of changed examination pattern in technical education system during COVID-19 in the survey questionnaire, the participants were requested to give their honest responses regarding their experiences of online education system and corresponding performances during the COVID-19 period.

4.2 Data Analysis Method

In this section, from the feedback, we have generated the parameter—"**Performance (*Δ*)**" and checked which factors have the influence on performance of participants.

Performance calculation (*Δ*) method: The performance of participant students (Δ) has been calculated on the basis of their average marks (in percentage form) obtained, before and during COVID-19 examination.

(*Δ*) = change of average marks of the participants (in grade) before and during COVID-19 pandemic in the following manner:

Increased. participant students' average marks (in %) before COVID-19 pandemic > participant students' average marks (in %) during COVID-19 pandemic.

Decreased. participant students' average marks (in %) before COVID-19 pandemic < participant students' average marks (in %) during COVID-19 pandemic.

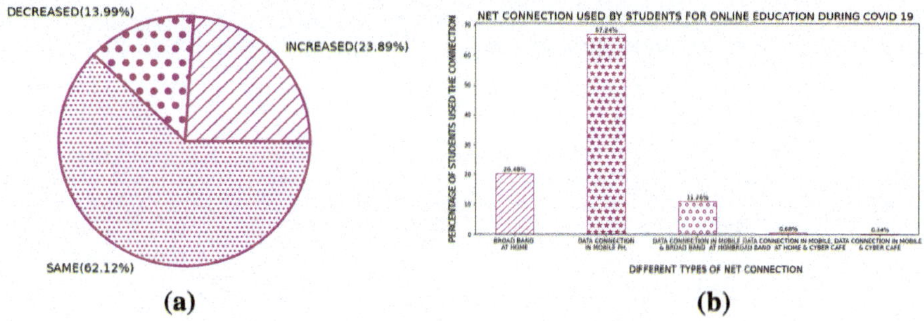

Fig. 2 Factors effect on performance. **a** Performance. **b** Different net connections

Same. participant students' average marks (in %) before COVID-19 pandemic = participant students' average marks (in %) during COVID-19 pandemic.

Graphs to show changed performance of participants and impact of different factors on performance (Fig. 2).

A correlation has been established with the performance of student and different factors by using the Chi-Square and Cramer's V rule.

Figure 2a shows the status of students' performance (Δ) which has been calculated through the change in average marks of the participants before and during COVID-19 pandemic. Out of 293 participants, change in performance were observed for 111 number of students (37.88%) and same performance were observed for rest 182 students (62.12%). Out of total 111 students, 63% responded increase in performance and 37% responded decrease in their performance.

4.3 Different Parameters Distributed in Survey

Background Demographic Profile. In this section, our survey studied on different demographic details of participants (age, gender, etc.). As shown in the Fig. 3, the male and female respondent ratio is 3:1. Here, we generated Table 1 for showing the range of marks obtained by male and female students before COVID-19 and during COVID-19.

We generated Table 1 which is describing the gender wise range of average marks of participants from 20 to 99%, before COVID-19 and during COVID-19. On the basis of the changed marks of participants, we calculated the median values to find the changes in marks before and during COVID-19 for male and female students separately.

The statistical median value of the above records calculated on the basis of the formula (1):

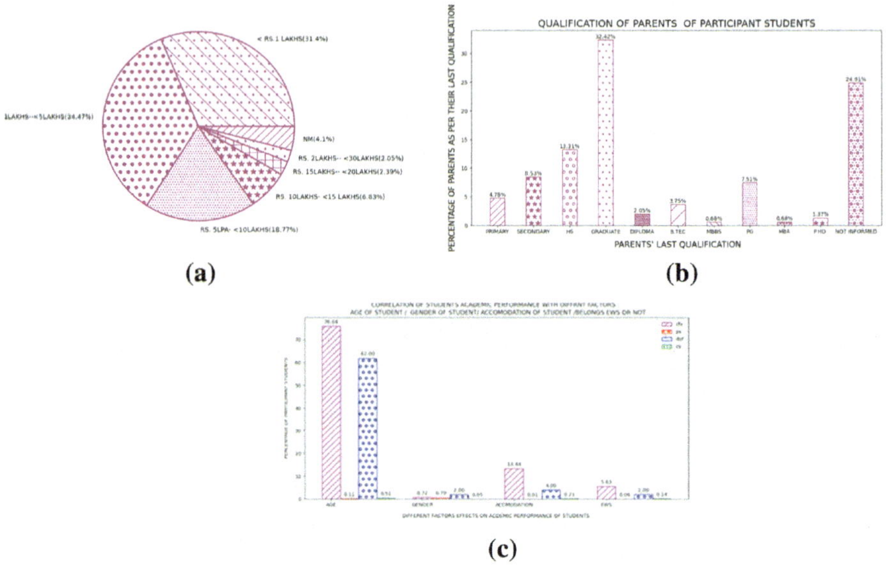

Fig. 3 Impact of different factors on performance. **a** Gross family income distribution. **b** Parents' highest qualification. **c** Effects on performance

Table 1 Table for showing the average marks (in percentage) of male and female participants before and during COVID-19

Marks (%)	Male		Female	
	Before COVID	During Covid	Before COVID	During Covid
20–39	4	1	–	–
40–59	8	9	3	4
60–79	98	77	38	30
80–99	112	135	30	37
Total	222		71	

$$\text{Median} = L + \frac{\left(\frac{N}{2}\right) - C}{f} \times h \tag{1}$$

where

L lower limit of the median class;
N total frequency;
C cumulative frequency up to the class preceding the median class;
f frequency of the median class;
h width of the median class.

The calculated result shown that median value of male before COVID-19 is 79.68 whereas the median value of male during COVID-19 is 83.06. Again the median value of female before COVID is 76.61 and the median value of female during COVID-19 is 80.31.

Therefore, it is clear that the median marks of male students increased by 3.38 (from 79.68 to 83.06) and the median marks of female students increased by 3.70 (from 76.61 to 80.3).

4.4 Family Income and Education

As per [14] the academic qualification and family background also play a vital role in education system. In this section, we prepared the questionnaire set to get the reviews of the participants regarding their family income, accommodation, EWS status, whether any family members related to technical education field and parents' highest qualification. Here, Fig. 3 is showing the details of above information and also showing an association of all the factors of this section with students' performance. From the feedback, it has been noticed that 71% of parents are from nontechnical background and 29% of students have their family members from the technical education system.

Figure 3 is representing the association of different demographic factors with student's performance (**Δ**).

4.5 Role of Online Technologies in Technical Education System During COVID-19

Impact of COVID-19 has caused closure of academic institutions all over the world. As a consequence, the teaching–learning process has shifted its mode radically to continue the learning process and accepted the digital online platform. Therefore, to cope with this system, applications of consistent network connection, necessary devices and the required knowledge became indispensable for both students and teachers. As per the survey feedback, it has been noticed that during COVID-19, the students used different online tools and technologies for continuing the online classes and examination system. We focused on the impact of different online technologies on the students' performance (Δ). As per the feedback report, most of the participant students have their supporting devices.

In our survey questionnaire set, when the students were asked regarding their supporting electronic devices for their online classes and examination, the feedbacks (Fig. 4.) are showing that 88% of total students already had their devices, 12% participants of total number were without required devices and 21% of that 12% purchased new devices while 11% continued online classes without own devices.

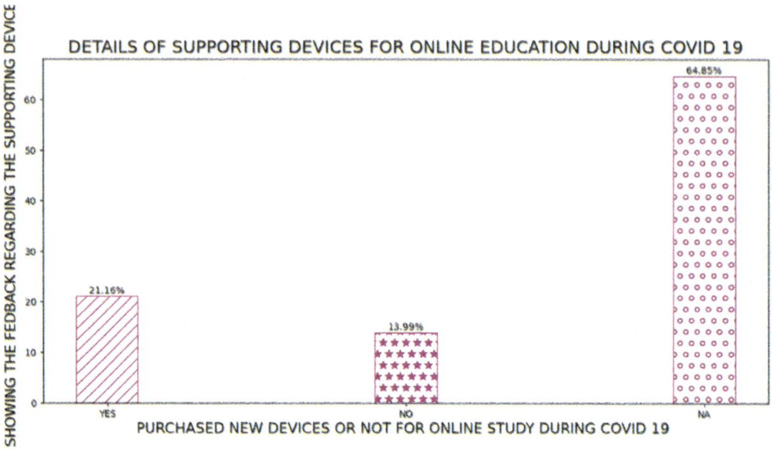

Fig. 4 Status of supporting devices of participants for continuing online study during COVID-19

Impact of online tools used by the participant: During survey, the participants were also asked to inform their online platforms, net connection and supporting devices they used during COVID-19. As per the participants' feedback, the most chosen tools for online mode education system for live as well as virtual theory and lab session were G-Meet, G-Classroom, Zoom, WhatsApp, Teams. Figure 5 is showing that out of total number of participants, 96% at least used the G-Meet, 32% at least used Zoom, 96% at least used G-Classroom, 15% at least used WhatsApp, 9% of participants at least used MS Teams, and 4% of participants at least used other tools during this pandemic phase to continue their online education.

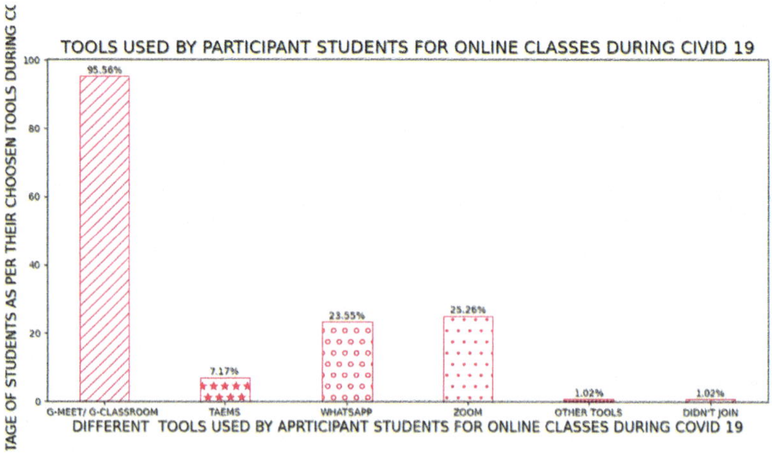

Fig. 5 Different online tools used by the participants for continuing online study during COVID-19

As per their feedback, we have got 17 sets of online tools combinations, they used during COVID-19, shown in Table 2.

The feedback report is showing that out of 293 students, total 290 students used different combination of tools (from set 1-set 17).

Table 3 is showing the percentage of students who used the 17 set of tools. Table 3 is displaying the percentage of average marks of participants they scored during COVID-19 phase by using 17 different sets of tools (Table 3) while appearing their online mode examination. It is observed that set 1 tool has been used by maximum number of participants (147) but the median result is showing that maximum number of students (59% of total students) scored 80–99% marks in the examination that held during COVID-19 phase.

On the basis of the record shown in Table 3, the calculation has been done by using Eq. 1. The result of calculation is showing that set 4, set 12, and set 13 are containing the highest median value. It indicates that students who used set 4, set 12, and set 13 as the tool for online study and online examination, got highest marks. Therefore set 4, set 12, and set 13 can be considered as most effective tools for online examination and they have major impact on performance of students.

Impact of Internet Connection, Online Class, and Study Materials: As per the feedback report, Fig. 2b is showing the details of different network connection used by the participants during the COVID-19.

Table 2 Different set of online tools used by the participant students during COVID-19 for their online exam

SET	Tool
SET 1	G-meet/ G-classroom
SET 2	G-meet/ G-classroom, Others
SET 3	G-meet/ G-classroom, Teams
SET 4	G-meet/ G-classroom, Teams, WhatsApp, Others
SET 5	G-meet/ G-classroom, Teams, Zooms
SET 6	G-meet/ G-classroom, Teams, Zooms, WhatsApp
SET 7	G-meet/ G-classroom, Teams, Zooms, WhatsApp, Others
SET 8	G-meet/ G-classroom, WhatsApp
SET 9	G-meet/ G-classroom, WhatsApp, Others
SET 10	G-meet/ G-classroom, Zooms
SET 11	G-meet/ G-classroom, Zooms, WhatsApp
SET 12	G-meet/ G-classroom, Zooms, WhatsApp, Others
SET 13	Others
SET 14	WhatsApp
SET 15	WhatsApp, Others
SET 16	Zooms
SET 17	Zooms, WhatsApp

Table 3 Percentage of marks scored the participants students by using different set of online tools during COVID-19

Marks (%)	Example																
	Set 1	Set 2	Set 3	Set 4	Set 5	Set 6	Set 7	Set 8	Set 9	Set 10	Set 11	Set 12	Set 13	Set 14	Set 15	Set 16	Set 17
20–39	0	0	0	0	0	0	0	1	0	0	0	0	0	0	0	0	0
40–59	6	0	1	0	0	0	1	2	0	0	1	0	0	0	1	0	0
60–79	70	1	1	0	1	1	0	12	2	9	4	0	0	1	0	0	2
80–99	71	0	3	1	5	8	3	21	1	33	18	3	2	0	0	3	0
Total	147	1	5	1	6	9	4	36	3	42	23	3	2	1	1	4	2

Figure 7 is presenting that factors which have influenced on the students' performance (Δ) are net connection, supporting devices to continue the online classes during the COVID-19 and study resources. In this section, the participants were asked for their reviews on the online theory and lab classes along with their attendance, mode of classes, their satisfaction level and their feedbacks on knowledge they acquired during COVID-19. The survey result is depicting through Fig. 6b that during the COVID-19 situation, 84.64% of total participants attended their lab and sessional classes and rest of 15.36% participants did not face any online lab or sessional classes.

In this survey, the participants' feedback is showing that during COVID-19 phase, they got the online study resources in different form. Figure 6a is displaying that 64.51% of total participants availed the study resources only in the form of soft copies, 2.73% of participants got the study resources only in hard copy form, 26.96% participants of total number received the study materials both in hard and soft copy form, and 5.80% participants did not get any study resources.

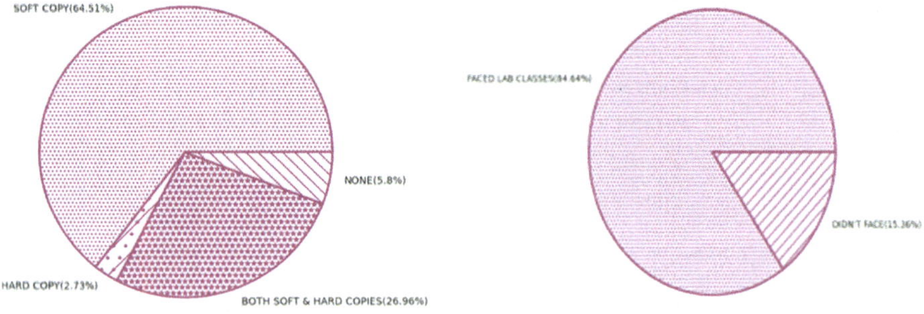

(a) **(b)**

Fig. 6 Ratio of study materials the participants availed and attendance of lab, sessional classes during COVID-19. **a** Study materials that the participants availed. **b** Attendance of lab classes

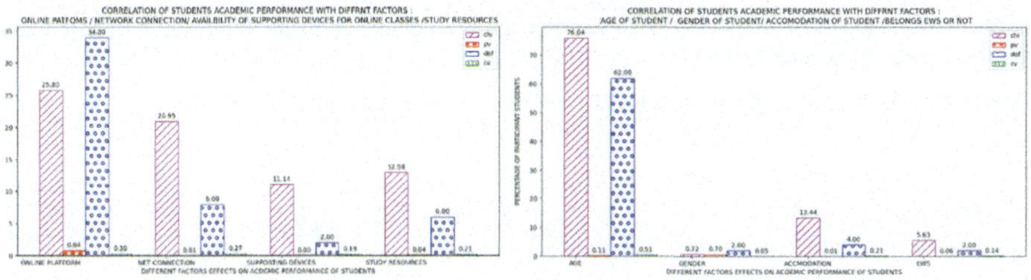

Fig. 7 Graph showing the *p*-value of the factors that have impact on the students performance

4.6 Role of Changed Examination Pattern in Technical Education System During COVID-19

In this section, the students were asked regarding their changed question pattern, they faced during COVID-19 phase. On the basis of their feedback, we got ten types of changed pattern (shown in Table 4).

We calculated here a statistical median value on the basis of the record shown in Table 5. We applied Eq. 1 to calculate the median of each set of changed equation. The result of calculation is showing that the question set 5 is containing the highest

Table 4 Ten sets of changed question pattern used by the participant students during their online examination in COVID-19

Set	Types of changed question pattern
SET 1	Decreased number of MCQ questions and corresponding marks increased
SET 2	Decreased number of subjective questions and corresponding marks increased
SET 3	Increased number of MCQ type questions and corresponding marks decreased
SET 4	Increased number of MCQ type questions and corresponding marks decreased creased, decreased number of MCQ questions and corresponding marks
SET 5	Increased number of MCQ type questions and corresponding marks decreased, decreased number of subjective questions and corresponding marks
SET 6	Increased number of MCQ type questions and corresponding marks decreased. creased, increased number of subjective type questions and corresponding
SET 7	Increased number of subjective type questions and corresponding marks decreased
SET 8	Increased number of subjective type questions and corresponding marks decreased, decreased number of MCQ questions and corresponding marks
SET 9	Increased number of subjective type questions and corresponding marks decreased, decreased number of subjective questions and corresponding marks
SET 10	Increased number of subjective type questions and corresponding marks decreased, decreased number of subjective questions and corresponding marks

Table 5 Percentage of marks scored the participant students by using different set of changed question paper during COVID-19

Marks	Marks obtained by participant students with changed question pattern										
	Before COVID-19	Set 1	Set 2	Set 3	Set 4	Set 5	Set 6	Set 7	Set 8	Set 9	Set 10
20–39%	4	0	0	1	0	0	0	0	0	0	0
40–59%	11	2	0	2	4	0	0	4	1	0	0
60–79%	136	27	2	17	27	0	1	6	24	2	1
80–99%	142	27	1	29	50	2	5	5	48	5	0
Total	293	56	3	49	81	2	6	15	73	7	1

median value. It indicates that the student who appeared the online examination with the changed question set 5, got highest marks. Therefore set 5 can be considered as the most effective question set for online examination and it plays important role on performance of students.

5 Survey Outcome Showing the Association Between Various Factors Reflected on Performance of Participated Student

In our survey study, the feedback report is emphasizing on the effects of different factors of online study, online examination, and performance of participants during COVID-19. To find the association among the performance of student (Δ) and various parameters, we have considered the p-value of factor from 0.00 to 0.5. In this range, we got the *p*-value of:

- Accommodation – 0.01
- Connectivity Problem – 0.00
- Need and availability of Supporting Devices for online classes – 0.02
- Delay during starting and completion of the online examinations – 0.04
- Attendance – 0.01
- Examination methods – 0.00
- Study resources students received – 0.04

Figure 7 is showing the factors that have impact on the students performance as per the generated graph.

On the basis of our survey study, the generated graphs depicted that factors which have influence on students assessment process are:

- Network Connectivity
- Availability of supporting devices for continuing online education
- Delay during starting and completion of the online examination

- Student attendance in online classes
- Study materials received during pandemic
- Online tools used during their examination

6 Survey Observation

Effects of Epidemic on Technical Education System. The pandemic greatly affected technical education system due to sudden change in mode of teaching–learning process from offline to online. The technical education system is fulfilled with some of the essential aspects like theoretical knowledge, practical experiments, laboratory training with relevant instruments and field work. But due to epidemic, technical students had to adopt to the changes from offline to online and as a result there was a lack in experimental and field-oriented knowledge. As per our survey it was observed that based on the knowledge acquired, 18% participants responded with "poor" as their satisfactory level, 39% participants responded with "moderately ok" as their satisfactory level, and the rest 43% participants responded with "satisfactory". Considering study materials, 21% students answered they were dissatisfied with the study materials they got, 73% students answered they were satisfied with the study materials they got, and rest 6% students answered they did not get any study materials.

As per the student feedback, factors like lack of proper presentation in virtual mode, network issues from faculty side and inadequate knowledge of new virtual mode operations created barrier in proper student-faculty interaction. Apart from that, the student feedback report had shown that a number of students discontinued their study due to sudden change in financial conditions as an effect of pandemic.

Problems Faced by Technical Students Since Online Education Began. During the pandemic situation when a worldwide online education system began, technical students faced various barriers and problems while attending their online classes and examinations. In that situation, network connection along with steady power supply was one of the major factor required to continue the education system. The different digital tools and platforms used by students also played a significant role in continuing their online education.

As per our survey report, some major problems that the participated students faced were poor network connection, lack of applicable devices, insufficient study resources. As per the feedback report, 60% of total students faced network issues and improper power supply during their online class and examination process, 12% of total students were without proper devices, and 6% of total students did not receive any study materials during online education system.

Survey-Based Study's Factors that Examined How Pandemic Conditions Affected Technical Education Evaluation. According to our survey-based study, some key factors that examined the effect of pandemic conditions on technical education evaluation are:

- Change in mode of teaching–learning process from offline to online.
- Change in student-faculty interaction process from traditional face-to-face interaction to virtual interaction.
- Transition from physical lab-oriented classes, practical experiments, and field work.
- Changed in question pattern during examination in the COVID-19 period.
- Network connectivity during ongoing online classes and examination.
- Proper handling of the online educational tools.

Effects of COVID-19 on Student Achievement. Overall achievement of any student depends on how they cope up with the new system, how they perform in the new system and how much satisfied they are with the changes. In this survey, students were asked whether they were able to easily handle the operations of the online platform they were using to which 76% of the total students replied as "YES". Regarding students performance, we set some questionnaire to know what were the percentage of average marks obtained by students during and prior to COVID-19 pandemic. Based on the data obtained, we analyzed the dataset (analysis given in Sect. 4.2), 62% of total students have no change in their average marks, 25% of total students have their average marks got increased during the pandemic and 13% of total students have their average marks got decreased during the pandemic.

Apart from these, the students were also asked to answer some questions regarding their satisfactory level on certain aspects like whether they were satisfied with the virtually lab-oriented class conducted, student-faculty interaction in online classes, study materials received, knowledge acquired in the online classes, and evaluation process. As a response to these questions, 21% students replied they were satisfied with virtually lab-oriented class conducted, 77% students replied they were satisfied with student-faculty interaction in online classes, 73% students replied they were satisfied with the study materials received, 43% students replied as "SATISFACTORY" to knowledge acquired in the online classes, and 56% replied as "FAIR" to evaluation process.

Components of Evaluation Process that Were Examined in This Survey

This survey have examined the following components of evaluation process:

- Comparison of students average marks they scored before and during COVID-19.
- Comparison of students highest marks they scored in a subject before and during COVID-19.
- Comparison of number of questions and question pattern at the time of examination before and during COVID-19.

7 Conclusion and Future Work

The sudden change in the technical education system due to the eruption of COVID-19 pandemic carries both positive as well as negative effects. The methods and policies applied by the institutions imparting technical education including introduction of technology-based new and innovative teaching–learning techniques brought a significant change in the traditional education system. At the same time, it posed difficulties to a group of students who were weak to handle this new system. Our survey reviewed a pragmatic study on education system examination process and post examination evaluation process to recognize the hindrances on fair and proper evaluation process during COVID-19 pandemic at different technical institutions in eastern India.

The outcome of survey revealed different factors influencing the academic performance including technical, behavioral, personal, demographical issues. This survey study also provide the experimental evidence to restructure the evaluation process and assessment techniques of engineering education system, examination process for offline as well as online process. It also highlights the need to introduce the active learning system for fair assessment process and appropriate judgment. Academic integrity is one of the fundamental and essential part of any education system and also a vital factor to sustain the reputation of any educational institute. It is very much related to evaluation and assessment process. Therefore, in any technical education system, the fair and perfect assessment process always brings the quality assurance and effectiveness. Our survey study also try to find out the factors that have great effects on engineering students career and try to focus on the rectification of the factors as a future study.

As a future work, impact of various remedial steps can be assessed: like how academic and industrial training session can change the student performance. Also, a study on the design of a fair and efficient learning and assessment system that can sustain future pandemic situations, would be beneficial for the society.

References

1. Chahrour, M., Assi, S., Bejjani, M., Nasrallah, A.A., Salhab, H., Fares, M.Y., Khachfe, H.H.: A bibliometric analysis of COVID-19 research activity: a call for increased output. Cureus. **12**, e7357 (2020). https://doi.org/10.7759/cureus.7357
2. UNICEF: 'What will a return to school during the COVID-19 pandemic look like?' UNICEF. Retrieved from https://www.unicef.org/coronavirus/what-will-return-school-during-covid-19-pandemic-look [2020]
3. Sahu, P.: Closure of universities due to coronavirus disease 2019 (COVID-19): Impact on education and mental health of students and academic staff. Cureus (2020)
4. Sim, G., Holifield, P., Brown, M.: Implementation of computer assisted assessment: lessons from the literature. ALT J. **12**(3), 215–230 (2004)
5. Hanson, J.M., Mohn, L.: Assessment trends: a ten-year perspective on the uses of a general education assessment. Assess. Update Progress Trends Practices High. Educ. **23**(5), 1–15 (2011)

6. Mislevy, R.J.: On measurement in educational assessment. In: Secolsky, C., Brian Denison, D. (eds.) *Handbook on Measurement, Assessment, and Evaluation in Higher Education*, pp. 11–31. Routledge, London (2017)
7. Tierney, R.: Fairness in Educational Assessment, 1st edn. Springer, Singapore (2020)
8. Holloman, T., et al.: The assessment cycle: insights from a systematic literature review on broadening participation in engineering and computer science. Re. J. Eng. Educ. **110**(4), 1027–1048 (2021)
9. Vattoy, K.D., Gamlem, S.M., Rogne, W.M.: Examining students' feedback engagement and assessment experiences: a mixed study. Stud. High. Educ. **46**(11), 2325–2337 (2021)
10. Lamprianau, I., et al.: A Teacher's Guide to Educational Assessment. Sense Publishers, Rotterdam (2009)
11. Tanrikulu, E.: What needs to be done for successful E assessment implementations. In: IFIP 2013, vol. 40, pp. 55–66. Spinger, Berlin (2012)
12. Tpaz, A.C., Saka, E., Fatsa, O.F., Kursun, E.: Emerging trends of online assessment system in the emergency remote teaching period. Smart Learn. Environ. **9**, 17 (2022)
13. Romeu, T., Carbonell, M.R., Catasús, M.G.: E-assessment process: giving a voice to online learners. Int. J. Educ. Technol. High. Educ. **13**, 20 (2016)
14. Li, Z., Qiu, Z.: How does family background affect children's educational achievement? Evidence from contemporary China. J. Chin. Sociol. **5**(1) (2018)

Chapter 9
Ensuring Energy Efficiency Using AI and Nudge Theory to Reach Sustainable Development Goals: A Research Framework

Anastasios Fountis and Anastasia Kiritsi

1 Introduction

In contemporary society, there exists a widespread apprehension regarding the ramifications of climate change on a worldwide scale. The discipline of behavioural economics is assuming a progressively significant role in influencing the development of environmental policies. The concept of "green nudges" has been suggested as a potentially effective strategy for promoting environmentally responsible behaviour among consumers, including the selection of renewable energy sources and the reduction of energy usage [1]. Presently, a substantial proportion of the global population, over 1.6 billion individuals, which accounts for approximately 25% of the total populace, face a lack of access to electricity. The numerical value in question has exhibited a notable degree of stability since the year 1970. Remarkably, the quantity of electrical power needed for tasks such as reading during the night, doing little water pumping, and engaging in radio listening is shown to be less than one percent of the global energy consumption. In the twenty-first century, emerging economies are confronted with a dual energy predicament. On one hand, they must address the pressing issue of delivering basic and contemporary energy services to billions of individuals who currently lack access. On the other hand, they must also actively engage in the worldwide transition towards clean and low-carbon energy systems. In

A. Fountis (✉)
Berlin School of Business and Innovation, Berlin, Germany
e-mail: anastasios.fountis@berlinsbi.com

A. Kiritsi
AUEB, Athens, Greece

J. M. R. S. Tavares et al. (eds.), *Proceedings of Second International Conference on Intelligent System*, Algorithms for Intelligent Systems,
https://doi.org/10.1007/978-981-99-8976-8_9

order to achieve these objectives, it is imperative to expedite development in key areas such as enhancing efficiency, decarbonizing processes, diversifying fuel sources, and mitigating emissions [2].

In addition, it is imperative to evaluate the implications of artificial intelligence (AI) on the attainment of the Sustainable Development Goals (SDGs) in the light of its increasing prominence in diverse sectors [3]. Historically, endeavours aimed at fostering energy efficiency frequently depended extensively on awareness campaigns that emphasised the possible economic benefits linked to adopting more energy-efficient practises. Nonetheless, research from a variety of countries suggests that the desire to live a morally upright life in comparison to one's peers may be more powerful than the prospect of lower utility costs [4]. This is consistent with the ideas of nudge theory, which is based on behavioural research, and emphasises the potency of this influence.

In order to attain optimal efficacy, it is very probable that behavioural methods will be most effective when implemented within a comprehensive policy framework that encompasses technical, market, and institutional barriers to the promotion of energy efficiency. Prior to the implementation of policies based on behavioural insights, it is imperative to build a complete research framework, undertake well-designed pre-tests, and engage in stakeholder talks. It is imperative to employ strategies aimed at mitigating the potential dangers and negative consequences associated with the practise of nudging [5]. In the year 2017, the concept of nudge theory gained significant acknowledgement, positing that subtly presented and indirect recommendations has the capacity to exert a comparable degree of effect on human behaviour as formal legislation, explicit directions, or obligatory mandates. The Nobel Prize in economics was awarded to Richard Thaler, an esteemed American economist, in recognition of his significant contributions in integrating economic and psychological perspectives in the study of individual decision-making processes. The researcher's innovative study went beyond the realm of economics, since it encompassed the utilisation of nudges to tackle the various issues presented by climate change [6].

2 The Research Framework

Both policy makers and economists hold the belief that alterations in relative prices have a significant impact on moulding consumer behaviour. This methodology encompasses the provision of financial assistance to environmentally conscious activities with the aim of cultivating interest among final customers. In order to substantiate the established correlations between artificial intelligence (AI) and diverse aims, a comprehensive literature search was carried out under the guidance of experts. During the search process, various sources were evaluated for their validity as evidence. These sources included publications that focused on real-world applications, with a specific emphasis on meeting certain quality standards. Additionally, evidence from controlled laboratory settings was examined, along with reports from reputable organisations such as the United Nations or government bodies [7]. Nevertheless, empirical studies have provided evidence indicating that knowledge pertaining to social norms can exert a significant influence on behaviour across diverse domains, encompassing activities such as voting and littering. The primary observation is that individuals often exhibit a contingent inclination towards conforming to societal conventions, wherein they modify their conduct in response to the behaviours and perceived intentions of others. The concept being referred to is the adherence to societal conventions and expectations.

2.1 Energy Trends

This section deals with the energy trends as follows:

- The rise in energy consumption can be attributed to the progressive industrialisation of societies, the accumulation of wealth, and the shift from traditional biomass-based fuels, such as wood, dung, and charcoal, to contemporary commercialised energy sources, primarily fossil fuels.
- There has been a continuous improvement in the power production and resource efficiency of energy-generating and energy-intensive technology.
- The mitigation of carbon emissions and the advancement of alternative fuels, specifically for application in the production of electric vehicles (EVs).
- The focus of this discussion centres around the concepts of decarbonisation and diversification, specifically in relation to the production of electricity.
- Throughout the bulk of the twentieth century, there was a notable decline in the per-unit electricity consumption.
- The utilisation of energy witnessed a reduction in the levels of conventional pollutants emitted [2].

2.2 The Problems Involved with Technology

Many scientists and scholars have analysed in detail the various energy supply systems that could be used in a future with carbon limitations. Typical examples of energy technologies include wind turbines, solar panels, biomass power plants, nuclear power plants, and sophisticated fossil fuel systems with carbon capture and sequestration capacities. Systems that utilise natural gas are usually regarded as a crucial "bridge" technology. In addition, energy efficiency is often regarded as an essential and often cheaper complement to supply side improvements [2].

2.3 Plans of Action and Procedures

- One potential strategy for improving energy efficiency is the implementation of measures aimed at enhancing the efficiency of various components, such as buildings, appliances, machinery, and vehicles. Additionally, the establishment of minimum efficiency standards can serve as a means to ensure that these components meet a certain level of efficiency.
- Conducting an assessment and reorientation of current energy subsidy programmes.
- The objective is to identify the most viable indigenous sources of renewable energy and establish policies that facilitate their continuous development.
- The goal is to garner assistance from developed nations in facilitating the widespread adoption of advanced energy technologies, while simultaneously fostering the growth of necessary skills and institutions for sustainable energy innovation in developing countries.
- The objective is to expedite the dissemination of environmentally sustainable, highly efficient, and cost-effective cooking stoves.

2.4 Some Statistics Regarding Energy

Despite the ongoing reduction in costs associated with renewable energy sources, a significant portion of the global population continues to face challenges in obtaining inexpensive and accessible clean energy alternatives. Moreover, the expeditious adoption of clean energy as a means to modify a nation's emissions trajectory presents several obstacles in numerous countries. Presently, there is a critical period that necessitates significant allocations of resources towards the field of energy. According to projections, energy is expected to account for approximately 30% of the total essential infrastructure expenditure over the next 15 years, which is estimated to be around $25 trillion US dollars. These investments possess the capacity to establish a pathway towards a future in which universal access to clean, affordable, and reliable energy is realised [8].

2.5 The Progression of the Consumption of Primary Energy for Each Energy Carrier

Having reliable access to energy is essential for promoting human health, fostering economic development, and combating poverty. It is a continuing and critical issue for global progress to solve the problem of insufficient access for all people. However, our energy generation practises also have far-reaching consequences for the natural world. Throughout the course of history and continuing into the contemporary era, fossil fuels such as coal, oil, and gas have consistently served as the primary sources of energy. Nevertheless, the combustion of these fuels results in the emission of carbon dioxide (CO_2) and other greenhouse gases, so serving as the primary drivers of the ongoing phenomenon of global climate change. In order to attain global climate objectives and mitigate the adverse effects of climate change, it is crucial to undertake a significant and synchronised transition in our energy resources [9].

3 Artificial Intelligence for Achieving the Sustainable Development Goals

Because of its rising popularity across numerous industries, analysing the influence of artificial intelligence (AI) on reaching the Sustainable Development Goals is critical. Researchers used an expert consensus-based process to discover that AI has the ability to advance 134 aims across all goals, but it also has the capacity to obstruct 59 targets. Current studies, however, may not capture all key factors. The rapid advancement of AI necessitates sufficient regulatory monitoring to ensure long-term development, as failure to do so may result in gaps in transparency, safety, and moral norms [3].

There are gaps in our understanding of AI's role in sustainable development. As we deploy AI-powered applications such as driverless vehicles, healthcare solutions, and smart grids to achieve the SDGs, investing in AI system safety research becomes critical to avoid malfunctions or security breaches. Investigating potential disasters caused by systemic AI failures is critical for safe AI integration. For example, the World Economic Forum's 2020 report expresses concern about the growing use of AI in the finance industry, which is inextricably linked to energy markets.

The notion of "nudge" change management includes several critical components that must be implemented. Offering a monetary incentive is one example, as is focusing on individuals and prioritising their needs to improve their life if they demonstrate a desire to do so [3, 10]. Nudges are basic, low-cost interventions that impact decision-making without the use of large monetary rewards or penalties. Nudges, which were popularised by Richard Thaler and Cass Sunstein's book "Nudge: Improving Decisions About Health, Wealth, and Happiness," seek to help individuals live happier, healthier, and more successful lives by aligning their decisions with their best interests. They can also encourage environmentally responsible behaviour, boosting demand for environmentally friendly products and services as consumer tastes shift towards sustainability.

Nudges can only be effective if they are well-designed as a whole, with appropriate safeguards built in for implementation and further refinement in the future. (see Fig. 1).

This demand can help to build marketplaces that attract investment and drive the shift to sustainable. Companies, whether intentionally or unintentionally, affect behaviours. Companies have the chance to promote ecologically and socially conscientious acts rather than reinforce existing behaviours. A travel website, for example,

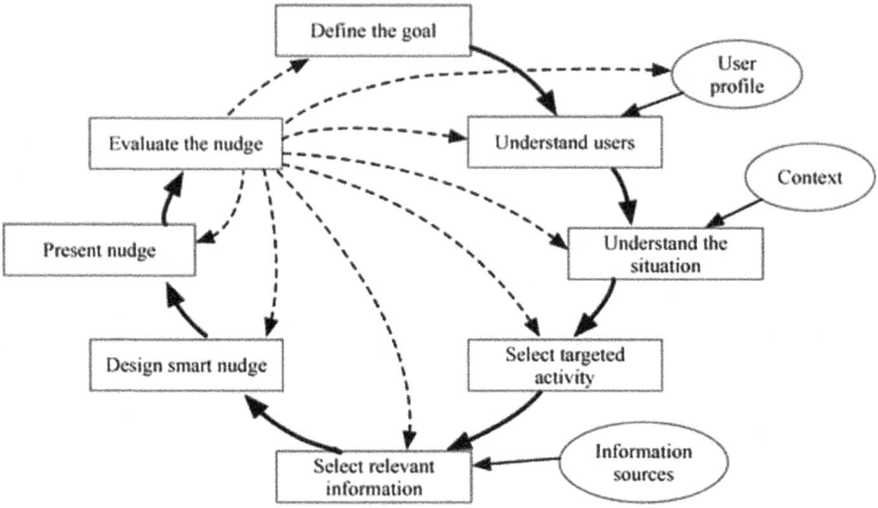

Fig. 1 Designing a smart nudge (*Source* Karlsen and Andersen [10])

may rank flights not only based on price and duration, but also on carbon emissions. Similarly, rather of promoting popular restaurants or their advertisers, meal delivery companies might highlight eco-friendly food options [11].

Energy firms may help their customers accomplish personal sustainability goals, boost their brand, create sustainable marketplaces, and help the environment. This can be accomplished by assessing current customer options and devising strategies to encourage long-term choices and behaviours. Before implementing a portfolio of sustainability nudges, it is necessary to define the most critical sustainability concerns for stakeholders and the company's performance. For example, if lowering a product's life-cycle emissions is a critical component of a company's sustainability plan, programmes might be designed to encourage customers to make choices that reduce emissions. This could include promoting the purchase of low-emission items or making use of the company's energy-optimisation service.

It's critical to assess how nudging for sustainability matches with broader social goals like improving customer satisfaction, expanding fairness and inclusion, and conducting business responsibly. Collaboration among corporations is critical for defining the larger ecosystem and pushing substantial changes, such as developing recycling infrastructure and influencing standards and legislation. The continual expansion of recycling infrastructure is one example of this. Sustainability nudges can also be used in corporate-led sustainability partnerships, particularly in the energy and raw materials industries. The burden of applying sustainability-promoting nudges, such as educating customers and reporting environmental costs, can be shared more equitably through such collaborations. In order to determine which prompts are appropriate in any given circumstance, several factors might be used. (see Fig. 2) [12].

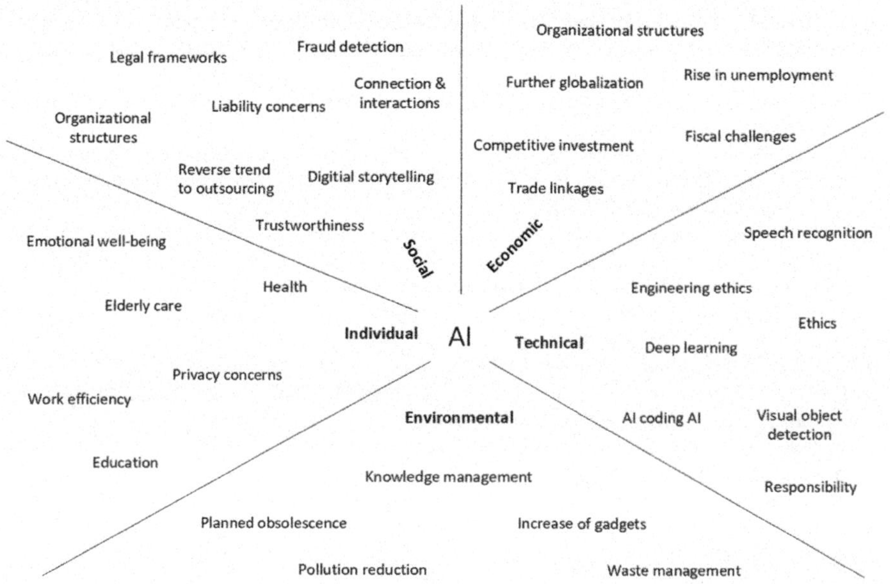

Fig. 2 The five pillars of sustainability applied to the field of artificial intelligence (AI) are depicted in this sustainability analysis diagram (*Source* Khakurel et al. [12])

4 Conclusion

Through this examination, the writers present their viewpoint on a relatively new topic. Some research on how AI might effect various SDGs may not have been widely published yet, needing a more comprehensive investigation. Within the strategy framework of energy firms, the concept of 'green nudges' and associated efforts can play a critical role. The energy sector's difficulties, exacerbated by the COVID-19 epidemic and the Russian invasion of Ukraine, have resulted in a temporary setback for renewable energies. Nonetheless, the referenced references attempted to reduce subjectivity in the review. Using nudge theory, the analysis contextualised and integrated the potential influence of AI on reaching the Sustainable Development Goals. This was accomplished by analysing the writings of numerous writers and examining multiple research for each interconnection.

References

1. Schubert, C.: Green nudges: do they work? Are they ethical? Ecol. Econ. **132**(2017), 329–342 (2017)
2. Ahuja, D., Tatsutani, M.: Sustainable energy for developing countries. S.A.P.I.EN.S Online **2**(1) (2009). http://journals.openedition.org/sapiens/823

3. Vinuesa, R., Azizpour, H., Leite, I., et al.: The role of artificial intelligence in achieving the sustainable development goals. Nat. Commun.Commun. **11**, 233 (2020). https://doi.org/10.1038/s41467-019-14108-y
4. Nudging energy saving behaviours (2020). https://cordis.europa.eu/article/id/415415-nudging-energy-saving-behaviours
5. Babette, N.: Making energy efficiency pro-poor: insights from behavioural economics for policy design. Discussion Paper, No. 11/2014. ISBN 978-3-88985-631-9. Deutsches Institut für Entwicklungspolitik (DIE), Bonn
6. Thaler, R.H., Sunstein, C.R.: Nudge: Improving Decisions about Health, Wealth, and Happiness. Yale University Press (2008). ISBN 978-0-14-311526-7. OCLC 791403664
7. https://www.un.org/sustainabledevelopment/. Accessed 24 Sept 2022
8. World Resources Institute: https://www.wri.org/our-work/topics/energy. Accessed 24 Sept 2022
9. Our World in Data: https://ourworldindata.org/energy. Accessed 24 Sept 2022
10. Karlsen, R., Andersen, A.: Recommendations with a nudge. Technologies **7**, 45 (2019). https://doi.org/10.3390/technologies7020045
11. Boston Consulting Group: https://www.bcg.com/publications/2022/nudging-consumers-to-make-sustainable-choices (2022)
12. Khakurel, J., Penzenstadler, B., Porras, J., Knutas, A., Zhang, W.: The rise of artificial intelligence under the lens of sustainability. Technologies **6**(4), 100 (2018). https://doi.org/10.3390/technologies6040100

Chapter 10
Configuration and Evaluation of Models for Ecological Systems the Case of Distribution of Koala

Yuting Zhao, Masoud Mohammadian, and Hamed Sarbazhosseini

1 Introduction

It is known that climate change has a strong influence on species distribution [1]. Species distributions are studied based on climate conditions. Ecology studies the relationship between environmental and living organisms [2]; species that survive in ecological niches relate to food sources and the survival environment [2]. Species distribution modeling (SDM) is a study to simulate species distribution under environmental elements. It analyzes species occurrence data and environmental raster layers by using computer algorithms to simulate species distribution probability in targeted geographic space and can display results on a map [3]. The process of SDM includes the following:

- It provides species occurrence data and absence data (or generates absence data).
- It provides environmental data.
- It allows to select algorithms and configuration settings with different algorithms having different configurations.

The SDM has been applied in many areas, particularly in decision-making when used in environmental protection and planning to evaluate the influence of climate change on biodiversity or indicate potential conservation areas [4–6]. Some SDM packages combine data analysis, big data, and cloud computing [6–8], which reduce the development time and provide the capabilities for complex analysis [4].

Y. Zhao (✉) · M. Mohammadian · H. Sarbazhosseini
ACT, School of It and Systems, University of Canberra, Canberra, Australia
e-mail: yuting.zhao@canberra.edu.au.com

M. Mohammadian
e-mail: masoud.mohammadian@canberra.edu.au

H. Sarbazhosseini
e-mail: hamed.sarbazhosseini@canberra.edu.au

J. M. R. S. Tavares et al. (eds.), *Proceedings of Second International Conference on Intelligent System*, Algorithms for Intelligent Systems,
https://doi.org/10.1007/978-981-99-8976-8_10

SDM has many challenges which could reduce its availability and reliability [8–10]. Researchers use sensitivity analysis to show a system's output reaction to input, to find out more about SDM uncertainty. It is found that SDMs are sensitive to climate data, pseudo-absence data inputs, or data size [3]. Some researchers also pointed out that configuration and parameter setting in SDM are challenging. Unfamiliarity with algorithms and configurations and the complexity hidden in SDM may lead to unsuitable configuration selection causing erroneous outputs [8, 9]. Several research studies [11] indicated that incorrect configuration of parameters can harm SDM. A clear strategy is required to set the configuration values of parameters for SDM; those researchers validated configurations could improve the performance of SDM more than the default value, but they have not analyzed the evaluation reactions to configuration changes [11, 12]. Evaluation is a necessary step to validate the performance of a model, particularly to compare different models [8, 12].

This research study employs Koala distribution data from the eastern part of Australia as a case study to model Koala distribution using Artificial Neural Network (ANN), Maxent, and Multivariate Adaptive Regression Splines (MARS). The results show the effect of parameter settings and configuration in the Koala distribution's modeling process. It evaluates each model's performance based on different parameter settings and configurations. Analyze the performance of selected evaluation methods, including the area under the receiver operating characteristic curve (AUC), True Skill Statistic (TSS), Kappa, and positive predictive value (PPV).

2 Methodology

2.1 Data

Koala (Phascolarctos cinereus) is a type of marsupial's native to Australia. The distribution covers a small area in South Australia and much of Queensland, New South Wales, and Victoria, and it is threatened by many issues, such as bush fire, tree clearance, and construction which threaten them. There is a particular need for urgent Koala protection in Australia [13, 14]. The Koala distribution dataset is obtained from The Atlas of Living Australia (ALA), a digital and open infrastructure [15]. The Koala occurrence data includes the longitude and latitude of an occurrence point, which is saved as a CSV-formatted file. The cleaned Koala data has 80,938 records and is applied to model Koala distribution in this research study.

In this research study, the environmental data is selected from WordClim's current climate (1976–2005) [16], which was collected from many weather stations worldwide during the noted period [11]. This research study uses biologically suggested bioclimatic variables, namely #5 Max Temperature of Warmest Month, #9 Mean Temperature of Driest Quarter, #10 Mean Temperature of Warmest Quarter, #12 Annual Precipitation, #14 Precipitation of Driest Month, and #17 Precipitation of Driest Quarter [11].

2.2 Algorithms and Configuration Selection

This study employs ANN, MARS, and Maxent to model and estimate Koala distribution. All selected algorithms are applied in SDM for Koala distribution modeling.

The configuration of SDM is different for different algorithms, which relate to each modeling technique. ANN, Maxent, and MARS configurations in SDM are commonly set based on expert experiences and elicitations in modeling [11]. This research proposed a set of test parameter values for each algorithm's configuration option based on a literature review and recommended settings for SDM [11]. Different parameter settings provide different performances of the Koala distribution model. This research study identifies the test values for configuration settings, which cover the gap between a setting range and experience setting values. The result of different test values is provided.

Table 1 shows the evaluated models and selected configurations with an explanation and their default settings available in SDM. This study applies the "one-at-a-time" (OAT) method, which changes one parameter value and keeps other configurations as default for Koala distribution modeling [17].

The model with the default parameter settings is the control model; the comparison model has only one configuration parameter changed for that model. There are three model sets, meaning a set of Koala distribution models generated based on different configurations for each algorithm. This research study provides the configuration settings change for each model that leads to different outputs and identifies good configuration.

2.3 Evaluation

This research study provides the result of the distribution map, the area under the receiver operating characteristic curve (AUC), True Skill Statistic (TSS), Kappa, and positive predictive value (PPV) to analyze. Receiver operating characteristic (ROC) is a method to evaluate the accuracy of ordinal score models, which generate a curve according to true positive rate and false positive rate; it shows that the AUC is the area under the ROC curve, which ranges between 0.1 and 1 [18]. A larger AUC score means a better performance; when it is larger than 0.9, it means that the target model has high accuracy [18]. Kappa is a statistical method used in machine learning and statistical analysis to measure the agreement between two raters or classifiers [18]; it ranges from -1 to 1, and large Kappa means more agreement between raters [6]. TSS considers both commission and omission errors without being affected by prevalence [18]. TSS with a value larger than 0.8 means high accuracy [38]. PPV is the probability that a site predicted as a present is a real present; the model is perfect if the PPV score is 1 [12]. Those evaluation methods are popular in SDM evaluation [8].

Table 1 Selected configuration test value

Model	Configuration	Description	Test value	Default
Maxent	WRW	Set the weight of observe	0.25, 0.33, 0.66, 0.75, 0.9	Null
	Maximum number of iterations (MINIs)	Set the maximum number of iterations	10, 100	200
	Linear/quadratic/ product/threshold/ hinge	Setting feature types	No	Yes
	Product feature threshold (PT)	Set the number of samples starting at the product	40	80
	Quadratic feature threshold (QT)	Set the number of samples starting at the Quadratic	5	10
	Hinge feature threshold (HT)	Set the number of samples starting at the Hinge	8	15
	Threshold feature regularization (TR)	Set the regularization parameter of threshold	2, 5, 10	1.0
	Categorical feature regularization (CR)	Set the regularization parameter of Categorical	2, 5,10	1.0
	Linear/quadratic/ product feature regularization (LQPR)	Set the regularization parameter this set	2, 5, 10	1.0
	Hinge feature regularization (HR)	Set the regularization parameter of	2, 5, 10	1.0
	Prevalence	Set the proportion of occupied location	0.3, 0.25, 0.1	0.5
ANN	WRW	Set the weight to observe action	0.25, 0.33, 0.66, 0.75, 0.9	Null
	Number of cross-validations (NbCV)	Set the number of cross-validations	10, 20	5
MARS	WRW	Set the weight of observe	0.5, 0.55, 0.6, 0.9, 0.99	Null
	Max no. of term (NK)	Used to consider at each stage until the model has the maximum number of terms set by this option	5, 10, 15, 20, 100	Null
	Penalty	Set the cost per degree of freedom change	2.5, 3, 4	2

(continued)

Table 1 (continued)

Model	Configuration	Description	Test value	Default
	Threshold	Set the forward stepping terminate	0.05, 0.01, 0.005, 0.001	0.001
	Prune	Set whether pruned in a backward stepwise fashion	False	Ture

3 Results

3.1 Maxent

The Maxent set has 27 Koala distribution models generated successfully (Table 2). This set has quite different values between TSS and Kappa, but the control model got 0 in both. The best performance setting at TSS is LQPR = 2; its score is 0.754, which means "good" condition. LQPR = 2 got lower scores, 0.720 in Kappa, and 0.953 in PPV. The highest Kappa score is 0.742 as set WRW = 0.25, LQPR = 10, or TR = 10, but they also got lower scores in PPV and TSS; LQPR = 10 has a lower score than TR = 10 and WRW = 0.25 in PPV and TSS, and WRW = 0.25 AUC score (0.940) is lower than other two settings (0.950). PPV shows that configuration setting TR = 5 or CR = 5 could generate the best model due to generating the highest PPV score (0.958). Only two values of AUC (0.95 and 0.94) are generated for all test parameters; the highest score models of PPV, TSS, and Kappa have got 0.950 in AUC evaluation. The control model performance is lower than many test-setting models.

3.2 ANN

This research study collected six estimated Koala models of ANN (Table 3), which shows the sensitivity for different configuration settings. This model was tested for sensitivity to non-default values of WRW and the best size and decay parameters (NbCV). The values of TSS and Kappa are the same; the highest score is 0.8 when configuration setting NbCV = 20, and the worst setting is WRW = 0.75. The highest PPV score is 0.873 as the setting is NbCV = 20, and the worst setting is WRW = 0.75. NbCV = 20 generates the highest AUC score (0.96), and WRW = 7.5 is the worst performance. The best configuration setting found was NbCV = 20, and the worst setting was WRW = 0.75 in the ANN model.

Table 2 Maxent evaluation results

Maxent model	PPV	TSS	AUC	Kappa
Control	0.801	0	0.94	0
CR = 10	0.958	0.744	0.94	0.684
CR = 2	0.948	0.726	0.94	0.709
CR = 5	0.953	0.737	0.94	0.697
HR = 10	0.95	0.743	0.94	0.734
HR = 2	0.956	0.731	0.94	0.667
HR = 5	0.951	0.744	0.95	0.728
Liner and quadratic = no	0.956	0.733	0.94	0.673
Liner and quadratic and threshold = no	0.957	0.749	0.94	0.7
Linear = *no*	0.955	0.733	0.94	0.677
LQPR = 10	0.947	0.738	0.95	0.742
LQPR = 2	0.953	0.754	0.95	0.72
LQPR = 5	0.948	0.739	0.94	0.738
MNI = 10	0.954	0.743	0.95	0.711
MNI = 100	0.954	0.751	0.95	0.725
pf = 40; qt = 5*;* ht = 8	0.947	0.735	0.95	0.736
Prevalence = 0.1	0.95	0.734	0.94	0.715
Prevalence = 0.25	0.953	0.738	0.94	0.703
Prevalence = 0.3	0.949	0.738	0.95	0.728
TR = 10	0.948	0.741	0.95	0.742
TR = 2	0.951	0.738	0.94	0.714
TR = 5	0.958	0.753	0.95	0.899
WRW = 0.25	0.948	0.741	0.94	0.918
WRW = 0.33	0.951	0.722	0.94	0.891
WRW = 0.66	0.947	0.727	0.94	0.91
WRW = 0.75	0.951	0.741	0.95	0.909
WRW = 0.9	0.951	0.741	0.94	0.909

Table 3 ANN evaluation results

ANN model	PPV	TSS	AUC	Kappa
Control	0.851	0.786	0.950	0.788
NbCV = 10	0.851	0.786	0.950	0.786
NbCV = 20	0.873	0.800	0.960	0.800
WRW = 0.66	0.872	0.799	0.960	0.799
WRW = 0.75	0.832	0.764	0.940	0.764
WRW = 0.9	0.847	0.779	0.930	0.779

Table 4 MARS evaluation results

MARS model	PPV	TSS	AUC	Kappa
Control	0.872	0.792	0.950	0.792
NK = 5	0.828	0.730	0.930	0.730
NK = 10	0.825	0.738	0.940	0.738
NK = 15	0.834	0.768	0.950	0.768
NK = 20	0.834	0.768	0.950	0.768
NK = 100	0.868	0.799	0.960	0.799
Penalty = 3	0.874	0.789	0.950	0.789
Penalty = 4	0.867	0.789	0.950	0.789
Thresh = 0.05	0.806	0.701	0.910	0.701
Thresh = 0.01	0.847	0.757	0.950	0.757
Thresh = 0.005	0.839	0.770	0.950	0.770
WRW = 0.5	0.870	0.791	0.950	0.791
WRW = 0.6	0.833	0.789	0.950	0.793
WRW = 0.9	0.844	0.763	0.940	0.789
WRW = 0.99	0.818	0.743	0.930	0.763
Resampling = 10	0.872	0.791	0.950	0.743
Resampling = 20	0.871	0.789	0.950	0.791
Resampling = 100	0.875	0.792	0.960	0.789

3.3 MARS

Table 4 lists the simulated Koala distribution models' evaluation results. TSS and Kappa's outputs are the same at MARS; the lowest score is 0.701 (Thresh = 0.05), and the highest is 0.799 (NK = 100). All PPV values show higher than 0.8 but lower than 0.9, which means a good performance; PPV outputs react as a configuration setting changes. Setting WRW = 0.5 get the highest value of PPV. NK = 100 and Resampling = 100, with the highest performance value of 0.96 in AUC. The second higher score of AUC is 0.950 for a set of generated models. Those configurations have higher performance than the control model in four evaluation methods.

4 Discussion

In this study, all selected evaluation methods identify the worst model without dispute in all model method sets. Still, it could conflict in some model sets in identifying the best model. The best configuration setting in the ANN model for Koala distribution is NbCV = 20, which achieves the highest scores of all evaluation methods; there is no conflict between evaluation methods in scoring the best and worst performance models in ANN. Maxent and MARS evaluation outputs show different results when identifying the best setting. In the Maxent model set, PPV, TSS, and Kappa have different results of the best performance model. PPV shows that the highest performance is 0.958 in setting *TR* = 5, TSS shows that the best configuration setting is LQPR = 2 (0.754), and setting TR = 10 outputs the highest Kappa score which is 0.742. LQPR = 2 could only get the best scores at TSS, but it could not achieve the highest or second higher score in Kappa (0.72) and PPV (0.953); setting TR = 5 and TR = 10 has the same issue. The PPV score of TR = 10 is 0.948 and lower than TR = 5, which people could not consider setting TR = 10 as better than TR = 5 if only applying Kappa as a reference; otherwise, people conclude that TR = 10 is better than TR = 5 if PPV is the reference. AUC only scores 0.950 and 0.940 for all test-setting models, and the three high-performance models all achieve 0.950 in AUC. AUC could not provide any further contributions to identify which model would be better in this scenario. As a result, we could obtain different results in the Koala distribution model if only one evaluation is applied in the modeling. We could obtain three different settings to be the better setting based on those three evaluation methods.

MARS has fewer disputes than Maxent, setting NK = 100 and achieving the highest score in TSS, AUC, and Kappa, but PPV's highest score is WRW = 0.55. AUC shows that two settings have the best performance at resampling = 100 and NK = 100; resampling = 100 also got a second higher score at TSS and Kappa; people could conclude that those two settings could be the better setting. The setting WRW = 0.55 only achieves the highest score at PPV but a much lower score in the rest of the evaluations. If people use PPV as a reference to evaluation models, tuning configuration WRW can obtain a better Koala distribution model.

AUC could not show all changes in configurations in Koala models. In Maxent, there are only two values in all AUC results, 0.95 and 0.94, which means that almost the configuration change could not affect the Koala model's performance; this conclusion conflicts with other evaluation results, which show that model performance is changed in many test configurations. MARS's AUC also has a limited reaction to configuration setting changes. Over 60% of simulated Koala models have the same AUC score of 0.950 in the MARS model set. All configuration *WRW* settings' AUC is 0.95, but their PPV, TSS, and Kappa scores are different as different configuration parameters are set in MARS.

In this case, the evaluation results make it hard to identify the best configuration setting for MARS and Maxent for Koala distribution. It increases the difficulty of configuration and tuning of parameters for these models. Only applying one evaluation method might let researchers ignore other configurations that could potentially improve the mode; meanwhile, it could lead to the incorrect director to model improvement if one selects unsuitable evaluation methods, particularly in some tuning functions of some SDM systems. One could conclude that TR could improve the Koala model if only considering any one PPV or Kappa [8].

5 Conclusion

This paper discusses the influence of configuration and parameter setting in SDM for three models (i.e., ANN, MARS, and Maxent) for the Koala distribution case study. Koala distribution data is used with three models with different configurations and parameter settings and generates three model sets for analysis. This study's results show that the selected configuration and parameter settings could improve the results obtained from the models, and some experience could improve the Koala distribution model performance. This study also finds that different evaluation methods have conflict scores in some Koala distribution cases, which could be a potential risk in improving SDM performance. The selected evaluation methods could avoid the worst performance model, but this issue could lead to simulated unreliable models. This issue needs further cases of other algorithms and evaluation methods to validate that the issue could repeat in other cases. SDM needs to look for a strategy for selecting suitable evaluation methods and validating the better model and its configuration. Particularly, select a suitable evaluation method in applying some configuration tuning functions with a single evaluation method.

References

1. Woodward, F.I., Woodward, F.I.: Climate and Plant Distribution. Cambridge University Press (1987)
2. Odum, E. P.: Fundamentals of Ecology, 3rd edn. Saunders, Philadelphia (1971)
3. Iturbide, M., Bedia, J., Gutiérrez, J.M.: Background sampling and transferability of species distribution model ensembles under climate change. Global Planet. Change **166**, 19–29 (2018)
4. Santana, F.S., Fonseca, R.R., Saraiva, A.M., Correa, P.L.P., Bravo, C., De Giovanni, R.: openModeller—an open framework for ecological niche modeling: analysis and future improvements. In: Computers in Agriculture and Natural Resources, 23–25 July 2006, Orlando Florida, p. 787. American Society of Agricultural and Biological Engineers (2006)
5. Santana, F.S., Saraiva, A.M.: SOC & SOA in Ecological niche modelling and agribusiness: discussion and case studies. In: 7th World Congress on Computers in Agriculture Conference Proceedings, 22–24 June 2009, Reno, Nevada, p. 1. American Society of Agricultural and Biological Engineers (2009)
6. Santana, F., Hallgren, W., Rehn, J., Cjiu, L., Holewa, H.: Implementing Best Practise and a Workflow for Modelling the Geospatial Distribution of Migratory Species, pp. 452–458. International Congress on Modelling and Simulation (2017)
7. Hallgren, W., Beaumont, L., Bowness, A., Chambers, L., Graham, E., Holewa, H., Laffan, S., Mackey, B., Nix, H., Price, J., Vamderwal, J., Warren, R., Weis, G.: The biodiversity and climate change virtual laboratory: where ecology meets big data. Environ Model Softw. **76**, 182–186 (2016)
8. Vignali, S., Barras, A.G., Arlettaz, R., Braunisch, V.: SDMtune: an R package to tune and evaluate species distribution models. Ecol. Evol. **10**(20), 11488–11511 (2020)
9. Araújo, M.B., Guisan, A.: Five (or so) challenges for species distribution modelling. J. Biogeogr. **33**(10), 1677–1688 (2006)
10. Li, X., Wang, Y.L: Applying various algorithms for species distribution modelling. Integr. Zool. **8**(2), 124–135 (2013)
11. Hallgren, W., Santana, F., Low-Choy, S., Zhao, Y., Mackey, B.: Species distribution models can be highly sensitive to algorithm configuration. Ecol. Model. **408**, 108719 (2019)

12. Zhao, Y., Romero, J., Mohammadian, M.: Issues related to modelling and parameter settings of models for ecological systems the case of distribution of thorny devil. In: 2022 2nd International Conference on Innovative Practices in Technology and Management (ICIPTM), vol. 2, pp. 669–674. IEEE (2022)
13. Koala Threats: Queensland Government, [online] Environment|Department of Environment and Science (2022). Available at: https://environment.des.qld.govau/wildlife/animals/living-with/koalas/threats
14. WWF: 60,000 koalas impacted by bushfire crisis. WWF Australia. [online] (2020). Available at: https://wwf.org.au/news/2020/wwf-60000-koalas-impacted-by-bushfire-crisis/?rd=1#gs.vmyq7c. Accessed 24 Apr 2023
15. Atlas of Livin Australia (ALA): [Online] (2018). https://bie.ala.org.au/species/urn:lsid:biodiversity.org.au:afd.taxon:cdbe8a3c-5de0-4140-aadc-2f0b48806e33. Accessed 25 Mar 2019
16. Worldclim.org.: WorldClim—Global Climate Data|Free climate data for ecological modeling and GIS. [online] (2019). Available at: https://www.worldclim.org/. Accessed 9 June 2018
17. Ochoa-Ochoa, L.M., Flores-Villela, O.A., Bezaury-Creel, J.E.: Using one vs. many, sensitivity and uncertainty analyses of species distribution models with focus on conservation area networks. Ecol. Model. **320**, 372–382 (2016)
18. Allouche, O., Tsoar, A., Kadmon, R.: Assessing the accuracy of species distribution models: prevalence, kappa and the true skill statistic (TSS). J. Appl. Ecol. **43**(6), 1223–1232 (2006)

Chapter 11
Design and Experiments of a Manta-Ray Robot for STEM Education

Radzi Ambar, Muhammad Firdaus Roslan, Chew Chang Choon, Mohd Helmy Abd Wahab, Muhammad Mahadi Abdul Jamil, and Suhazlan Suhaimi

1 Introduction

Science, Technology, Engineering, and Mathematics (STEM) education is aimed at preparing students for future professions by assisting students in obtaining a solid grasp of several topic domains and improving their recollection of new knowledge. STEM educations are fundamentally involving hands-on activities that encompass a variety of academic areas. STEM education can be effectively supported by educational robotics because it gives students' access to an enjoyable, interesting, and active hands-on education environment. Robotics requires students to learn the process of learning and solving problem by engaging in utilizing, constructing, and reconstructing objects, in contrast to conventional education environments where students pay closer attention to instructors' teachings in a more organized and controlled manner. The focus of learning STEM with robotics is to improve learning by putting more emphasis on the learning process than the result.

R. Ambar (✉) · M. F. Roslan · C. C. Choon · M. H. A. Wahab · M. M. A. Jamil
Universiti Tun Hussein Onn Malaysia, Batu Pahat, Johor, Malaysia
e-mail: aradzi@uthm.edu.my

C. C. Choon
e-mail: chewcc@uthm.edu.my

M. H. A. Wahab
e-mail: helmy@uthm.edu.my

M. M. A. Jamil
e-mail: mahadi@uthm.edu.my

S. Suhaimi
Universiti Pendidikan Sultan Idris, Tanjung Malim, Perak, Malaysia
e-mail: suhazlan@meta.upsi.edu.my

J. M. R. S. Tavares et al. (eds.), *Proceedings of Second International Conference on Intelligent System*, Algorithms for Intelligent Systems,
https://doi.org/10.1007/978-981-99-8976-8_11

In conventional or traditional teaching techniques, students are not appropriately engaged in hands-on activities mixed with lack of interest toward STEM subjects which eventually causes students to fall behind in their academic performance or even boosting the decreasing trend of student enrollment in STEM courses in universities [1]. In conventional techniques, students are required to find information only from textbooks, looking up words and figures to understand things, and experience random demonstrations and discussions. Perhaps, this method was effective for earlier generations, but may not be effective for students nowadays. STEM education with robotics or educational robotics deviates from traditional methods by exploring the blend of various fields of STEM. Robotics incorporates the use of additional fields, including physics, mechanical, electronics, and communication [2]. It is an engaging field that enables students to feel as though they are learning by doing as they plan, construct, program, and carry out their projects.

Educational robotics sparks students' interest in STEM by exposing them to real-world applications that exceed classroom education. There are various educational robotics platforms that have been proposed around the globe. LEGO Mindstorms/ NXT/EV3 educational kits have been used in primary school's robotic education to give a first-hand experience on programming concept [2, 3]. Some institutions also utilize VEX Cortex platform from VEX robotics to increase STEM learning [4, 5]. Other popular platforms include Fischertechnik [6], Parallax Boe Bot [7], and Bioloid [8]. These platforms are popular but highly expensive and not affordable for schools, especially in rural areas. Furthermore, these platforms are mainly used for robotic activities on land such as line following robot, sumo robot, and obstacle avoiding robot. However, these platforms are not suitable for underwater robot activities which require additional modifications on the structure and the type of components used in the robot design. A major significant difference between an underwater robot and land robot is the electronic component for underwater robot which requires waterproofing to prevent damage.

This project proposes a low-cost solution by demonstrating the design and development of a manta-ray underwater robot to increase the interest of students toward STEM education via hands-on learning process. In this paper, the hardware components used in the development and control algorithm for the robot are described. The body structure of the robot is 3D printed and the propulsion mechanism is utilizing two waterproof servomotors used for pectoral fins movements. At the end of the paper, experiments to show the functionality of the robot for STEM education are demonstrated.

2 Methodology

2.1 Algorithm

Figure 1 shows the algorithm flowchart for controlling the manta-ray robot using the remote controller. The user provides the robot with input in the form of joystick movement; then, the servos move the robot's fins accordingly. If the joystick is pushed up, the left servo rotates counterclockwise, but the right servo rotates clockwise. This produces forward propulsion for forward motion. If the joystick is pushed down, the left servo rotates clockwise, and right servo rotates counterclockwise, which produces backward propulsion for backward motion. On the other hand, motions to the left and right are possible by reducing servo motor speed on either side.

2.2 Hardware Design

The proposed manta-ray robot overall circuit design is shown in Fig. 2. As shown in the figure, the design consists of the remote controller part and the robot part. For the remote controller part, it consists of an Arduino Nano microcontroller, a thumb joystick, and an HC-05 Bluetooth module. The robot can move horizontally (front, back, left, and right) using the signal from the thumb joystick. All signals from the thumb joystick are sent to the robot part via Bluetooth communication.

As shown in Fig. 2, the robot part is mainly consisting of an Arduino Mega microcontroller that receives the transmitted data from the remote controller part using another HC-05 Bluetooth module connected to the robot's microcontroller which is connected to two waterproof servomotors. The two servomotors are used for moving the manta-ray's two pectoral fins to make the robot move horizontally according to the input from the user via the remote controller.

Figure 3 depicts the internal electronic circuit for the robot part. An Arduino Uno microcontroller is utilized to process signal received from the remote controller via an HC-05 Bluetooth module. Two HITEC HS-5646WP waterproof servomotors are used to produce propulsion for the robot to move. There is also a DC–DC voltage regulator used to keep the power source voltage within the servomotor's permitted limits. Finally, the figure also shows an ESP32 camera module, but this is an optional part for the final design. The ESP32 sends underwater images captured by the camera to the Arduino Uno microcontroller.

Figure 4 shows the actual circuit for the remote controller. Arduino Nano microcontroller is utilized to process data from the thumb joystick. The thumb joystick is used to control the servomotors that control the pectoral fins movement. As shown in the figure, the remote controller is also connected with an HC-05 Bluetooth module for wireless communication with the robot part.

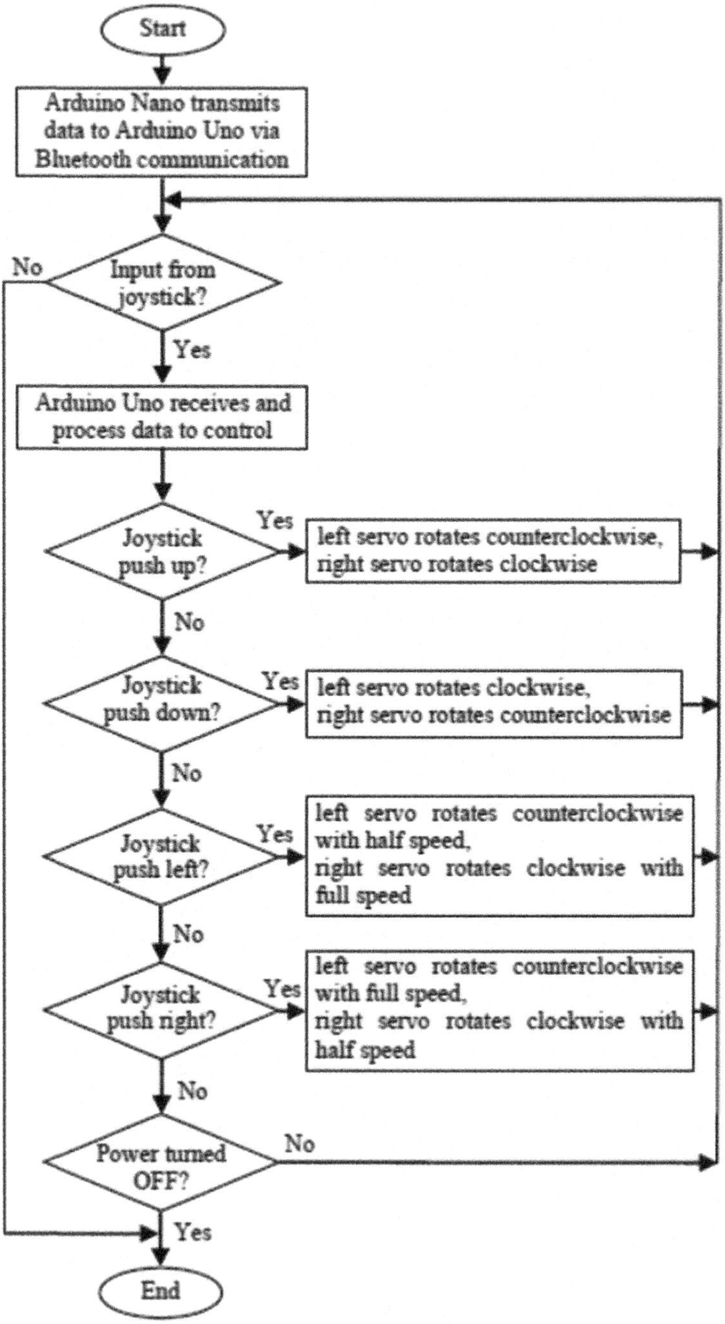

Fig. 1 Flowchart of the algorithm to control robot motions

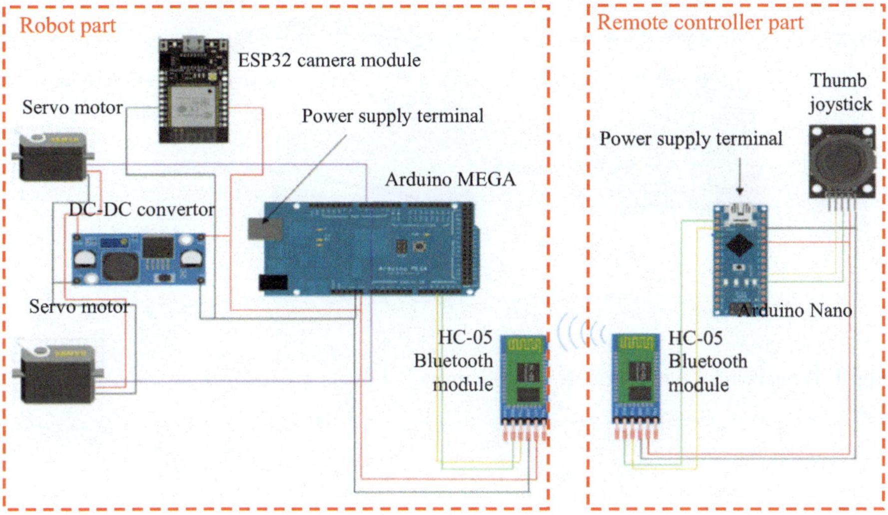

Fig. 2 Manta-ray robot overall circuit design

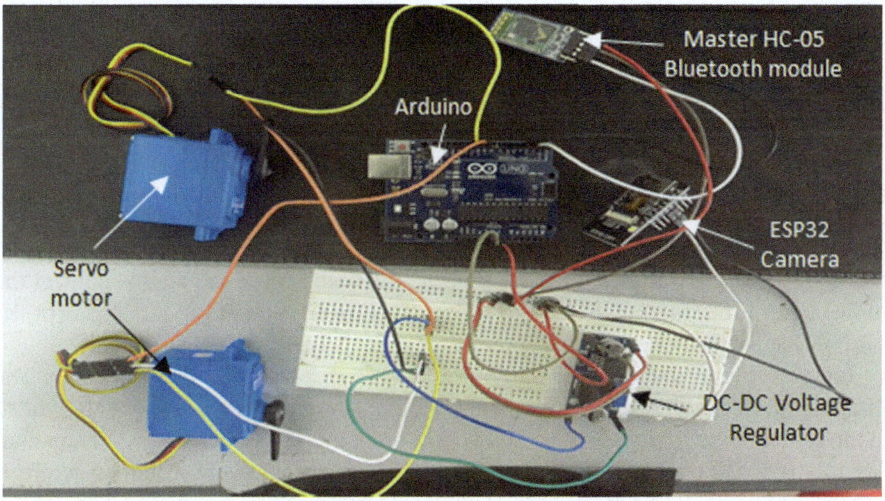

Fig. 3 Actual circuit for the robot part

Figure 5 shows various views of the preliminary design for the robot. Initially, this design was sketched on the SketchUp software before the actual hardware design was developed. Manta-ray structure was used as the main design of the robot's body due to its stable motions in underwater environment. In the proposed design,

Fig. 4 Actual circuit for the remote controller part

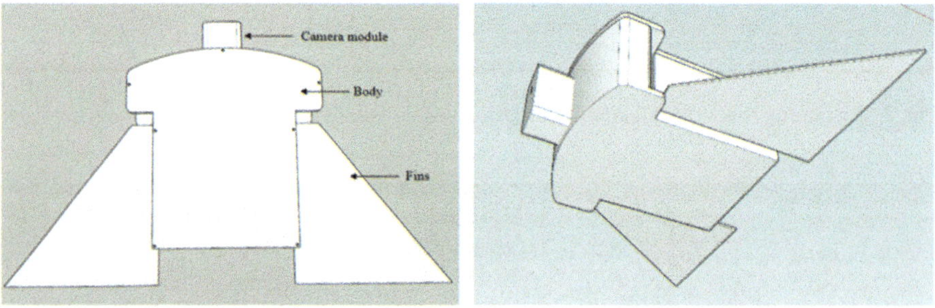

(a) Top view (b) Side view

Fig. 5 Various views of the proposed manta-ray robot

the servomotors are attached on the sides of the body, while the servos control the movement of the fins. Figure 5 also shows the ESP32 camera module used for the vision system positioned at the center of the robot's head.

Figure 6 shows the actual image of the proposed robot without a top cover. The image shows a waterproof compartment that stores the electronic components located at the center of the robot between the two servomotors. The dimensions of the manta-ray robot are 30 cm length × 15 cm width × 6 cm height. The robot structure is constructed using 3D-printed material, while the pectoral fins are created using cheap polyvinyl chloride (PVC) sheets that were cut in rectangular shape.

On each pectoral fin, a 20 cm length stick is attached to one side of the rectangular-shaped fin as depicted in Fig. 7. One end of the stick is attached to the servomotor horn with hot glue. So, when the servo rotates, the stick follows it which in turn makes the fin flaps. Figure 7 also shows how the two servo motors are attached inside the body.

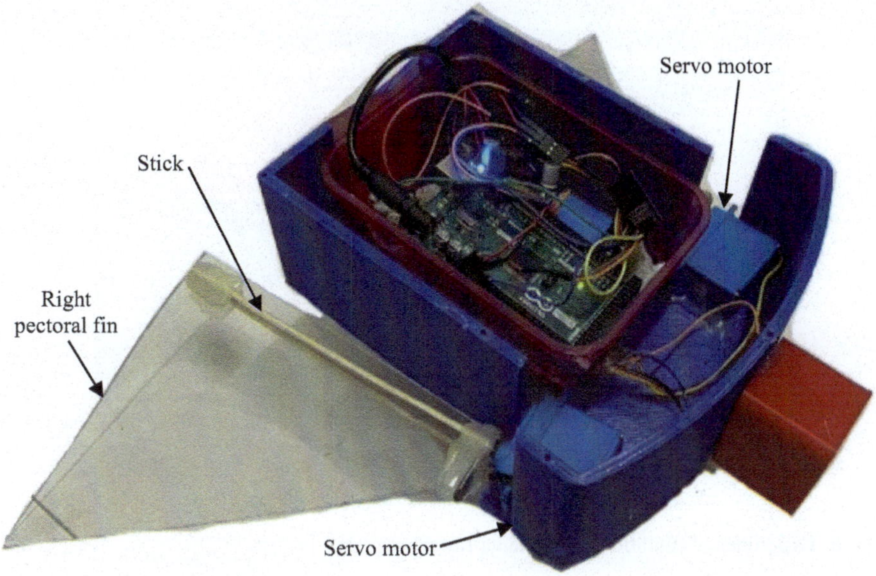

Fig. 6 Actual manta-ray robot (without top cover)

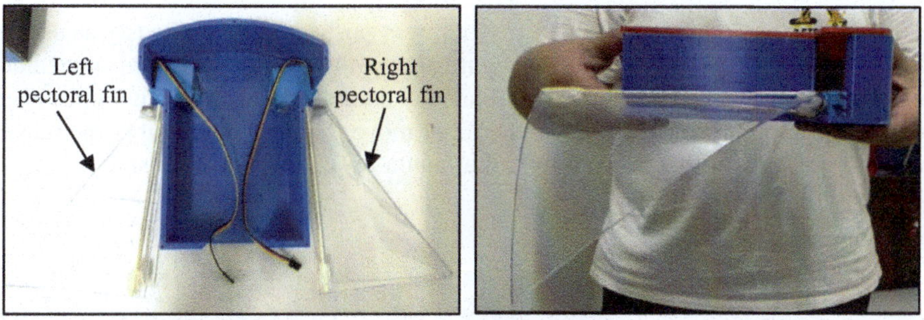

Fig. 7 Servomotor's location (left) and pectoral fins design (right)

3 Methodology

Experiments have been implemented to test the manta-ray robot functionality. The tests consist of movement tests to the right, to the left, and forward. The tests also determine how long the manta-ray robot can swim underwater, and how far the robot can travel underwater before the battery is fully drained. Before testing the manta-ray robot, some weights were placed inside the manta-ray body. The weights help to submerge the manta-ray underwater. In this project, small bearings were used as weights. The weight of the bearing is 150 g/pcs. A total of seven bearings were needed to submerge the manta-ray completely. Figure 8 shows the robot underwater.

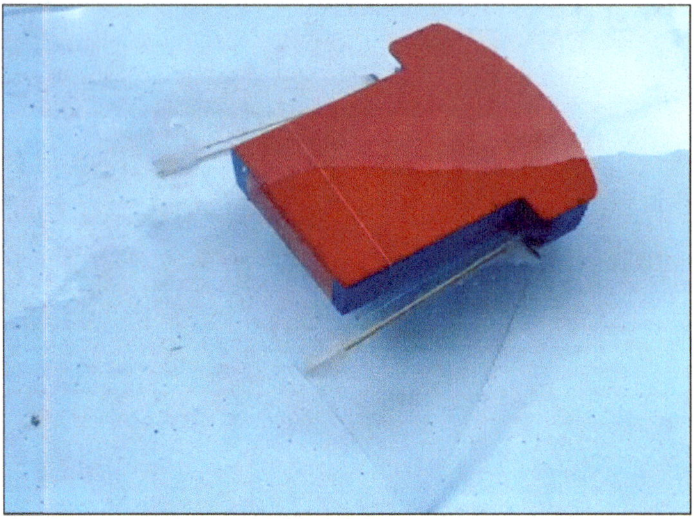

Fig. 8 Experimental condition of the manta-ray robot

3.1 Forward Movement

For forward movement, the thumb joystick was pushed up continuously as shown in the top left figure in Fig. 9. The figure shows the result of forward movement experiment, where the robot's left fin rotated counterclockwise, and right fin rotated clockwise simultaneously, to make the robot move forward. In addition, the movement of the fins in this test prevents the manta-ray fish robot from sinking to the bottom of the pool.

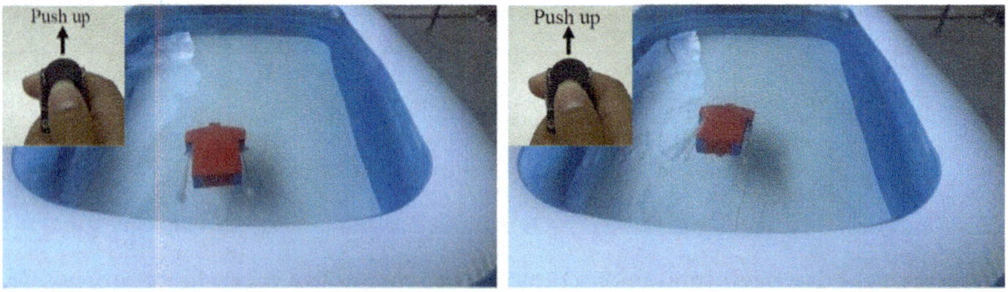

Fig. 9 Manta-ray robot forward movement experiment. Robot starts to move forward (left image). Robot continues swimming forward (right image)

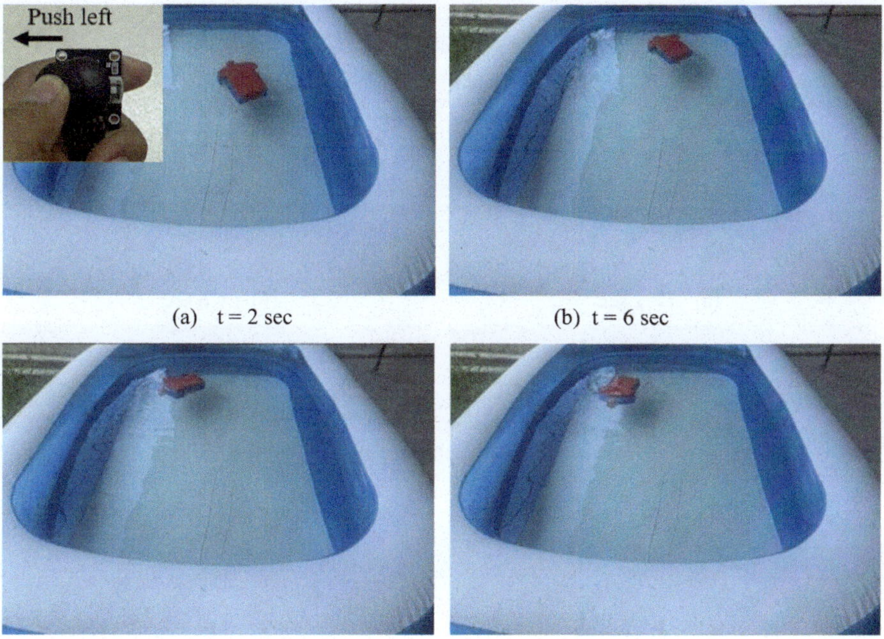

(a) t = 2 sec (b) t = 6 sec

(c) t = 10 sec (d) t = 15 sec

Fig. 10 Manta-ray robot left movement experiment

3.2 Left Turn Movement

This experiment was conducted to test whether the manta-ray robot can turn left or not. For left turn movement, the joystick was pushed to the left as shown in the top left figure in Fig. 10a. When the joystick was pushed to the left, only the right fin rotated clockwise continuously, while the left fin rotated with half speed. Both fins continuously rotated until the joystick was released. Figures 10a–d show the time histories for the movement of the robot turning to the left. As shown in the figure, the manta-ray robot can turn to the left. It was also found that the robot needs more time to rotate its body to the left, which caused the battery life to drain fast due to the high load.

3.3 Right Turn Movement

This test was conducted to test whether the manta-ray robot can turn right or not. For right turn movement, the joystick was pushed to the right as shown in the top right figure in Fig. 11a. When the joystick was pushed to the right, the left fin rotated at full speed counterclockwise continuously, while the right fin rotated with half speed.

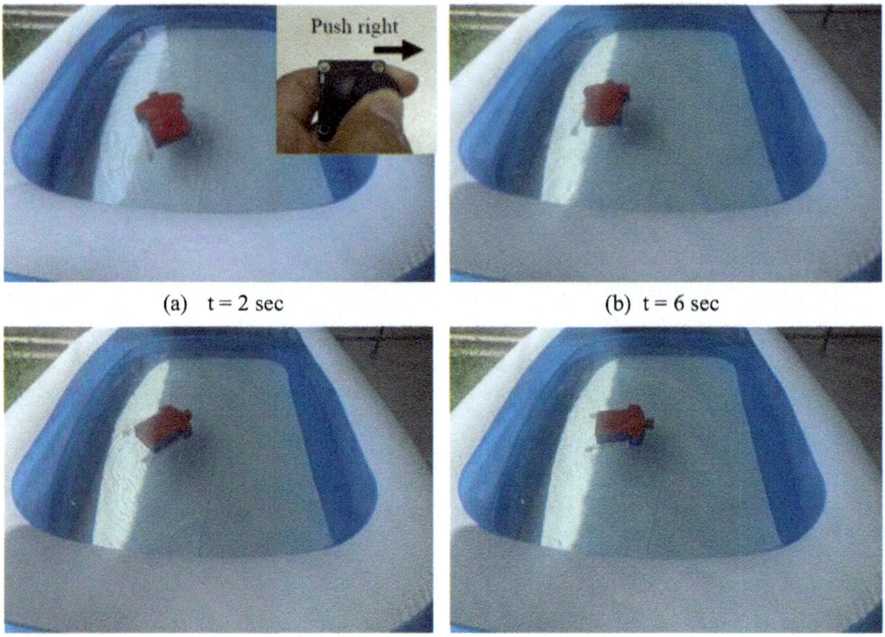

(a) t = 2 sec (b) t = 6 sec

(c) t = 10 sec (d) t = 15 sec

Fig. 11 Manta-ray robot right movement experiment

The fins continuously rotated until the joystick was released. Figures 11a–d show the time histories for the movement of the robot turning to the right. As shown in the figure, the manta-ray robot can turn to the right.

Based on all the experiments that have been conducted, the manta-ray robot was able to be operated underwater without any problem. The robot was able to move forward and turned to the left and right. Both servomotors were able to run smoothly without any issue. Both servomotors rotate clockwise or counterclockwise from 0° to ± 180°. The waterproofing of the manta-ray body and ESP32 camera was also successful. All the components such as the microcontroller, HC-05 Bluetooth module, DC–DC voltage regulator, and ESP32 camera were secured inside the waterproof compartments. No leakage occurs inside the compartments and can be operated without any issues.

Furthermore, a battery drain test was conducted to observe the battery life of the robot. The tests were conducted three times and the average results were taken. The distances traveled by the robot during the battery drain test were also recorded. Table 1 shows the results for the average battery life of the manta-ray before the battery was completely drained. It was also found during the experiment that the average distance traveled before the battery completely drained was approximately 200 m.

Table 1 Test results for battery life

Test item		Average battery life results
Battery life	Surface test	1 h 24 min
	Underwater test	
	• Forward motion	36 min
	• Right movement	42 min
	• Left movement	44 min

4 Conclusion

This paper presents the design and development of a manta-ray robot for STEM education. It is strongly believed that the integration of robotics in classroom learning can enhance students' interest in STEM education through a hands-on learning process. This paper provides a detailed description of the robot's hardware design. Preliminary experiments have demonstrated the functionality of the robot. This is a preliminary work for implementing underwater robotics in STEM education; therefore, several improvements can be made.

In addition to improving the current design, future work will also focus on enhancing user-friendliness and determining the extent of students' knowledge required to learn about the proposed robot. A module or manual will be prepared to assist users in assembling the robot. In terms of design, the current challenge is reducing the production cost, as a remote controller is used to control the robot's motions, adding additional expenses. However, for future work, there is a plan to utilize smartphone application as a controller for the robot. Additionally, in the current design, Bluetooth wireless communication is implemented, which raises questions about its use underwater. Based on the work, communication was possible, but the robot must be submerged near the water's surface. For the authors, this is still the best communication method compared to a wired remote controller, which can be troublesome due to long wires. Nevertheless, considering the budget constraints, the authors will also explore other wireless communication methods such as semi-tether, acoustic, or optical communication methods.

The potential for scalability of the manta-ray robotic system is significant. The robot's modular design makes it easily customizable and adaptable to different teaching environments. A variety of features can be added or modified to meet specific needs, making it adaptable for larger class sizes or educational settings. In addition, the principles underlying the design of the manta-ray robot can be applied to other robotic assemblies, allowing it to be generalized to other educational robotics projects. In addition, robotic hardware and software can be tailored to specific STEM fields such as physics, biology, and environmental sciences, by integrating sensors or devices for data collection and data analysis. Hopefully, this initiative will increase students' interest in STEM education.

Acknowledgements This research was supported by Ministry of Higher Education (MOHE) through Fundamental Research Grant Scheme for Research Acculturation of Early Career (FRGS-RACER) (RACER/1/2019/TK04/UTHM//5) and Universiti Tun Hussein Onn Malaysia (UTHM) through Tier 1 (vot Q368).

References

1. Graham, M.J., Frederick, J., Byars-Winston, A., Hunter, A., Handelsman, J.: Increasing persistence of college students in STEM. Science 27(341), issue 6153, 1455–1456 (2013)
2. Wang, W.: A mini experiment of offering STEM education to several age groups through the use of robots. In: Proceedings of the 2016 IEEE Integrated STEM Education Conference (ISEC), Princeton, NJ, USA, 5 Mar 2016, pp. 120–127
3. Grandi, R., Falconi, R., Melchiorri, C.: Robotic competitions: teaching robotics and real-time programming with LEGO. IFAC Proc. Vol. **47**(3), 10598–10603 (2014)
4. Nugenta, G., Barkera, B., Grandgenett, B., Welcha, G.: Robotics camps, clubs, and competitions: results from a US robotics project. Robot. Autonom. Syst. **75**(B), 686–691 (2016)
5. Sullivana, F., Heffernana, J.: Robotic construction kits as computational manipulatives for learning in the STEM disciplines. J. Res. Technol. Educ. **48**(2), 105–128 (2016)
6. Fishertechnik: http://www.fischertechnik.de/en/Home.aspx
7. Parallax: https://www.parallax.com/boe-bot-robot/
8. Bioloid: http://www.robotis.com/xe/bioloid_en

Chapter 12
House Price Forecast for Imbalanced Label Data by Metaheuristic Model

V. Anjana Devi, Ahmed J. Obaid, Vithya Ganesan, V. Sri Anima Padmini, Subrata Chowdhury, and Saurabh Adhikari

1 Introduction

Real estate sector is one of revenue-based industries with more supports from stakeholders for common man, private and public companies, and government bodies with a high demand and better understanding of operational and driving mechanism.

Prediction of prices has been executed by using machine learning techniques. According to the requirements of the clients, an appropriate machine learning algorithm will be used to identify best prices of current real estate trends. For example, it is observed that apartments close to shopping areas, bus stop, railway station will be preferred by people and the value of the property in this area will be quite high than other area. Businesspeople need places in the central part of the city. Based on customer requirement, the price for the house will be predicted.

V. A. Devi
Rajalakshmi Institute of Technology, Chennai, Tamil Nadu, India
e-mail: anjanadevi.aby06@gmail.com

A. J. Obaid
Faculty of CS and Mathematics, University of Kufa, Kufa, Iraq
e-mail: ahmedj.aljanaby@uokufa.edu.iq

V. Ganesan (✉) · V. S. A. Padmini
Koneru Lakshmaiah Education Foundation, Vaddeswaram, Andhra Pradesh, India
e-mail: vithyamtech@gmail.com

S. Chowdhury
Sreenivasa Institute of Technology and Management Studies, Murukambattu, India

S. Adhikari
School of Engineering, Swami Vivekananda University, Kolkata, India
e-mail: saurabhadhikari@svu.ac.in

J. M. R. S. Tavares et al. (eds.), *Proceedings of Second International Conference on Intelligent System*, Algorithms for Intelligent Systems,
https://doi.org/10.1007/978-981-99-8976-8_12

1.1 Objective of the Proposed System

1. To analyse various features of a house and apply variables for house price forecast.
2. To help people who plan to buy a house within their budget.
3. To improve real estate efficiency.
4. To better predict the house prices.
5. Introduce lottery system to select the house based on features and customer customization.

1.2 Scope of the Proposed System

House price prediction with specific attributes has wide application prospects. Some of them are,

1. Using this proposed model, people can buy houses or any other land with fair prices and want to ensure that they were not cheated by agents.
2. This model also helps big companies by giving accurate predictions for them to set the pricing and save a lot of precious time and money
3. This model ensures the accurate real estate prices that are the essence of the market.

2 Literature Survey

A method for prediction of house price in the market by prediction process .it uses historical data in real estate market. This data is analysed by applying machine learning techniques to predict the price which is very useful for buyers and sellers to improve the prediction accuracy [1]. In this paper [2], the authors proposed a solution for house price forecasting using regression techniques which was held on Kaggle platform. House's sale price will be predicted by using several attributes such as house area, year of building. House prices are going to be predicted by using classic machine learning algorithms having into account all these attributes.

Real estate is one of the major areas to apply machine learning algorithms to forecast the price with better accuracy. Factors such as physical conditions, concepts, and location are used as parameters in prediction process. Customer's financial needs and requirements are considered for the house price [3].

An unsupervised neuron model (DNM) for the prediction of house price by considing the data from Chinese housing market. The effectiveness is analysed by comparing the traditional versus statistical model. The performance metrics are normalised mean square error, absolute percentage of error, and correlation coefficient [4]. Several parameters may influence price variation in the real estate market.

Bankers and other investors keep monitoring the price changes in this field. Macroeconomic parameters are responsible for price variation by mean absolute error and root mean square error which are considered for analysing the variation in prices [5].

House price prediction is the major source of employment. In the real estate field, housing is considered as major domain [6]. Finding fair prices in major cities is a challenging process for the buyer to buy their desired property. A better way to predict the prices is essential for thr buyer and seller.

Due to insufficient information about housing features, the performance and applications have been limited [7]. Spatial–temporal interactions behind the evolution of prices removes the need for a historical transaction.

This study proposes a comprehensive framework for an intelligent smart home that enables individuals with disabilities to reside in the familiar environment of their own homes while receiving healthcare services [8]. The proposed system is an Internet of Things (IoT)-enabled mobile gateway designed for the purpose of monitoring and controlling an interactive home wireless system. This gateway serves as a central hub for data collecting inside the system.

The utilisation of a flexible spatiotemporal model (FSTM) is proposed for the purpose of predicting housing values in middle-small cities, as an alternative to the conventional linear models commonly employed for larger cities [9]. The Faculty of Science, Technology, and Medicine (FSTM) takes into account both spatial and temporal dimensions, effectively reflecting the spatial diversity and temporal patterns in housing costs. The residential housing prices in middle-small cities are greatly influenced by government policies as well as several elements, including road density, bank density, and supermarket density. It is recommended that local governments undertake the implementation of socio-economic development strategies in order to enhance water usage facilities, hence fostering the advancement of environmentally sustainable agricultural production [10]. The analysis of spatial autocorrelation and space–time pattern mining examines the distinguishing traits and evolutionary patterns of residential, commercial service, and industrial land concession prices [11]. The dimension reduction method known as stepwise support vector machine (SVM) is employed to decrease the dimensions of datasets characterised by a large number of features (p) and a relatively limited number of observations (n). In the context of analysing home prediction datasets, many dimension reduction approaches are commonly employed. These methods include the Pearson product difference correlation coefficient (PCCs), recursive feature elimination based on random forest (RF-RFE), and principal component analysis (PCA) [12].

In this study [13], it is postulated that fluctuations in housing values are associated with the timing of their initial sale and subsequent resale, as well as the characteristics of the dwelling unit and neighbourhood, including those of neighbouring properties.

This research [14] examines various models for the forecast of house prices. This study compares and analyses three distinct types of machine learning algorithms, namely Random Forest, XGBoost, and Light GBM, together with two methodologies in machine learning, Hybrid Regression, and Stacked Generalisation Regression, in order to identify the most optimum answers. Despite the favourable outcomes attained by all of these strategies, it is important to acknowledge that diverse models possess their own distinct advantages and disadvantages.

In the referenced research [15], the primary aim is to predict the future prices of cohesive residential properties for individuals who do not currently own a house. This prediction is based on an analysis of their financial provisions and personal objectives. Through a comprehensive analysis of the aforementioned items, fare ranges, and anticipatory trends, it is possible to make estimations for predicted prices. This study explores the application of various regression techniques, including multiple linear regression, ridge regression, LASSO regression, elastic net regression, gradient boosting regression, and AdaBoost regression, for the purpose of making predictions. The process of predicting house prices on a given dataset has been conducted by employing all of the aforementioned methodologies in order to determine the most effective approach.

3 Proposed System

The goal is to find out how much a house is worth by looking at a number of factors that the user inputs into the machine learning model. It makes a guess based on how these traits change the label. The information will be cleaned up and each piece of raw data will be saved as a .csv file after it is collected. It is used to deal with lost data during data preprocessing. Soon after being cleaned up, the data will be transformed into a NumPy array so that it can be sent to support vector machine and Decision Tree to train the models. The accuracy of the results from both models was then checked.

3.1 System Architecture

The proposed system has three Generative Adversarial Algorithms. The algorithms are support vector machine, Decision Tree, XGBoost. Figure 1 describes the architecture of our system.

3.2 Data Collection

Dataset includes parameters such as areas in square metres, builders ranking, quality location, year of built, Garage area and number of cars that can fit in garage, swimming pool area, selling year of the house and price at which house is sold. In addition, six forecasts such as global recession, construction cost declines, property price

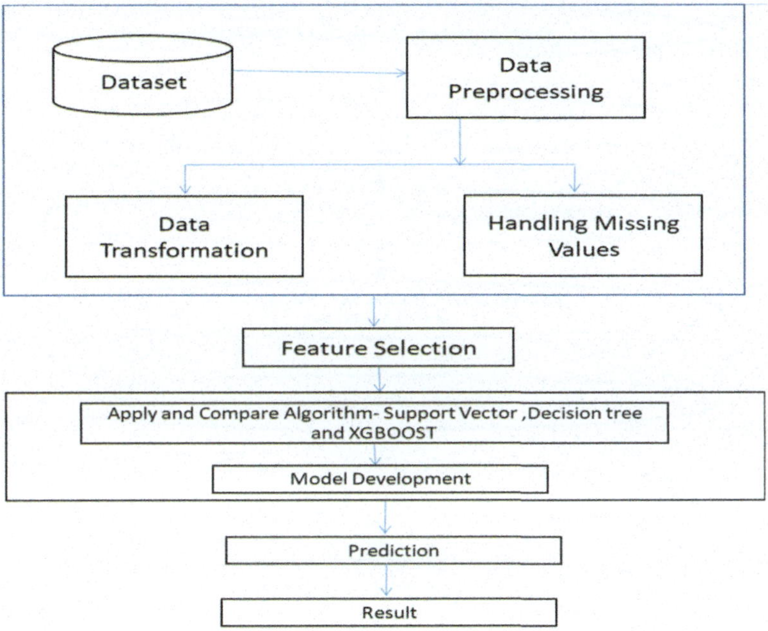

Fig. 1 System architecture

falls, office vacancy rate rise, gold price rise, and global deflation may affect the selling price. The lottery system may uses phase wise anaysis to improve selling and customer satisfaction. The selling price of a house is dependent on six of these forecasts.

3.3 Features Extraction

3.3.1 Importing Dataset

The dataset will be downloaded from Kaggle and stored in our project folder. The size of the dataset is about 900 Kb. The dataset will contain the house price and details of other parameters which are required for our prediction process (Fig. 2).

3.3.2 Data Analysis

The .csv file which has been created has 20,315 rows. Each row contains the parameters and their respective values. There are 182 different areas present in the dataset. The CSV file has nine columns first columns represent the index number the remaining Price, Sqft, Location, No. of Bedrooms, Swimming Pool, Jogging Track,

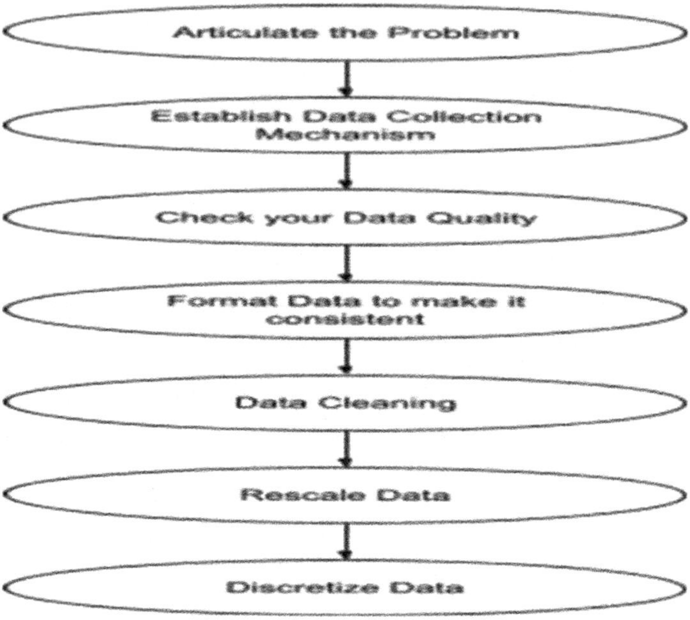

Fig. 2 Collection of dataset

Rain Water Harvesting, Car. Parking CSV file will be 0 or 1 . 0 for parameter is not present for the corresponding area, 1 for the parameter is present.

3.4 *Data Preprocessing*

It is a process of transforming raw, complex data into systematic understandable knowledge. It involves finding missing and redundant data in the dataset. In this dataset, there were no missing values found. But in our dataset, there were records which had values other than the ones required for our model. Those values will be changed to a uniform value of 0 (Fig. 3).

3.5 *Import Libraries and Dataset*

Import all the basic libraries which will be required to perform basic operations and other libraries that are needed for implementing the algorithm. Data is normally set in CSV formats. By calculations, the categorical data must be encoded. Data is normally set in CSV formats. Dataset is split for training and the other for testing purposes.

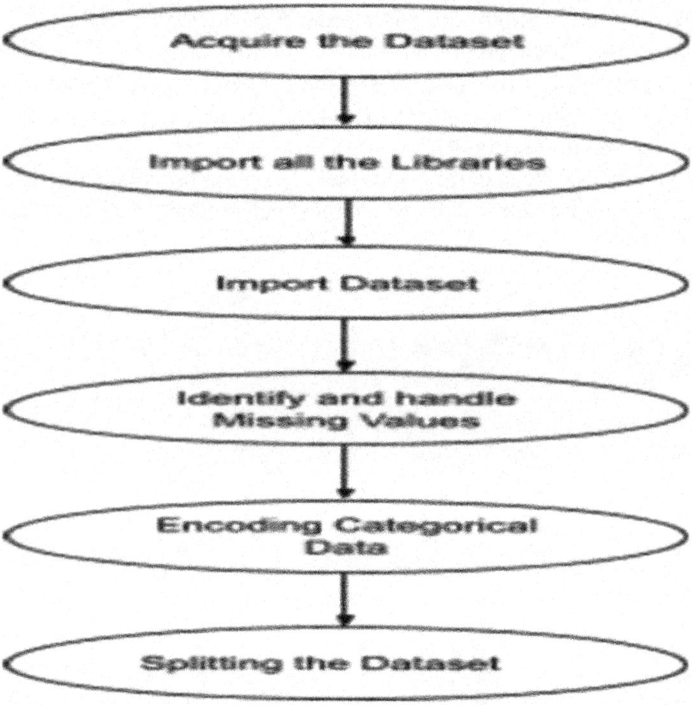

Fig. 3 Data preprocessing

Normally 80% of the dataset is used for training and 20% is used for testing. For this test, import test_train_split from model_selection by library of scikit.

4 Implementation and Result

Three machine learning algorithms such as SVM, Decision Tree, and XGBoost are used for the prediction process and the results were compared in terms of accuracy.

4.1 Working of XGBoost

"Extreme Gradient Boosting" is what XGBoost means for. It is an optimised distributed gradient boosting library that is made to be very flexible, portable, and efficient. Gradient Boosting is a machine learning algorithm that gives you a parallel tree boosting system to quickly and correctly solve data science tasks. When you make a decision tree, the steps are put in order. Variables were given weights, which were then put into a decision tree to make predictions about what would happen.

Variables are fed into the next step of the decision tree to give it more weight. Then, different classifiers and predictors are put together to make a model that is better and more accurate (Fig. 4).

Figure 5 compares the accuracy of model evaluation process using three different algorithms, and XGBoost provides the highest percentage of accuracy prediction.

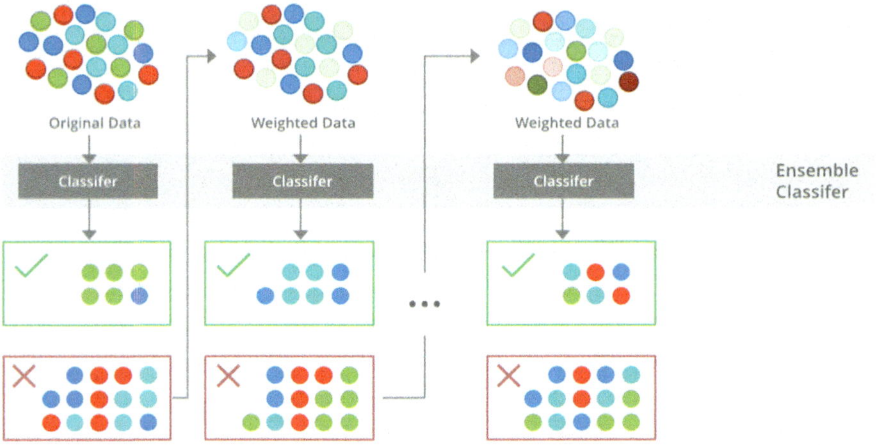

Fig. 4 XGBoost working

Fig. 5 Accuracy comparison model

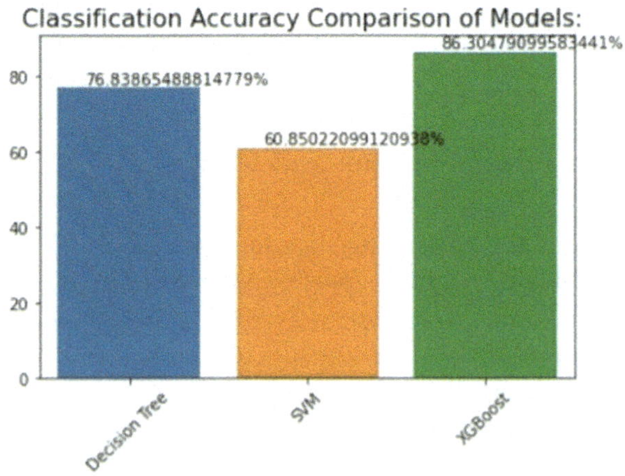

5 Conclusion

In conclusion, this study aimed to determine house prices by utilising a Machine learning model that takes several user-input features into account. The collected dataset underwent a data cleaning process, transforming the raw data into a .csv file. This prepared the data for training the model using support vector machine and Decision Tree algorithms. The accuracy of both models was compared to evaluate their performance.

The results of this study highlight the potential of machine learning in predicting house prices based on user-provided features. By employing the XGBoost algorithm, this study demonstrated its effectiveness in accurately predicting house prices. The data cleaning and preprocessing steps ensured the quality and reliability of the dataset. Overall, this study contributes to the field of house price prediction using machine learning and provides insights for future advancements in this domain.

References

1. The Danh Phan: Housing price prediction using machine learning algorithms: the case of Melbourne City, Australia. In: International Conference on Machine Learning and Data Engineering (ICMLDE) (2018)
2. Viktorovich, P.A., Aleksandrovich, P.V., Igor, K., Vasilevna, P.I.: Predicting Sales price of the houses using regression methods of machine learning (2018)
3. Velthorst, M., Güven, C.: Predicting housing market trends using Twitter data. In: 6th Swiss Conference on Data Science (SDS) (2019)
4. Yu, Y., Song, S., Zhou, T., Yachi, H., Gao, S.: China house price index forecasting using dendritic neuron model. In: International conference on advances in informatics and computing (ICPIC) (2016)
5. Li, L., Chu, K.-H.: Real estate price variation prediction based on economic parameters. In: International Conference on Applied System Innovation (ICASI) (2017)
6. Sawant, R., Jangid, Y., Tiwari, T., Jain, S., Gupta, A.: Comprehensive analysis of housing price prediction. In: Fourth International Conference on Automatic Computing ICCUBEA) (2018)
7. Tan, F., Cheng, C., Wei, Z.: A Time-oriented latent hierarchical model for domestic price prediction. In: IEEE International Conference on Data Mining (ICDM) (2017)
8. Rakshanasri, S.L., Naren, J., Vithya, G., Akhil, S., Dinesh Kumar, K., Sai Krishna Mohan Gupta, S.: A framework on health smart home using IoT and machine learning for disabled people. Int. J. Psychosoc. Rehabil. 1–9. https://doi.org/10.37200/IJPR/V24I2/PR200304
9. Wang, L., Wang, G., Yu, H., Wang, F.: Prediction and analysis of residential house price using a flexible spatiotemporal model. J. Appl. Econ. **25**(1), 503–522
10. Wang, Q., Wang, Y., Chen, W., Zhou, X., Zhao, M., Zhang, B.: Do land price variation and environmental regulation improve chemical industrial agglomeration? A regional analysis in China. Land Use Policy **94**, 104568
11. Yang, S., Hu, S., Wang, S., Zou, L.: Effects of rapid urban land expansion on the spatial direction of residential land prices:evidence from Wuhan, China. Habitat Int. **101**, 102186
12. Chou, E.P., Ko, T.W.: Dimension reduction of high-dimensional datasets based on stepwise SVM (2017)
13. Phipps, A.G.: Inner-city neighbourhood changes predicted from house prices in Windsor, Ontario, since the early- or mid-1980s. J. Build. Construct. Plann. Res. **08**(02), 138–160. https://doi.org/10.4236/jbcpr.2020.82009

14. Truong, Q., Nguyen, M., Dang, H., Mei, B.: Housing price prediction via improved machine learning techniques. In: International Conference on Identification, Information and Knowledge in the Internet of Things (2019)
15. Raga Madhuri, C.H., Anuradha, G., Vani Pujitha, M.: House price prediction using regression techniques: a comparative study. In: 2019 International Conference on Smart Structures and Systems (ICSSS)

Chapter 13
Unmanned Aerial Vehicles (UAVs): Performance Analysis of Routing Protocols for Optimized Operations

Vinti Gupta and Dambarudhar Seth

1 Introduction

FANETs are acquiring prevalence because of their effortless deployment, flexibility, simplicity in arrangement, high dynamic nature, and low functional cost. FANETs are, for the most part, outlined by UAVs, which can fly freely or be controlled from a distance [1]. FANET is the new family part of the ad hoc environment which has significant interest from many industries and academic persons. Possessing the fantasy, formative, generally little working costly, and accessible to the establishment, the use of FANETs has drawn in more consideration and significance as of late in the different military, business, and regular citizen applications like disaster management, the urgency of control, and unfriendly environment, loss of life operation, border lookout, forest fire operation, relaying networks, assessment of wind, common security, cultivating distant perception, vehicles monitoring activity [2].

FANETs environment always supports different ad hoc network architectures. This permits information to convey to the base station in a split second and gives the capacity to interface different UAV nodes if connection loss occurs during the operation, then instantly connect to the other UAV node [3]. Although FANET follows some of the same characteristics as MANET, VANET is like some other ad hoc networks. Besides this, it always faces lots of design issues. FANET is an enhancement above typical system administration frameworks: In contrast to traditional systems administration frameworks, FANET enables the establishment of dynamic and mobile networks that may be swiftly installed and altered as needed. Uavs in a FANET may cover wide geographical regions, allowing communication linkages to be established over vast and difficult terrains where traditional systems may not be

V. Gupta (✉) · D. Seth
Department of Computer Science and Engineering, SRM Institute of Science and Technology, Delhi-NCR Campus, Ghaziabad, India
e-mail: gupta.vinti1@gmail.com

J. M. R. S. Tavares et al. (eds.), *Proceedings of Second International Conference on Intelligent System*, Algorithms for Intelligent Systems,
https://doi.org/10.1007/978-981-99-8976-8_13

viable or cost-effective. Barely any qualities like distance coverage by UAVs, node speed, limit, and kinds of power supply utilized separately from different ad hoc networks (VANETs, MANETs) [4]. FANETs have highly dynamic nodes in speed and 3D environment, and instantly changes occur in network topology or energy and weight limitations compared to other ad hoc nodes [5]. As a result, the FANET environment can be especially effective in time-sensitive scenarios, where it can finish tasks faster and with less investment.

Similarly, these are useful for a variety of purposes. Many routing protocols are implemented for these challenging scenarios by developing different mobility models. The best suite routing protocols are necessary to establish UAV- to UAV communication. Researchers always search the routing protocols with minimum overhead end-to-end delay for data transmission [6]. One of the UAVs from a group of UAVs nodes of a FANET connected to the base station, which causes them to turn into an appealing innovation in the marine mission [7]. UAVs fly in totally independent design and can convey products starting with one spot and then onto the following and, surprisingly, paired with High Definition (HD) cameras for rural field monitoring, manufacturers, and so on [1]. Besides 2D network topology, FANET follows a 3D environment where the movement of UAVs node continuously changes the shape of the network and disconnections of different nodes occur, and communication between nodes also halts [8]. As a reason, designing routing protocols for traditional networks is inappropriate for FANET [9]. UAVs in FANETs move at fast speeds all the time, causing frequent connection interruptions and topological changes. Due to battery restrictions, mission constraints, or navigation modifications, UAVs in FANETs may enter or quit the network at any moment. UAVs often have a restricted communication range that varies with environmental conditions and altitude variations. As a result, network architecture might change quickly, necessitating regular route modifications. Variable signal quality might cause sporadic connectivity and make determining the most reliable pathways difficult for routing techniques. Continuous data transmission is critical for various reasons, for being the dynamic nature of UAVs and the real-time applications they serve. FANETs are frequently used for applications that need real-time data interchange for important missions. Continuous data transmission guarantees that information reaches its destination on time, allowing for quick decision-making and action. Many FANET applications require mission-critical operations, such as aerial surveillance or emergency communication. Continuous data transmission is critical in such settings to ensure a steady flow of information between UAVs and base stations.

From the literature survey, many researchers reported the performance analysis of metrics through different methods such as OPNET, QUALNET, and NS-2. When all nodes in MANET, VANET, and FANET are in motion, few works are published on the NS-3 network simulator. In two scenarios, researchers analyzed the performance metrics of MANET routing protocols in a FANET context. We investigated different densities of UAVs node in the first condition and varying speeds of UAVs node in the second condition.

The main contribution reported in this area is as follows.

- Some famously utilized routing protocols are analyzed considering the fundamental FANET network conditions and necessities.
- Provide a related research paper summary based on Routing Protocols for Unmanned Aerial Vehicles (UAVs) performance discussed and compare the various routing protocols applicable in FANET.
- Compare the performance concerning different metrics by using the network simulator-3
- Suggest the best routing protocol with the help of resulted model by concerning different densities of UAVs as well as the extra speed of UAVs. This is suitable for use in FANETs.

The layout of the paper is summarized below.

Area II of this paper covers some related examination papers of UAV routing protocols. In this section, we present the written work on analyzing the performance of several routing protocols. In Sect. 3, the fundamentals of AODV, OLSR, and DSDV are discussed. Simulation parameters and performance metrics are discussed in Sect. 4. Results in the form of a table and graph for different conditions, as various numbers, and speeds of the presence of UAVs node, are discussed in Sect. 5. In area V we also give a fundamental analysis of the results for these routing protocols in UAV conditions concerning the different performance metrics discussed. Section 6, concludes our study.

2 Related Work

A few different routing protocol studies have sparked a lot of interest in FANET. Routing protocols are responsible for finding, making, and remaining mindful of communication routes between two nodes. The vertical and data transfer speed utilization of these protocols should be kept to an absolute minimum. Because of the incredible versatility of UAVs, network topology could shift over time, so the routing protocols, route discovery, and route maintenance are both the main challenges to solve [10]. In the paper [11], the authors proposed the utilization of CHNN to enhance the route discovery of DSR protocols. With this adjustment, the suggested work enhances the rapid development of the FANET node. The NS-3 simulator is used for simulation, and the suggested approach increases the performance of routing methods, but the author did not focus on QoE and QoS metrics. In the paper [12], the author gives the utilization of OLSR and MP-OLSR routing protocols in FANET applications, particularly in rapidly spreading fires. For chances of protecting casualties, the author thought about the situation with various speed scenarios but still needs to address the UAVs' independence issues. In a highly dynamic condition, OLSR delivered a greater PDR than other routing protocols, according to the paper [13]. A paper [14] examined the PKT delivery ratio of AODV using multiple mobility models, including RW, RWP, PURSUE, and RDIR and found that PDR declined more severely in the RWP model. The author of the paper [15] gives a detailed study of geographic FANET routing protocols, compares their path and packet (PKT) forwarding mechanism, and then discusses the issues related to the routing protocols. Paper [16] gives the performance examination of various routing protocols regarding different mobility-model and finally proposes which routing protocols provide the best execution for agriculture applications. Paper [17] presents a comparative analysis. The paper [18] thoroughly examined the various classes of position-based routing algorithms. The originality of this work is that it adds to previous evaluations on a similar topic by providing more information on various performance measures. Paper [19] proposed five classifications of FANET routing protocols and presented a correlation of routing procedures considering the specific measures with critical issues such as route selection and data transmission quality in FANET research. Paper [20] presented the order of UAVs from two perspectives: set-up and working level. The uses of the different communication models in an advanced wireless network are introduced. The author gives the classification of routing protocols and then thinks about it regardless of whether the base station is involved. [21–23] concentrated on the energy element in the FANET environment and documented several routing methods given energy. Paper [20] not only displays the FANET network structure in detail but also examines UAV mobility concepts and then examines various application scenarios.

3 Routing Protocols

Unmanned aerial vehicles (UAVs) are increasingly utilized for a variety of army, civic functions, and applications. Ad hoc wireless interconnection of UAVs can build a UAV network for cooperation and engagement. The dynamic behaviors of UAV networks, such as rapid mobility and topology changes, make designing a routing protocol difficult. When UAVs node communicates with one another, routing protocols are very significant. The significance of UAV routing protocols derives from their ability to properly handle communication, data transmission, and decision-making inside the network, which influences many aspects of FANET functioning such as security and privacy, load balancing, energy efficiency, and so on. Due to nature of communication environment, designing routing protocols for UAV networks presents several specific challenges, such as dynamic network topology, limited energy and resources, unpredictable link quality, high mobility, three-dimensional space, and so on. These are important conceptual issues because they necessitate novel methods to provide effective and reliable communication among UAVs.

UAV routing protocols are categorized into four different routing techniques as location-based, topology-based, energy-based, and swarm intelligence routing-based protocols.

3.1 Location-Based Routing Protocols

The location-based protocol uses node location information to increase performance and scalability. Three basic facts must be provided. First, each node in the network is aware of its position, whether via GPS or another method. Second, each node knows its neighbors' position, which is only one hop away. Third, the position of the base station was sent to the source node.

3.2 Topology-Based Routing Protocol

These routing methods employ user accounts to indicate nodes in the networks that utilize topology-based routing.

Figure 1 illustrates the FANET topology-based routing protocol subgroups of proactive routing, reactive routing, and hybrid routing.

Proactive Routing Protocols: These proactive protocols keep most of the network's related routing information in the routing table. This information is regularly revised as well as transferred among all UAVs nodes, and these tables need to be amended as topologies change. The proactive routing systems often have the latest information. Routing commands are used to deliver to all network elements to keep the navigation records completely maintained. As a result, proactive routing algorithms no longer provide routes to destinations, and data deliver directly, resulting in dramatically decreased delivery latency. Both OLSR and DSDV fall within this categorization.

Optimized link state routing (OLSR).

OLSR follows the approach of proactive routing. It utilizes two types of messages for communication.

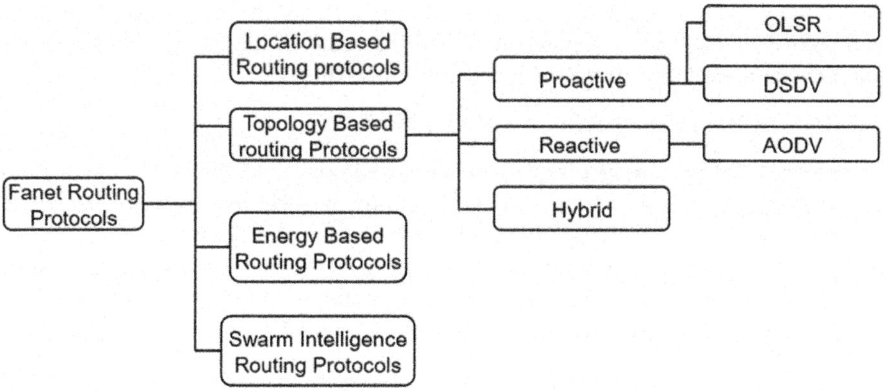

Fig. 1 UAVs routing protocol classification

I. Hello Message.
II. Topology control (TC) message.

For direct communication set-up, OLSR transmits the Hello message to identify the neighbor nodes. The second message is the topology control message used for getting the network topology information. TC messages are communicated occasionally on altogether network to illuminate the other nodes of the network to update their routing tables. For large-scale wireless ad hoc networks, OLSR gives the best solution as it provides optimal route information. Multi-point relaying is a significant feature of OLSR for decreasing control messages.

Destination-Sequenced-Distance Vector Routing (DSDV)

DSDV is a member of the proactive routing protocols family. DSDV is based on Distributed Bellman-Ford algorithm. In DSDV, each route is labeled with a sequence number (Sq. no.) which is begun by the destination, demonstrating how old the route is. Every node deal with its Sq. No. by assigning it two more prominent than the former one without fail, after getting the higher Sq. No., old route information is updated. If the same sequence number is assigned to the different routes, the route is selected with better performance metrics. After any updating, the information is transmitted over the network. To stay away from fluctuations in route updating, DSDV utilizes "setting-time" information which is used to figure out the time when the route becomes stable. This protocol has the advantage of providing loop-free pathways to each destination and selecting the best path depending on an average settling delay, which is the time between advertising a route and it is reached. But when the network is inactive, the routing tables require a small quantity of bandwidth, periodic updating, and battery power.

Reactive Routing Protocols: These protocols reply to routing requests on the fly. A route between two nodes is maintained when they interact. The primary design aim of RRP is to eliminate the overhead involved with proactive routing options. Whenever a UAV desires to communicate, a route selection phase begins, during which the maximum number of feasible routing alternatives are assessed, declared, and recorded. The time necessary to locate the route inside the routing stage may add unnecessary delay. AODV is included in this category.

Ad Hoc On-Demand Distance Vector (AODV)

This is applicable routing protocol in wireless communication networks. Robust connectivity between the nodes is where this protocol works best, with less utilization of the network, long computation of data, in case of the burden of memory, and disclosure of routing path in ad hoc network. AODV is a reactive routing protocol that maintains all information of all nodes in a suitable routing table. For maximizing the bandwidth of the web, it is considered just the next hop.

For data transmission in an ad hoc network, AODV follows the three steps.

(1) This is a discovery process in which it gathers information on routes that apply from source to a destination node.

(2) In this phase, AODV recovers any failure if it occurs in the network and updates the routing table.
(3) This is a loop-free protocol. As a new route is found, this utilization of a sequence number at each progression will be updated with network nodes.

AODV provides overhead and performance on a range of data traffic levels while consuming less battery power, reducing data traffic overhead, and improving scalability and performance.

Hybrid Routing Protocols: Hybrid protocols solve their shortcomings by integrating the traits of proactive and reactive procedures. Indeed, these protocols need extensive complexity to manage the whole network, as well as adequate time to identify and select the best routes. Hybrid protocols, like a conventional way, focus on the idea of regions, delivering the proactive method only within the regions and so lowering overhead. The reactive method is only required for inter-zone connectivity across at-location nodes. These protocols apply to massive networks with several inter and intra networks.

3.3 Energy-Based Routing Protocols

Energy-efficient routing protocols aim to select the most suitable paths for transmitting packets from the source to the base station, considering factors such as the energy level of the sensor node or the energy expended by the transmitter.

3.4 Swarm Intelligence Routing Protocols

Swarm intelligence involves solving and controlling problems by leveraging the behaviors observed in biologically inspired species. Swarm-based intelligence algorithms draw inspiration from the actions of various organisms such as ants, bees, fireflies, wasps, bats, and fish. When it comes to optimizing routing for Unmanned Aerial Vehicles (UAVs), swarm-based algorithms are considered highly effective in delivering optimal solutions.

4 Simulation

In this segment, we test and evaluate the routing protocols cleared up above for choosing the best presentation between AODV, OLSR, and DSDV. We utilized the NS3 network simulator through this review to understand this performance comparison. The support of NS3 is C++ and Python with a set-up limit.

NS3 is a fantastic network simulation software. It promises an easy graphical interface, and a strong graphic appeal, and may be used to signify action. When it arrives to software processes, NS3 has a relatively ideal one in that it can confirm the data forwarding time difference, packet length distribution, network node type, and connection type in detail, and it can layout its simulation model using network tools and implementations provided by various computer manufacturers.NS3 is built on C++ and Python and supports writing the code. The NS3 simulator works with a system that includes attributes for handling default values and simulation parameters; this tool is commonly used to model and test the behavior of various types of networks, such as mobility models, topology, and node energy consumption. There is no assumption about NS3's advantages in terms of ease of use, as it can produce more detailed and realistic simulation results with fewer operations. As a side effect, the simulation tool in this paper is NS3. The usage of the NS3 simulator in research offers researchers a great tool for exploring, analyzing, and optimizing UAV routing protocols in a virtual environment. It aids in overcoming the practical constraints and expenses involved with real-world trials, and it speeds up the development of efficient and stable routing systems for UAV networks.

The network simulator-3 capacities with a framework that incorporates characteristics to deal with default values as well as a simulation matrix. This simulator is primarily used to display and test the behavior of various sorts of networks, for example, different mobility models, topology, and energy utilization of nodes.

4.1 Parameters for Simulation

A range of parameters was utilized, with variations in both the quantity and velocity of UAVs. Details of these parameters can be found in Table 1.

Table 1 Simulation Parameter

Simulator Parameter	Values
No. of UAVs node	20, 25, 40, 60, 80
Simulator	NS-3
Simulation time	600 s
Data payload	512bytes/packet
Wireless range	27 m
Speed of UAVs	20, 40, 60, 80 m per second
Protocols	AODV, OLSR, DSDV
Link layer	802.11 b-11 Mbps
Channel type	Channel/wireless channel

4.2 Performance Metrics

The presentation measurement thought about for the assessment of our three dynamic routing protocols are as follows: -

Packet Loss Ratio: Metrics addresses the proportion of lost data PKTs to all out numbers of sent data PKTs. This measurement is equivalent to the number of data packets not partitioned by the total number of data packets sent. It should be noticed that fault packets are not commonly viewed as outside the lost packages in communication networks because most applications require the authority of data.

$$\text{PKT Loss Ratio} = (\text{PKT}_{\text{nt}} - \text{PKT}_{\text{nr}}/\text{PKT}_{\text{nt}}) * 100\% \tag{1}$$

PKT_{nt} = Total no. of transmitted data packets.
PKT_{nr} = Total no. of received data packets.

Packet Delivery Ratio: PDR can be estimated as the proportion of delivered several packets at the destination altogether to transmit PKTs from the source in the network. The greatest number of data PKTs must be delivered to the destination. As the value of PDR increases, the performance of the network is directly dependent on the PDR of the network.

$$\text{Delivery ratio} = \text{Total Delivered data packet/Total Transmitted data PKT.} \tag{2}$$

End-to-End Delay: End-to-end delay is the total amount of time it takes for PKT to go from source to destination over the network. So, this metric somewhat relies upon the PDR. As the distance increases between the source and the goal, the packet drops probability also increases.

$$E\_D = 1/N \sum_{i=1}^{N} (\text{Rti} - \text{Sti}) * 1000(\text{ms}) \tag{3}$$

E_D End-to-End Delay.
N Successful delivered packets.
i Packet identifier.
Rti Received Time.
Sti Send time.

Throughput: Network throughput is the number of effectively sent data PKTs from the origin to the destination per second in wireless sensors. Throughput is the average amount of data (kbps) provided per unit of time by each pair of nodes in the network.

$$\text{Throughput} = (\text{received PKT Size}/(\text{stop Timing} - \text{start Timing})) * (8/1000). \tag{4}$$

Jitter Delay: Packet Delay statistical variation is called jitter. It's used to figure out how long it takes for all packets to transit from a source to a destination. It is determined in the following way:

$$\text{Jitter rate} = \text{Delay in Total Variant} / \text{No. of packet sent} - 1. \quad (5)$$

5 Evaluation Result

Following the simulation, we obtained the following results, which reveal the diversity of the beginning-to-end concede PKT loss ratio, PDR, End-to-End delay, throughput, and jitter in different speeds of UAVs and various node densities.

5.1 Performance of AODV, OLSR, and DSDV

Case I: In the first case, the speed of UAVs in the network is changed from 20 m per sec to 80 m per second, leading the network's multiple routing algorithms to perform differently. In this case, the no. of UAVs is fixed as 25 nodes.

Case 2: In the second case, the No. of UAVs connected in a FANET network vary from 20 to 80 UAVs nodes, resulting in modifying the number of connections.

5.2 Graphs After Simulation

Both cases are simulated, and the requisite trace file is generated, as illustrated in Figs. 2, 3, 4, 5, 6, 7, 8, 9, 10 and 11. The UAV nodes were simulated in this scenario using AODV, OLSR, and DSDV protocols, with different parameters such as PKT Loss Ratio, PDR, End-to-End Delay, Throughput, and jitter delay being checked. X-axis represents UAV's node speed in case 1 (Figs. 2, 3, 4, 5 and 6). In contrast, the x-axis denotes the number of UAV nodes in case 2 (Fig. 7, 8, 9, 10 and 11), whereas the y-axis represents different performance metrics.

5.3 Analysis and Result

To construct the various scenarios, the NS-3 was employed. We simulate for 600 s before creating the trace file, which we utilize to save the graphs for analysis and calculation, as seen above. These visuals substantially enhance the statistical analysis

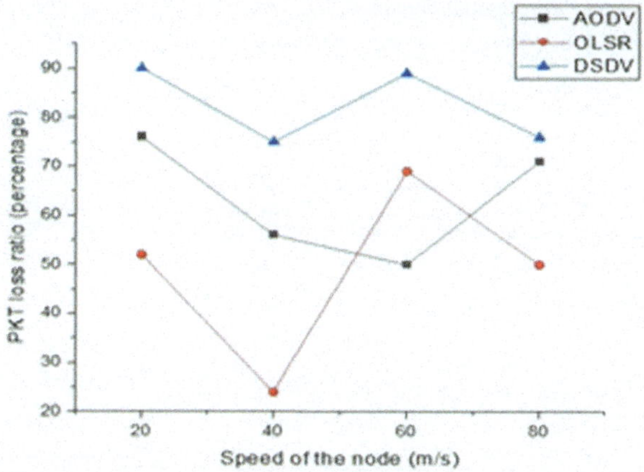

Fig. 2 PKT loss ratio versus speed of UAVs

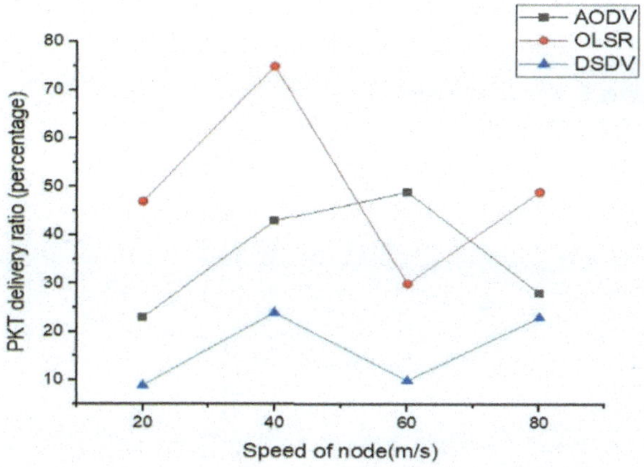

Fig. 3 PDR versus the rate of UAVs

of the performance of these routing protocols. The necessary graphs were saved as a bitmap graphic for statistical analysis.

Case 1: Speed of UAVs Nodes Varied.

PKT Loss Ratio: As per Fig. 2, DSDV shows the lowest performance as the speed of UAV nodes increases. It remains the same for all lower rates of UAV nodes, but as the speed of UAVs increases, DSDV shows the most packet loss because each node must spend a significant amount of time processing any control data received to get routing information, even if it is not the intended destination.

PKT Delivery Ratio: The graphical analysis of Fig. 3 demonstrate that AODV's performance in terms of boosting the speed of UAVs is poor, whereas OLSR outperforms both DSDV and AODV because with the fast changes of the source and destination pairs, OLSR improves the protocol's appropriateness for an ad hoc network.

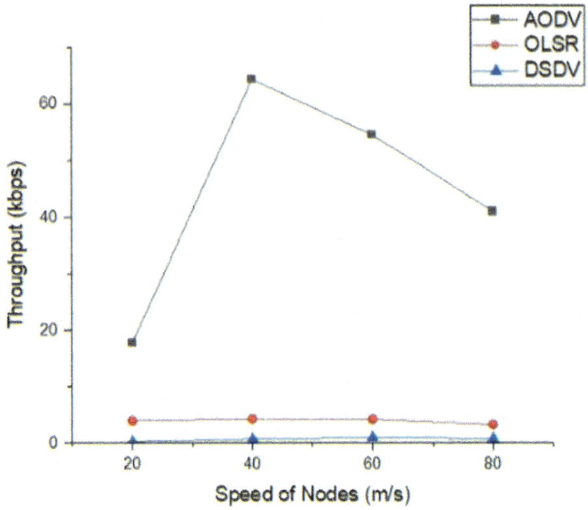

Fig. 4 Throughput versus speed of UAVs

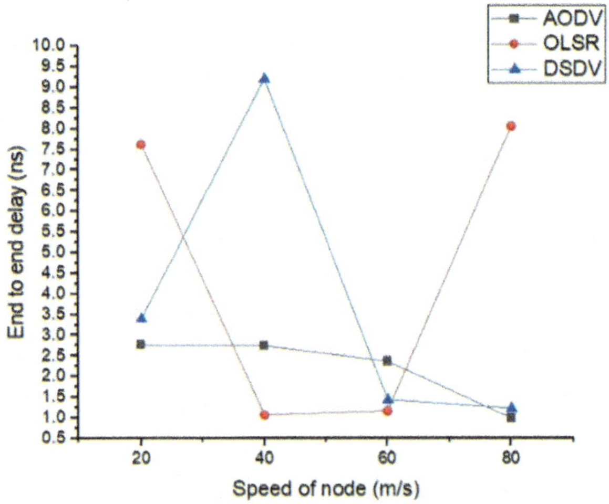

Fig. 5 End-to-end delay versus speed of UAVs

Throughput: The graphical presentation in Fig. 4 shows that when it comes to increasing the speed of UAVs, DSDV, and OLSR perform roughly identically, whereas AODV beats both because AODV offers a faster connection set-up time and detects the most recent path to the destination.

End-to-End Delay: Fig. 5 shows a graphical overview of the situation. As the speed of UAVs rises, the DSDV end-to-end delay increases, then drops. As the speed of UAVs rises, the AODV lowers. When the nodes are moving at 80 m per second, the OLSR has the longest end-to-end delay because OLSR keep a routing table with

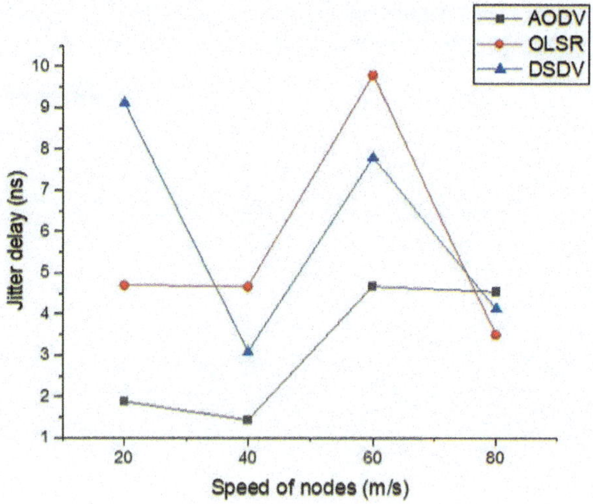

Fig. 6 Jitter versus speed of UAVs

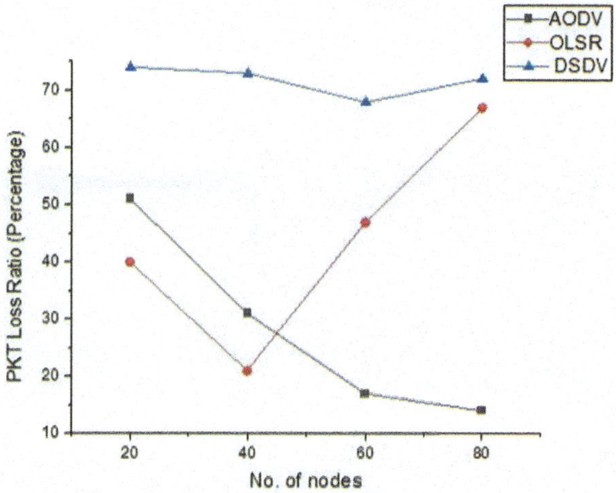

Fig. 7 PKT loss ratio versus no. of UAVs

all possible routes and overhead from control messages grows in proportion to the number of mobile hosts.

Jitter Delay: Fig. 6 shows a graphical overview of the situation. AODV has the most extended jitter delay of the two routing protocols when the node speed increases because AODV takes up a significant amount of bandwidth.

Case 2: Number of UAVs Nodes Varied.

PKT Loss Ratio: Fig. 7 illustrates the scenario graphically. When it comes to expanding the density of UAV nodes, AODV excels in all other protocols. In DSDV

Fig. 8 PDR versus no. of UAVs

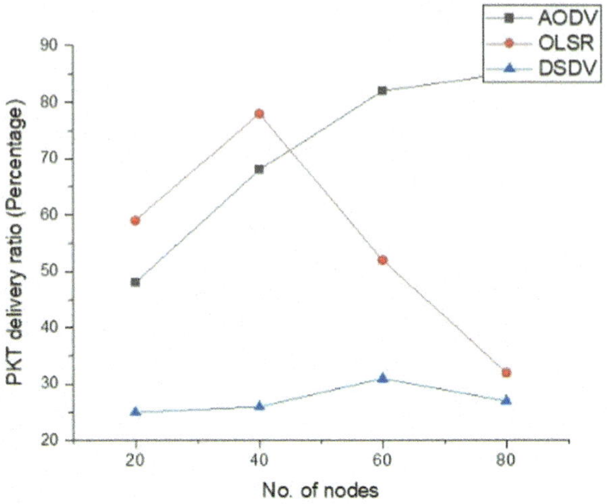

Fig. 9 Throughput versus no. of UAVs

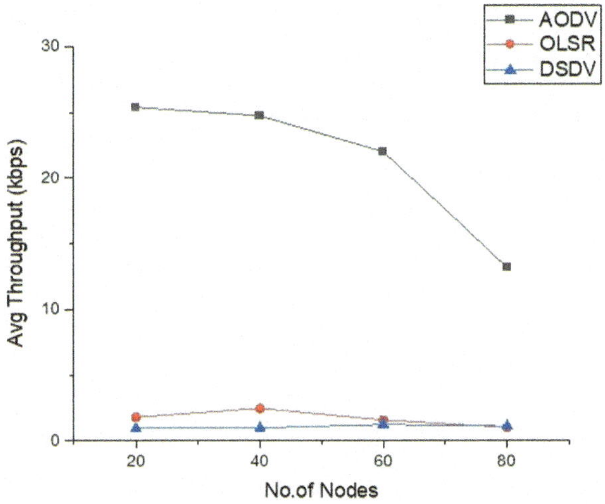

routing protocols, the highest PKT loss ratio is recorded because a broken connection is not repaired locally by the route maintenance mechanism.

PKT Delivery Ratio: The scenario is presented visually in Fig. 8. AODV beats DSDV and OLSR routing protocols because even for nodes in continual motion, AODV can handle both unicast and multicast packet transfers.

Throughput: Fig. 9 represents the situation effectively. In this scenario, AODV performs better than the two routing protocols because AODV protocol is a flat routing mechanism. It does not require a centralized administration system to conduct the routing operation.

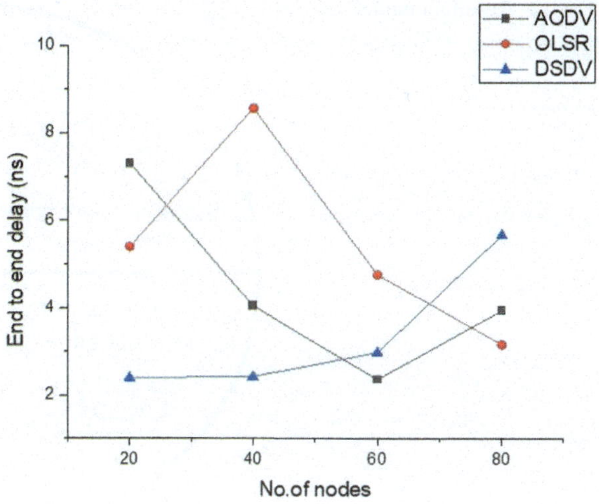

Fig. 10 End-to-end delay versus no. of UAVs

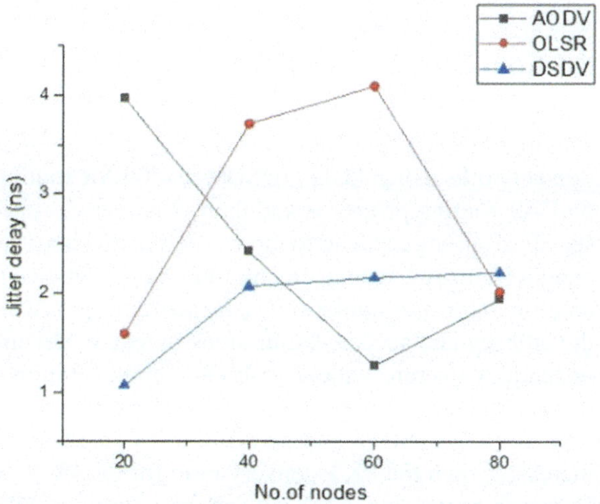

Fig. 11 Jitter delay versus No. of UAVs

End-to-End Delay: Fig. 10 effectively shows the case scenario where DSDV has the greater value for the parameter End-to-End delay than OLSR. OLSR performs much better on this parameter because the messages are transmitted on a regular basis and do not have to be delivered sequentially, the OLSR protocol does not require that the link be trustworthy for the control messages.

Jitter Delay: Fig. 3 represents the case scenario effectively. The jitter latency of DSDV increases as the density of UAV nodes increases because of the source routing technique used in DSDV, there is routing overhead. This routing overhead is proportional to the path length and data load (Fig. 12).

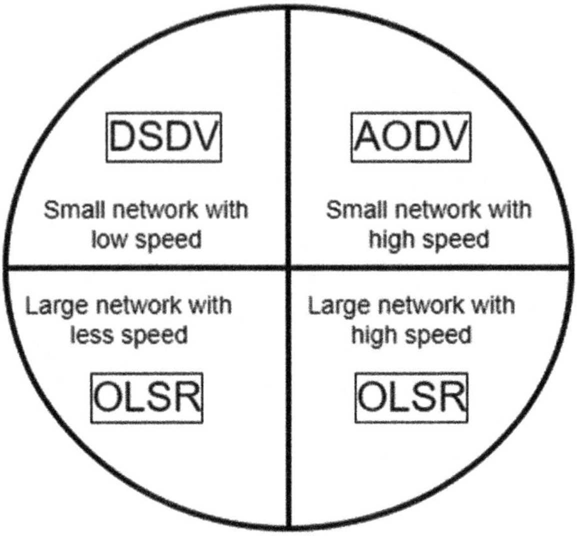

Fig. 12 Resulting model

6 Conclusion

This manuscript uses simulation to investigate the efficiency of UAV telecommunication networks using AODV, DSDV, and OLSR routing protocols The study findings on UAV routing protocols and their effectiveness under varying UAV densities and speed variations can lead to more reliable, efficient, and adaptable UAV-based applications in disaster management, military operations, and patrols. By exploiting this data, operators may make more educated decisions while deploying UAVs, increasing the efficacy and success of missions in real-world circumstances. Different routing techniques operate widely in UAV communication networks depending on node density and node movement speed. As the number of UAVs grows, so does the PKT Loss Ratio of DSDV and speed of UAV increase. When the speed of UAVs has increased, then the PKT delivery ratio for OLSR is the highest compared to others. However, as the size of the network increased, AODV has the greatest PDR comparable to others. In both cases, AODV provides the highest throughput. When we speed up the UAVs OLSR's end-to-end latency is increased. In the case of DSDV, the value of this performance indicator grows as increases the nodes. The jitter delay of AODV is enhanced with the increase in the speed of nodes. But in the other scenario, the density of UAVs is increased, and the jitter delay of DSDV is increased. To improve communication assurance for the UAV cluster, the performance of the UAV communication networks should be optimized, and the best appropriate routing protocol would be chosen depending on the node density and movement speed of the UAV. Based on analysis, the conclusions are.

1. For high network density, AODV is advised. Because AODV is a self-starting, loop-free protocol that can handle many mobile nodes.
2. OLSR is good for highly dynamic networks. Because OLSR has a reduced average end-to-end latency, it is used for applications that require as little delay as possible.
3. The packet overhead of a DSDV protocol increases the total amount of nodes in the ad hoc network. Because of this, DSDV is ideal for small networks.

References

1. AlKhatieb, A., Felemban, E., Naseer, A.: Performance evaluation of ad-hoc routing protocols in (FANETs). In: 2020 IEEE Wireless Communications and Networking Conference Workshops (WCNCW), pp 1–6. IEEE (2020). https://doi.org/10.1109/AECT47998.2020.9194201
2. Erdelj, M., Natalizio, E., Chowdhury, K.R., Akyildiz, I.F.: Help from the sky: Leveraging UAVs for disaster management. IEEE Pervasive Comput. **16**(1), 24–32 (2017). https://doi.org/10.1109/MPRV.2017.11
3. Alghamdi, Y., Munir, A., La, H.M.: Architecture, classification, and applications of modern unmanned aerial vehicles. IEEE Consum. Electron. Mag. **10**(6), 9–20 (2021). https://doi.org/10.1109/MCE.2021.3063945
4. Mariyappan, K., Christo, M.S., Khilar, R.: Implementing FANET energy-efficient AODV routing protocols for flying ad hoc networks [FEEAODV]. Mater. Today Proc. (2021). https://doi.org/10.1016/j.matpr.2021.02.673
5. Guillen-Perez, A., Montoya, A.M., Sanchez-Aarnoutse, J.C., Cano, M.D.: A comparative performance evaluation of routing protocols for flying Ad-Hoc networks in real conditions. Appl. Sci. **11**(10), 4363 (2021). https://doi.org/10.3390/app11104363
6. Al-Emadi, S., Al-Mohannadi, A.: Towards enhancement of network communication architectures and routing protocols for FANETs: a survey. In: 2020 3rd International Conference on Advanced Communication Technologies and Networking (CommNet), pp. 1–10. IEEE (2020). https://doi.org/10.1109/CommNet49926.2020.9199627
7. Xie, P.: An enhanced OLSR routing protocol based on node-link expiration time and residual energy in ocean FANETS. In: 2018 24th Asia-Pacific Conference on Communications (APCC), pp. 598–603. IEEE (2018). https://doi.org/10.1109/APCC.2018.8633484
8. Wubben, J., Aznar, P., Fabra, F., Calafate, C.T., Cano, J.C., Manzoni, P.: Toward secure, efficient, and seamless reconfiguration of UAV swarm formations. In: 2020 IEEE/ACM 24th International Symposium on Distributed Simulation and Real-Time Applications (DS-RT), pp. 1–7. IEEE (2020). https://doi.org/10.1109/DS-RT50469.2020.9213669
9. Cheriguene, Y., Djellikh, S., Bousbaa, F.Z., Lagraa, N., Lakas, A., Kerrache, C.A., Tahari, A.E.K.: SEMRP: an energy-efficient multicast routing protocol for UAV swarms. In: 2020 IEEE/ACM 24th International Symposium on Distributed Simulation and Real-Time Applications (DS-RT), pp. 1–8. IEEE (2020). https://doi.org/10.1109/DS-RT50469.2020.9213700
10. Souza, J., Jailton, J., Carvalho, T., Araújo, J., Francês, R.: A proposal for routing protocol for FANET: a fuzzy system approach with QoE/QoS guarantee. Wirel. Commun. Mob. Comput. (2019). https://doi.org/10.1155/2019/8709249
11. Yang, H., Liu, Z.: An optimization routing protocol for FANETs. EURASIP J. Wirel. Commun. Netw. **2019**(1), 1–8 (2019). https://doi.org/10.1186/s13638-019-1442-0
12. Radu, D., Cretu, A., Parrein, B., Yi, J., Avram, C., Aştilean, A.: Flying ad hoc network for emergency applications connected to a fog system. In: International Conference on Emerging Internetworking, Data & Web Technologies, pp. 675–686. Springer, Cham (2018). https://doi.org/10.1007/978-3-319-75928-9_60

13. Mohapatra, S., Kanungo, P.: Comparative performance analysis of MANET routing protocols using nS2 simulator. In: International Conference on Computational Intelligence and Information Technology, pp. 731–736. Springer, Berlin (2011). https://doi.org/10.1007/978-3-642-25734-6_127
14. Katiyar, S., Gujral, R., Mallick, B.: Comparative performance analysis of MANET routing protocols in military operation using NS2. In: *2015 International Conference on Green Computing and Internet of Things (ICGCIoT)*, pp. 603–609. IEEE (2015). https://doi.org/10.1109/ICGCIoT.2015.7380536
15. Agrawal, J., Kapoor, M.: A comparative study on geographic-based routing algorithms for flying ad-hoc networks. Concurr. Comput. Practice Exper. **33**(16), e6253 (2021). https://doi.org/10.1002/cpe.6253
16. Kakamoukas, G.A., Sarigiannidis, P.G., Economides, A.A.: FANETs in agriculture—a routing protocol survey. Internet Things **18**, 100183 (2022). https://doi.org/10.1016/j.iot.2020.100183
17. Sang, Q., Wu, H., Xing, L., Xie, P.: Review and comparison of emerging routing protocols in flying ad hoc networks. Symmetry **12**(6), 971 (2020). https://doi.org/10.3390/sym12060971
18. Khan, M.A., Safi, A., Qureshi, I.M., Khan, I.U.: Flying ad-hoc networks (FANETs): A review of communication architectures and routing protocols. In: 2017 First International Conference on the Latest Trends in Electrical Engineering and computing Technologies (INTELLECT), pp. 1–9. IEEE (2017). https://doi.org/10.1109/INTELLECT.2017.8277614
19. Nadeem, A., Alghamdi, T., Yawar, A., Mehmood, A., Siddiqui, M.S.: A review and classification of flying Ad-Hoc network (FANET) routing strategies. J. Basic Appl. Sci. Res. **8**(3), 1–8 (2018)
20. Lakew, D.S., Sa'ad, U., Dao, N.N., Na, W., Cho, S.: Routing in flying ad hoc networks: a comprehensive survey. IEEE Commun. Surv. Tutor. **22**(2), 1071–1120 (2020). https://doi.org/10.1109/COMST.2020.2982452
21. Oubbati, O.S., Atiquzzaman, M., Lorenz, P., Tareque, M.H., Hossain, M.S.: Routing in flying ad hoc networks: survey, constraints, and future challenge perspectives. IEEE Access **7**, 81057–81105 (2019). https://doi.org/10.1109/ACCESS.2019.2923840
22. Li, K., Ni, W., Wang, X., Liu, R.P., Kanhere, S.S., Jha, S.: Energy-efficient cooperative relaying for unmanned aerial vehicles. IEEE Trans. Mob. Comput. **15**(6), 1377–1386 (2015). https://doi.org/10.1109/TMC.2015.2467381
23. Khelifi, F., Bradai, A., Singh, K., Atri, M.: Localization and energy-efficient data routing for unmanned aerial vehicles: Fuzzy-logic-based approach. IEEE Commun. Mag. **56**(4), 129–133 (2018). https://doi.org/10.1109/MCOM.2018.1700453

Chapter 14
Processing the Missing Value Based on the Linear Regression Approach

Pham Le Son, Nguyen Ngoc Dung, and Doan Van Thang

1 Introduction

Data has changed the face of our world over the years. Many businesses rely heavily on the ability to collect meaningful information, interpret statistical data and turn it into useful information that businesses can use to make important decisions. Data analysis is a process from data collection, mining, management, and processing, in which data preprocessing is the most important step because the collected data can be considered unclean because of the following reason:

- Incomplete data: This can happen for a number of reasons: (1) Some important attributes are not provided, (2) Data at the time of data entry and analysis are different, and (3) data inconsistent with previously saved data may be deleted resulting in data loss.
- Inconsistent data: This can be because the data is gathered from many different sources.

Data cleaning is an important step in the knowledge search process because if the data is not of good quality, the mining results will also be of poor quality. For example, duplicate or missing data can cause erroneous statistics. Clearly, quality decisions must be based on quality data.

P. Le Son
Thu Dau Mot University, Ho Chi Minh City, Vietnam

N. N. Dung · D. Van Thang (✉)
Industrial University of Ho Chi Minh City, Ho Chi Minh City, Vietnam
e-mail: doanvanthang@iuh.edu.vn

N. N. Dung
e-mail: nguyenngocdung@iuh.edu.vn

J. M. R. S. Tavares et al. (eds.), *Proceedings of Second International Conference on Intelligent System*, Algorithms for Intelligent Systems,
https://doi.org/10.1007/978-981-99-8976-8_14

Currently, there are many approaches to solve such as: similarity-based model [12], probability distribution model [11], similarity reasoning [10]. All approaches are intended to fully capture and handle incomplete, inaccurate, or uncertain information.

The paper, based on the correlation between the attributes, then builds a regression equation based on the maximum correlation with the attribute whose data is incomplete, inaccurate, or uncertain. Finally, apply the regression equation to fill in the data for the attribute whose data is incomplete, incorrect, or uncertain. The content of the article is presented as follows: Related studies are presented in Sect. 2. Correlation coefficients, linear regression in statistical theory are presented in Sect. 3. Section 4 presents procedures for handling missing attribute values and finally concludes.

2 Related Studies

In 1894 Sir Francis Galton et al. introduced the concept of linear regression for the first time. Linear regression is applied to a data set to determine and quantify the relationships between the variables under consideration [7, 8]. Univariate statistical tests, exact tests, t tests, and analysis of variance did not allow the calculations to affect covariates/other confounding factors during the analysis (Chang 2004). However, correlation and regression are tests that allow studies to control the effect of confounding factors in understanding the relationship between two variables (Chang 2003).

Some recent works theoretically analyze linear models Belkin et al. [1], Bibas et al. [2], Mei et al. [9], Liang et al. [4]. At a high level, these studies analyze the experimental risk of over-analyzed linear regression estimators with different assumed values on the variables.

Breusch and Pagan [3] introduce two least squares regressions instead of performing the probability ratio test as a conventional condition. However, it can be calculated using two least squares regressions. Eliminating the use of repetitive calculations is necessary to obtain maximum likelihood estimates of the constants throughout the model given in the paper.

Doan et al. [5, 6], use fuzzy dependencies to handle missing attribute values. The concept of fuzzy attribute dependency and fuzzy modal dependency is an extension of the fuzzy functional dependency concept in fuzzy relational databases. On that basis, the paper has applied these fuzzy dependencies to find the correct answer for the Null query.

Wang et al. [12], using the same argument theory, if A is called the source feature set, and B is the target feature set, then the source and target feature sets have similar properties P. Then, if A has the attribute P′, then that B can also have P′ based on the properties P present in both A and B. The same reasoning can be applied to dealing with the missing value and finding the answer approximation for Null query is quite efficient.

The authors in [6] have proposed a theoretical model that uses the theory similar argument to answer Null queries for a fuzzy relational database model based on the probability distribution. However, in this model, author is only interested in the similarity relationship and ignores the discrete similarity domains, where the data can be modeled through similarity relationships.

The paper study regression model to handle missing value. The theoretical basis will be presented in the next section.

3 Theoretical Basis

3.1 Simple Linear Regression

Given X and Y is two random variables, experimentally observed two samples of size n: X: $X_1, X_2, \ldots, X_n$; Y: $Y_1, Y_2, \ldots, Y_n$.

Y has a linear relationship with X, if $Y_i = b_1 + b_0 X_i + \varepsilon_i$, $i = 1, 2, \ldots, n$ with: ε_i is a random variable according to the normal distribution $N(0;\sigma 2)$, b_1: call intercept, and b_0: call gradient.

These coefficients are estimated from the data. The estimation method is the least squares method. This method finds b_1, b_0 to $\sum_{i=1}^{n} [y_i - (b_1 + b_0 x_i)]^2$ to get the minimum value.

Then

$$\widehat{b_0} = \frac{\sum_{i=1}^{n} (x_i - \overline{x})(y_i - \overline{y})}{\sum_{i=1}^{n} (x_i - \overline{x})^2}, \quad \widehat{b_1} = \overline{y} - \widehat{b_0}\overline{x}$$

Note that: $\widehat{b_1}$, $\widehat{b_0}$ is approximately of, b_0. From $\widehat{b_1}$, $\widehat{b_0}$, we has $\widehat{y_i} = \widehat{b_1} + \widehat{b_0}\widehat{x_i}$, then, the quantity $\left(y_i - \widehat{y_i}\right)$ call is residual. The variance of the residual can be estimated by:

$$s^2 = \frac{\sum_{i=1}^{n} (y_i - \hat{y}_i)}{n - 2}$$

Example 1: The study data on blood cholesterol levels of 18 male subjects were consider as follows (BMI: ratio of weight (kg) with height squared (cm^2). Estimating the correlation coefficient between age and cholesterol.

id	Age	BMI	Chol	Id	Age	BMI	Chol
1	47	24.4	3.5	10	42	24.3	3.8
2	19	19.6	1.9	11	56	22.2	4.1
3	51	25.2	4	12	32	20.8	3

(continued)

(continued)

id	Age	BMI	Chol	Id	Age	BMI	Chol
4	29	21.6	2.6	13	21	19.9	2.5
5	56	24.4	4.5	14	62	25.7	4.6
6	24	22.1	3.0	15	39	25.4	3.2
7	27	21.7	2.9	16	47	20.2	4.2
8	35	23.9	3.8	17	27	20.2	2.3
9	21	18.8	2.1	18	48	21.8	4

For simple linear regression analysis for two quantities age, chol. We have correlation between age and chol: 0.9336382192613233, and linear equation $\hat{y}_i = 37.75 + 0.0573 * \hat{x}_i$.

3.2 Correlation Analysis

To evaluate the strong and weak correlation of the linear relationship between two random variables, we use correlation analysis. The strength of this relationship is expressed through the coefficient of correlation.

a. *Correlation Coefficients Pearson r*

Given two random variables X and Y obeying the normal distribution, experimentally observed by two samples of size n:

$$[X : X_1, X_2, .., X_n; Y : Y_1, Y_2, .., Y_n]$$

Correlation coefficients Pearson r determined:

$$r_{XY} = \frac{\sum_{i=1}^{n} \left(X_i - \overline{X}\right)\left(Y_i - \overline{Y}\right)}{\sqrt{\sum_{i=1}^{n} \left(X_i - \overline{X}\right)^2 \sum_{i=1}^{n} \left(Y_i - \overline{Y}\right)^2}}$$

Correlation coefficients $r_{XY} \in [-1, +1]$, conventional reality.

If linear correlation is very strong then r_{XY} is larger 0.8.

If linear correlation is strong then r_{XY} is in interval 0.6–0.8.

If linear correlation has correlation then r_{XY} is in interval 0.4–0.6.

If linear correlation has weak correlation then r_{XY} is in interval 0.2–0.4.

If linear correlation has very weak correlation then r_{XY} is in less then 0.2.

Based on the correlation coefficient, we will know the relationship between two variables. Thereby helping us to know the strength and weakness in the relationship

Table 1 Correlation matrix between attributes

	Age	BMI	Chol
Age	1.000000	0.691420	0.936726
BMI	0.691420	1.000000	0.693392
Chol	0.936726	0.693392	1.000000

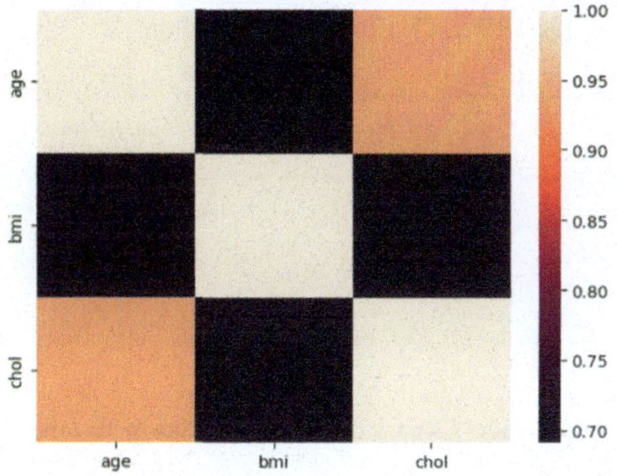

Fig. 1 Correlation matrix shown in heat map

of the two variables under consideration. We find that the closer the correlation coefficient value is to 1, the stronger the relationship between the two variables.

b. *Correlation matrix*

In the data set to show the affinity between attributes, we use the affinity matrix, also known as the correlation matrix. Each cell in the table shows the correlation between two variables. In the multivariable linear regression model, to consider the correlation between factors and diagnose multicollinearity, the correlation matrix is also used. However, the diagnosis of multicollinearity is also relative, since variables can have multicollinearity even if there is no significant correlation (Table 1; Fig. 1).

It can be seen that the age and chol attributes are strong linear correlation.

4 Apply the Regression Model to the Missing Value Treatment

To apply the research method, in this paper we use the Employee database (with 9 attributes and 92 records) with the following structure (Table 2).

We noticed that the salary attribute currently has two employees E006, E028 missing data (Fig. 2).

So the question arises how to fill data the salary value for these two employees? As we mentioned in Sect. 2 of the article, there are many research directions proposed

Table 2 Employee database structure

ENO	Age	Dept.	Gender	Skill	Work in year	Salary	Trained year	Office city
E001	29	HR	Female	SQL	3	833.2380613	3	Danang
E002	39	IT	Male	Java	7	1459.629839	7	Danang
E003	32	IT	Male	SQL	6	1239.044884	4	Danang
E004	38	HR	Female	C#	6	1328.732047	5	Danang
E005	28	IT	Male	Python	2	902.1951618	3	Hanoi
…	…	…	.	…	…	…		…

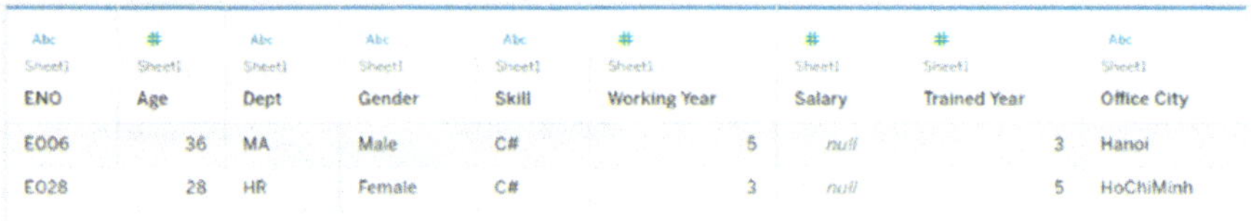

ENO	Age	Dept	Gender	Skill	Working Year	Salary	Trained Year	Office City
E006	36	MA	Male	C#	5	null	3	Hanoi
E028	28	HR	Female	C#	3	null	5	HoChiMinh

Fig. 2 Show salary of two employees with missing data

to solve this problem. In this section, the paper presents a new processing solution based on a simple linear regression model.

Algorithm RMV (Replace Missing Values)

Input: *Table E with attribute set* (a_1, a_2,..., a_n), *where* a_i *is the attribute with the missing value.*

Output: *Table E with attribute set* (a_1, a_2,..., a_n), *with* a_i *attribute has been filled in with the full value.*

Method:

Step 1: Determine the correlation between the attributes in E with a_i.
Step 2: Determine which attribute a_j has the greatest correlation with attribute a_i.
Step 3: Build the linear regression equation $a_i = \alpha + \beta a_j$.
Step 4: Fill in the missing value in a_i using the equation developed in step 3.

Apply the **RMV** algorithm to fill in the missing data value for the **Salary** attribute in the Employee database (Table 3; Fig. 3).

Step 1: Determine the correlation between the attributes in Employee database with Salary.
Step 2: Determine attribute a_j has the greatest correlation with attribute Salary. From correlation graph, we find that the attribute **TrainedYear** has the greatest correlation with attribute **Salary**.
Step 3: The linear regression equation is built as follows: Salary $= \alpha + \beta$TrainedYear.

Table 3 The salary data of two employees has been filled in

ENO	Age	Dept.	Gender	Skill	Work in year	Salary	Trained year	Office city
E006	36	MA	Male	C#	5	981.183	3	Hanoi
E028	28	HR	Female	C#	3	1320.423	5	Ho Chi Minh

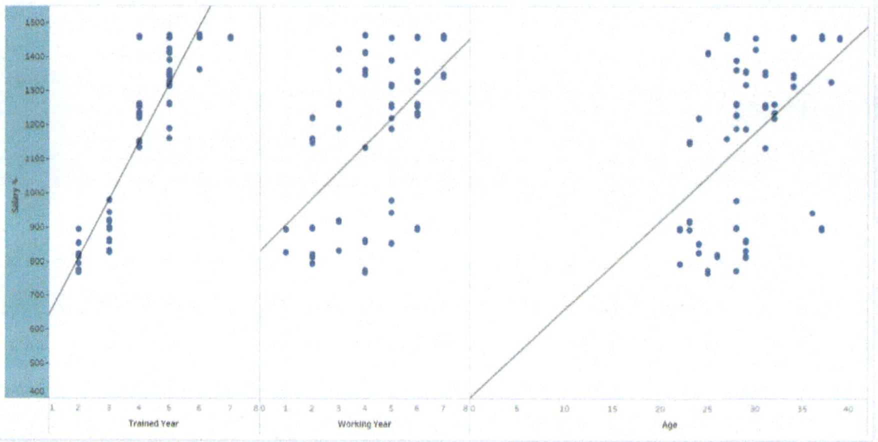

Fig. 3 Correlation graph

Linear regression equation: **Salary** = 169.62 * **Trained Year** + 472.323.

Step 4: Fill in the missing Salary value for two employees E006 and E028.

Theorem RMV algorithm always stops and is correctness.

Proof

1. *Proof of stationarity*: In a data table with a fixed set of attributes, it should be considered finite, so the algorithm stops when it has traversed all the attributes.
2. *Proof of correctness*: RMV algorithm is based on correlation theory and linear regression to handle missing values. The concepts of correlation and linear regression in statistical theory, proposed by Sir Francis Galton in 1894, were gradually withdrawn into an increasingly complete scientific theory. That is the basis for determining the correct RMV algorithm.

The **RMV** algorithm complexity is $O(n)$ with n attributes of table E.

5 Conclusion

The paper have proposed a new method to deal with missing data in the database according to the linear regression approach in statistical theory. In this solution, the biggest problem is to determine the correlation between the attributes and from there to build the regression equation. The paper also built the RMV algorithm to fill in missing data values from an incomplete database. The algorithm has also been proven correct and stable.

References

1. Belkin, M., Hsu, D., Xu, J.: Two models of double descent for weak features. SIAM J. Math. Data Sci. **2**(4), 1167–1180 (2020)
2. Bibas, K., Fogel, Y., Feder, M.: A new look at an old problem: A universal learning approach to linear regression. In: 2019 IEEE International Symposium on Information Theory (ISIT), pp. 2304–2308 (2019)
3. Breusch, T.S., Pagan, A.R.: A simple test for heteroscedasticity and random coefficient variation. Econometrica J. Economet. Soc. **47**(5), 1287–1294 (1979)
4. Dereziński, M., Liang, F., Mahoney, M.W.: Exact expressions for double descent and implicit regularization via surrogate random design. In: Conference on Neural Information Processing Systems (NeurIPS 2020), Vancouver, Canada, pp 1–26 (2020)
5. Van Ban, D., Tuan, T.C., Van Thang, D.: The fuzzy object dependences based on semantic approximation by the hedge algebraic approach. J. Inform. Cybern. **29**(1), 66–78 (2013)
6. Van Ban, D., Tuan, T.C., Van Than, D.: Approximate answer to NULL queries in the fuzzy object-oriented database model short of information. J. Inform. Cybern. **27**(4), 329–340 (2011)
7. Jay Kens, G.: Introduction to Probability and Statistics Using R, 1st edn (2010)
8. Wickham, H.: Ggplot2: Elegant Graphics for Data Analysis. Springer, Berlin (2016)
9. Mei, S., Montanari, A.: The generalization error of random features regression: precise asymptotics and double descent curve. Commun. Pure Appl. Math. **75**(4), 665–891 (2022)
10. Dutta, S.: Approximate reasoning by analogy to answer null queries. Int. J. Appr. Reasoning **5**, 373–398 (1991)
11. Biazzo, V., Giugno, R., Lukasiewiez, T., Subrahmanian, V.S.: Temporal probabillistic object bases. IEEE Trans. Knowl. Eng. **15**(4), 921–939 (2023)
12. Wang, S.L, Huang, T.J.: Analogical reasoning to answer null queries in fuzzy object-oriented data model. In: Proceedings of the 6th IEEE International Conference on Fuzzy Systems, Barcelona, Spain, 1–5 July 1997, pp 31–36 (**1997**)

Chapter 15
Improving Object Detection Versatility with 6G in VANETs

Vithya Ganesan, R. Dhanalashmi, Ahmed J. Obaid, Viswanathan Ramasamy, Sri Anima Padmini, and Subrata Chowdhury

1 Introduction

Enhanced mobile broadband (eMBB), ultra-reliable low-latency communications (URLLC), and massive machine-type communications (mMTC) are the three use cases that the International Telecommunication Union (ITU) has identified for the 5G era. Since the commercial release of 5G systems is on the horizon, there will be a lot of attention paid to 6G technologies in the academic and industry communities. Artificial intelligence (AI) technology, a cornerstone of 6G networks, is excellent at uncovering previously undiscovered relationships among massive datasets. Cloud computing (CC) has emerged as a possible solution, although it is not the optimal paradigm for time-critical traffic conditions. Some data processing and storage functions can be moved from the network's central node to peripheral devices and infrastructure in the form of "edge computing" (EC) paradigms. The computer, storage, and caching needs of applications and services across numerous smart terminals can

V. Ganesan · V. Ramasamy (✉) · S. A. Padmini
Department of CSE, Koneru Lakshmaiah Education Foundation, Vaddeswaram, A.P, India
e-mail: rvnathan06@gmail.com

V. Ganesan
e-mail: vithyaganesan@kluniversity.in

S. A. Padmini
e-mail: vsapadmini@kluniversity.in

R. Dhanalashmi
Department of AI & DS, Madanapalle Institute of Technology & Science, Madanapalle, A.P, India

A. J. Obaid
Faculty of CS and Mathematics, University of Kufa, Kufa, Iraq

S. Chowdhury
Sreenivasa Institute of Technology and Management Studies, Chittoor, AP, India

J. M. R. S. Tavares et al. (eds.), *Proceedings of Second International Conference on Intelligent System*, Algorithms for Intelligent Systems,
https://doi.org/10.1007/978-981-99-8976-8_16

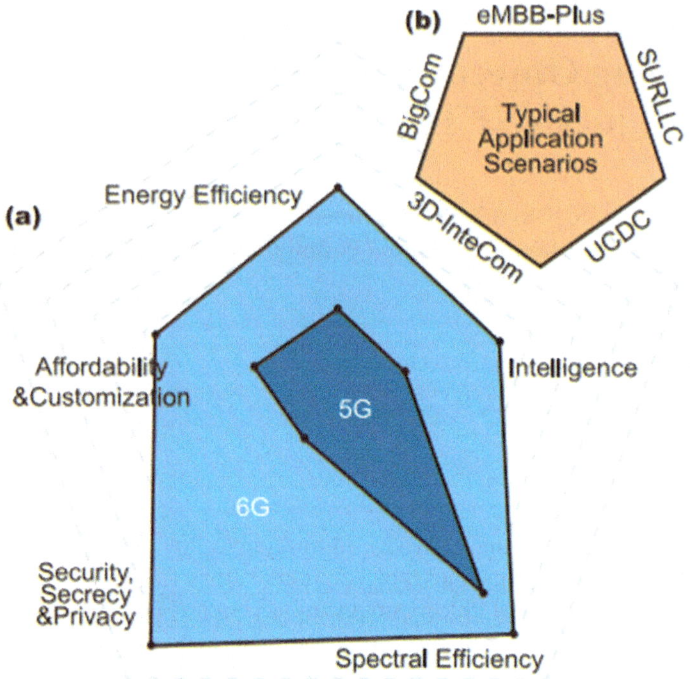

Fig. 1 Enhanced mobile broadband plus (eMBB-Plus), big communications (BigCom), three-dimensional integrated communications (3D-InteCom), unconventional data communications (UCDC), and secure ultra-reliable low-latency communications (SURLLC) are just some of the typical 6G application scenarios

be met by vehicular edge computing (VEC), leading to enhanced in-vehicle entertainment, safety, and efficiency. Figure 1 depicts some typical 6G use scenarios and provides a qualitative comparison to 5G networks.

The basis of investigation into 6G systems is VEC. Combining the VEC with BVLoS detection technology is offered as a method of deployment for such systems. This approach strikes a great mix between efficiency and computing volume, allowing for outstanding performance despite restricted resources. This approach will help the Internet of Vehicles (IoV) shift its complicated computing activities to the edge, freeing up resources for use by both in-vehicle and off-road infrastructure. Feedback on the 6G system's performance will also increase the effectiveness with which tasks are completed. The numerical findings presented in this study illustrate that our approach has the ability to greatly improve the computational economy of the model in a vehicle-mounted system, while still maintaining a high level of accuracy and recall in detection. This technique presents a framework that outlines a pathway for achieving seamless deployment of algorithms relevant to various Internet of Vehicles (IoV) tasks onto edge nodes. The subsequent findings presented in this study hold paramount importance:

The deployment approach for the VEC model entails the relocation of the model from the desktop system to the edge device installed within the car. By significantly reducing resource expenditure and meeting the criteria of the VEC, it is possible to preserve the original effectiveness.

A methodology is presented for enhancing the stability of VEC models. The utilization of model stabilization techniques aims to ensure that the process of model adaption does not have a detrimental impact on performance. The enhancement of the model's stability can be achieved through the implementation of minor modifications to the network architecture. Upon completion of this process, the model reaches its ideal state. Figure 2 demonstrates the significance of edge data preprocessing in 6G systems for alleviating network congestion and improving decision-making quality.

An approach for modifying VEC models process involves modifying the model to fit the specifics of the intended device. Reducing the amount of computing required

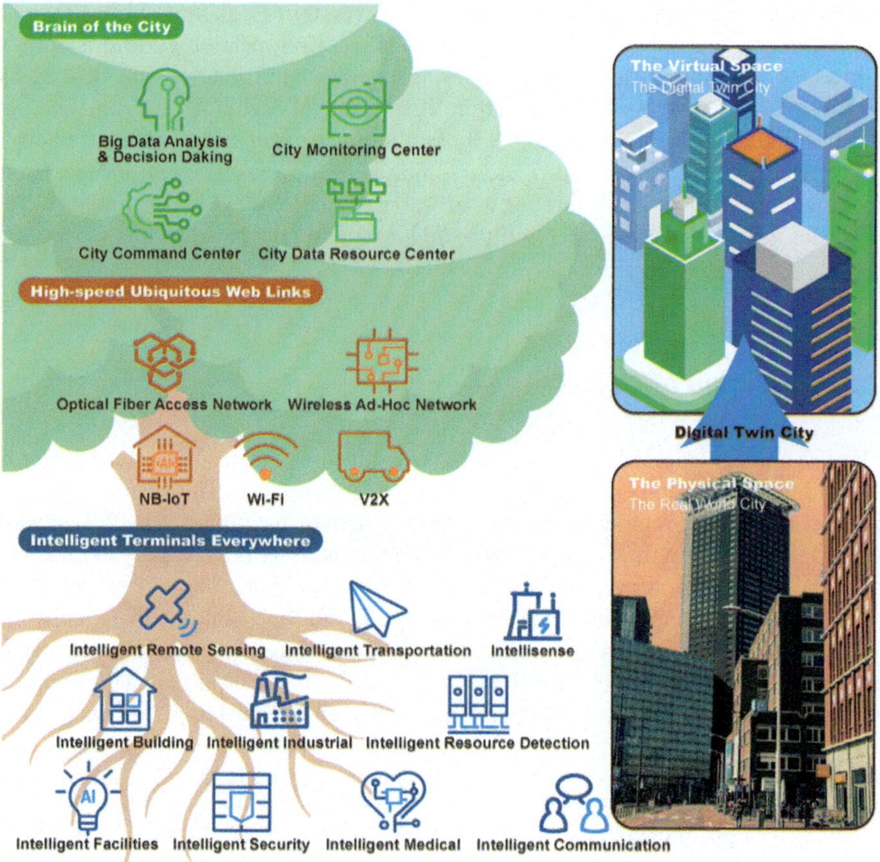

Fig. 2 Digital twin (DT) technology on a 6G system can create digital copies of real-world objects

by a model through adaptation saves time and money. For this, a channel-pruning strategy draws on the concept of knowledge distillation.

2 Literature Survey

2.1 Integrating VECs into 5G Networks

Using VECs in a 5G network technology for a wide variety of onboard devices to operate at high speeds and densities promotes the growth of ITS and autonomous driving systems. It aids in the development of smart cities. Scheduling and programming of heterogeneous networks, such as vehicle-to-vehicle (V2V), vehicle-to-infrastructure (V2I), vehicle-to-pedestrian (V2P), and vehicle-to-everything (V2X) connections, can be difficult due to the wide diversity of onboard applications [1, 2]. Expanding the driver's perceptual horizons, BVLoS technology in this context allows for the gathering and broadcast of road condition information outside the Line of Sight (LoS) [3, 4]. As a result, Within Visual Line of Sight (WVLoS) technology has become increasingly popular [5]. By facilitating drivers' access to more relevant information, WVLoS can contribute to smarter traffic management and decision-making [6]. As such, it can be seen as an early example of 6G DT technology. The opposite is true, though, as 6G will provide a tremendous boost to this technology [7].

2.2 Detection of Objects

One can classify target detection algorithms as either one-stage or two-stage detectors depending on the number of stages at which they are executed. The former begins the process of creating regional proposals with the help of the region proposal network (RPN) [8]. Based on these suggestions, the computer assigns a category to the target and makes a location prediction. For the latter, a backbone network is employed to get to feature maps within the input image without any intermediate steps [9]. The process of target detection is executed rapidly through the utilization of feature maps. The categorization of target localization methods can be based on the presence or absence of an anchor box, distinguishing between anchor-based and anchor-free detectors. Anchor-based approaches are frequently constructed manually due to their need on anchor boxes for target selection. It is worth noting that the utilization of distinct datasets may necessitate varying anchor boxes for the same algorithm. The user's text is already academic and does not require any rewriting [10]. The latter group necessitates a restricted workforce due to their enhanced capacity in item classification and informed speculation regarding object placement. The user's text does not contain any information to rewrite in an academic manner [11].

2.3 Compression in a Network

Pruning models, quantizing networks, low-rank sparse decomposition, and distilling knowledge are the four most common forms of model compression [12]. It is essential to the model pruning procedure to identify irrelevant model parameters and prune them from the network in accordance with predetermined criteria. Network quantization employs quantization and coding techniques to decrease the number of bits required to represent the original network characteristics [13]. When the precision of the parameters is decreased, the model shrinks. The sparsity of neural network parameters is used by low-rank sparse decomposition techniques [14]. Parameter reduction or computation speedup can be achieved by low-rank decomposition [15]. In general, the introduction of some procedures can add sparsity to the parameters even if the original parameters themselves did not have sparsity [16, 17]. To facilitate the training of simpler models, the knowledge distillation techniques employ complicated models [18, 19]. Soft targets for the output layer, feature maps for the intermediate layer, and relationships between instances are all examples of the various shapes this knowledge [20]. By combining various strategies for model compression, a better compression ratio may be achieved.

3 Methodology

3.1 Model 6G VEC Deployment Strategy

It is important to consider the resource cost of the model while implementing the VEC algorithm on a 6G system. Overhead in terms of model resources is quantified in this paper by counting the number of model parameters and floating-point operations. Deployment and use of the model in the VEC system improves as its resource cost falls. It is well knowledge, however, that decreasing the resource cost of a model typically leads to a corresponding decrease in performance. That is why our solution takes a dual-pronged approach to model deployment on VEC devices. To adapt the model to VEC devices, it is necessary to first stabilize the model's performance. Figure 3 explains how WVLoS and BVLoS work together within the traffic system, which makes use of the IoT to set up a comprehensive target sensing system (Table 1).

Latency, mobility, traffic area capacity, efficiency of network energy, and efficiency of spectrum are considered as performance comparison between 5G and 6G.

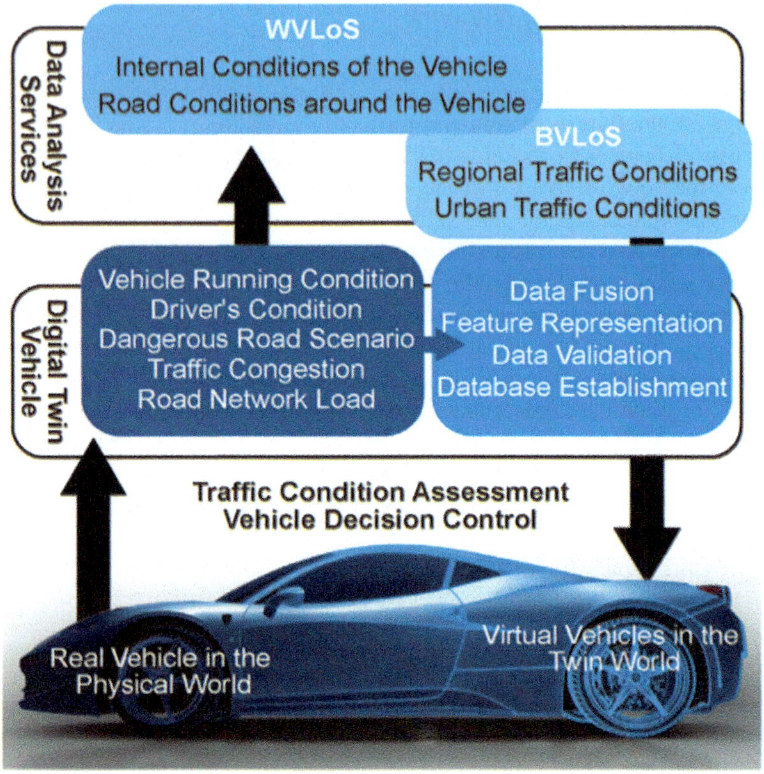

Fig. 3 When used together, WVLoS and BVLoS technologies not only aid the vehicle in making judgements about its driving behavior, but also gather data on the traffic system as a whole and how it is functioning. When they work together, they can ease congestion on the roads and make travel more efficient and safer. Table 1 shows the estimated performance comparison between 5 and 6G systems

Table 1 Estimated performance comparison between 5 and 6G systems

	Generation		
	5G	6G	Improvement
Peak (Tbps)	0.01	≥ 1	$\geq 60\times$
Experienced (Gbps)	0.11	1	$10\times$
Latency (ms)	1	0.11	$10\times$
Mobility (km/h)	600	1000	$2\times$
Density (dev/km^2)	10^6	$\geq 10^7$	$\geq 10\times$
Traffic area capacity (Gbps/m^2)	0.01	1	$100\times$
Efficiency of network energy	$1\times$	$100\times$	$100\times$
Efficiency of spectrum	$1\times$	$\geq 3\times$	$\geq 3\times$

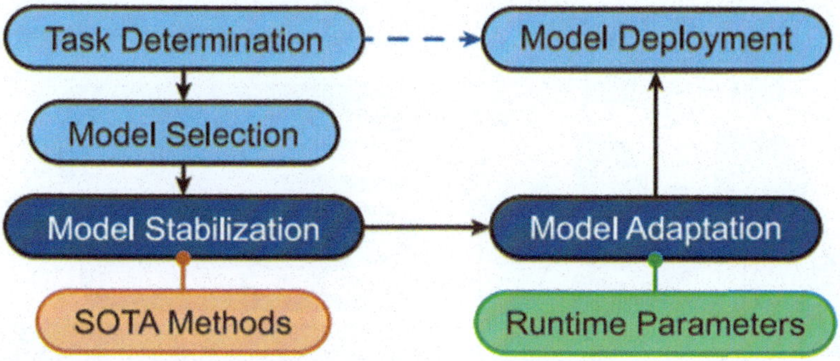

Fig. 4 Our planned rollout timeline for the 6G VEC architecture

3.2 Stabilization of the Model

The criteria for selecting models are as follows: (1) Go with the most up-to-date, cutting-edge model available. To rephrase, the model is guaranteed to have adequate performance and can be integrated with the 6G system to acquire edge intelligence in a hierarchical setting. The model should be simple to install on the automobile. The algorithm execution environment of a 6G system is more restrictive than that of a desktop computer.

This research uses the YOLOv5 and YOLOX models with respect to these two tenets, and it is shown in Fig. 4. The object in this model is a traffic participant, and it can help with the implementation of autonomous driving and traffic control in 6G-oriented IoV systems.

The fact that YOLOv4/5's open-source code is still being updated and expanded upon a year after its initial release is cause for optimism. The schematic layout of our model is depicted in Fig. 5.

Backbone: To solve the issues with repeated gradient information during network feature extraction, YOLOV4/5 uses the cross-stage partial (CSP) module as its central network. A portion of a neural network's redundant computation can be eliminated with a CSP structure. YOLOv5 makes use of the CSP module along with the residual module. The residual module helps mitigate gradient vanishing and overfitting issues in deep learning model training by allowing feature reuse.

However, it fails to perform as well as densely linked networks at detecting and representing novel features. Deep network can learn from the experiences of shallow network, and the feature map in a densely linked network combines the data from all the feature maps that came before it. Group convolution is used by the dual-path network (DPN) to take advantage of both types of networks. As a result, the CSP and DPN modules are used as the primary and secondary networks, respectively, in this paper. The three networks discussed above are depicted in Fig. 6.

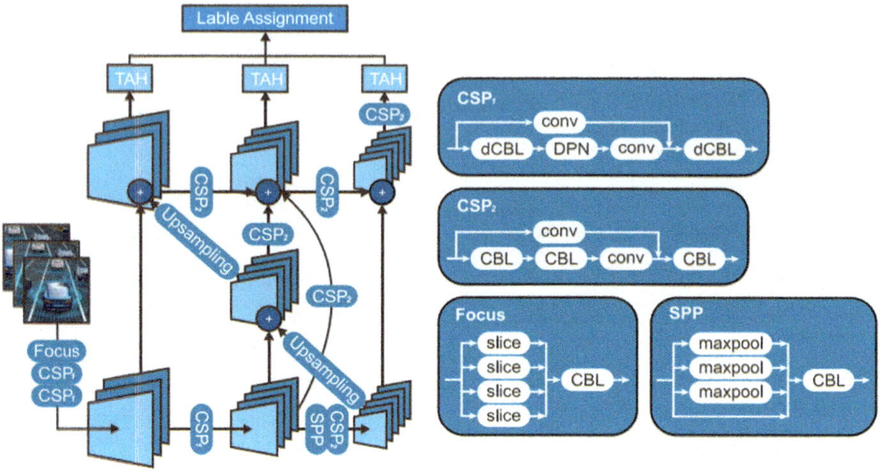

Fig. 5 General design of our model

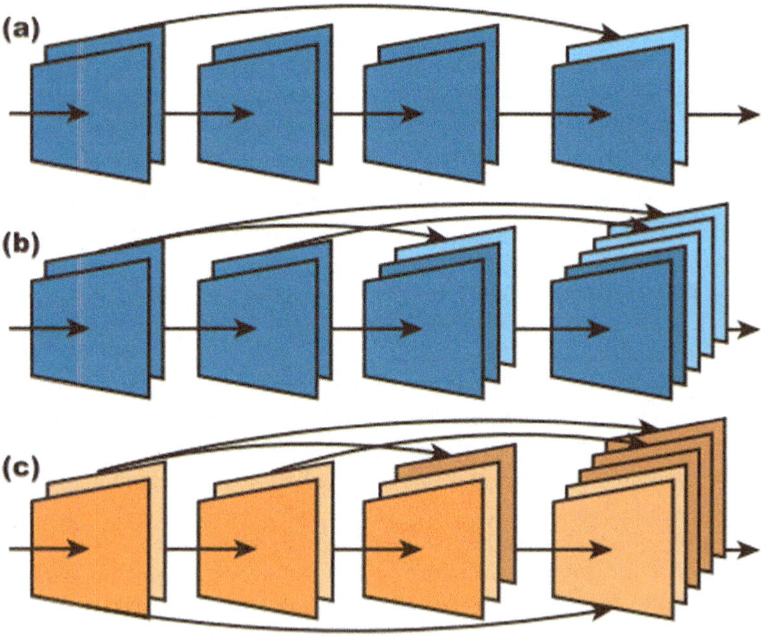

Fig. 6 Network can be broken down into three distinct types of blocks: **a** residual, **b** dense, and **c** DPN. Different feature transmission paths are represented by different tones of color

4 Experiments and Results

In order to provide further information regarding the dataset for analyzing the authenticity of KITTI features in various settings such as city, country, and highway, it is necessary to offer extra details. The entire collection consists of around 200,000 object images labeled with 3D information, sampled and synced at a data frequency of 10 Hz. Additionally, the collection includes visual range sequences spanning a distance of 39.2 km, as well as 389 pairs of stereo images and corresponding optical flow diagrams.

There can be up to 15 cars and 30 people in one picture. The Eoscannon 550D was used to capture the 25 FPS video and JPEG photographs that make up the UA-DETRAC dataset on a road flyover in Beijing and Tianjin. The 60 movies included in the collection were shot on days with clear skies, overcast skies, and rain. About 8250 cars and 1.21 million targets were tagged among the 82,085 pictures. About 15 h worth of driving data from Boston and Singapore make up the nuScenes collection. Table 2 listed the parameters (Table 3).

About 1.4 million camera photographs and 1.4 million 3D bounding boxes, annotated by hand for 23 object classifications, make up the collection. You can see how pruning works in Fig. 7.

Performance (speed/accuracy) can be tuned by adjusting the pruning ratio. The model with the best time-to-solution and an accuracy of 80% or above is chosen. When the model pruning ratio is at or above 80%, the inference time strangely increases, as seen by the inference time graph. All predicted items currently have low scores, making it hard for NMS to select the best result and invalidating our current selections. This compromise can be used with various models. Among models with enough accuracy, the one with the quickest execution time can be selected, depending on the specifics of the application scenario at hand. On the flip side, among the

Table 2 Parameters of the platform

	System	TX2
CPU	Core i and -10,700	Dual (Core) Denver 2 and Quad Core ARM Cortex
GPU	3090 24 GB	256 CUDA cores
Memory	32 GB DDR4	8 GB LPDDR4, 128 bit
Storage	3 TB TOSHIBA32 GB DDR4	32 GB

Table 3 Parameters and their components

Components	mAP (%)	Parameters (M)	GFLOPs	Latency (ms)
EfficientDet	82.9	9.5	11.7	1.5
YOLOv5s	74.9	8.3	13.8	1.2
YOLOv5s adapted	78.9	9.1	138	1.2
Proposed adapted	79.3	8.7	11.9	1.2

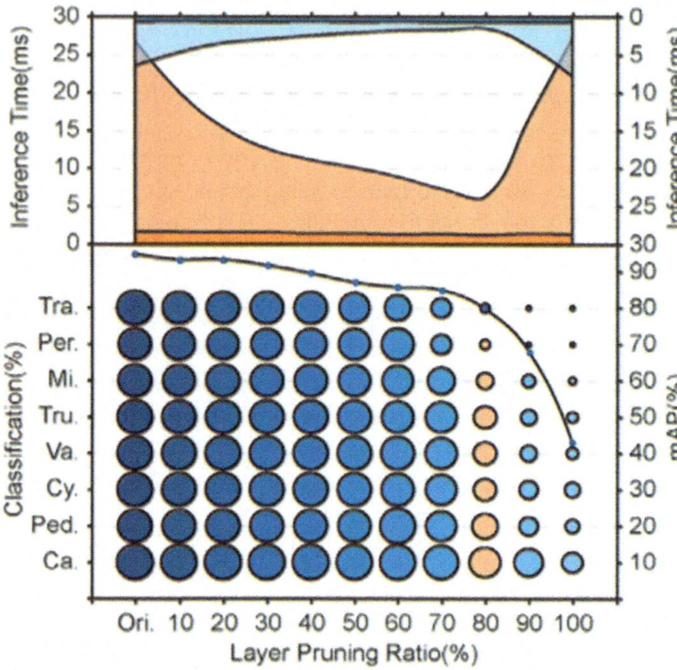

Fig. 7 As the pruning ratio is adjusted, the performance of the model shifts. The inference time for the neural network (dark) and the NMS time (light) required to run the model are shown in the top section, along with the total time it took to run the model on the desktop system (blue) and Jetson TX2 (orange). The model's AP and mAP are shown for different classes as a function of the circle's radius and the value of the line graph, respectively, in the figure's lower section

models that meet the satisfactory computation time, the most accurate one can be chosen. To accomplish the same goals as the network compression method, several lightweight networks take a similar approach. However, the scene complexity may cause the intended network to be overfitted or underfitted depending on the task at hand. Our approach is distinct because it is more malleable. It significantly decreases the computational cost compared to the artificially constructed lightweight networks. In addition, when compared to lightweight networks created by simplifying big networks for accuracy stabilization.

5 Conclusions

The proposed 6G VEC-based approach improves the stability of AI model deployment. Consequently, model stabilization is conducted to optimize the model, and then model adaption is conducted to strike a compromise between performance and speed within a given set of restrictions. It performs well in both desktop and smart edge terminals. Therefore, similar models in subsequent research are used to test the strategy's viability and efficiency.

References

1. Li, Y., Liao, C., Wang, Y., Wang, C.: Energy-efficient optimal relay selection in cooperative cellular networks based on double auction. IEEE Trans. Wireless Commun.Commun. **14**(8), 4093–4104 (2015)
2. Zhang, S., Zhu, D.: Towards artificial intelligence enabled 6G: state of the art, challenges, and opportunities. Comput. Netw. 107556 (2020)
3. You, X., Wang, C.-X., Huang, J., Gao, X., Zhang, Z., Wang, M., Huang, Y., Zhang, C., Jiang, Y., Wang, J., et al.: Towards 6G wireless communication networks: vision, enabling technologies, and new paradigm shifts. Sci. China Inf. Sci. **64**(1), 1–74 (2021)
4. Latva-aho, M., Lepp€anen, K., Clazzer, F., Munari, A.: Key drivers and research challenges for 6G ubiquitous wireless intelligence
5. Alsharif, M.H., Kelechi, A.H., Albreem, M.A., Chaudhry, S.A., Zia, M.S., Kim, S.: Sixth generation (6G) wireless networks: vision, research activities, challenges and potential solutions. Symmetry **12**(4), 676 (2020)
6. Austin, M., Delgoshaei, P., Coelho, M., Heidarinejad, M.: Architecting smart city digital twins: combined semantic model and machine learning approach. J. Manag. Eng.Manag. Eng. **36**(4), 04020026 (2020)
7. Ma, X., Xu, H., Gao, H., Bian, M.: Real-time multiple-workflow scheduling in cloud environments. IEEE Trans. Netw. Service Manag. **18**(4), 4002–4018 (2021)
8. Raza, S., Wang, S., Ahmed, M., Anwar, M.R.: A survey on vehicular edge computing: architecture, applications, technical issues, and future directions. Wireless Commun. Mobile Comput. **2019**
9. Chen, C., Zeng, Y., Li, H., Liu, Y., Wan, S.: A multi-hop task offloading decision model in MEC-enabled internet of vehicles. IEEE Internet Things J. (2022)
10. T. Xiao, C. Chen, Q. Pei, H.H. Song, Consortium Blockchain-Based Computation Offloading Using Mobile Edge Platoon Cloud in Internet of Vehicles, IEEE Transactions on Intelligent Transportation Systems, 2022.
11. Xia, S., Yao, Z., Li, Y., Mao, S.: Online distributed offloading and computing resource management with energy harvesting for heterogeneous MEC-enabled IoT. IEEE Trans. Wireless Commun. (2021)
12. Huang, Y., Xu, H., Gao, H., Ma, X., Hussain, W.: Ssur: an approach to optimizing virtual machine allocation strategy based on user requirements for cloud data center. IEEE Trans. Green Commun. Netw. **5**(2), 670–681 (2021)
13. Saad, W., Bennis, M., Chen, M.: A vision of 6G wireless systems: applications, trends, technologies, and open research problems. IEEE Network **34**(3), 134–142 (2019)
14. Cong, W., Chen, C., Qingqi, P., Zhiyuan, J., Shugong, X.: An information centric in-network caching scheme for 5G-enabled internet of connected vehicles. IEEE Trans. Mobile Comput. (2021)
15. McCrink, M.H., Gregory, J.W.: Design and development of a high-speed UAS for beyond visual line-of-sight operations. J. Intell. Rob. Syst.Intell. Rob. Syst. **101**(2), 1–16 (2021)
16. Li, Y., Ma, H., Wang, L., Mao, S., Wang, G.: Optimized content caching and user association for edge computing in densely deployed heterogeneous networks. IEEE Trans. Mobile Comput. (2020)
17. Yin, Y., Huang, Q., Gao, H., Xu, Y.: Personalized APIS recommendation with cognitive knowledge mining for industrial systems. IEEE Trans. Ind. Inf. **17**(9), 6153–6161 (2020)
18. Jiang, X., Yu, F.R., Song, T., Leung, V.C.: A survey on multi-access edge computing applied to video streaming: some research issues and challenges. IEEE Commun. Surv. Tutorials **23**(2), 871–903 (2021)

19. Huang, T., Yang, W., Wu, J., Ma, J., Zhang, X., Zhang, D.: A survey on green 6G network: architecture and technologies. IEEE Access **7**, 175758–175768 (2019)
20. Gao, H., Liu, C., Yin, Y., Xu, Y., Li, Y.: A hybrid approach to trust node assessment and management for VANETs cooperative data communication: historical interaction perspective. IEEE Trans. Intell. Transp. Syst. (2021)

Chapter 16
Analysis of Human Behaviour on Movie Review

Saksham Sharma, V. Vipin Deval, Vidushi, Mohd Anas Wajid, Bharat Bhushan, and Mustafa Al-Asadi

1 Introduction

There is a huge amount of text available online now, and it keeps expanding every day. Due to how simple it is to create and distribute, text is the most common data type on the Internet. Nowadays, obtaining vital information is less of a challenge than sorting through the vast amount of information and placing it in the proper perspective. Given that it would take too much time and effort for humans to manually sort through all the data, automatic categorization and organisation of the data is clearly a scientific difficulty. Through a variety of channels, such as Facebook, Twitter, LinkedIn, and Instagram, Web 2.0 has allowed users to communicate their ideas, opinions, and

S. Sharma · V. Vipin Deval
KIET Group of Institutions, Delhi-NCR, Ghaziabad, India
e-mail: saksham.2123mcse1002@kiet.edu

V. Vipin Deval
e-mail: vipin.deval@kiet.edu

Vidushi
CHRIST (Deemed to be University), Delhi-NCR, Ghaziabad, India
e-mail: vidushi@christuniversity.in

M. A. Wajid (✉)
School of Computer Application & Technology, Galgotias University, Greater Noida, India
e-mail: anaswajid.bbk@gmail.com

B. Bhushan
Department of Computer Science and Engineering (CSE), School of Engineering and Technology, Sharda University, Greater Noida, India

M. Al-Asadi
Faculty of Engineering and Natural Sciences, Computer Engineering Department, KTO Karatay University (KTO Karatay Üniversitesi), Konya, Turkey
e-mail: mustafa.alasadi@karatay.edu.tr

J. M. R. S. Tavares et al. (eds.), *Proceedings of Second International Conference on Intelligent System*, Algorithms for Intelligent Systems,
https://doi.org/10.1007/978-981-99-8976-8_17

judgements on a range of topics, from entertainment to education. There is a plethora of information available on these platforms, such as tweets, blogs, status updates, and articles.

From the articles, reviews, and postings on these platforms, individual character analysis aims to extract the polarity of emotions like happiness, sorrow, sadness, fury, rage, and affection as well as viewpoints [1]. Opinion mining assesses the sentiment of a text in respect to a particular information source. Slang, misspellings, abbreviations, repetitious characters, regional language use, and the addition of new emoticons make it difficult to analyse human behaviour. Determining the appropriate behaviour for each phrase is therefore a very difficult task [2]. Data mining is actively investigating the continuous study topic of human behaviour analysis. Since views underpin most human actions and behaviours, human behaviour analysis is used in practically every facet of economic and social life.

In the recent past, there have been lots of websites that let users add, edit, and rate information. On a variety of topics, users can express their own opinions. Such websites include blogs, forums, social networks, and product review sites. It has many uses because it is an effective process with high precision:

- **Research on consumer behaviour**: It may draw on the findings of human behaviour analysis approaches. This method can be used to examine how consumers feel about a new government strategy as well as a product or service.
- **Recommendation Systems**: The algorithm can anticipate which item should be suggested and which should not by studying along with putting thoughts into categories based on people's preferences and interests.
- **Buying Products or Services**: Making the right choice while buying a product or service is no longer a difficult task. Using human behaviour analysis, people may quickly analyse reviews and opinions about any product or service, as well as the ability to contrast various businesses.
- **Product or service improvement**: Opinion mining enables producers to gather consumer input on their goods or services, whether favourable or unfavourable, and then upgrade and boost the quality of their goods or services.
- **By using human behaviour analysis**, newsgroup, blog, and social media monitoring is made simple. This method may be used to identify haughty, conceited, or highly emotional language in online tweets, comments, forums, and blogs.

The analysis of movie reviews provides valuable insights into how people communicate their thoughts, preferences, and attitudes about movies. It helps in sentiment analysis, identifying recurring themes, and understanding factors influencing decision-making. By studying language, cultural influences, and constructive feedback, filmmakers and the entertainment industry can improve their productions and marketing strategies. Movie reviews also reflect evolving tastes and engagement within movie enthusiast communities, offering valuable information for understanding audience perceptions and shaping future cinematic experiences.

The study enhances our understanding of human behaviour in film criticism by analysing movie reviews. It reveals movie preferences, communication patterns, and the impact of social influences. Using machine learning algorithms, the research builds predictive models and provides feedback to filmmakers. By understanding how people express opinions about movies, the study contributes valuable insights to the entertainment industry and academic research.

To identify the components of subjective style and sentence polarity, we employ statistical techniques. At the sentence level, statistical analysis is performed. In order to categorise a collection of communications, we apply ML techniques. Here, three distinct machine learning algorithms—NB, K-NN, and Random Forest—were used to analyse the data set. These algorithms were chosen in order to assess how well they addressed the research issue and to determine which algorithm gave the most accurate outcomes. The study report offers helpful insights into the efficacy of various machine learning approaches for the provided data set by emphasising the superiority of the NB algorithm [3, 4]. These results increase the amount of knowledge in the field and may help direct future study or real-world applications in related fields.

The Naive Bayes algorithm's outstanding accuracy in this research report acts as a strong incentive to continue the investigation. Due to its exceptional performance, people are curious in how it works and want to know what fundamental components make it effective. Researchers can learn important lessons from examining the subtleties of the algorithm's effectiveness that could progress the use of machine learning in a variety of industries. Additionally, understanding the factors that led to its high accuracy might motivate changes and advancements, spurring innovation and boosting the dependability of predictive models. The rest of the paper is divided into following sections. Section 2 gives a thorough literature survey of the previous studies conducted in this direction. Section 3 presents the methodology adopted in current research. Section 4 is focused on results and analysis, while Sect. 5 concludes the paper.

The main goal of the project, with the application of machine learning, is to leverage computational techniques to analyse movie reviews and gain insights into various aspects of human behaviour related to movie preferences and attitudes. Through sentiment analysis using machine learning algorithms, the project aims to identify the overall sentiment (positive, negative, neutral) expressed in the reviews. Natural Language Processing (NLP) techniques can be used to analyse the language and expressions used in the reviews, providing a deeper understanding of how people communicate their thoughts. Additionally, machine learning models can help identify recurring themes and patterns in the reviews, which shed light on the specific aspects of movies that resonate with viewers. By correlating these findings with demographic information, cultural influences, and historical trends, the project can provide valuable insights into how human behaviour shapes movie preferences and decision-making processes.

2 Literature Review

Behavioural analysis has gained a lot of popularity in recent past this could be inferred from the recent work [5]. Machine learning plays a very vital role in this regard [6]. Machine learning algorithms are extensively utilised not only in behavioural analysis but also gaining popularity in sentiment analysis [6]. In this regard, the work by [7] Joscha et al. devised and investigated a number of methods, including n-grams and Words in a Bag models, to add purely semantic data to accelerate the effectiveness of human behaviour assessment. Previous methods did not take into account the semantic connections between phrases or other document components. The best strategy for integrating disclosure units was not compared by researchers, and they did not offer one [8]. In an effort to improve human behaviour analysis, RST, which was used offers a document-level hierarchical representation. They recommended merging the grid search and weighting to get the Rhetoric Structure Theory (RST) tree's average human behaviour scores. They were able to considerably reduce the complexity of the original RST tree using feature engineering while encoding binary data into the Random Forest. They found that machine learning improved balancing accuracy and gave them a stellar F1 rating of 71.9%.

The research provided a novel interpretation of the human behaviour analysis problem in this study [9] by using numerical data in medicine assessments. They employed fuzzy set theory to assess the polarity conveyed and checked whether or not sentences had quantitative words to see if they contained any opinions. To build a fuzzy knowledge base, several doctors from various medical facilities were questioned. Despite the fact that numerous research in this field have been carried out, despite acknowledging the polarity of emotion, they don't consider the quantitative (numerical) information from the reviews. Furthermore, it cannot be applied to different domains due to the training data's heavy domain dependence. They concluded that their proposed knowledge engineering method that utilised fuzzy sets was substantially more straightforward, effective, and accurate, having an F1 grade of over 72%.

Dhiraj Murthy outlined the purposes that tweets serve in political campaigns in his study [10]. He stressed that despite several efforts to research and analyse, no attempt was made to identify whether these tweets were predictive or reactionary when it came to politics on Twitter. In this article, he came to the conclusion that tweets are more reactive than predictive. He learned that a candidate's popularity on various social media platforms was increased by spreading news about them. He also learned that political success was not in any way related to Twitter success.

Ahmad Kamal developed an opinion mining framework in his research [11] that makes activities like feature extraction, review summary, and objectivity or subjectivity analysis simpler. He used supervised ML to classify the subjectivity along with objectivity of assesses. He used a variety of techniques, including Decision Trees, Bagging, Multilayer Perceptron, and Naive Bayes. He increased mining output by minimising extraction as well as noise, like in Kamal's study [12, 13].

Researchers used the WEKA tool to classify movie reviews for human behaviour analysis in this study [14]. They enhanced earlier research on the classification of human behaviour, which labels thoughts as positive or negative depending on their emotional content. In this research, the idea that a single review, as well as reviews that incorporate the opinions of numerous people, can contain both positive and negative emotion was also taken into consideration. They conducted their experiment using WEKA and discovered that for both written and visual assessments, NB outperformed SVM. 85.1% of all the times, NB is accurate.

Akshay Amolik et al. created a dataset for their study [15] using tweets from reviews of films and tweets linked to those flicks. These tweets are subjected to sentence-level research of human behaviour. It involves three steps. The first step is pre-processing. After that, a feature vector is produced utilising pertinent features. Tweets were finally identified as positive, negative, or neutral using a variety of classifiers, including NB, SVM, ensemble classifiers, k-means, and ANN. The findings demonstrate that we achieve 75% accuracy using SVM. A claim made in the Wu et al. research [16] that the @username in a tweet increases likelihood and influences an action was challenged by him. However, Akshay Amolik removed hashtags and modified @username in this study, which forced us to utilise SVM rather than NB, increasing accuracy with 10%.

The study sources information from online review platforms, social media, movie forums, and surveys. It analyses the data using sentiment analysis, topic modelling, statistical analysis, word frequency analysis, and clustering to gain insights into human behaviour, movie preferences, and attitudes.

3 Methodology

3.1 Methods

The methodology adopted to perform current work consists of various methods that are widely used in machine learning for carrying out effective work in various domains of research such as healthcare, sentiment analysis, information retrieval and bioinformatics, etc. Further sections explain the methods of machine learning which have formed the methodology of current research.

By applying these feature extraction techniques and utilising machine learning algorithms such as Naive Bayes, K-Nearest Neighbours, and Random Forest, the project can analyse the textual data and gain insights into human behaviour and preferences related to movie reviews [17].

- **Bag-of-Words (BoW) Representation**: The reviews are converted into numerical vectors using the BoW representation, where each review is represented as a set of word frequencies. This technique ignores the order of words but captures their occurrence in the reviews.

- **Feature Engineering**: Custom features may be created from the review text, such as the presence of specific keywords, mentions of movie-related entities (actors, directors), or indicators of genre.
- **Handling Categorical Data**: If reviews are associated with specific metadata (e.g. movie genre, release year, reviewer rating), techniques like one-hot encoding or label encoding may be used to represent categorical data numerically.

The bag-of-words (BoW) model is a text representation technique that converts movie reviews into numerical vectors by counting the frequency of each word in the review. It treats each review as an unordered set of words, disregarding word order and context. By using BoW, one can analyse the distribution of words in the reviews to gain insights into the reviewer's sentiments and feelings about a specific movie. Comparing word frequencies across reviews helps identify commonly mentioned positive or negative aspects. However, BoW has limitations as it lacks contextual understanding and cannot capture complex linguistic nuances. Despite its simplicity, it remains a useful tool for sentiment analysis and providing quantitative understanding of feelings towards a movie based on word occurrences.

The knowledge gathered in this research offers valuable applications for the entertainment sector and decision-making. It provides audience insights, aiding content creation, and marketing strategies. Predictive models help forecast box office success, guiding release schedules, and resource allocation. Filmmakers benefit from constructive feedback, improving storytelling, and filmmaking techniques. Segmenting audiences based on sentiments tailors content and enhances promotional efforts. Understanding social and cultural influences assists in adapting to trends. Engaging online communities strengthens connections with fans. Leveraging this knowledge enables data-driven decisions, higher audience satisfaction, and increased success in the dynamic entertainment industry.

3.1.1 Random Forest

RF is the preferred learning technique we use for classification as well as regression. While training, a variety of DTs are generated. New cases are added to each tree for classification. A class is produced by the categorising process of each tree. The Random Forest output class will be chosen by majority vote as the most equivalent class that might be produced by several trees. Since minimal programming or research is necessary, both specialists and laypeople may quickly learn Random Forests. Even individuals without a background in statistics can use it easily (Fig. 1).

3.1.2 Naïve Bayes

The Bayes theorem is the basis for it. The NB classifier is based on the notion that each characteristic in a class exists separately from one another. This model is simple to construct and performs very well with big datasets. In addition being

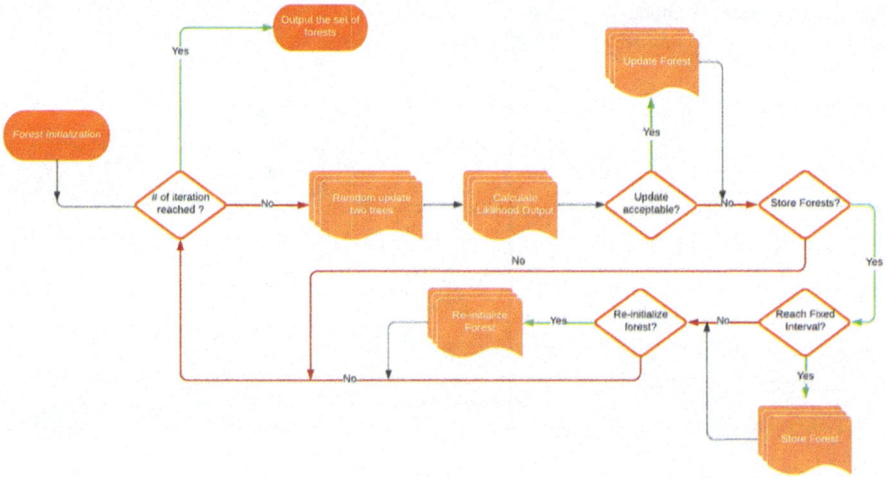

Fig. 1 Random Forest flow chart

simple to use, Naive Bayes has been demonstrated to perform better than even the most sophisticated categorization systems (Fig. 2).

$$P(C|X) = P(X|C) * P(C)/P(X). \tag{1}$$

$P(X)$ predictor prior probability
$P(X|C)$ predictor's likelihood given the class.
$P(C|X)$ Class C posterior probability.
$P(C)$ Class C prior probability.

Fig. 2 Naive Bayes

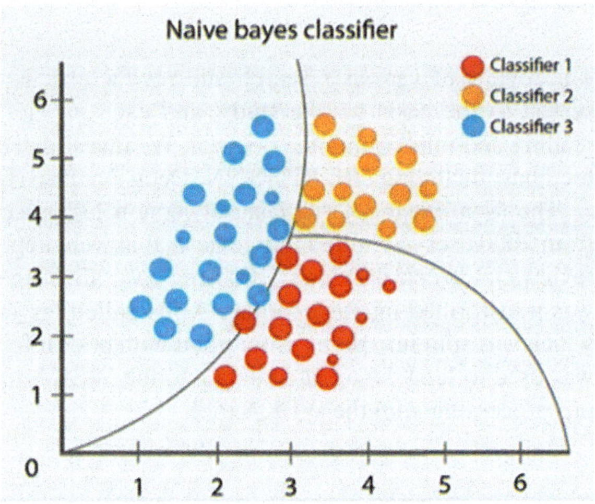

Fig. 3 K-Nearest Neighbour

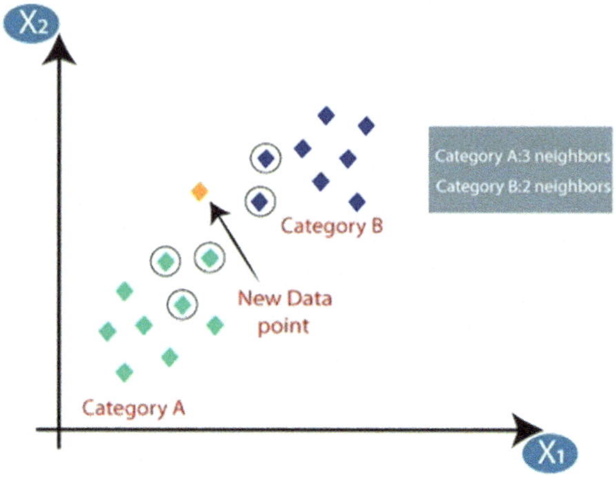

3.1.3 K-Nearest Neighbour

The most fundamental ML algorithm is K-NN. The basic idea of the technique is to anticipate a set of training samples that are physically close to the new place are used to label the area. The number of samples may be a user-defined constant or be dependent on the local point density. It is acceptable to measure length in any metric unit. The standard Euclidean distance is the most often used technique for calculating the distance between two places. Several classification and regression problems, like those involving handwritten digits or the processing of satellite pictures, have been addressed using the Nearest Neighbours technique (Fig. 3).

3.2 Testing Setup

The review of various research and review articles on human behaviour analysis served as the basis for the study, and each paper's summary was produced after reading and interpreting the original. We investigated a wide range of classification methods, including NB, RF, K-NN, Decision Tree (DT) Induction, and SVM.

The "Sentiment Polarity Dataset version 2.0" is a collection of information drawn from 2000 user-submitted film reviews that were kept on the IMDB website at http://reviews.imdb.com/Reviews: 1000 reviews, all of them favourable. Then, this data was transformed into arff format. A file called txt_token included the information, which was split into portions for good with poor information. Text Directory Loader was used to feed the updated data into the WEKA tool. The next step involved text pre-processing with the WEKA tool.

WEKA is free software as defined by the GNU General Public Licence. The system is being developed in New Zealand at the University of Waikato. The acronym "WEKA" is used to refer to the Waikato Environment for Knowledge Analysis. Java was used to create the programme. It offers ML algorithm implementation. It has components for processing data.

Weka concurs with a variety of data mining (dm) techniques, including feature selection, regression, clustering, and DM. There is further software that can join numerous connected tables of data into a single table for WEKA processing, despite the fact that it cannot perform multi-relational data mining. The categorization issue is a task for supervised learning, is to assign a an unclassified tuple with a class label in accordance with a group of examples that have already been classified and are being used as an instructional set for the method.

Our three classification techniques are NB, K-NN ($k = 3$), and RF.

As a quality indicator, the percentage of cases that are correctly classified will be considered. For the validation stage, we used tenfold cross validation. These are the procedures used to conduct experiments:

Step 1: The data is imported into WEKA (Fig. 4). The dataset was initially imported into the WEKA tool. The Text Directory Loader component, a straightforward import method for textual datasets, was used to complete this phase.
Step 2: The available dataset is saved in the ARFF format after being imported.
Step 3: Following that, a connection is made with two thousand instances that have the "Text" and "Class" properties. The graph displays the consistent distribution of the Class characteristic. Individuals with a negative polarity are represented by blue, while those with a positive polarity are represented by red.
Step 4: Next, the StringToWordVector filter is used.
Step 5: Next, the AttributeSelection filter is used.
Step 6: The required outcomes are attained with the use of the AttributeSelection filter.
Step 7: The information produced by the aforementioned processes is subjected to three algorithms. The three algorithms are RF, NB, and K-NN.

3.3 Challenges

- Data collection and cleaning from various sources can be time-consuming and complex.
- Integrating diverse data formats and structures may lead to inconsistencies.
- Scalability issues with large volumes of textual data require efficient processing and storage solutions.
- Textual ambiguity, sarcasm, and colloquial language may affect sentiment analysis accuracy.
- Lack of context in the bag-of-words representation may result in information loss.
- Imbalanced data distribution can introduce bias in the analysis.

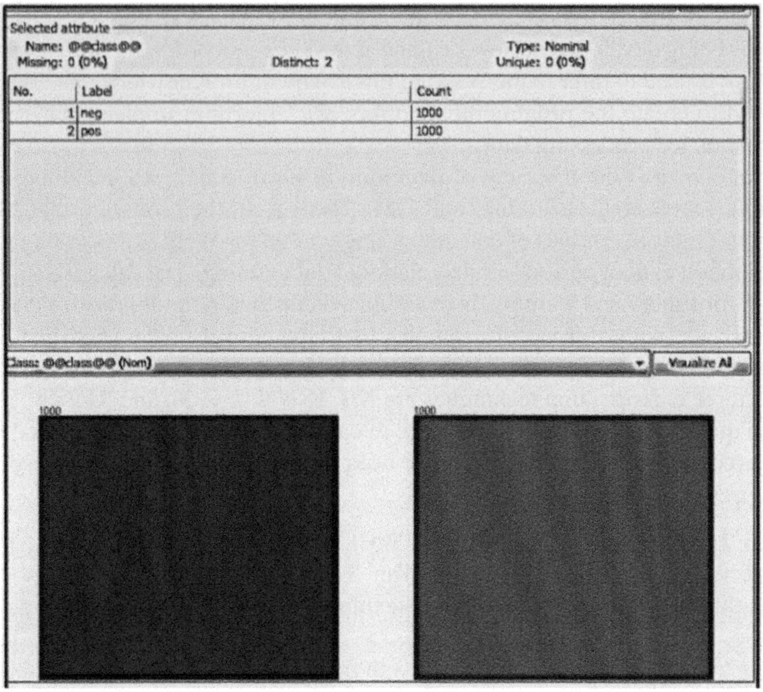

Fig. 4 Weka interface following data loading

- Ethical considerations for user-generated content and data privacy must be addressed.
- Generalisation and bias challenges when applying findings to broader populations.
- Domain-specific challenges like slang usage and movie jargon need consideration.
- Interpreting subjective aspects of reviews, such as individual tastes, is difficult.
- Continuous data updates require robust collection and updating mechanisms.

4 Results

The NB, K-NN, and RF algorithms were applied to the data set. Figures 5, 6, 7, and 8 and Tables 1, 2, and 3 show how the results are presented. The NB algorithm provided the most accurate results which can be referred from the figures and tables.

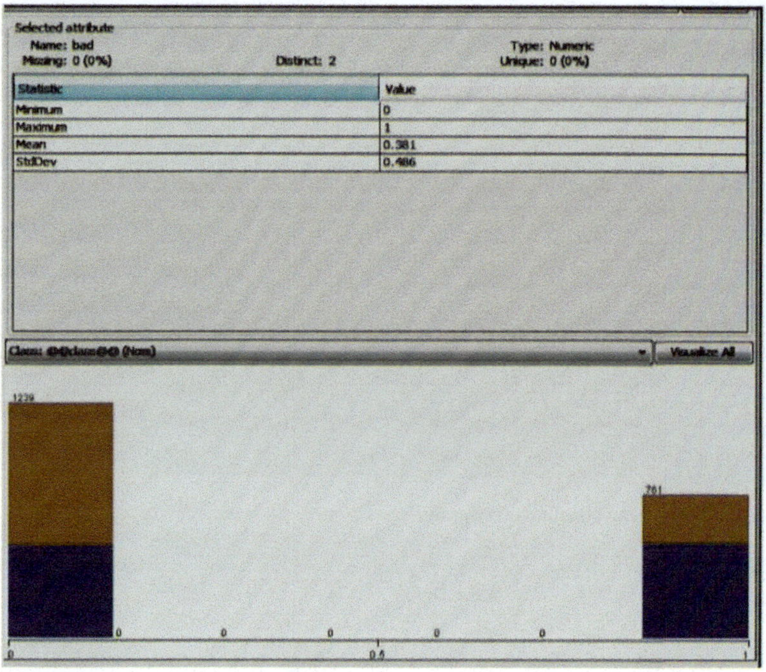

Fig. 5 Following AttributeSelection Filter, the Weka interface

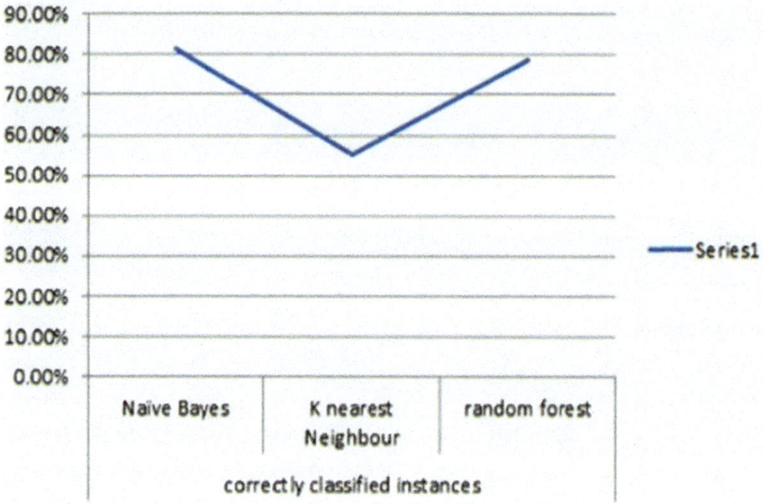

Fig. 6 Graph of different algorithms' accuracy

Fig. 7 Algorithm mean absolute error graph

Fig. 8 Graph of algorithms' root mean squared error

Table 1 Accuracy ratios for different ML methods

NB	K-NN	RF
81.4%	55.3%	78.65%

Table 2 Accuracy ratios for different ML methods on one dataset

NB	K-NN	RF
0.2112	0.4475	0.3278

Table 3 Accuracy ratios for different ML methods on another dataset

NB	K-NN	RF
0.3793	0.6554	0.3925

5 Conclusions

In this study, a variety of methods were used to determine the polarity of the tweets. RF, K-N N, and NB were the algorithms employed. The best outcomes were provided by the NB classifier. The NB classifier had an accuracy of 81.45%, the RF classifier had an accuracy of 78.65%, and the K-NN classifier had an accuracy of 55.30%. Only a few algorithms were tested; thus, more algorithms need to be tested or hybrid approaches need to be developed in order to improve the results' accuracy. To find the review polarity could be beneficial in a number of circumstances. Intelligent systems may be created to give customers comprehensive reviews of films, goods, and services without requiring them to read individual reviews. Without having to read individual evaluations, the user may then form opinions based on the outcomes that the intelligent systems deliver. The future work in this direction could be done by employing larger datasets on movie reviews or recommendations. Also applications of deep learning could be explored in this domain.

References

1. Wajid, M.A., Zafar, A., Wajid, M.S., Terashima-Marín, H.: Neutrosophic-CNN-based image and text fusion for multimodal classification. J. Intell. Fuzzy Syst. (Preprint), 1–17
2. Zafar, A., Wajid, M.A.: A mathematical model to analyze the role of uncertain and indeterminate factors in the spread of pandemics like COVID-19 using neutrosophy: a case study of India, vol. 38. Infinite Study (2020)
3. Wajid, M.A., Zafar, A.: Multimodal fusion: a review, taxonomy, open challenges, research roadmap and future directions. Neutrosophic Sets Syst. **45**(1), 8 (2021)
4. Wajid, M.A., Zafar, A., Bhushan, B., Khanday, A.M.U.D., Wajid, M.S.: Artificial intelligence (AI) and internet of things (IoT): application in detecting and containing the spread of COVID-19. In: AI Models for Blockchain-Based Intelligent Networks in IoT Systems: Concepts, Methodologies, Tools, and Applications, pp. 373–392. Springer International Publishing, Cham (2023)
5. Wajid, M.S., Terashima-Marin, H., Paul Rad, P.N., Wajid, M.A.: Violence detection approach based on cloud data and neutrosophic cognitive maps. J. Cloud Comput. **11**(1), 1–18 (2022)
6. Wajid, M.S., Wajid, M.A.: The importance of indeterminate and unknown factors in nourishing crime: a case study of South Africa using neutrosophy. Neutrosophic Sets Syst. **41**(2021), 15 (2021)
7. Markle-Huß, J., Feuerriegel, S., Prendinger, H.: Improving sentiment analysis with document-level semantic relationships from rhetoric discourse structures. In: Proceedings of the 50th Hawaii International Conference on System Sciences (2017)
8. Hogenboom, A., Frasincar, F., de Jong, F., Kaymak, U.: Using rhetorical structure in sentiment analysis. Commun. ACM **58**(7), 69–77 (2015)
9. Yazdavar, A.H., Ebrahimi, M., Salim, N.: Fuzzy based implicit sentiment analysis on quantitative sentences. J. Soft Comput. Decis. Support Syst. **3**(4), 7–18 (2016)
10. Murthy, D.: Twitter and elections: are tweets, predictive, reactive, or a form of buzz? Inf. Commun. Soc. **18**(7), 816–831. https://doi.org/10.1080/1369118X.2015.1006659
11. Wajid, M.S., Terashima-Marin, H., Najafirad, P., Wajid, M.A.: Deep learning and knowledge graph for image/video captioning: a review of datasets, evaluation metrics, and methods. Eng. Rep. e12785 (2023)

12. Kamal, A.: Subjectivity classification using machine learning techniques for mining feature-opinion pairs from web opinion sources. Int. J. Comput. Sci. Issues **10**(5), 191–200 (2013)
13. Wajid, M.A., Zafar, A.: Pestel analysis to identify key barriers to smart cities development in India. Neutrosophic Sets Syst. **42**, 39–48 (2021)
14. Shaziya, H., Kavitha, G., Zaheer, R.: Text categorization of movie reviews for sentiment analysis. Int. J. Innov. Res. Sci. Eng. Technol. **4**(11)
15. Amolik, A., Jivane, N., Bhandari, M., Venkatesan, M.: Twitter sentiment analysis of movie reviews using machine learning techniques. Int. J. Eng. Technol. **7**(6), 2038–2044 (2016)
16. Wajid, M.A., Zafar, A., Wajid, M.S.: A deep learning approach for image and text classification using neutrosophy. Int. J. Inf. Technol. 1–7 (2023)
17. Mishra, A., Wajid, M.S., Dugal, U.: A comprehensive analysis of approaches for sentiment analysis using Twitter data on COVID-19 vaccines. J. Inf. Electr. Electron. Eng. (JIEEE) **2**(2), 1–10 (2021)

Chapter 17
Artificial Intelligence (AI) in Education—Current Trends

Anastasios Fountis and Christos Lemonakis

1 Introduction

Artificial intelligence (AI) is a crucial pillar for the growth of the educational process in the Internet age. Its constant growth and incorporation into numerous aspects of life, such as education, have given rise to the need to analyze not just its benefits, but also the perspectives it can offer for all levels and types of education. The successful use of information technology in education, whether in live classes or remotely, enhances the quality of instruction and learning [1].

Intelligent education systems are the most important AI-based application in education. With such intelligent education systems, step-by-step teaching approaches are used based on the individual's demands and learning pace [2]. It is divided into a domain model, a pedagogical model, and a learner model. The subject model provides the background knowledge of the subject to be taught to the learner. The learner model describes the student's abilities, particularly his or her knowledge [3].

The learning environments in which AI can be used can thus be divided into two types:

(A) **Face-to-face classroom education with AI**

Face-to-face education with AI (see Fig. 1) is concerned with real-time education. All pupils are present in a classroom setting during instruction. The instructor is a humanoid robot or system that permits conversation through an interface. Students and the surroundings are real, while professors are virtual. The system that serves as the guide can be constructed using either supervised or unsupervised learning

A. Fountis (✉)
Berlin School of Business and Innovation, Berlin, Germany
e-mail: anastasios.fountis@berlinsbi.com

C. Lemonakis
Mediterranean University, Agios Nikolaos, Crete, Greece

J. M. R. S. Tavares et al. (eds.), *Proceedings of Second International Conference on Intelligent System*, Algorithms for Intelligent Systems,
https://doi.org/10.1007/978-981-99-8976-8_18

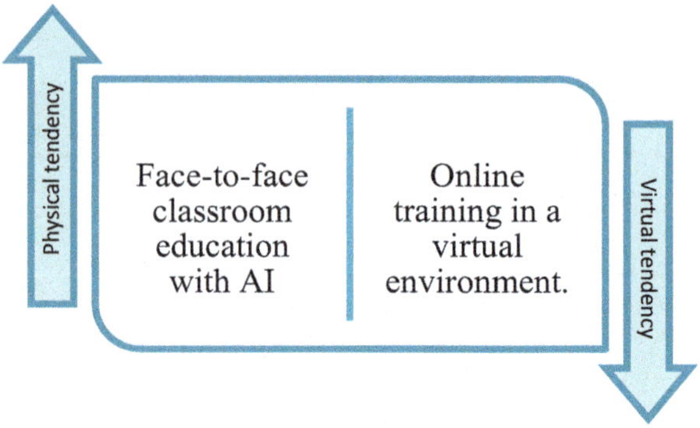

Fig. 1 Physical versus virtual tendency in education with AI use. *Source* Seren and Oezcan [2]

techniques [2]. The selected algorithms can function independently as well as based on the class and general requirements.

(B) **Online training in a virtual environment**

Remote education can be considered as simulation-based instruction (see Fig. 1). Moreover, AI simulation training can be conducted on two independent platforms. The first pertains to individuals who can participate in the simulation without the use of a device, while the second refers to assistive technology. This device is comparable to the virtual reality goggles, gloves, and clothes that make the simulation more accessible [2].

2 AI Application Categories

AI can be applied to the following main categories:

- **Face-to-face education backed by AI**

It is the human to human and physical environment design. It supports students by analyzing their profiles, offering suggestions and solutions to their problems, or supporting instruction by assigning specific tasks [4].

- **Face-to-face education by means of AI**

The design concerns in this case. By using animations or an anthropomorphic talking robot, clear and controllable algorithms are implemented that act as trainers completely independent or dependent on deep learning [2].

- **Training with simulation and use of AI**

We can include distance learning in the design since participants do not need to relocate because the setting is mimicked. The user can be engaged in two ways by employing augmented reality:

(1) using an assistive device, and
(2) connecting computers with neural networks [2].

Applications that use the cutting edge of AI in education are as follows:

(A) **Automated grader**

This sophisticated computer technology is utilized to assess and grade students in an educational context. Automatic grading utilizing AI is one of the most recent and widely adopted advanced technologies in schools and colleges. This technique not only decreases excessive latency, but also simplifies the grading process. By reducing instructors' workload, more time is available for grading [5].

(B) **Adaptive learning**

Adaptive learning concentrates on a single source of individualized data, such as learning style. It is an educational technology that can respond in real time to student interactions by giving automatic responses and tailored support. Adaptive learning systems collect data about each student's behavior by monitoring how they respond to questions [5].

(C) **Chatbots**

The term chatbots describes chat and communication programs. Currently, it is one of the most basic programs used to imitate humans. Chatbots are programs that enable conversation via text or audio systems. The programs are designed to realistically simulate the behavior of a human and show how they would behave as a conversational partner [5].

(D) **Virtual Facilitator**

Virtual facilitators can act and respond like humans naturally and understand both nonverbal and verbal cues. They have been developed in education, health, and all other areas of human life [5]. In addition, virtual facilitators can use gesture and facial expression recognition technologies to detect the mood of students and assess how easy or difficult the lesson is for them. The interaction with all previous applications constitutes a radical way to progress a unique way of learning style in education with the use of AI technics (see Fig. 2).

3 Learning Styles Educational Systems and Adaptive Learning in AI

Kids learn through a variety of approaches. Some individuals prefer ideals and speculations to facts, data, and experiments. Others would rather solve problems, while others would rather read written content. Individual preferences determine a person's learning style, which forms a "learning circle" with the following characteristics:

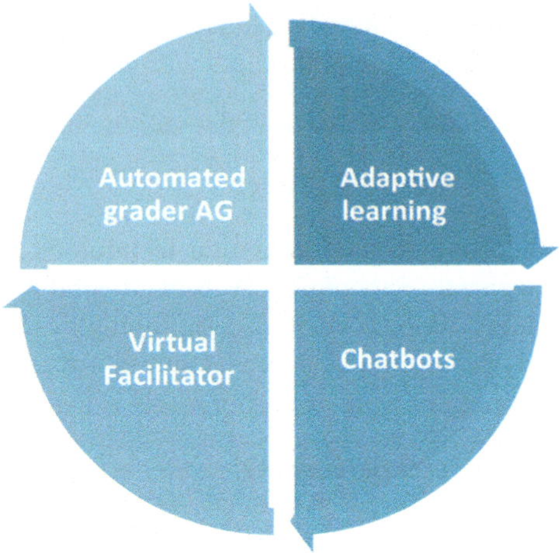

Fig. 2 Cutting edge applications of AI in education. *Source* Hashimu and Anka [5]

concrete experience (CE), reflecting observation (RO), active experimentation (AE), and abstract conceptualization (AC). The constant collaboration and information sharing between the four aspects of the learning cycle create a dynamic setting. Kolb's [6] theory of experience learning was significantly influenced by experiential learning theorists. Early learning management systems were created with a "one-size-fits-all" mentality, which caused students to get confused and suffer from information overload, hence reducing their effectiveness. Every student has their own learning style. Knowing a student's learning style is essential for tailoring e-learning or traditional education to their individual needs [7]. Applications of artificial intelligence are deemed significant due to their ability to duplicate and evolve the human decision-making process. Several approaches have been utilized in adaptive learning systems. Such examples include fuzzy logic, decision trees, neural networks, Bayesian networks, genetic algorithms, and hidden Markov models. Bajaj and Sharma provide a solution for intelligent education that enables tailored learning at all times and in all locations. Learning environments include conventional schools in urban centers or isolated villages, higher education institutions or training centers in cities, and distance learning institutions [7].

AI enables teachers to assess the learning profiles of their students and identify those who are slow learners. Whenever a student demonstrates deficiencies in particular subject areas or fails to reach learning objectives, the application is responsible for giving a report to the teacher or him in order to increase education [1].

A comparable study done at the College of Canterbury across two years, from 2018 to 2019, with 201 and 193 participants, respectively, assessed the impact of employing a mature Intelligent Tutoring System, referred briefly as ITS for conceptual database building over 12 weeks [8]. As a model, Enhanced Entity-Relationship (EER) data were used. The participants were undergraduate computer science students who volunteered willingly [8]. Due to insufficient self-regulation abilities and motivation, instructional technology use was limited, according to the survey results. As problem-solving in the EER-Tutor was optional, students who solved more problems may have been more motivated. Nonetheless, the number of solution attempts and the amount of time spent with the EER-Tutor were revealed to be significant predictors of task performance. In 2018, students who completed at least three tasks on EER-Tutor had considerably higher task scores than the rest of the class.

Hence, learners' voluntary participation in the usage of ITS can promote learning if participants have high levels of self-regulation and motivation, regardless of whether the ITS is suited to the learner's specific needs [8]. Even if the study's small sample size makes it impossible to extrapolate the findings to the entire population, it is nevertheless interesting since it suggests that kids with high levels of self-regulation can perform better. An important prediction that can be highlighted while assessing the learning profile on the ITS website is the level of self-regulation.

In addition, AI can highlight courses that require improvement, as professors are not always aware of gaps in their teaching materials that may confuse pupils. Courses can be tailored to the profile and interests of individual students.

Dropout is one of the most complex aspects of student careers that may be investigated using IT, both in terms of research design and outcomes management. Dropping out is a potentially terrible event in a student's life and has a detrimental effect on the institution's finances. It can also be an indicator of possible course organization and quality issues. Using machine learning approaches that have already been validated in assessing student performance, we can anticipate student dropouts.

Del Bonifro et al. [9] created a machine learning-based approach for predicting dropout likelihood. In one academic year, 15,000 students from the University of Bologna participated in the experimental study [10]. Students between the ages of 19 and 23 from both genders participated. The following three design principles were followed in their study: First, the danger of withdrawal was discovered early, either before the trial began or within the first year. According to statistics, this is one of the most crucial times for dropouts. Owing to the focus on first-year students, the data utilized to train the prediction models comprised exclusively of personal information and academic records from high school, including gender, age, high school education, and final grade, as well as the number of credits obtained in the first few months. Second, the research did not focus on a particular predictive model. Instead, a rigorous examination of several machine learning algorithms was conducted to provide a benchmark. Using post-enrollment data such as student performance and conduct at the college, the model was deployed at the time of admissions application. Yet, gaps exist when the researcher considers the likelihood that students may take extra courses or change academic tracks after required courses have already begun. The study's categorization algorithms provide a reasonably good prediction of the phenomenon of dropout. Random Forest, Linear Discriminant Analysis (LDA), and Support Vector Machine (SVM) are three machine learning algorithms (RF). LDA is a dimensionality reduction technique that aims to reduce the complexity of the data while keeping vital information, e.g., by projecting the real feature space onto a space with less dimensions. This alternative technique permits the use of the tool to actively enhance the academic standing of students from the beginning, as opposed to simply making forecasts and tracking them throughout their academic career. In any case, the obtained results indicate that, beginning with data devoid of educational or didactic value, the measurement instrument can effectively assist in minimizing the dropout problem. Although secondary education is structured differently in each country compared to higher education, which is more standard, the planned and implemented prediction model was designed for higher education. In addition, the anticipated quantity of participants may generate generalizable findings. In contrast, Kadar et al. [10] attempted to forecast dropout using an entirely different method. In a smart classroom, it collects data from webcams, eye trackers, and other similar devices [11]. This information allows for the analysis and detection of students' emotions in the classroom, which can then be utilized to forecast dropout.

4 Students' Readiness, Digital Literacy, and Acceptance of AI Applications

AI in education provides a system that can monitor and display the learning stages of any subject, as well as provide personalized recommendations and guidance to pupils. Some AI applications can assist educators in understanding their students' capabilities and developing future learning experiences based on current outcomes. AI supports students by motivating them, evaluating their success and improvement, and collecting classroom participation data. In addition, it can review and grade work supplied by students [3].

Engagement and personality traits determine performance in online learning [11]. For decades, the purpose of technology-related education has been to prepare students [12]. AI courses and other AI applications in education should not only equip students with IT-related material knowledge, but also enable them to feel confident and secure in the workplace. Yet, AI education in schools is still in its infancy [13]. In addition, research on student readiness at many levels, including higher education and online education, has yielded inconsistent results regarding the potential influence of gender differences on student readiness. For instance, it was found that male students were more confident and equipped to utilize computers than their female counterparts. It was revealed that female students were considerably better equipped for technological developments than male students [13]. Another component is digital literacy, which refers to the user's ability to acquire, comprehend, and utilize information to achieve the intended objective. Several forms of digital literacy, such as information literacy and media literacy, have emerged in the context of technical education. Occasionally, there is no clear distinction between these terms. Given the expanding importance of AI, we refer to students' access to and application of AI-related information and abilities [13].

When there is a sense of object relevance between the learning object and the student's internal needs, future goals, or living environment, students apply and develop what they have learned in a context that goes beyond the development of intrinsic personal attributes. During instruction, relevance can be established in a variety of ways [13]. Much use is made of students' everyday experiences and relatable examples when teaching and learning about Tennessee. For example, how image recognition can be utilized to unlock mobile phones and diagnose cancer. Connecting TN knowledge to students' everyday lives may help them establish relevance, which defines the purpose of learning.

Ultimately, the effect of technology on potential users spans from optimism for the technology's use to uneasiness with the technology [13]. Users' interactions with a technology can influence whether they accept or reject its use. Wang and Wang [14] conducted a survey utilizing a scale they developed to gauge anxiety toward the implementation of AI in education [10]. Anxiety is a conscious emotion of apprehension. It includes negative ideas about technology, as well as skepticism, unease, anxiety, intimidation, and reluctance. Negative emotions related with technology can hinder technological learning experiences and adoption in the future. Anxiety, confusion, frustration, and anger, as well as other related emotional states, can affect not only the learning process, but also an individual's productivity, social interactions, and overall well-being [14]. The Artificial Intelligence Anxiety Scale (AIAS) was used in an online survey conducted in Taiwan with 301 participants from around the country. Participants ranged in age from 15 to over 60, were of both genders, and had high school to doctoral degrees. According to the findings of the study, reducing users' perceived fear through fostering extended usage of IT systems or products through enhancing learning channels is vital for successfully promoting user adoption.

5 The Applications of AI in Higher Education

The application of artificial intelligence in higher education is progressing rapidly and has a significant impact on the services given by universities. Australian universities such as Deakin use the IBM Watson supercomputer, for instance. Watson provides guidance to students 24/7, 365 days per year [15]. Two tiers define personalized instruction for students in higher education. Levels of conceptual theory and practical application co-exist [16]. In the traditional sense, the level of concept theory of personalized education involves teachers classifying pupils according to their learning profile. Personalized education hinges on the relationship between teachers and students [16]. Furthermore, personalized subject matter learning for pupils comprises involvement, feedback, adaptation, and improvement. In terms of alignment, there are also concepts that direct engagement and emphasize interaction [16].

AI can answer questions and deliver information, much like humans. The chatbot, in which a computerized information technology system interacts with a human, is one of the most well-known forms of communication [17]. It is characterized as "software that simulates human communication using text, speech, graphics, or a combination of spoken and visual heuristics." It is used on many commercial and educational college websites to respond to visitors as if they were human. Chatbots are said to as virtual assistants powered by technology and developed from established/scripted scripts or AI. They operate 24 h/7 days a week and provide the necessary knowledge and answers on the topic of admission. They not only serve the tourists or information seekers but also relieve the secretarial staff. The application of Big Data and AI in higher education can facilitate individualized teacher development, personalized institutional administration, and personalized student learning. Mixing them can lead to improvements:

(A) **Mining of data**

The system provides educators with the necessary information to properly comprehend students' backgrounds in order to give correct and effective instruction [18].

(B) **Emphasis on the Process and the Result**

AI and Big Data enable instructors to realize that they can "teach to the best of their abilities." By observing and analyzing instruction, teachers play a much larger role in the educational process of their pupils. In addition, instructional design, methodologies, and materials are optimized, and instructional content is modified in a timely manner [18]. Teachers are able to monitor the completion time, duration, and accuracy of students' assignments, as well as determine whether the material has been retained [18].

The marriage of AI and Big Data can not only enhance instructional procedures, but also reduce student weariness and boost learning. Many students experience frustration and despair during their studies because they do not comprehend or are uninterested in the supplied knowledge [18]. At this moment, instructors can intervene promptly to resolve potential issues. Thereafter, individualized advising plans are developed for each student requiring assistance. By merging AI and Big Data, it is possible to lower the dropout rate [18].

Despite strong support for AI in education, research on its efficacy, safety, integration, and ethics is lacking. As previously said, the research community has done a number of evaluations over the past several decades, but they have typically been small scale, conducted over brief time periods, and focused on specific scientific themes. There are insufficient robust, independent reviews of AI systems used in educational settings, with few exceptions, to determine policy. Current unbiased evaluations sometimes compare an AI tool against "business as usual" (i.e., when no tool is employed), so any success may be attributable to the use of the technology in general as opposed to the specific technique. Additionally, these "evaluations" never assess the tool's impact on students' cognitive skills or mental health, instructional techniques, or the teacher's role; only students' academic achievement is considered (e.g., whether they received a better grade on a standardized test or exam).

Manufacturers of artificial intelligence (AI) tools utilized in state education systems, the vast majority of whom are private commercial entities, may continue to sell or deliver their goods to schools without accreditation or direct accountability to students.

6 The New Enhanced Role of Teachers

Due to the application of AI in education, many aspects of instructors' work, such as evaluating trainees' work and providing constructive comments, may now be accomplished rapidly [1]. Professors can use the given time for professional development, student interaction, and class preparation. In addition to assisting with grading, AI provides further assistance to teachers. AI is able to do routine tasks and communicate with students [18]. Throughout the semester, one university instructor utilized an AI chatbot as a teaching assistant without the pupils being aware that they were conversing with a computer.

The LearnEven app and virtual lectures are also capable of employing a robot in place of an instructor. With gesture recognition technology, existing guides and facilitators may think, act, and react intuitively and respond to vocal and nonverbal instructions [18].

MentorPal [19] is a research project that uses virtual mentoring and is based on the following design (see Fig. 3):

(a) Rapid real-time responses to user queries, including the ability to interrupt mentors and advisors who consider previous queries when responding.
(b) Authentic video chat graphics and sounds that resemble a high-quality video conference (such as Skype MS-Teams) with the mentor.
(c) Recording devices must be inexpensive, reusable, and use open-source software.

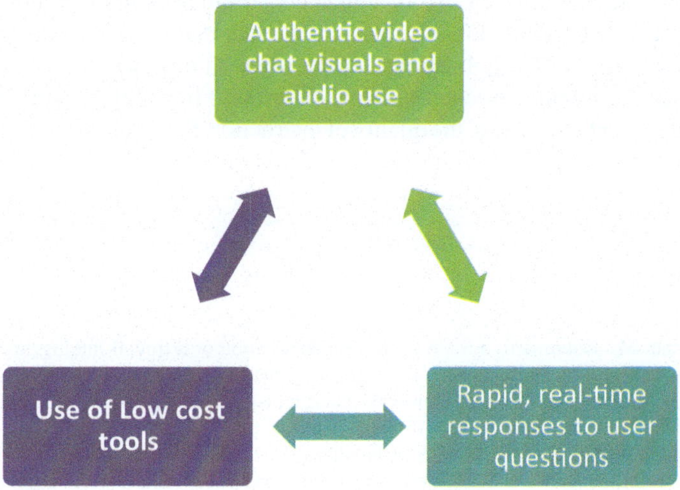

Fig. 3 Virtual mentoring triangle in AI. *Source* Subrahmanyam and Swathi [18]

In a study [19] of high school students participating in a summer internship program, the majority of respondents were male and Caucasian. Throughout the interview, researchers observed students and provided support as needed. Teachers can teach students skills that artificial intelligence systems cannot, such as critical thinking and creativity.

7 Conclusions

Increased engagement in education and continuous economic expansion can make AI an attractive answer. Examining the ways in which AI has benefited education, we conclude that its uses can have a wide range of effects, the most of which are beneficial. After doing a literature analysis on the effect of applications on the quality of instruction, we reached the following conclusions: Technology offers chances to enhance both distant education, the subject of this study, and everyday instruction. Automatic grading has a significant effect on the quality of instruction by minimizing delays in outcomes and facilitating instructor evaluation. Automated grading can provide generic feedback for each student action, which might result in ambiguity without enhancing learning and necessitates human intervention to interpret the grades. TN can resolve this issue by delivering adaptive learning for each student's unique learning style. Human–machine interaction is crucial teachers will continue to assist their students.

As many may have believed, teachers will not retreat from the educational process; rather, their role will shift to one of guidance, stimulating their pupils to become independent, collaborative thinkers. Despite the fact that the nature of their work is changing as a result of IT applications, teachers benefit from having cobots (collaborative robots), or robotic colleagues, in the classroom. In addition to technical capabilities and student assistance, the quality of instruction depends on the tools available to instructors to conduct individualized, needs-based instruction in the classroom.

References

1. Fahimirad, M., Kotamjani, S.S.: A review on application of artificial intelligence in teaching and learning in educational contexts. Int. J. Learn. Dev. **8**(4), 106–118 (2018)
2. Şeren, M., Özcan, Z.E.: Post pandemic education: distance education to artificial intelligence-based education. Int. J. Curr. Instr. (Special Issue) **12**(1), 212–225 (2021)
3. Chaudhari, S., More, P., Bhadak, S., Chaudhari, S., Gawali, T.: A survey on applications of artificial intelligence for enhancement in learning experience. Asian J. Converg. Technol. **6**(3), 86–89 (2020)
4. Zadrozny, W.W., Gallagher, S., Shalaby, W., Avadhani, A.: Simulating IBM Watson in the classroom. In: Proceedings of the 46th ACM Technical Symposium on Computer Science Education, pp. 72–77 (2015)
5. Hashimu, M., Anka, N.A.: An overview of the application of artificial intelligence in education. Int. J. Sci. Res. Publ. (IJSRP) **9**(9), 837–843 (2019)

6. Kolb, D.-A. Experiential Learning : Experience as the Source of Learning and Development. Upper Saddle River, New Jersey: Pearson Education Inc. (1984)
7. Bajaj, R., Sharma, V.: Smart education with artificial intelligence-based determination of learning styles. Procedia Comput. Sci. **132**(2018), 834–842 (2018)
8. Mitrović, A., Holland, J.: Effect of non-mandatory use of an intelligent tutoring system on students' learning. In: Bittencourt, I., Cukurova, M., Muldner, K., Luckin, R., Millán, E. (eds.) Artificial Intelligence in Education, vol. 12163, pp. 386–397. Springer International Publishing (2020)
9. Del Bonifro, F., Gabbrielli, M., Lisanti, G., Zingaro, S.P.: Student dropout prediction. In: Bittencourt, I.I., Cukurova, M., Muldner, K., Luckin, R., Millán, E. (eds.) Artificial Intelligence in Education, vol. 12163, pp.129–140. Springer International Publishing (2020)
10. Kadar, M., Sarraipa, J., Guevara, J.C., Restrepo, E.G.: An integrated approach for fighting dropout and enhancing students' satisfaction in higher education. In: Proceedings of the 8th International Conference on Software Development and Technologies for Enhancing Accessibility and Fighting Info-Exclusion, pp. 240–247 (2018)
11. Dray, B.J., Lowenthal, P.R., Miszkiewicz, M.J., Ruiz-Primo, M.A., Marczynski, K.: Developing an instrument to assess student readiness for online learning: a validation study. Distance Educ. **32**(1), 29–47 (2011)
12. Soulé, H., Warrick, T.: Defining 21st century readiness for all students: what we know and how to get there. Psychol. Aesthet. Creat. Arts **9**(2), 178–186 (2015)
13. Dai, Y., Chai, C.-S., Lin, P.-Y., Jong, M.S.-Y., Guo, Y., Qin, J.: Promoting students' well-being by developing their readiness for the artificial intelligence age. Sustainability **12**(16), 1–15 (2020)
14. Wang, Y.-Y., Wang, Y.-S.: Development and validation of an artificial intelligence anxiety scale: an initial application in predicting motivated learning behavior. Interact. Learn. Environ. **30**(4), 619–634 (2022)
15. Popenici, S.A.D., Kerr, S.: Exploring the impact of artificial intelligence on teaching and learning in higher education. Res. Pract. Technol. Enhanced Learn. **12**(1)22, 1–13 (2017)
16. Lan, Q.: Construction of personalized education model for college students driven by big data and artificial intelligence. J. Phys. Conf. Ser. **1744**(032022), 1–7 (2021)
17. Ahmad, S.F., Alam, M.M., Rahmat, M.K., Mubarik, M.S., Hyder, S.I.: Academic and administrative role of artificial intelligence in education. Sustainability **14**(22), 1–11 (2022)
18. Subrahmanyam, V.V., Swathi, K.: Artificial intelligence and its implications in education. [Paper Presentation] International Conference on Improved Access to Distance Higher Education focus on Underserved Communities and Uncovered Regions, Kakatiya University, Warangal, Telangana, India, 11–12 Aug 2018
19. Nye, B.D., Davis, D.M., Rizvi, S.Z., Carr, K., Swartout, W., Thacker, R., Shaw, K.: Feasibility and usability of MentorPal, a framework for rapid development of virtual mentors. J. Res. Technol. Educ. **53**(1), 21–43 (2021)

Chapter 18
The Study of Randomness Properties Exhibited by LAO-3D Lightweight Block Cipher Algorithm

Nik Abdullah Nik Azura, A. H. Azni, Zakaria Nur Hafiza, and Zakaria Abdul Alif

1 Introduction

In an increasingly digitized world where the exchange and storage of sensitive data have become universal, cryptographic algorithms are essential in safeguarding data confidentiality, integrity, and authenticity. These algorithms employ complex mathematical techniques to convert plaintext into ciphertext which makes it extremely difficult for unauthorized entities to access the data. Numerous fields, including communication systems, financial transactions, and data storage, find extensive use of cryptography. In communication systems, these algorithms secure the transmission of messages, protecting them from interception and eavesdropping. In financial transactions, the algorithms secure online transactions, preventing unauthorized access and fraud. Additionally, in data storage, the algorithms encrypt sensitive information so that anyone lacking the right key cannot read it. By understanding the foundational

N. A. Nik Azura (✉) · Z. Abdul Alif
CyberSecurity Malaysia, 63000 Cyberjaya, Selangor, Malaysia
e-mail: azura@cybersecurity.my

Z. Abdul Alif
e-mail: alif@cybersecurity.my

N. A. Nik Azura · A. H. Azni · Z. Nur Hafiza · Z. Abdul Alif
Faculty of Science and Technology, Universiti Sains Islam Malaysia, 71800 Nilai, Negeri Sembilan, Malaysia
e-mail: ahazni@usim.edu.my

Z. Nur Hafiza
e-mail: mzhafiza@usim.edu.my

A. H. Azni · Z. Nur Hafiza
CyberSecurity and System Research Unit, Faculty of Science and Technology, Universiti Sains Islam Malaysia, 71800 Nilai, Negeri Sembilan, Malaysia

J. M. R. S. Tavares et al. (eds.), *Proceedings of Second International Conference on Intelligent System*, Algorithms for Intelligent Systems,
https://doi.org/10.1007/978-981-99-8976-8_19

role of cryptographic algorithms in securing sensitive information, we can appreciate the significance of conducting a thorough analysis of randomness properties to ensure their effectiveness and resilience against potential threats.

The integrity and security of cryptographic algorithms heavily rely on the quality of randomness integrated into their design. However, the presence of inherent weaknesses in random number generation mechanisms can expose these algorithms to potential vulnerabilities. When there is insufficient randomness, biases or patterns might arise that attackers can use to their advantage, jeopardizing the confidentiality and integrity of sensitive data. Although there has been a lot of research on cryptographic algorithms, there is still a need for more research on the full understanding of randomness features and how they affect algorithmic security. Therefore, the research gap lies in the need for a thorough investigation of the randomness properties within cryptographic algorithms, including the identification of sources of bias, potential patterns, and inadequate entropy levels. By filling in this gap, our study aims to provide insightful information into the vulnerabilities arising from insufficient randomness.

Over the recent years, numerous lightweight algorithms have been suggested in academic research, encompassing BRIGHT [1], CRAFT [2], ACT [3], LRBC [4], LWE [5], PriPresent [6], T-TWINE [7], LBC-IoT [8], and an enhanced version of SM4 [9]. The randomness of several block cipher algorithms such as AES, RC6, Twofish, MARS, and Serpent [10] have been analyzed using several randomness tests from the NIST Statistical Test Suite. Therefore, LAO-3D lightweight algorithm should also be analyzed using the same approach.

The primary aim of this research is to perform a thorough analysis of the randomness properties exhibited by LAO-3D lightweight block cipher [11]. By conducting rigorous testing and evaluation, we aim to analyze the strength of random number generation properties employed within this algorithm and identify any potential weaknesses or vulnerabilities. The research will involve a range of statistical tests to quantitatively assess the characteristics associated with randomness in LAO-3D lightweight block cipher. Through this research objective, we aim to contribute to the field of cryptography by providing insightful analysis and recommendations for improving the randomness aspects of LAO-3D lightweight block cipher, thereby enhancing its resistance to attacks, and strengthening its overall security.

The research paper is structured in the following manner to offer a thorough investigation of the randomness characteristics associated with the LAO-3D lightweight block cipher being examined: Sect. 1, Introduction provides an overview of the research topic, emphasizing the significance of randomness analysis in cryptographic algorithms, and highlights the research objectives and the research gap being addressed. In Sect. 2, LAO-3D lightweight block cipher, we present a comprehensive review of LAO-3D algorithm. This review serves as a foundation for our research and helps identify research gaps. Section 3, Randomness Tess, outlines the experimental setup utilized to conduct the randomness analysis. We also describe the dataset used, including the size of the data, as well as any specific configurations or variations used in the experiments. In Sect. 4, Results and Analysis, we present the results and the findings obtained from the randomness analysis performed toward LAO-3D lightweight block cipher. The results section seeks to offer a thorough evaluation of the properties of randomness and forms the groundwork for subsequent discussion and interpretation. Section 5, Conclusion, summarizes the main findings of the study and highlights the contributions of this research.

2 LAO-3D Lightweight Block Cipher

LAO-3D is a block cipher that is crafted to be lightweight, utilizing 20 iterated rounds of encryption. The algorithm comprises a 64 bits block size and a 128 bits key length. The construction of this algorithm implemented the 3D rotation method. LAO-3D lightweight block cipher aims to provide security through its encryption technique and is efficient for implementation due to its design simplicity. Each iteration of LAO-3D block cipher incorporates three primary components in its encryption round. Firstly, the linear layer assures high diffusion across the encryption iterations. Secondly, the nonlinear layer incorporates a parallel implementation of the S-box, which exhibits high nonlinearity properties. Thirdly, the key addition layer employs an XOR function between the round key and its cipher state, adding confusion property to the algorithm. Implementation of the linear and nonlinear layers in LAO-3D lightweight block cipher would provide diffusion and confusion properties in the algorithm which are the fundamental requirements of a secure lightweight block cipher [12].

2.1 Encryption Algorithm

LAO-3D encryption algorithm is inclusive of four primary components including the *AddRoundKey*, *SubColumn*, and *Double3DRotation* functions that are applied to the *State* and *RoundKey*. Figure 1 illustrates the encryption process of LAO-3D.

Firstly, in the *AddRoundKey* operation, the plaintext defined as $B = b_{3,15} \dots b_{3,0} || b_{2,15} \dots b_{2,0} || b_{1,15} \dots b_{1,0} || b_{0,15} \dots b_{0,0}$ performed a bitwise XOR operation with the round key denoted as $K = k_{3,15} \dots k_{3,0} || k_{2,15} \dots k_{2,0} || k_{1,15} \dots ka_{1,0} || k_{0,15} \dots k_{0,0}$ to generate the *State*. The 64 bits round key is produced by the key schedule algorithm.

Secondly, a nonlinear bit substitution called *SubColumn* is operated on each *State* column. This operation implemented PRESENT S-box [13] as presented in Table 1 due to its satisfactory cryptographic properties as well as its small hardware footprint. The employed S-box of the *SubColumn* operation receives its input from the *AddRoundKey* operation defined as $\mathrm{Col}_j = h_{3,j} || h_{2,j} || h_{1,j} || h_{0,j}$ and the resulting output is $S(\mathrm{Col}_j) = q_{3,j} || q_{2,j} || q_{1,j} || q_{0,j}$ for $15 \geq j \geq 0$.

After that, the *Double3DRotation* function is constructed by combining the *3DBitRotation_X-axis*, *AddRoundKey*, and *3DBitRotation_Z-axis* operations. The function enhanced the security by improving its confusion and diffusion properties

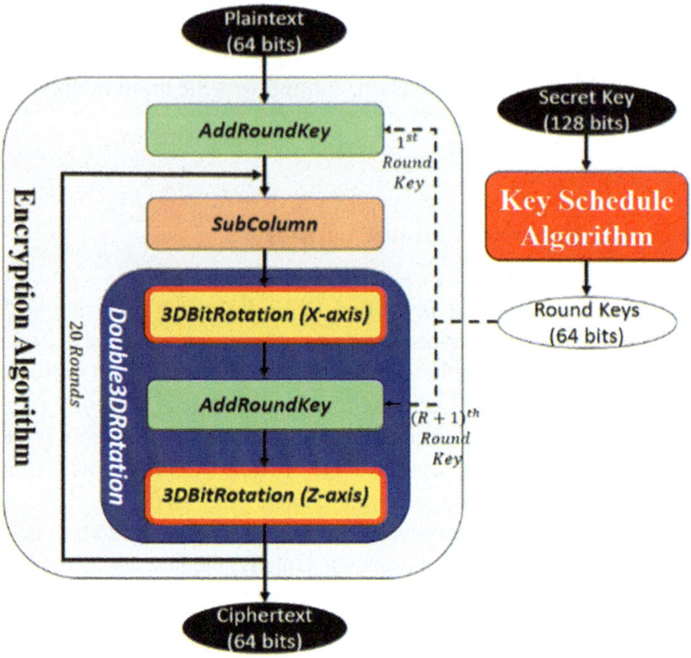

Fig. 1 Encryption algorithm of LAO-3D

Table 1 S-box

x	0	1	2	3	4	5	6	7	8	9	A	B	C	D	E	F
$S(x)$	C	5	6	B	9	0	A	D	3	E	F	8	4	7	1	2

[11]. Bit permutation is performed by this function on the quadruple 4 by 4 cipher state that consists of four *Slices* of 16 plaintext bits. The rotation formation follows a specific clockwise direction. While $Slice_0$ remains unaltered, $Slice_1$, $Slice_2$, and $Slice_3$ rotate 90°, 180°, and 270°, respectively. The *Double3DRotation* function introduced significant improvements by permuting the cipher bits in various orientations, rather than the standard rotation method, thus enhancing the randomness and its overall security.

Table 2 displays the permutation table representing the values associated with the *3DBitRotation_X-axis* function. Output from the *SubColumn* operation assigned as $Q = q_{3,15} \ldots q_{3,0}||q_{2,15} \ldots q_{2,0}||q_{1,15} \ldots q_{1,0}||q_{0,15} \ldots q_{0,0}$ is used as the input of the *3DBitRotation_X-axis* operation. Meanwhile, output for this operation is assigned as $P(Q) = e_{3,15} \ldots e_{3,0}||e_{2,15} \ldots e_{2,0}||e_{1,15} \ldots e_{1,0}||e_{0,15} \ldots e_{0,0}$.

The output of the *3DBitRotation_X-axis* assigned as $E = e_{3,15} \ldots e_{3,0}||e_{2,15} \ldots e_{2,0}||e_{1,15} \ldots e_{1,0}||e_{0,15} \ldots e_{0,0}$ is XOR using the *AddRoundKey* operation with the round key defined as $K =$

Table 2 *3DBitRotation_X-axis* permutation table

Permutation table																
x	0	1	2	3	4	5	6	7	8	9	10	11	12	13	14	15
$P(x)$	0	1	2	3	4	5	6	7	8	9	10	11	12	13	14	15
x	16	17	18	19	20	21	22	23	24	25	26	27	28	29	30	31
$P(x)$	28	24	20	16	29	25	21	17	30	26	22	18	31	27	23	19
x	32	33	34	35	36	37	38	39	40	41	42	43	44	45	46	47
$P(x)$	47	46	45	44	43	42	41	40	39	38	37	36	35	34	33	32
x	48	49	50	51	52	53	54	55	56	57	58	59	60	61	62	63
$P(x)$	51	55	59	63	50	54	58	62	49	53	57	61	48	52	56	60

Table 3 *3DBitRotation_Z-axis* permutation table

Permutation table																
x	0	1	2	3	4	5	6	7	8	9	10	11	12	13	14	15
$P(x)$	0	13	62	51	4	29	58	35	8	45	54	19	12	61	50	3
x	16	17	18	19	20	21	22	23	24	25	26	27	28	29	30	31
$P(x)$	16	9	46	55	20	25	42	39	24	41	38	23	28	57	34	7
x	32	33	34	35	36	37	38	39	40	41	42	43	44	45	46	47
$P(x)$	32	5	30	59	36	21	26	43	40	37	22	27	44	53	18	11
x	48	49	50	51	52	53	54	55	56	57	58	59	60	61	62	63
$P(x)$	48	1	14	63	52	17	10	47	56	33	6	31	60	49	2	15

$k_{3,15} \dots k_{3,0} || k_{2,15} \dots k_{2,0} || k_{1,15} \dots k_{1,0} || k_{0,15} \dots k_{0,0}$ to produce the new cipher *State*.

Next, the output of the *AddRoundKey* operation is used as the input for the *3DBitRotation_Z-axis* defined as $F = f_{3,15} \cdots f_{3,0} || f_{2,15} \cdots f_{2,0} || f_{1,15} \cdots f_{1,0} || f_{0,15} \cdots f_{0,0}$ is permutated using Table 3 that will produce the ciphertext defined as $P(F) = g_{3,15} \cdots g_{3,0} || g_{2,15} \cdots g_{2,0} || g_{1,15} \cdots g_{1,0} || g_{0,15} \cdots g_{0,0}$.

2.2 Key Schedule Algorithm

LAO-3D employed its key schedule algorithm to produce the round keys. Overall, the computation of the round key bits involves multiplying the block size by the round number and adding one to the result. A block size of 64 bits with 20 encryption rounds requires a total of 1344 round key bits. After performing the *NonceXOR*, *KeySubColumn*, and *RowTransformation* functions, the *RoundKeyExtraction* process extracts a 64-bit round key from the modified key as depicted in Fig. 2.

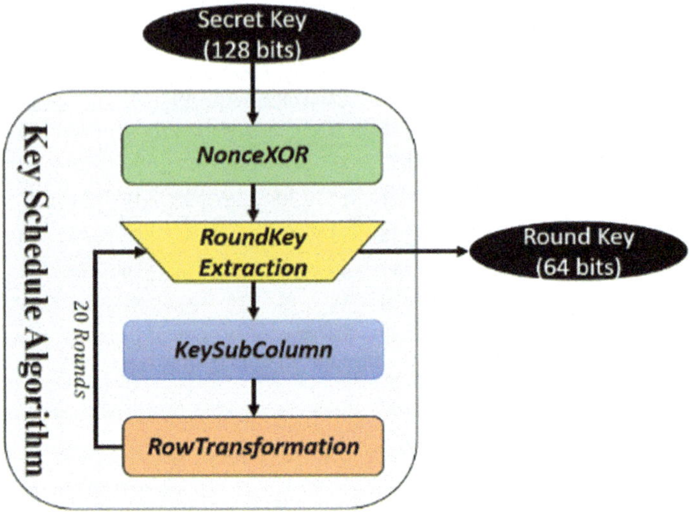

Fig. 2 Key schedule algorithm of LAO-3D

Firstly, the secret key denoted as $S = s_{127}||\ldots||s_1 s_0$ is subjected to an XOR operation with a nonce referred to as $N = n_{127}||\ldots||n_1 n_0$ to obtain the key state. The nonce is a random 128-bit data that is modifiable and embedded in the algorithm. The purpose of adding the nonce in the key schedule algorithm is to enhance confusion property during round keys generation. Furthermore, the nonce would eliminate the correlation of the generated round keys with the original secret key.

LAO-3D employed a 128-bit secret key to produce 21 round keys. The key schedule algorithm needs a 64-bit portion from the key state. Assign $V = v_{127}||\ldots v_1||v_0$ as the key state and each row is notated as the *RowKeys*. To produce the round key K_i at round i for $0 \leq i \leq 20$, the concatenation takes place by combining the 16 columns located on the right side of each *RowKey*, resulting in a 64-bit value. The round keys are used in the *AddRoundKey* operation during both encryption and decryption executions.

Upon completion of *RoundKeyExtraction* in each round, the round key value is modified through the utilization of the *KeySubColumn* function. The S-box from the *SubColumn* operation in the encryption algorithm is employed to construct the key state's four uppermost rows and eight rightmost columns, i.e., $k'_{3,j}||k'_{2,j}||k'_{1,j}||k'_{0,j} = S(k_{3,j}||k_{2,j}||k_{1,j}||k_{0,j})$ for $7 \geq j \geq 0$.

The *KeySubColumn* operation generates an updated round key, which is then rearranged into the *RowKeys* format to form the new key state. All *RowKeys* are modified by applying the *RowTransformation* function, i.e., $RowKey'_0 = (RowKey_0 <<< 8) \oplus RowKey_1$, $RowKey'_1 = RowKey_2$, $RowKey'_2 = (RowKey_2 <<< 16) \oplus RowKey_3$, and $RowKey'_3 = RowKey_0$. The above processes will be repeated until the completion of 21 round keys.

3 Randomness Tests

The examination of randomness was conducted by utilizing a group of fifteen statistical tests adopted from the NIST Statistical Test Suite [14]. It is an open-source application designed for evaluating randomness that focuses on different characteristics associated with non-randomness in ciphertext generated by LAO-3D lightweight block cipher.

Non-parameterized tests consist of eight distinct tests. These tests do not require parameter values to be included. The purposes of these tests are as follows:

i. Random Excursion Variant—to detect any deviations from the anticipated frequency of visits to distinct states in the random walk.
ii. Random Excursion—to assess whether the number of visits to a specific state within a cycle deviates from the expected occurrence in a random sequence.
iii. Cumulative Sums—to evaluate if the cumulative sum of partial sequences within the tested sequence is significantly larger or smaller than what is typically expected in random sequences.
iv. Runs—to establish if the number of runs of ones and zeros, varying in length, conforms to the anticipated pattern for a random sequence.
v. Spectral DFT—to detect repetitive patterns in close proximity, known as periodic features, within the examined sequence. The presence of such patterns would indicate a departure from the assumption of randomness.
vi. Frequency—to determine if the quantity of ones and zeros in a sequence closely aligns with the expected distribution for a genuinely random sequence.
vii. Longest Runs of Ones—to assess if the length of the longest run of ones in the tested sequence aligns with what is typically anticipated for the longest run of ones in a random sequence.
viii. Binary Matrix Rank—to examine the presence of linear dependence among fixed-length substrings of the original sequence.

Parameterized tests consist of seven tests. These tests demand the insertion of specific parameter values into the application. The focuses of these tests are as follows:

i. Non-overlapping—to detect generators that produce an unusually high frequency of a particular non-repeating (aperiodic) pattern.
ii. Serial—to examine the occurrence rate of every conceivable overlapping m-bit pattern throughout the entire sequence.
iii. Block Frequency—to establish if the occurrence rate of ones within an M-bit block aligns closely with the expected value of $M/2$, which is anticipated in a random sequence.
iv. Approximate Entropy—to conduct a comparison of the occurrence rates of all potential overlapping m-bit patterns throughout the whole sequence.
v. Maurer's Universal—to assess if the sequence can undergo substantial compression without sacrificing any information.

vi. Overlapping Templates—to identify generators that produce an unusually large number of occurrences of a specific non-repeating (aperiodic) pattern.
vii. Linear Complexity—to ascertain if the sequence possesses enough complexity to qualify as random.

For randomness assessment, it is necessary to establish the level of significance, denoted by α. The value of α must be set to a minimum of 0.1% (0.001) and should not surpass 1% (0.01), whereby the minimum required samples are equal to the inverse of α (100 samples) [15]. When the p-value is bigger than or equal to α, the ciphertext is considered to exhibit randomness. Conversely, it is classified as not random.

Randomness of a cryptographic algorithm is determined according to the generated test samples using the subsequent equation [14].

$$p_\alpha = (1 - \alpha) - 3\sqrt{\frac{\alpha(1 - \alpha)}{s}}, \quad (1)$$

where the significance level α is defined as 0.001, while s indicates the 1000 sample size. Should the rejections exceed the proportion denoted by p_α, these test samples are considered to be non-random.

To generate the input data for the LAO-3D, a total of nine data categories were utilized that include Ciphertext Block Chaining (CBC), Plaintext/Ciphertext Correlation (PCC), Random Plaintext/ Random Key (RPRK), Strict Key Avalanche (SKA), Strict Plaintext Avalanche (SPA), Low-Density Key (LDK), High-Density Key (HDK), Low-Density Plaintext (LDP), and High-Density Plaintext (HDP). Table 4 details the format of data categories input based on 64-bit plaintext and 128-bit key implemented by LAO-3D lightweight block cipher. From each data category, 1000 ciphertext samples were produced where the number of blocks for each sample is to be determined by the block size and key size [16]. These derived blocks were combined, resulting in the construction of a large number of bit sequences used as the test samples.

Multiple ciphertext lengths were generated for each data category. All of the 15 statistical tests are executable for the evaluation of five data categories in LAO-3D, namely CBC, RPRK, SKA, SPA, and PCC. These categories are capable of generating a minimum of 1,000,000 bits of data, meeting the requirements for NIST statistical tests. In contrast, LDK and HDK can undergo evaluation with only 11 tests, whereas LDP and HDP are limited to ten tests due to insufficient data length.

4 Results and Analysis

The test results for $\alpha = 0.1\%$ are presented in Tables 5 and 6. Findings from the analysis show that LAO-3D passed all 15 statistical tests from the nine data categories. Furthermore, the p-values generated from the NIST Statistical Test Suite exhibit a uniform distribution, as they are all greater than 0.0001, as depicted in Fig. 3. The

Table 4 Data categories of LAO-3D lightweight block cipher input data

Data category	Key	Plaintext	Derived blocks	Derived bits
CBC	1 random key (128-bit)	1 all zero plaintext (64-bit)	15,625	1,000,000
PCC	1 random key (128-bit)	15,625 random plaintexts (64-bit)	15,625	1,000,000
RPRK	1 random key (128-bit)	15,625 random plaintexts (64-bit)	15,625	1,000,000
SKA	123 random keys (128-bit)	1 all zero plaintext (64-bit)	15,744	1,007,616
SPA	1 all zero key (128-bit)	245 random plaintexts (64-bit)	15,680	1,003,520
LDK	3241 specific keys (128-bit)	8257 random plaintexts (64-bit)	8257	528,448
HDK	3241 specific keys (128-bit)	8257 random plaintexts (64-bit)	8257	528,448
LDP	2081 specific keys (128-bit)	2081 random plaintexts (64-bit)	2081	133,184
HDP	2081 specific keys (128-bit)	2081 random plaintexts (64-bit)	2081	133,184

fact that the randomness and uniformity evaluation results achieved a 100% passing rate serves as confirmation for the random characteristics of the encryption output.

The assessment observes the influence of changing any parameter values on the algorithm randomness attributes. The fact that the tests show a 100% passing rate indicates characteristics of randomness and proves that LAO-3D has the capability to function as a generator of pseudorandom bits. Random ciphertext can be generated by a combination of good substitution and permutation implementation. The *SubColumn* and *Double3DRotation* functions are able to optimize the algorithm's confusion and diffusion, thereby contributing to the randomness of the resulting ciphertext.

Table 5 Randomness results

	Statistical test	Data category				
		CBC	PCC	RPRK	SKA	SPA
		No. of passed samples (out of 1000) (acceptable rejection range: [0, 4])				
1	Runs	997	999	999	1000	999
2	Frequency	1000	999	998	997	1000
3	Spectral DFT	999	1000	996	1000	1000
4	Block frequency	998	999	1000	999	1000
5	Binary matrix rank	999	997	1000	1000	999
6	Approximate entropy	1000	1000	998	997	999
7	Longest runs of ones	998	999	1000	1000	997
8	Serial	999	1000	1000	1000	1000
9	Cumulative sums	1000	999	999	997	1000
10	Non-overlapping templates	999	999	999	999	999
11	Maurer's universal	999	998	998	999	1000
12	Linear complexity	998	998	999	1000	999
13	Overlapping template	999	999	999	1000	999
		No. of passed samples (acceptable rejection range: [0, 3])				
14	Random excursion	583/584	616/617	595/595	620/622	632/633
15	Random excursion variant	583/584	616/617	594/595	621/622	633/633

Table 6 Results of randomness (continued)

		Data category			
	Statistical test	LDK	HDK	LDP	HDP
		No. of Passed Samples (out of 1,000) (acceptable rejection range: [0, 4])			
1	Runs	999	998	1000	1000
2	Frequency	1000	997	999	998
3	Spectral DFT	998	999	997	998
4	Block frequency	1000	1000	998	1000
5	Binary matrix rank	999	998	1000	1000
6	Approximate entropy	999	999	999	1000
7	Longest runs of ones	1000	999	999	999
8	Serial	999	999	999	999

(continued)

Table 6 (continued)

		Data category			
	Statistical test	LDK	HDK	LDP	HDP
9	Cumulative sums	1000	997	999	997
10	Non-overlapping templates	999	999	999	999
11	Maurer's universal	1000	999	X	X
12	Linear complexity	X	X	X	X
13	Overlapping template	X	X	X	X
		No. of passed samples (Acceptable rejection range: not available)			
14	Random excursion	X	X	X	X
15	Random excursion variant	X	X	X	X

X = No tests were conducted due to inadequate data length

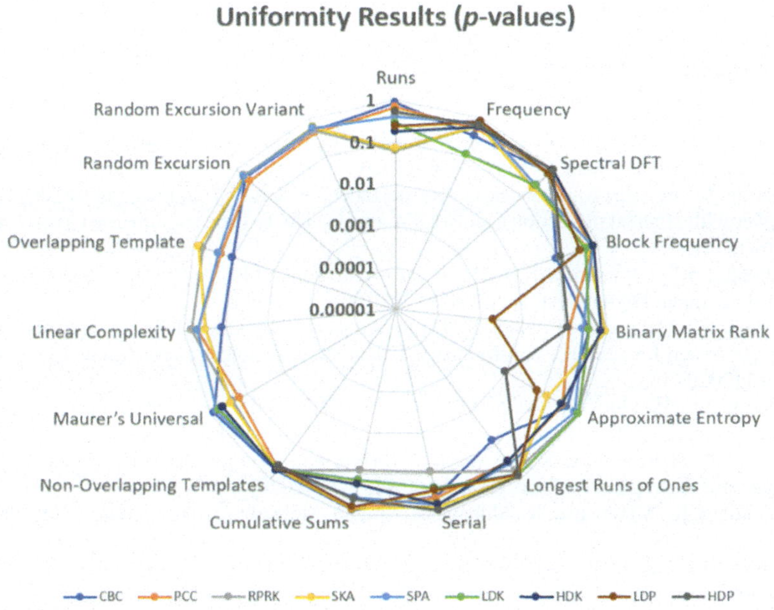

Fig. 3 Uniformity results

5 Conclusion

The goal of this exploration is to analyze the randomness attributes of LAO-3D algorithm. We aimed to assess LAO-3D's resistance against statistical tests and evaluate its suitability for secure data transmission and implementation. We first analyzed LAO-3D design principle and its underlying mechanisms. Next, we generated a significant number of ciphertexts in which the corresponding plaintext was generated using nine different data categories. Then, we analyzed the ciphertext statistical characteristics by utilizing the NIST Statistical Test Suite. Finally, LAO-3D's adherence to randomness criteria was analyzed. From the findings, LAO-3D passed all fifteen statistical tests, and it is determined as random according to the 0.1% significance level. This proves that the ciphertext generated by LAO-3D exhibited excellent adherence to randomness criteria and indicated a high degree of uniformity and unpredictability. The algorithm's excellent performance ensures its suitability for robust cryptographic applications. LAO-3D meets the required standards, such as those outlined in the FIPS 140, making it a viable choice for secure data encryption and implementation in cryptographic applications. The algorithm has been integrated into mobile applications and will further be integrated into resource-constrained devices. Therefore, as part of future work, our intention is to undertake a thorough analysis of the algorithm's suitability for resource-constrained devices commonly found in IoT ecosystems.

References

1. Sehrawat, D., Gill, N.S.: BRIGHT: a small and fast lightweight block cipher for 32-bit processor. Int. J. Eng. Adv. Technol. **8**, 1549–1556 (2019)
2. Beierle, C., Moradi, A., Leander, G., Rasoolzadeh, S.: CRAFT: lightweight tweakable block cipher with efficient protection against DFA attacks. IACR Trans. Symmetric Cryptol. **1**, 5–45 (2019)
3. Jithendra, K.B., Kassim, S.T.: ACT: an ultra-light weight block cipher for Internet of Things. Int. J. Comput. Digital Syst. **9**, 921–929 (2020)
4. Biswas, A., Majumdar, A., Nath, S., Dutta, A., Baishnab, K.L.: LRBC: A lightweight block cipher design for resource constrained IoT devices. J. Ambient Intell. Humanized Comput., 1–15 (2020)
5. Toprak, S., Akbulut, A., Aydın, M.A., Zaim, A.H.: LWE: an energy-efficient lightweight encryption algorithm for medical sensors and IoT devices. Electrica J. **20**, 71–81 (2020)
6. Girija, M., Manickam, P., Ramaswami, M.: PriPresent: an embedded prime lightweight block cipher for smart devices. Peer Netw. Appl. **14**, 1–11 (2020)
7. Sakamoto, K., Minematsu, K., Shibata, N., Shigeri, M., Kubo, H., Funabiki, Y., Bogdanov, A., Morioka, S., Isobe, T.: Tweakable TWINE: building a tweakable block cipher on generalized Feistel structure. IEICE Trans. Fundam. Electron. Commun. Comput. Sci. **103**, 1629–1639 (2020)
8. Ramadan, R.A., Aboshosha, B.W., Yadav, K., Alseadoon, I.M., Kashout, M.J., Elhoseny, M.: LBC-IoT: lightweight block cipher for IoT constraint devices. Comput. Mater. Continua **67**, 3563–3579 (2021)
9. Chen, B.W., Xia, X., Liang, Q.M., Zhong, W.D.: Lightweight design of SM4 algorithm and realization of threshold scheme. J. Phys. Conf. Ser. **1871**, 012124 (2021)
10. Aljohani, M., Ahmad, I., Basheri, M., Alassafi, M.O.: Performance analysis of cryptographic pseudorandom number generators. IEEE Access **7**, 39794–39805 (2019)
11. Zakaria, A.A., Azni, A.H., Ridzuan, F., Zakaria, N.H., Daud, M.: LAO-3D: a symmetric lightweight block cipher based on 3d permutation for mobile encryption application. Symmetry **14**(10), 2042 (2022)
12. Shah, I.N.M., Ismail, E.S., Samat, F., Rahman, N.N.A.: Modified generalized Feistel network block cipher for the Internet of Things. Symmetry **15**(4), 900 (2023)

13. Bogdanov, A., Knudsen, L. R., Leander, G., Paar, C., Poschmann, A., Robshaw, M. J. B., Seurin, Y., Vikkelsoe, C.: PRESENT: An ultra-lightweight block cipher. In: Proceedings of the Cryptographic Hardware and Embedded Systems-CHES 2007: 9th International Workshop; Springer Berlin Heidelberg, Vienna, Austria, pp. 450–466 (2007)
14. Rukhin, A., Soto, J., Nechvatal, J., Smid, M., Barker, E., Leigh, S., Levenson, M., Vangel, M., Banks, D., Heckert, A.: A statistical test suite for random and pseudorandom number generators for cryptographic applications. NIST Special Publication 800–22 Revision 1a. http://csrc.nist.gov/groups/ST/toolkit/rng/documents/SP800-22rev1a.pdf
15. Imdad, M., Ramli, S.N., Mahdin, H.: An enhanced key schedule algorithm of PRESENT-128 block cipher for random and non-random secret keys. Symmetry **14**(3), 604 (2022)
16. Zakaria, A.A., Azni, A.H., Ridzuan, F., Zakaria, N.H., Daud, M.:. Randomness analysis on RECTANGLE block cipher. In: Proceedings of the International Cryptology and Information Security Conference, pp.133–142 (2020)

Chapter 19
Association Rule Mining-Based Food Preferences Analysis Using FP-Growth Method

Hamidah Jantan, Nurhidayah Supardi, Hayati Adilin Mohd Abd Majid, and Ummu Fatihah Mohd Bahrin

1 Introduction

With the rapid advancements in science and technology, we can tap into extensive volumes of structured, unstructured, and semi-structured data. This dataset holds valuable knowledge that can be uncovered using data mining technology. Various data mining technologies have emerged, including Association Rules Mining (ARM) and sequence pattern discovery. ARM, initially developed to extract valuable information from large datasets, is capable of detecting and extracting useful insights [1, 2]. ARM is used in various applications, including recommender systems, customer relationship management (CRM), and cross-selling [1, 3–5].

H. Jantan (✉) · N. Supardi · U. F. M. Bahrin
College of Computing, Informatics and Mathematics, Universiti Teknologi MARA, Terengganu, Malaysia
e-mail: hamidahjtn@uitm.edu.my

U. F. M. Bahrin
e-mail: ummufatihah@uitm.edu.my

H. A. M. A. Majid
Faculty of Hotel and Tourism Management, Universiti Teknologi MARA, Terengganu, Malaysia
e-mail: hayati2959@uitm.edu.my

J. M. R. S. Tavares et al. (eds.), *Proceedings of Second International Conference on Intelligent System*, Algorithms for Intelligent Systems,
https://doi.org/10.1007/978-981-99-8976-8_20

The FP-Growth algorithm utilizes an extended prefix-tree structure called the FP-tree, which efficiently stores customer transactions in a compressed format. This algorithm is known for its speed and scalability, outperforming other well-known techniques for association rule mining, including Apriori and Tree Projection algorithms, and delivering superior performance compared to Eclat and Relim [6, 7]. Another fascinating method worth mentioning is FP-Growth, which employs a divide-and-conquer approach. Its initial step builds a compact data structure known as a frequent pattern tree (FP-tree). This FP-tree is an extended prefix-tree structure that efficiently stores essential quantitative data regarding frequent patterns. Only frequent items with length-1 will have nodes in the tree [8–10].

This study explores the use of ARM to analyze food preferences and promote nutrition awareness among teenagers. By applying ARM techniques, the study aims to uncover strong relationships [1, 2] and associations among various food preferences, thereby aiding in developing personalized and healthy diet plans. The research focuses on leveraging data mining methodologies to extract valuable insights that can inform teenagers about their food choices and nutritional needs.

Good nutrition awareness is important for the community, especially among teenagers. However, most teenagers face problems eating good nutrition to maintain their healthy diet plan [11, 12]. Traditional statistical analysis methods have been predominantly used in previous studies. Still, adopting machine learning and data mining techniques [11, 13, 14], such as ARM, offers a promising approach to uncovering hidden relationships and extracting meaningful knowledge from large datasets [15]. By analyzing the confidence and support metrics of generated rules, ARM enables the identification of significant associations between different food preferences. The application of ARM in the context of food preferences analysis highlights its significance in understanding the food consumption patterns among teenagers.

The structure of this paper is as follows: The second section delves into the discussion of related food preferences analysis, ARM applications, and techniques. The third section provides an overview of the research methodology. Subsequently, the fourth section presents the analysis of results and corresponding discussions, leading to the conclusion and prospects for future work in the final section.

2 Related Work

2.1 Food Preferences Analysis

Food preferences analysis plays a crucial role in formulating a health-conscious diet that aligns with the recommended nutrient proportions outlined in the food pyramid. For instance, a descriptive, questionnaire-based, cross-sectional analysis study on the relationship between fast-food preferences and BMI among female university students has been carried out [16]. A separate study examined the association between the density of fast-food restaurants near residences and the proximity of fast-food establishments to homes about BMI [17]. The method that has been used was descriptive statistics.

Recently, several studies on food preferences analysis using a data mining approach have been conducted for recommending healthy diet planning using predictive analysis techniques in data mining [13–15, 18, 19]. These studies aim to suggest beneficial dietary practices and eating routines that can assist individuals in understanding both calorie expenditure and macronutrient consumption through data mining techniques. For example, a study used predictive analysis in data mining to discover hidden patterns and customer eating habits from different sources [18, 20]. The outcome will aid in mitigating the risk of illness since the system offers monitoring capabilities, thereby enhancing an individual's health and the quality of their dietary choices.

Predictive analysis in the data mining approach can be implemented using classification, clustering, or association tasks [13, 16, 21]. Association rule mining is a data mining technique employed to identify the likelihood of specific patterns occurring within a set of items. It accomplishes this by generating association rules that reveal patterns, relationships, and correlations among these items [2, 13, 15]. This method can discover hidden knowledge in food preferences analysis, especially for healthy diet planning.

2.2 *Association Rule Mining (ARM)*

Association mining extracts linkages, recurrent patterns, and association structures between items or artifacts. It uses support and confidence statistical measures to measure the frequency of events and the priority of rules. The association's regulations are significant if they meet a minimum support threshold and a minimum level of confidence [5, 19]. Association rule mining, often referred to as ARM, is a frequently employed technique used to identify relationships between item co-occurrences in datasets through the creation of association rules. These rules consist of both antecedent and consequent components, and their importance is determined by metrics such as support and confidence.

Previous research has used various standard ARM algorithms, such as Apriori, Eclat, and the FP-Growth algorithms [4]. The Apriori algorithm identifies frequent item sets by relying on a principal property where all subsets of an item set are considered frequent. In contrast, Eclat utilizes binary integer programming and implements a satisfied constraint approach for modeling. On the other hand, the FP-Growth algorithm efficiently compresses and stores quantitative data related to frequent patterns within a database, employing a divide-and-conquer strategy [6]. This approach has attracted many researchers in food preferences analysis, such as the study on the association between eating habits and BMI of adolescents [19], comparison of factors associated with disordered eating between male and female Malaysian university students [11], to find relationships between the dishes offered by the restaurant for the elaboration of combos based on the preferences of the diners [15] and many others.

2.3 FP-Growth

The FP-Growth algorithm is known for efficiently storing quantitative information about frequent patterns within a database. It accomplishes this by employing a divide-and-conquer strategy. The FP-Growth method finds the most common itemset without needing candidate generation [7]. This method involves two steps: constructing the frequent pattern tree (FP-tree) from a transaction database and mining the FP-tree by creating conditional pattern bases. The second step focuses on generating rules from the list of frequent patterns, considering their support and confidence values. The method offers advantages such as generating a compact conditional pattern base with minimal support and requiring less memory since it avoids candidate generation. The FP-Growth method offers benefits such as completeness, preserving complete information for frequent pattern mining, and compactness, reducing irrelevant information and items in frequency descending order. This method ensures transaction patterns remain consistent and avoids breaking long patterns. FP-Growth is a divide-and-conquer structure. It decomposes the mining task and database scanning according to the frequent patterns obtained, leading to a focused search of smaller databases. In addition, there is no candidate generation, no candidate test, and a compressed database in the FP-tree structure, where there is no repeated scan of the entire database. FP-Growth algorithm is faster, more efficient, and uses small memory, but struggles with large pattern datasets compared to the Apriori algorithm [6, 8–10]. Many studies used the FP-Growth method, such as in preventing maverick buying [7], discovering customer patterns [7, 9], and predicting sales. In food preferences analysis, this method has attracted researchers in this field due to its ability to produce good decision-making results.

3 Research Method

In this study, the Cross-Industry Standard Process for Data Mining (CRISP-DM) model is employed, encompassing six primary phases for analyzing food preferences, which also incorporates BMI data through the utilization of the FP-Growth algorithm. Figure 1 shows the six phases involved: The first phase is business understanding to identify the business goals gathered from the literature study and interviews with experts in food nutrition. The second phase is data understanding, which aims to understand the dataset on food preferences among the students. This study's dataset is obtained from the survey regarding student food preferences based on the food pyramid. The food preference criteria were chosen based on some issues that raise health problems, especially among teenagers. In this phase, several interviews with the experts were conducted to strengthen and verify the proposed criteria. The dataset comprises 160 respondents' feedback and attributes on demographic information and food preference criteria. Twenty-five (25) attributes have been chosen, encompassing factors such as BMI, the food group with the highest preference, the preferred food within each group, the preferred cooking method, favored food sources, and the frequency of food consumption. The dataset sample, featuring metadata for BMI and food preference criteria, is illustrated in Table 1 and depicted in Fig. 2.

The third phase is data preparation to handle some issues in the dataset, such as missing values and noisy and inconsistent data. In this study, data preparation was conducted using Rapid Miner software. Several steps are involved in data preparation, such as data consolidation, cleaning, transformation, reduction, and arrangement. In the data preparation process, some attribute's data type must be changed to a suitable data type that depends on the algorithm requirement. It can be performed by using features in Rapid Miner, as an example, from nominal to numerical data type, where the non-numeric attributes will be changed to a numeric type. Besides that, it can also map all attribute values to numeric values. Furthermore, attributes can be renamed by replacing parts of the attribute's name with a specified replacement. The numerical values can also be changed to binomial by selecting the numeric attributes to a binomial type, as shown in Fig. 3. Figures 4 and 5 show the sample of demography and food preferences datasets after data preparation processes. The data

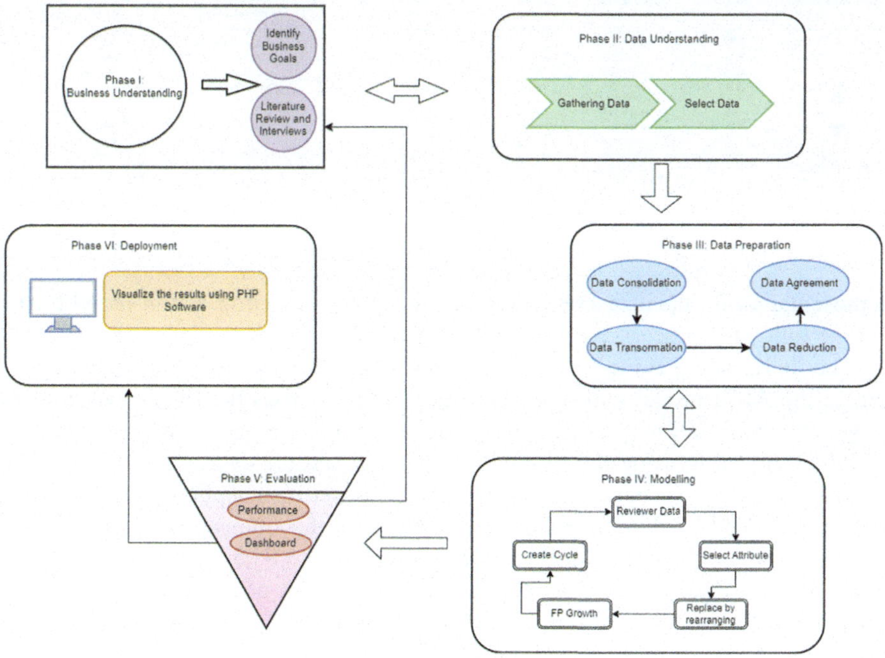

Fig. 1 CRIPS-DM-based food preferences analysis framework

Table 1 Metadata of demography (BMI) and food preferences

Attribute	Description	Value
Body mass index (BMI)	A measure of body fat based on height and weight that applied to the student	Normal, underweight, overweight, obese
Type of food in carbohydrate	The type of food mostly likes for food in carbohydrate	White Rice, brown rice, bihun, pasta
Type of food in fat, oil, sugar, and salt	The type of food mostly likes for food in fat, oil, sugar, and salt	Butter, coconut milk, granulated sugar
Type of food in protein	The type of food mostly likes for food in protein	Chicken, fish, lamb
Type of food in vegetable/fruit	The type of food mostly likes for food in vegetable/fruit	Watermelon, apple, kangkung
Type of fast food	The type of food mostly likes for food in fast food	Fried chicken, beef burger, KFC

	A	B	C	D	E	F	G
1	BMI	Type of Carbo	Type of Food in Carbo	Method of Cooking Carbo	The Preferred Places to Find Carbo	Frequency of Consuming Carbo	Type of Vegetables/Fruits
2	Overweight (25-29.9)	Rice	White Rice	Frying	Cafeteria	2-3 times per day	Fruits
3	Overweight (25-29.9)	Rice	White Rice	Boiling	Cafeteria	4 times per week	Fruits
4	Normal (18.5-24.9)	Rice	White Rice	Frying	Cafeteria	4 times per week	Fruits
5	Normal (18.5-24.9)	Rice	White Rice	Steam	Others	2-3 times per week	Vegetables
6	Overweight (25-29.9)	Noodle	Bihun	Boiling	Cafeteria	1 time per day	Fruits
7	Normal (18.5-24.9)	Rice	White Rice	Frying	Cafeteria	2-3 times per day	Fruits
8	Normal (18.5-24.9)	Noodle	Bihun	Boiling	Cafeteria	1 time per day	Fruits
9	Normal (18.5-24.9)	Rice	White Rice	Boiling	Cafeteria	1 time per day	Fruits
10	Normal (18.5-24.9)	Rice	White Rice	Boiling	Cafeteria	1 time per week	Vegetables
11	Overweight (25-29.9)	Rice	White Rice	Frying	Cafeteria	1 time per day	Fruits
12	Underweight (Below 18.5)	Rice	White Rice	Boiling	Others	2-3 times per day	Fruits
13	Normal (18.5-24.9)	Rice	White Rice	Boiling	Others	2-3 times per day	Fruits
14	Normal (18.5-24.9)	Rice	White Rice	Boiling	Others	2-3 times per day	Fruits
15	Normal (18.5-24.9)	Rice	White Rice	Boiling	Groceries	1 time per day	Fruits
16	Normal (18.5-24.9)	Rice	White Rice	Boiling	Cafeteria	2-3 times per day	Fruits
17	Normal (18.5-24.9)	Rice	White Rice	Boiling	Cafeteria	4 times per week	Fruits
18	Normal (18.5-24.9)	Rice	White Rice	Frying	Cafeteria	1 time per day	Fruits
19	Normal (18.5-24.9)	Rice	White Rice	Frying	Cafeteria	1 time per day	Fruits
20	Normal (18.5-24.9)	Rice	White Rice	Boiling	Cafeteria	2-3 times per day	Fruits
21	Overweight (25-29.9)	Rice	White Rice	Boiling	Others	2-3 times per day	Vegetables
22	Normal (18.5-24.9)	Rice	White Rice	Boiling	Cafeteria	2-3 times per day	Fruits
23	Overweight (25-29.9)	Noodle	Yellow mee	Boiling	Cafeteria	2-3 times per day	Vegetables
24	Overweight (25-29.9)	Rice	White Rice	Boiling	Cafeteria	2-3 times per day	Vegetables
25	Overweight (25-29.9)	Rice	White Rice	Boiling	Cafeteria	4 times per week	Fruits

Fig. 2 Sample of dataset

representation for the dataset needs to be transformed to a transaction-based format for the following process: association rule mining modeling.

The fourth phase is model building, which will address a specific business need where the ARM model will be embarked. This study uses the FP-Growth method to generate association rules from frequent item mining. Figure 6 shows the process of ARM modeling using the FP-Growth method, and the dataset is retrieved after the data preparation process is completed. In the experimental phase, the FP-Growth operator determines all frequent itemset for the items that often appear together in the data. This study chooses minimum support of 0.01 and 0.05 for model evaluation. Then, the association rule operator will generate a set of association rules from the given frequent itemset. The minimum confidence of 0.6 and 0.5 was chosen to determine the strongly generated rules for future use.

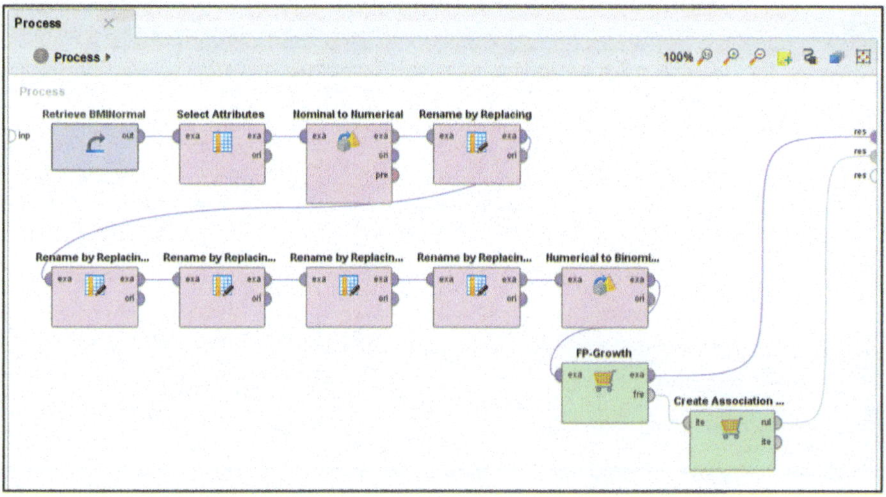

Fig. 3 Data preprocessing and ARM modeling

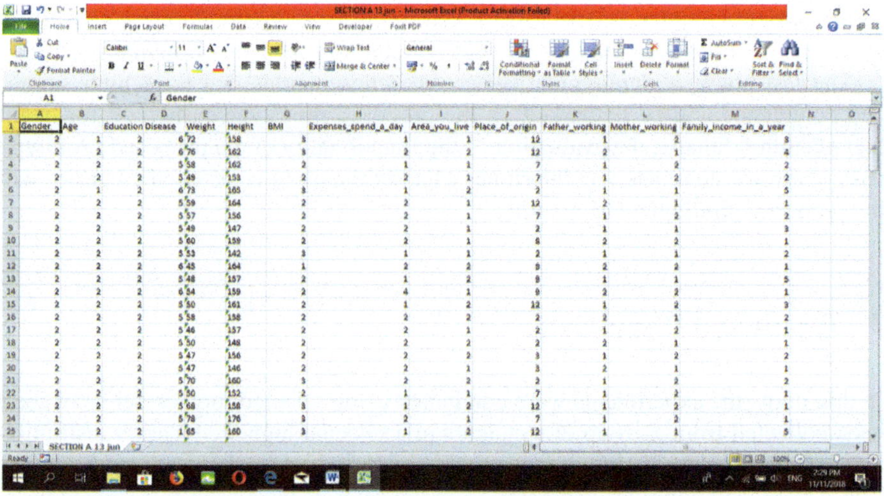

Fig. 4 Demographic information in dataset

The ARM rules performance and proposed prototype/dashboard evaluation were conducted in the evaluation stage. The performance of rules is analyzed based on the confident and support metrics to determine the strong association rules for future use. The minimum support and confidence can vary and need to be experimented with. In the study, the first experiment used a minimum support of 0.01 (1%), and a minimum confidence of 0.60 (60%) was chosen. The minimum support 0.05 (5%)

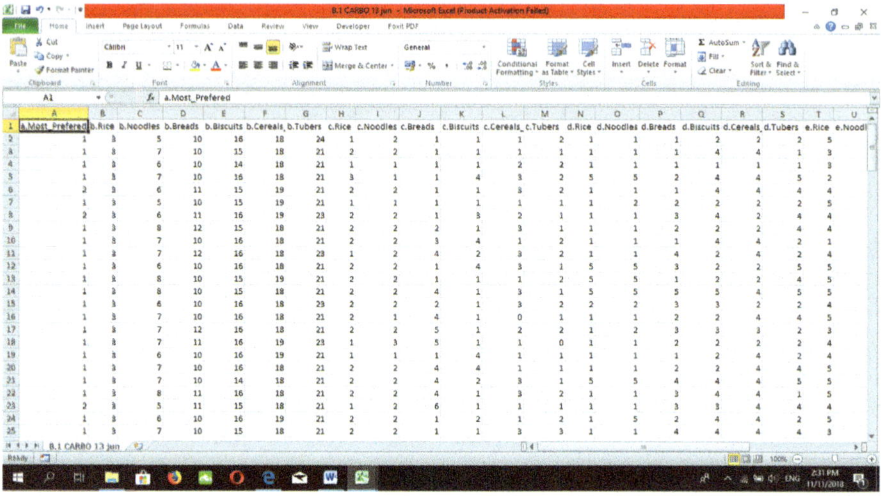

Fig. 5 Food preferences criteria in dataset

and minimum confidence 0.50 (50%) were used for the second experiment. If the confidence in rules is high, that rule will be used for future decision making.

The last phase is deployment, where the chosen rules that have high support and confidence will be embarked on the proposed prototype/dashboard for health diet planning. Finally, the evaluation of the functionality of the proposed prototype/ dashboard was conducted to determine whether the users were satisfied or not with the proposed prototype/dashboard.

4 Result and Discussion

In this study, the experiments were conducted using selected food preferences feedback collected from a group of teenagers in a selected higher education institution. Several attributes are identified for future business and data understanding phase analysis. The association rules generated from the datasets have been categorized based on BMI metrics, which are underweight, normal, overweight, and obese. Tables 2 and 3 show the analysis of selected rules based on BMI category from Experiment 1 and Experiment 2.

In the first experiment, most rules for each category have a support value greater than minimum support values. This indicates that each rule's itemset is significant enough to produce a strong rule. On the other hand, most of the rules generated in the second experiment have support values lower than the minimum support value for each category. The ARM rules for deployment are selected based on the higher confidence values for each rule. Figure 7 shows the sample of association rules for food preferences analysis.

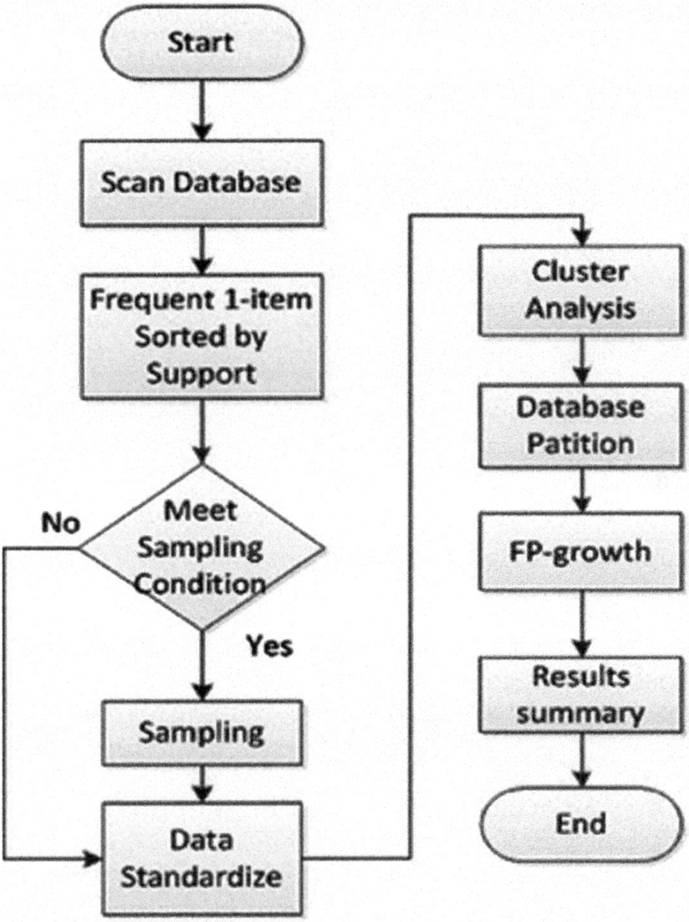

Fig. 6 ARM modeling using FP-Growth method

Table 2 Experiment 1: rules for minimum support 0.01 and minimum confidence 0.60

BMI category	Sample of rule	Support	Confidence
Underweight	[Apple] → [White rice]	0.03	0.67
Normal	[Orange] → [White rice]	0.01	0.67
Overweight	[Chicken] → [White rice]	0.05	0.64
Obese	[White rice, beef] → [Burgers]	0.33	0.67

Table 3 Experiment 2: rules for minimum support 0.05 and minimum confidence 0.50

BMI Category	Sample of Rule	Support	Confidence
Underweight	[Chicken] → [White rice]	0.03	0.57
Normal	[Chicken, apple] → [Granulated sugar]	0.02	0.56
Overweight	[Apple] → [White rice]	0.03	0.56
Obese	[White rice] → [Beef]	0.05	0.60

No.	Premises	Conclusion	Support	Confidence
11	Chicken, Apple	Granulated sugar	0.015	0.562
12	White Rice, Fried Chicken	Chicken	0.020	0.571
13	White Rice, Chicken, Apple	Granulated sugar	0.014	0.571
14	White Rice, Granulated sugar	Chicken	0.026	0.577
15	Apple	Chicken	0.027	0.593
16	Granulated sugar, Apple	White Rice, Chicken	0.014	0.615
17	Pizza	White Rice	0.020	0.632
18	White Rice, Burgers	Chicken	0.020	0.632
19	Prawn	Watermelon	0.012	0.636
20	White Rice, Apple	Chicken	0.024	0.667
21	Butter	Chicken	0.015	0.692
22	Granulated sugar, Apple	Chicken	0.015	0.692
23	White Rice, Butter	Chicken	0.012	0.700
24	Others	White Rice	0.014	0.727
25	Gula melaka	White Rice	0.014	0.727

Fig. 7 Sample of association rule for food preferences

In the deployment phase, the proposed dashboard consists of user BMI information, health diet planning, and the description of food preferences analysis. Several menus contain information on the food group and several facts that help users to decide on their food consumption. Figure 8 represents the primary dashboard interface, where users are prompted to provide their weight and height to access information about their food preferences and healthy diet planning. Before accessing the diet planning menu, users are required to input their weight and height to determine their BMI category.

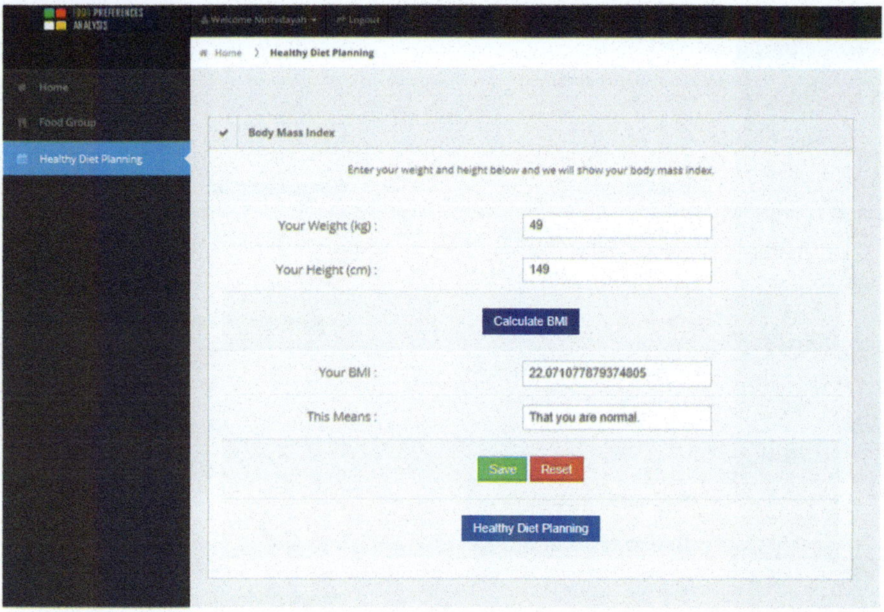

Fig. 8 BMI category information

Additionally, users can view their healthy diet plan by selecting the 'Healthy Diet Planning' button. Then, it shows the type of food usually consumed for that category of BMI. In the recommendation menu in Fig. 9, users can view suitable healthy food options based on their BMI. For instance, if a user has a normal BMI, the menu will display information related to a normal BMI and vice versa.

Users can also access information on appropriate food consumption for their healthy diet planning. The recommended food choices are determined by considering the food pyramid servings and the association rules generated on this page. Additionally, users have the option to organize their food intake for a week. This is an easier way for them to eat healthy meals in their schedule. Besides, the user can also see the analysis result based on the association rules. For example, the description of each food the user consumes is shown in Fig. 10.

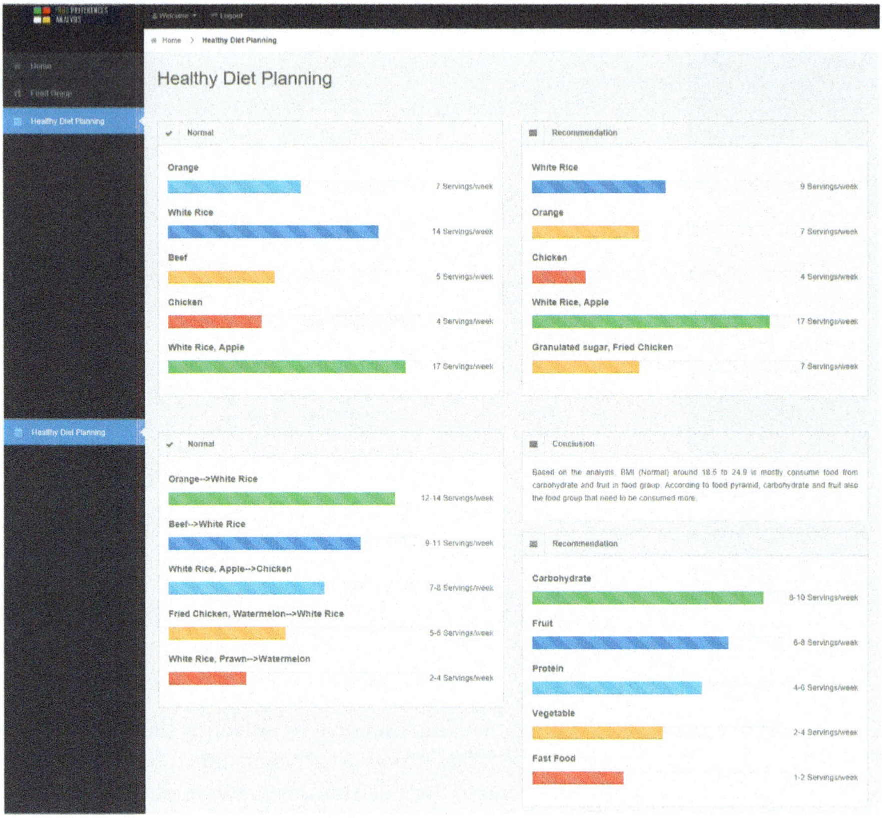

Fig. 9 Healthy diet planning information

It is helpful for them to be aware of their diet planning toward healthy eating practices. In the evaluation phase, most users are satisfied with the result of the food preferences analysis, as shown in the dashboard. Besides that, it can also display food recommendations based on each BMI category. It is helpful for users to keep track of their healthy diet planning in handling several issues, such as obesity and health problems among teenagers.

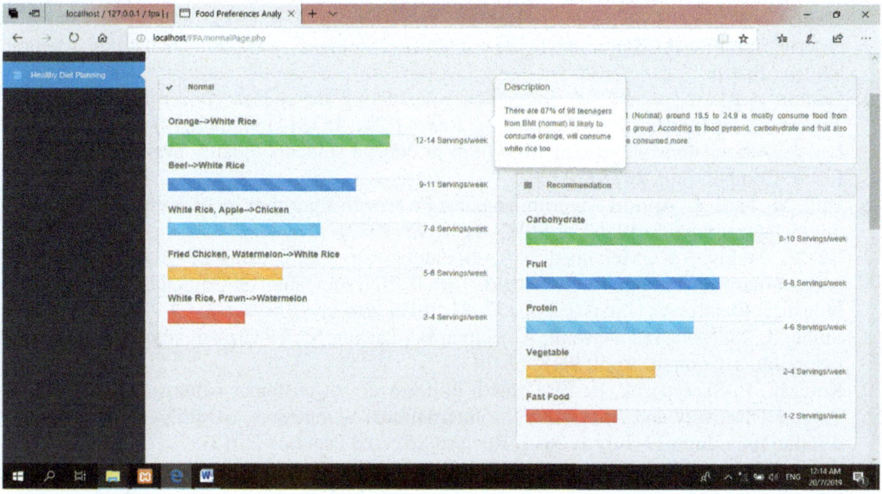

Fig. 10 Description of analysis

5 Conclusion

This study demonstrates how using the FP-Growth algorithm and ARM techniques can help teenagers make healthier food choices. This proposes a system that analyzes teenagers' food preferences and BMI types. The results show that the FP-Growth algorithm effectively provides valuable insights for food preferences analysis. These findings can guide teenagers in planning healthier diets. Exploring other ARM algorithms to strengthen these results and expand the understanding of this field is also recommended. Overall, the study paves the way for using ARM algorithms in food preferences analysis with different datasets.

Acknowledgements The registration fee is funded by Pembiayaan Yuran Prosiding Berindeks (PYPB), Tabung Dana Kecemerlangan Pendidikan (DKP), Universiti Teknologi MARA (UiTM), Malaysia.

References

1. Mittal, K., Aggarwal, G., Mahajan, P.: A comparative study of association rule mining techniques and predictive mining approaches for association classification. Int. J. Adv. Res. Comput. Sci. **8**(9) (2017)
2. Shahin, M., Arakkal Peious, S., Sharma, R., Kaushik, M., Ben Yahia, S., Shah, S.A., Draheim, D.: Big data analytics in association rule mining: a systematic literature review. In: 2021, the 3rd International Conference on Big Data Engineering and Technology (BDET) (2021)

3. Chen, C.-W., Tsai, C.-F., Tsai, Y.-H., Wu, Y.-C., Chang, F.-R.: Association rule mining for ordering traditional Chinese medicine container placement: an experimental study. Medicine **99**(18) (2020)
4. Guntur, A.P.I.: A review on association rule mining techniques concerning their privacy preserving capabilities. Int. J. Appl. Eng. Res. **12**(24), 15484–15488 (2017)
5. Lei, J.: Association rule mining algorithm in college students' quality evaluation system. J. Electr. Comput. Eng. **2022** (2022)
6. Patil, M., Patil, T.: Apriori Algorithm against Fp growth algorithm: a comparative study of data mining algorithms. Available at SSRN 4113695 (2022)
7. Isa, N., Neddy, S.K., Mohamed, N.: Association rule mining using FP-growth algorithm to prevent maverick buying. In: 2021 IEEE 11th IEEE Symposium on Computer Applications & Industrial Electronics (ISCAIE), pp. 77–81. IEEE (2021)
8. Sidhu, S., Meena, U.K., Nawani, A., Gupta, H., Thakur, N.: FP Growth algorithm implementation. Int. J. Comput. Appl. **93**(8) (2014)
9. Korczak, J., Skrzypczak, P.: FP-Growth in discovery of customer patterns. In: Data-Driven Process Discovery and Analysis: First International Symposium, SIMPDA 2011, Campione d'Italia, Italy, June 29–July 1, 2011, Revised Selected Papers 1 (2012)
10. Zhang, W., Liao, H., & Zhao, N. (2008). Research on the FP growth algorithm about association rule mining. 2008 international seminar on business and information management.
11. Chin, Y.S., Appukutty, M., Kagawa, M., Gan, W.Y., Wong, J.E., Poh, B.K., Mohd Shariff, Z., Mohd Taib, M.N.: Comparison of factors associated with disordered eating between male and female Malaysian university students. Nutrients **12**(2), 318 (2020)
12. Pike, K.M., Dunne, P.E.: The rise of eating disorders in Asia: a review. J. Eat. Disord. **3**(1), 1–14 (2015)
13. Jaiswal, V.: A new approach for recommending healthy diet using predictive data mining algorithm. Int. J. Res. Anal. Rev. **6**(2), 58–65 (2019)
14. Bahrin, U.F., Jantan, H., Sofian, M.A.S.M., Ismail, I.S., Samsudin, S.H.A.: Classifying body type based on eating habits and physical condition using decision tree technique. In: 2022 International Visualization, Informatics and Technology Conference (IVIT), pp. 95–102. IEEE (2022)
15. Vazquez, R.M., Bonilla, E., Sanchez, E., Atriano, O., Berruecos, C.: Application of data mining techniques to find relationships between the dishes offered by a restaurant for the elaboration of combos based on the preferences of the diners. Appl. Comput. Sci. **15**(2), 73–88 (2019)
16. Vijayalakshmi, V., Anuratha: A comparative study on mining the healthy food preferences of women clusters. Int. J. Sci. Eng. Res. **8**(7), 126–131 (2017)
17. Alfawaz, H.A.: The relationship between fast food consumption and BMI among university female students. Pak. J. Nutr. **11**(5), 406 (2012)
18. De Decker, A., Sioen, I., Verbeken, S., Braet, C., Michels, N., De Henauw, S.: Associations of reward sensitivity with food consumption, activity pattern, and BMI in children. Appetite **100**, 189–196 (2016)
19. Benazeera, U.J.: Association between eating habits and body mass index (BMI) of adolescents. Int. J. Med. Sci. Public Health **3**(8), 940–943 (2014)
20. Reitzel, L.R., Regan, S.D., Nguyen, N., Cromley, E.K., Strong, L.L., Wetter, D.W., McNeill, L.H.: Density and proximity of fast food restaurants and body mass index among African Americans. Am. J. Public Health **104**(1), 110–116 (2014)
21. Dharkar, S., Rajavat, A.: Performance analysis of healthy diet recommendation System using Web data mining. Int. J. Sci. Eng. Res. **3**(5), 1–4 (2012)

Chapter 20
Studies on Potential Conflicts of Network Densification in 6G

Viswanathan Ramasamy, Saikat Maity, N. Mohana Priya, Vithya Ganesan, Sri Anima Padmini, Subrata Chowdhury, and Saurabh Adhikari

1 Introduction

One innovative strategy that has arisen to address the needs of 5G ultra-dense networks is for dense massive installations of tiny base stations (BSs). It is envisioned in to boost data throughput and coverage. Dense, massive installations of tiny base stations (BSs) are envisioned in UDNs to boost data throughput and coverage [1]. However, there has been a lot of research into how higher base station (BS) densities affect the efficiency of wireless networks [2, 3]. Compared to the 5G system, the 6G version of ICT is expected to be able to accommodate the capacity and latency. Sixth generation (6G) wireless networks already face challenges from factors including interference and coverage gaps caused by the proliferation of base stations (BSs).

V. Ramasamy · V. Ganesan (✉) · S. A. Padmini
Koneru Lakshmaiah Education Foundation, Vadeeswaram, AP, India
e-mail: vithyamtech@gmail.com

V. Ramasamy
e-mail: rvnathan06@gmail.com

S. A. Padmini
e-mail: sapadmini@kluniversity.in

S. Maity
Department of CSE, Sister Nivedita University, Kolkata, India

N. Mohana Priya
Department of AI & DS, Madanapalle Institute of Technology & Science, Madanapalle, AP, India
e-mail: mohanapriyan@mits.ac.in

S. Chowdhury
Sreenivasa Institute of Technology and Management Studies, Murukambattu, AP, India

S. Adhikari
School of Engineering, Swami Vivekananda University, Kolkata, India
e-mail: saurabhadhikari@svu.ac.in

J. M. R. S. Tavares et al. (eds.), *Proceedings of Second International Conference on Intelligent System*, Algorithms for Intelligent Systems,
https://doi.org/10.1007/978-981-99-8976-8_21

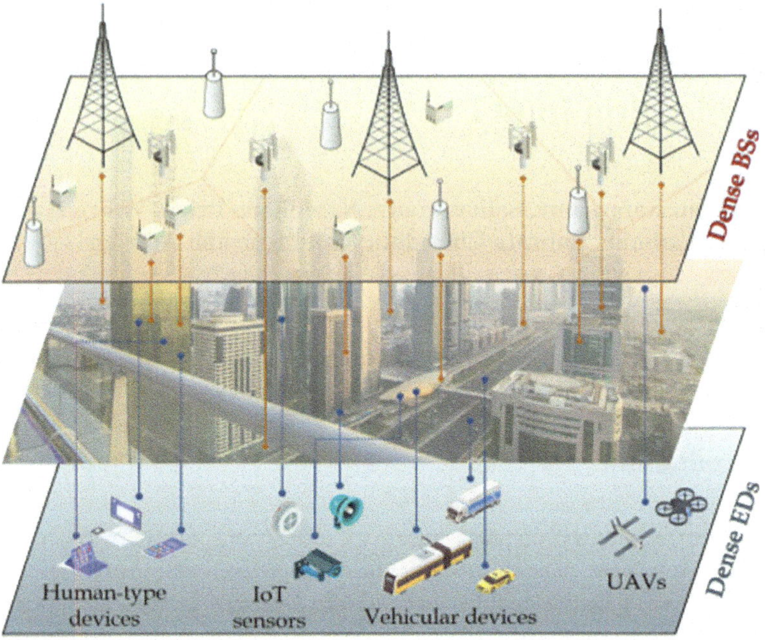

Fig. 1 Densification of networks using several base stations and end nodes

Further, it is expected that the dense network environment will make use of numerous layers of BSs to facilitate various coverage areas, functionality, and connectivity with preexisting cellular networks.

The benefits of 6G wireless network densification, as well as its accompanying challenges and technology, are the primary focus of this essay [4]. Densification of the network is described in terms of the densification of the base station and the edge node [5]. There is now the option for regular or ad hoc traffic in MTD communications. When the distance between EDs is minimised, spatial correlations in interference and main link channels become substantial [6]. Finally, the advantages of a more dense 6G wireless network are explored and shown in Fig. 1. Densification of networks: key features and difficulties are shown in Fig. 2.

2 Literature Survey

2.1 Density in Network

The forthcoming 6G wireless network will have distinctive characteristics in comparison to the current network, owing to the increased density of base station (BS) and end device (ED) nodes. Key features of the 6G wireless network, as determined by

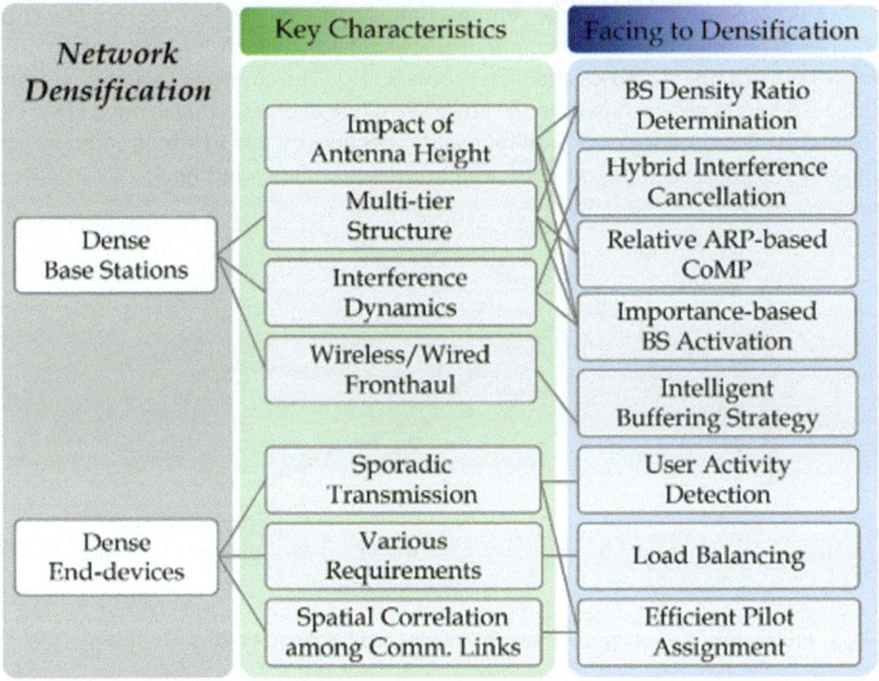

Fig. 2 Densification of networks: key features and difficulties

the density of base station (BS) and end device (ED) deployments, are discussed below.

2.2 *Important Network Properties Derived from BS Density*

Antenna height has a significant effect

The benefits of 6G wireless network densification, as well as its accompanying challenges and technology, are the primary focus of this essay. The densification of the network is described in terms of the densification of the base station and the edge node. There is now the option for regular or ad hoc traffic in MTD communications. When the distance between EDs is minimised, spatial correlations in interference and main link channels become substantial. Finally, the advantages of a more dense 6G wireless network are explored.

$$P_{\mathrm{CON}} = P\left[\frac{\mathrm{SS}}{1+\sigma^2} > \theta_{\mathrm{SINR}}\right]. \tag{1}$$

In Fig. 3, SS and I denote the primary link signal strength and interference from other transmitting BSs. As the height of BSs increases, the coverage probability in a dense network first increases and then decreases [7]. The coverage probability grows as the BS height grows; however, the coverage likely decreases due to huge I [8]. The optimal BS height decreases with increasing BS density due to strong interference, hence lowering BSs improves network performance shown in Fig. 4.

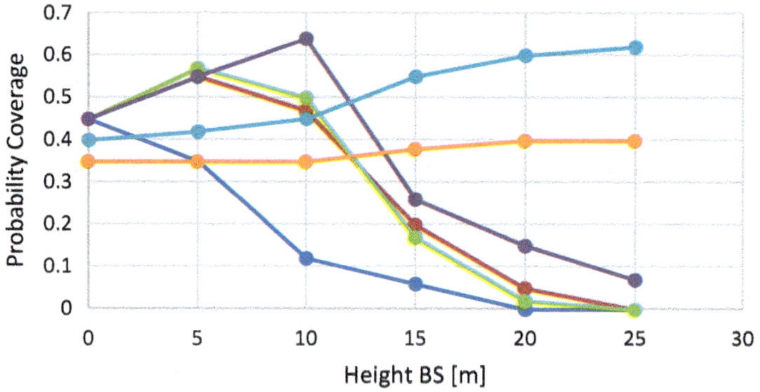

Fig. 3 Probability of coverage at various BS heights and BS densities λT

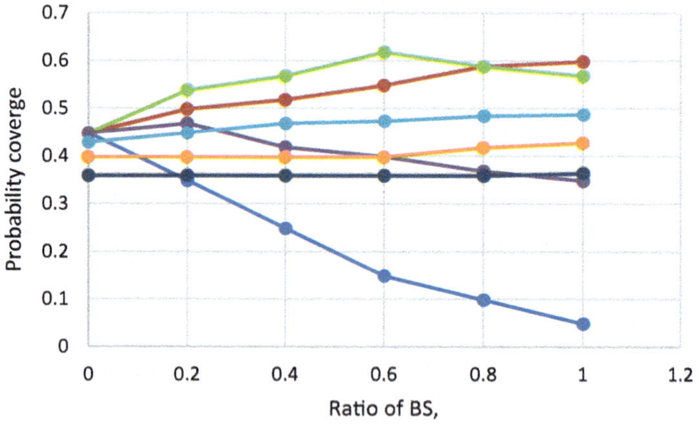

Fig. 4 Probability of coverage based on the ratio of first tier to total base station densities for various cell densities

3 Methodology

Layered network architecture

In this research, it is analysed the two-tier network's coverage probability using 35 dBm and 24 dBm base station transmission strengths and 6 m and 3 m antenna heights, respectively [9]. Nakagami-m fading was utilised as the channel model, and a diagram of the two-tier network's coverage probability is shown in Fig. 4. The optimal BS density ratio in the HetNet setting was 1, suggesting the use of only primary BSs [10]. Because of the normally enormous distance between the ED and its associated BS in a HetNet scenario, the effect of modifying the BS density ratio is limited [11].

The impact of the BS density ratio changes more gradually as network density increases. Using second tier BSs is best in highly dense networks, while using first-tier BSs is optimal in sparse networks [12]. When using the optimal ratio of two tiers, however, a multi-tier network structure outperforms a single-tier network structure. This is due to the fact that the best deployment of two-tiered BSs takes full advantage of the benefits offered by both the first-tier BSs, which boost main link performance, and the second-tier BSs, which mitigate interference. As a result, the ideal combination of two types of BSs must be used, and this is determined by the overall BS density.

Dynamics of Interference

The following is a summary of the interference characteristics brought on by the increased density of BSs:

Enhanced interfering strength: As the number of BSs in a network increases, the connection distance between an interfering BS and an ED decreases, increasing the possibility of a LoS interfering link. The tiers of competing BSs are the cause of interference in a multi-tiered network.

Main link channels and spatially correlated interference: As the BSs get closer together, the serving BS and its interfering BSs will have spatial channels that are more similar to one another. This will cause a high degree of spatial correlation between the interfering link channels and the main link channels.

Interfering node sets tend to fluctuate often: Due to the greater variety of BSs and unpredictable traffic patterns, the network is more dynamic, allowing for more frequent handovers of interfering BSs.

Fronthaul can coexist with wireless and wired connections

Increasing the BS density has the potential to enhance the BS-ED link quality and throughput. However, as network traffic increases, high-capacity low-latency backhaul is required to connect the many BSs to the backbone network. To connect to the backbone network, BSs (especially smaller BSs) should set up fronthaul links with macro-BSs. Densification of BSs, however, makes it unlikely that wired fronthaul will be sufficient for managing all small BSs. Due to channel fluctuations and fading, wireless fronthaul capacities are lower than cable fronthaul. The issue of data corruption can be resolved and the consequences of the limited fronthaul capacity reduced by placing buffers at the small BSs. Small BSs can temporarily store data or information in the buffer before discarding it.

4 Implementation

Important features of networks uncovered via ED density

Occasionally Transmitted Devices

Unlike typical HTDs, MTDs only see irregular bursts of traffic. In other words, while ED densification has led to a large total number of EDs, including HTDs and MTDs, only a small percentage of the MTDs are active at any given time. Because of this trait, it is challenging for BSs to identify the EDs doing the actual transmission. Receiving and correctly decoding the messages provided by the active ED set requires the receiver to employ an active ED detection approach and a suitable decoding technique in order to achieve optimal network performance.

Concurrent operation of multiple traffic modes

Densifying EDs leads to the emergence of both traditional HTDs and newer, more complex MTDs. It's possible that each of these EDs has slightly different standards. Some MTDs, like vehicles, need constant, high-speed connections, while others, like simple Internet of Things (IoT) sensors, can get by with best-effort connections and relatively recent data with a low age of information (AoI). To maximise network speed, optimal resource allocation and scheduling across EDs must take into account the presence of many traffic categories with varying service requirements.

Setting in which EDs are spatially connected to one another

However, when the number of EDs increases in a given area, the possibility of any given ED associating with a BS that also serves other nearby EDs likewise rises. Using techniques like channel estimation substitution and inference, which take use of the spatial correlation between neighbouring EDs, network performance can be further improved.

The Challenge of Increasing Network Density

Here, we highlight some of the major obstacles inherent in the process of BS and ED densification.

In the face of increasing BS: Issues like Base Station (BS) activation control, interference management, and the CoMP Design can arise as BS densities increase. These problems have been solved by HetNets, but the densification of BSs introduces new ones.

CoMP gearbox based on the relative average of the received power

Because of the CoMP's potential to boost per-user throughput, it has been explored for inclusion in 3GPP LTE-Advanced (Release-11) and 6G wireless technologies. In a multi-tier dense network, however, contemporary CoMP systems have serious limitations. The average received power (ARP) is used in traditional CoMP algorithms to select a predetermined number (N) of collaborating BSs. The authors recommend a relative ARP-based architecture to choose the CoMP set for downlink transmission in a dense network, where the $(N + 1)$th BS can still cause substantial interference due to BS densification. The other BSs will work with it if and only if the BS with the highest ARP is part of the CoMP set:

$$\frac{\mathrm{SS}_i}{\mathrm{SS}_{\mathrm{strong}}} \geq \eta. \tag{2}$$

This graph is the same as Fig. 4 in terms of the simulation settings used. The relative ARP-based CoMP that we propose outperforms the conventional CoMP method by a significant margin. To make the suggested system more manageable, we normalise the CoMP set so that it consists on average of two BSs. In a multi-tier dense network environment, the proposed method is demonstrated to outperform the standard CoMP strategy.

Managing interference in a hybrid system

By combining interference cancellation with interference coordination, hybrid interference management weakens networks by neutralising potentially disruptive nodes. It relies on precise channel knowledge and uncorrelated channels among competing links, and it's been the focus of a lot of recent investigations. The beam-managed interference coordination method combines interference cancellation and coordination by way of reinforcement learning in Fig. 5.

Priority-based activation of BS

In order to account for the number of collaborative BSs, the authors of this work offer an importance-based BS activation control technique. The number of connected EDs and the number of cooperative BSs for each connected ED are both considered by this method. It is important to consider both the number of affiliated EDs and the number of collaborating BSs when determining a BS's relative importance. Then, the suggested system would selectively turn off the BSs. In Fig. 1, the authors provide the coverage probability for different total BS densities in a two-tier network.

$$d_i = \sum_{u=1}^{n_i} \frac{1}{c_{i,u}}. \tag{3}$$

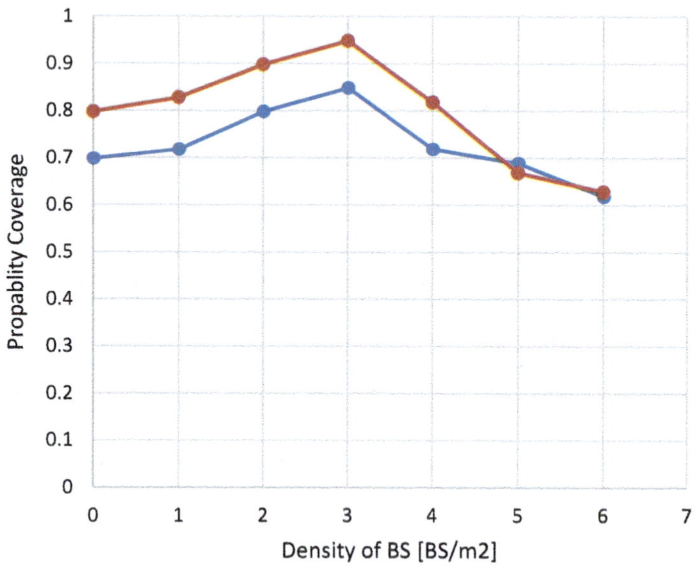

Fig. 5 Probability of coverage for various CoMP architectures as a function of the total number of BSs in an open, two-tier dense network where the SIR threshold is set to zero. Tier-1 and tier-2 base station transmission powers are 35 and 24 dBm, respectively; base station antenna heights are 6 and 3 m; and user antenna height is 1.5 m

Six, the antenna heights of a BS are 25 m and 10 m, the transmission power of a tier-1 BS is 44 dBm, the transmission power of a tier-2 BS is 33 dBm, and the antenna height of an ED is 1.5 m. The authors used mL = 10 and L = 3.5 for a line-of-sight (LoS) channel and utilised Nakagami-m fading as their channel model.

$$P_{\text{cov}}^{\text{CoMP}} = P\left[\frac{\sum_{i\in\phi} \text{SS}_i}{1+\sigma^2}\right] > \theta_{\text{SINR}}. \tag{4}$$

Facing the increasing density of EDs

One of the major difficulties in massive MIMO, involves many users and many inputs and outputs at once, is correct channel estimation. Because there may not be enough orthogonal pilot sequences pilot to split uplink channel estimation from different EDs, ED densification can result in pilot contamination. Since MTC traffic is rare, it is important to precisely identify or estimate user activity so that scarce orthogonal resources may be directed to the actual transmitting EDs. Probability of coverage in a two-tier network is shown in Fig. 6.

Associating BSs based on their fronthaul bandwidth availability

To counteract the effects of the limited capacity of the wireless fronthaul and greatly improve network performance, we suggest a BS association method depending on

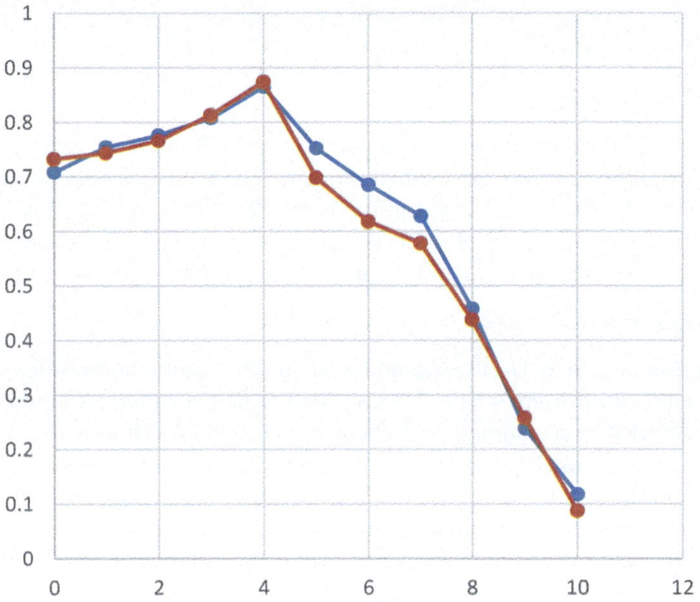

Fig. 6 Probability of coverage in a two-tier network as a function of the total number of base stations for various activation strategies for the base stations

the capacity of the available fronthaul. Common practices for grouping BSs give preference to those that are geographically close together or have the highest ARP. To solve this problem, we pick the serving BS according to its fronthaul capacity. The ED might pair up with the BS, which is expected to provide superior and reliable communication despite constrained fronthaul resources.

The Possibilities of a More Densely Connected Network

Making BSs Service-Functional

As described the network's BS density will result in the deployment of multiple BS kinds. Each base station (BS) may have many purposes to maximise the use of these densely deployed BSs, including but not limited to BSs with computation and caching capabilities and BSs geared toward aerial devices.

BSs that can perform computational tasks

Edge computing is a dense network with multiple tiers that may process data independently of the network's central node. The CPU frequency and transmission power, among other computing and communication resources, are not uniform between BS tiers. While higher-tier BSs have more computing and transmission power, lower-tier BSs may provide better communication performance. Computing performance and the ED's quality of services and experiences would both benefit from an intelligent offloading method that makes use of the multi-tier dense network structure.

Cached-data-capable BSs

By storing frequently requested data on BSs and allowing EDs to get it from neighbouring BSs, the caching method dramatically decreases latency. By caching frequently used service algorithms at the network's edge, we can improve our hit rate. The performance of a multi-tier dense network can be improved with the help of data and service cache techniques that are adapted to the specifics of each BS.

BSs geared for aerial devices

Airborne users (such as UAVs) should be taken into account alongside more traditional ground-based users when planning for ED densification. A new BS service scheme is needed to accommodate users on the ground as well as in the air. When it comes to base stations (BSs), there are two main types: exclusive-service BSs (which only serve ground users) and inclusive-service BSs (which also support air users). Better network performance can be attained through BS densification, but only if we methodically plan the deployment of BSs using each scheme.

Pilot allocation optimization by spatial correlation of communication lines

The channel estimate at the BS was traditionally carried out with the aid of an orthogonal pilot signal broadcast by the EDs. Densification of ED shows that there are more ED linked to the BS than there are orthogonal pilot sequences. A unique channel estimate scheme derived from network densification can be used to address the pilot shortage. Using the EDs' inherent spatial correlation, conventional orthogonal pilot signals will be used for channel estimation of those EDs connected to the BS. Since EDs are spatially correlated, the channel information of nearby EDs can be approximated based on the estimation result. Using the spatial correlation function, we can perform channel estimate of widely dispersed EDs without needing as many pilots, thus solving the pilot shortage issue.

Planning a front-loading limit reduction buffer technique

In a multi-tier dense network, the capacity of the wired backhaul may be less than that of the access connection if the wireless fronthaul is of low quality. One proposed solution involves using buffers to prevent data from being lost between the tiny BSs. Modern tools, such as a reinforcement learning algorithm, can efficiently handle this buffer technique.

5 Conclusion

The benefits and drawbacks of network densification are discussed here. In this paper, we focus on recent developments in BS densification and ED densification as they pertain to network densification. We then describe some of the major obstacles inherent in BS densification, such as interference management, BS activation control, and CoMP design. This research offers helpful design insights for the future multi-tier dense network by taking into account the densification of both BS and ED nodes. It draws attention to the pilot contamination problem, user activity detection, and base station association, in addition to functionalized base stations, spatial correlation exploitation, and an intellectualised buffer technique.

References

1. Andrews, J.G., Buzzi, S., Choi, W., Hanly, S.V., Lozano, A., Soong, A.C., Zhang, J.C.: What will 5G be? IEEE J. Sel. Areas Commun. **32**(6), 1065–1082 (2014)
2. Ge, X., Tu, S., Mao, G., Wang, C.X.: 5G ultra-dense cellular networks. IEEE Trans. Wirel. Commun. **23**(1), 72–79 (2016)
3. Zhang, H., Lee, J., Quek, T.Q.S., I, C.-L.: Ultra-Dense Networks: Principles and Applications. Cambridge University Press (2020)
4. Bhushan, N., et al.: Network densification: the dominant theme for wireless evolution into 5G. IEEE Commun. Mag. **52**(2), 82–89 (2014)
5. Romanous, B., Bitar, N., Imran, A., Refai, H.: Network densification: challenges and opportunities in enabling 5G. In: 2015 IEEE 20th International Workshop on Computer Aided Modelling and Design of Communication Links and Networks, CAMAD, Guildfood, UK (2015)
6. Gupta, A.K., Sabu, N.V., Dhillon, H.S.: Fundamentals of network densification. In: Lin, X., Lee, N. (eds.) 5G and Beyond, pp. 129–163. Springer (2021)
7. Jiang, W., Han, B., Habibi, M.A., Schotten, H.D.: The road towards 6G: a comprehensive survey. IEEE Open J. Commun. Soc. **2**, 334–366 (2021)
8. C.V.N. Index: Cisco visual networking index: Global mobile data traffic forecast update, pp. 2017–2022 (2019)
9. Yang, P., Xiao, Y., Xiao, M., Li, S.: 6G wireless communications: vision and potential techniques. IEEE Network **33**(4), 70–75 (2019)
10. Sharma, S.K., Wang, X.: Toward massive machine type communications in ultra-dense cellular IoT networks: current issues and machine learning-assisted solutions. IEEE Commun. Surveys Tuts. **21**(4) (2019)
11. Wu, Y., Gao, X., Zhou, S., Yang, W., Polyanskiy, Y., Caire, G.: Massive access for future wireless communication systems. IEEE Wireless Commun. **27**(4), 148–156 (2020)
12. Recommendation ITU-R P.1410-5, propagation data and prediction methods for the design of terrestial broadband millimetric radio access systems, P series, radiowave propagation (2012)

Chapter 21
Machine Learning Framework for Detecting Fake News Over Social Media Platforms

Vidushi, Zainab Zubair, Muskan Agrawal, Shivansh Srivastava, Mohd Anas Wajid, and Bharat Bhushan

1 Introduction

The world is constantly changing. There are several benefits and drawbacks to living in the digital age. In our digital age, various issues occur. One of them is the fake news. Anyone can easily distribute false information. Spreading false information can harm a person's or an organization's reputation [1]. The propaganda may have as its objective a political party or other organization. There are several online sites where incorrect information can be disseminated. These include, among others, Instagram, Facebook, and Twitter. As we spend more and more time online speaking on these platforms, more and more people are choosing to search for and consume news on social media platforms rather than from traditional news agencies. On social media, news is frequently cheaper and more timely than it is in conventional media, such newspapers as well as television and it's easier to share, debate, and discuss news with friends or other users on social media. Fake news is untrue information that is independently verifiable. This helps a country maintain a false statistic or inflates the cost of a certain service, which may lead to upheaval in some countries, as the

Vidushi
CHRIST (Deemed to be University), Delhi-NCR, Ghaziabad, India
e-mail: vidushi@christuniversity.in

Z. Zubair · M. Agrawal · S. Srivastava
KIET Group of Institutions, Delhi-NCR, Ghaziabad, India

M. A. Wajid (✉)
School of Computer Application & Technology, Galgotias University, Greater Noida, India
e-mail: anaswajid.bbk@gmail.com

B. Bhushan
Department of Computer Science and Engineering (CSE), School of Engineering and Technology, Sharda University, Greater Noida, India

J. M. R. S. Tavares et al. (eds.), *Proceedings of Second International Conference on Intelligent System*, Algorithms for Intelligent Systems,
https://doi.org/10.1007/978-981-99-8976-8_22

Arab Spring [2]. The Crosscheck project as well as the House of Commons are two organizations that deal with difficulties like establishing the authorship of works. However, their reach is severely constrained since they rely on manual identification by individuals, this, in a world where millions of things are withdrawn or released every minute, is neither practicable nor reliable. The development of a system would offer a precise automated index scoring, or rating, for the standing of different media and the news's context.

Ongoing study and adaptation of learning algorithms are necessary to address the dynamic nature of fake news due to several reasons:

1. Evolving Tactics: Fake news creators constantly adapt their tactics to bypass existing detection algorithms. As machine learning algorithms are trained on historical data, they might become less effective in identifying newer, more sophisticated forms of misinformation. Continuous study allows researchers to identify these changing tactics and update the algorithms accordingly.
2. Data Distribution Changes: The distribution of fake news and genuine news in online platforms can change over time. What was once a rare type of fake news might become more prevalent, and vice versa. This shifts the balance of the training data, making it essential to continuously monitor and update the learning algorithms to maintain their accuracy.
3. Contextual Changes: The context in which fake news emerges and spreads can change rapidly. Political events, societal trends, and other contextual factors impact the way misinformation is crafted and disseminated. Keeping up with these contextual changes requires ongoing research and adaptation of algorithms to recognize fake news in different situations.
4. Adversarial Attacks: Fake news creators may deliberately attempt to deceive algorithms using adversarial attacks. They can tweak the content or structure of fake news to exploit vulnerabilities in the detection algorithms. Regular adaptation and improvements are necessary to mitigate these adversarial attacks and maintain the efficacy of the algorithms.
5. Platform and Algorithm Updates: Online platforms and search engines frequently update their algorithms and interfaces. These changes can inadvertently affect the performance of fake news detection systems. Continuous study ensures that the learning algorithms remain compatible with the platform changes and deliver reliable results.

There are several ethical considerations related to the use of learning algorithms to combat fake news. Some of the key ethical considerations include

1. Bias and Fairness: Learning algorithms may inadvertently perpetuate biases present in the training data, leading to unfair treatment of certain individuals or groups. For example, if the training data is biased towards specific perspectives or demographics, the algorithm might discriminate against certain sources or viewpoints, affecting the diversity of information available to users.
2. Freedom of Speech and Censorship: The use of learning algorithms to combat fake news raises concerns about potential censorship and restrictions on freedom

of speech. Striking a balance between preventing the spread of misinformation and upholding freedom of expression is a challenging ethical dilemma.

3. Privacy and Data Use: Learning algorithms often require access to user data to function effectively. Ensuring the privacy and security of user data is essential, as mishandling personal information can have serious ethical implications.
4. Misclassification and Consequences: False positives and false negatives in fake news detection can have real-world consequences. Misidentifying legitimate news as fake news or failing to detect actual misinformation can harm individuals, reputations, and businesses.
5. Algorithmic Manipulation: The use of learning algorithms can be exploited by bad actors who intentionally create content to deceive or manipulate the algorithms. This can lead to adversarial behaviour, further complicating the fight against fake news.

Using learning algorithms to detect fake news presents several challenges and limitations that researchers and developers must address to build effective and reliable systems. Some of the key challenges include

1. Lack of High-Quality Labelled Data: Building accurate fake news detection models requires large, high-quality, and diverse labelled datasets. However, obtaining such data can be challenging due to the subjective nature of labelling and the limited availability of reliable sources.
2. Dynamic and Evolving Nature of Fake News: Fake news tactics are continuously evolving to evade detection. Learning algorithms need to keep pace with these changes to remain effective. Developing algorithms that are resilient to new forms of fake news is a constant challenge.
3. Data Imbalance: The distribution of fake news and genuine news in the real world is often imbalanced, with significantly fewer instances of fake news compared to genuine news. Imbalanced data can lead to biased classifiers and impact the model's performance.
4. Adversarial Attacks: Fake news creators may actively try to manipulate learning algorithms by crafting content to deceive the system (adversarial attacks). Developing algorithms that are robust against such attacks is a complex task.
5. Interpretability and Explainability: Learning algorithms, especially deep learning models, can be highly complex and challenging to interpret. The lack of interpretability and explainability may hinder the understanding of how the algorithms make decisions and raise concerns about accountability.

We can all see the harm that fake news can do, which is why it is urgently needed for a tool that can verify specific news, whether it is fake or real, and give people a sense of authenticity based on which they can decide whether or not to take action. With so much noise from fake news and fake data, if people lose faith in information, they will no longer be able to access even the most important information, which can even occasionally be life-changing or game-changing. Our strategy is to create a model that uses long short-term memory (LSTM) and other machine learning ideas like NLP, word embedding, one hot representation, etc., to determine whether the provided news is accurate or not. The model will provide us with the outcomes for the given information [3]. To successfully stop dissemination of incorrect information, it will be essential to explore the creation of more complex algorithms, the integration of contextual data, and real-time analysis. Several key factors can significantly affect the performance of learning algorithms in detecting fake news. These factors influence the accuracy, precision, recall, and overall effectiveness of the algorithms. Some of the key factors includes

1. Quality and Diversity of Training Data: The quality and diversity of the labelled training data used to train the learning algorithms are crucial. A well-balanced dataset with a representative mix of fake and genuine news examples is essential for the algorithms to generalize well to new, unseen data.
2. Feature Selection and Representation: The choice of features and their representation plays a vital role in capturing the relevant information for fake news detection. Properly selected and well-engineered features, such as word frequencies, TF-IDF scores, sentiment analysis, and named-entity recognition, can significantly improve the algorithm's performance.
3. Algorithm Selection: Different learning algorithms have different strengths and weaknesses. The choice of the appropriate algorithm for fake news detection depends on the specific characteristics of the dataset and the problem at hand. Some algorithms might be more effective at handling certain types of fake news than others.
4. Model Complexity and Overfitting: The complexity of the learning model can impact its ability to generalize to new data. Complex models might memorize the training data (overfitting) and perform poorly on unseen data. Balancing model complexity to avoid overfitting is critical for better performance.

The rest of the paper is divided into following sections. Section 2 gives a thorough literature survey of the previous studies conducted in this direction. Section 3 presents the methodology adopted in current research. Section 4 is focused on results and analysis, while Sect. 5 concludes the paper.

2 Literature Review

Due to the increasing need of fake news identification over social media platforms, a lot of research has been carried out by researchers using various data mining techniques in collaboration of machine learning algorithms. In order to understand the importance of current research, we have divided literature in two different sections. The one is determining the news article's trustworthiness, while the other is evaluating the reliability of tweets. Let us understand each of them one by one.

- **Determining the news article's trustworthiness**: The majority of recent studies on evaluating the truthfulness of false information have analysed portions of social media sites. Investigative journalists and expert panels are used by websites like Politifact.com and Snopes.com to distinguish between reputable and unreliable information sources. In 2016, authors questioned whether human involvement is required to discern between information pieces that are authentic and those that are false and advised us to take the validity of expert judgements into account [4]. Reputable informational websites including Reddit.com, Dig.com, and Newstrust.com allegedly provide visitors the chance to rank and comment on informational articles, according to a study by Mukherjee and Weikum (2015). Since they believed that personal feelings and acknowledgment also have an impact on the dissemination of news, they also emphasized the need to joint evaluate for the credibility and dependability of clients, articles, and information reviews [5]. Due to the speed at which false information spreads through social networks and the amount of content on them, the guide evaluation of fake information isn't always scalable. Using the cognitive (message and comments), visual (images), affective (various emojis), and behavioural (sharing and tagging) cues of the news posts, researchers in 2017 used convolution neural network (CNN), SVM, and DT-rank to distinguish between true and false news postings by mainstream media pages on Facebook. In 2016, authors used a classification model to analyse news claims' credibility using a variety of characteristics, including the language of the study, subjectivity, implicative verbs, and article sources. Additionally, they tested their plan on Amazon Mechanical Turk. An end-to-end neural network model with bidirectional LSTM is presented in another research by Popat et al. (2019) to benefit from generated past and new features for the evaluation of news article veracity [6]. The authors gleaned characteristics like the language used in the articles, their position on a claim, and the reliability of the sources. To find false material on Wikipedia, Kumar et al. (2016) used ML methods like logistic regression (LR), SVM, and RF. They concentrated on elements like text length, markup ratio, and link density that affect article structure and content.

- **Evaluating the reliability of tweets**: Twitter was developed in a way to communicate information in brief bursts of text and quickly rose to prominence as a streaming medium. Researchers have developed a method for spotting fake tweets that involves using systems and deep learning techniques. Utilized a range of extracted capabilities, such as tweet length, account age, the number of followers, the number of tweets, the number of hash tags, the number of retweets, etc., in 2018 to uncover fraudulent information on Twitter [7]. Qazvinian et al. (2011) applied Bayes class based purely lexical patterns, part-of-speech, retweets, hash tags, and universal resource locators to discover tweets that supported rumours. Castillo (2011) verified their method by using threefold cross-validation and choice trees for automatic trending information tweet categorization. They took into account a variety of factors, including the number of friends, retweets, URLs, and hash tags, as well as manual labelling of data by human raters. Similar to this, Alrubaian et al. (2018) discovered tweets having malicious statistics by combining RF, NB, and choice trees [8]. In 2018, studied the problem of false news tweets as well as four related problems, such as rumour tracking, stance class, rumour detection, and rumour verification. To resolve the problems, they applied the department LSTM method. Agrawal and co. (2019) expanded a technique for identifying fake information using logistic regression and investigated how well this kind of approach performs when used to the data that was tweeted over the course of some months. With reference to the present literature, our study aims to expand a version for the automatic detection of false news stories and tweets. Our version is intended to be applicable when readers encounter false information in both lengthy series and short series writings [9]. In order to benefit from deep learning's advantages in analysing continuous representations of textual input, we employed an LSTM neural network, as did researchers in 2016, and in 2018, 2019. In this work, we have employed a variety of textual, tagging, and syntactic capabilities in addition to the procedures mentioned above, e.g. N-grams. Our method was emphasized on the text-based data found in news tweets and articles. The suggested version regularly extracts all mentioned capabilities to train the LSTM-based version to comprehend the unique contextual meaning of a phrase. (Example—the phrase Bank refers to the place in which we maintain money or a riverside) [10]. Word embedding is a feature that associates every word with exceptional meanings by developing awesome vectors. Therefore, avoid relying on the manual extraction as done in 2018 and in 2011. Additionally, our method is able to catch the spread of false information in its early phases. Additionally, the suggested method effectively handles any information-related textual material, including both lengthy articles and brief tweets [11].

3 Methodology

Fake news detection involves the use of various methodologies and techniques to identify and classify news articles or information that is intentionally misleading, false, or fabricated [12]. Here is an overview of a typical methodology used in fake news detection:

1. **Data collection**: Compile a sizable database of news articles, both legitimate and potentially false. This dataset needs to be diverse and inclusive of a range of sources, subjects, and writing elaborations.
2. **Extrapolate**: Pertinent details from news stories that can be utilized to discern between authentic and false news. These features may include linguistic features (e.g. grammar, vocabulary), content-based features (e.g. sentiment, readability), and metadata (e.g. source reputation, publishing date).
3. **Preprocessing**: Sort the gathered information by eliminating any extraneous details like HTML elements, punctuation, or stop words. Reduce word variations by using text normalization techniques like stemming or lemmatization.
4. **Training and validation data preparation**: Create training and validation sets from the dataset. The validation set aids in assessing the performance of the fake news detection model, which is trained using the training set.

5. **Model choice**: Pick the best ML or NLP model for identifying fake news. LR, SVM, decision trees, RF, and deep learning models like CNNs or recurrent neural networks (RNNs) are examples of models that are frequently employed.
6. **Feature representation**: Convert the textual data into numerical representations that can be understood by the selected model. Popular techniques include the word embedding Word2Vec and GloVe, the term frequency-inverse document frequency (TF-IDF), bag-of-words (BoW), and word frequency.
7. **Model training**: Apply the training dataset to the chosen model. The retrieved attributes and the labels (false or actual news) are analysed by the model, which learns to spot trends and connections.
8. **Evaluate Model**: Utilizing the validation dataset, evaluate the trained model. A few common evaluation measures are F1 score, recall, accuracy, and precision. This process evaluates the model's effectiveness and identifies areas for development.
9. **Fine Tuning**: Consider fine-tuning the model if its performance is subpar by changing its hyperparameters, attempting various feature representations, or testing out various models.
10. **Deployment and testing**: Once the model demonstrates good performance, use it to determine whether recently published news articles are true or fraudulent. Test and assess the model's performance on new data continuously to verify its accuracy and efficacy.

We selected two datasets to test the efficiency of the algorithm on lengthy and less textual posts. A sizable number of online news articles are included in the first dataset, and tweets regarding current events are present in the second. Using Python's Natural Language Toolkit (NLTK) package, we first performed text data preprocessing by removing errors and abnormalities from the datasets and cleaning any remaining text. Then, groups for training and testing were created using our data. Our long short-term memory (LSTM)-based model for text categorization was created using NB and Keras, an open-source deep learning Python package [13, 14]. Finally, we assessed the effectiveness of our model by well-known metrics like accuracy, precision, recall, and F1 scores. The integration of contextual data and real-time analysis can significantly enhance the detection of fake news by providing a more comprehensive and dynamic approach to identifying misinformation. Here's how these factors contribute to improving fake news detection

1. Contextual Understanding: Fake news detection systems that take contextual data into account can better understand the broader context in which news articles are presented. Contextual information includes the source of the news, the time and location of publication, the related events or topics, and the social and political environment. By considering these factors, the algorithms can better assess the credibility and reliability of the news content.
2. Identifying Misleading Content: Contextual data can help identify potentially misleading content, such as headlines or images that don't match the actual content of the article. For example, if a headline is sensationalized or misleading

but the article itself is factual, the system can differentiate between sensationalism and actual misinformation.

3. Temporal Analysis: Real-time analysis allows the algorithms to analyse news articles as they are published, capturing the most recent developments and trends in fake news dissemination. By incorporating temporal analysis, the system can identify emerging fake news stories quickly, respond to rapidly changing narratives, and stay up-to-date with the latest misinformation tactics.
4. Social Media and Network Analysis: Real-time analysis can include monitoring social media platforms and online networks to track the spread and virality of fake news. By analysing user interactions, sharing patterns, and engagement metrics, the system can identify suspicious content and potential sources of misinformation.
5. Incorporating User Behaviour: Real-time analysis can leverage user behaviour data to assess the credibility of information based on user engagement and reactions. For instance, if an article is receiving a high volume of shares and engagement, the system may flag it for further scrutiny.

The integration of contextual data and real-time analysis provides a more nuanced and agile approach to fake news detection. By incorporating dynamic information and staying up-to-date with the rapidly evolving nature of misinformation, these systems can better equip individuals and platforms to combat the spread of fake news and foster a more informed and responsible digital information environment.

Order reliance can be learned using RNNs of the LSTM variety. Finally, we assessed the performance of our model using well-known metrics like accuracy, precision, recall, and F1 scores. The LSTM is one type of RNN design. RNNs are widely used for text and speech recognition and are efficient at processing sequential data, but they struggle to comprehend long-term dependencies [15]. To get over the limitations of traditional RNNs and maximize efficiency, LSTM was developed. In order to learn long-term contextual dependencies, LSTM employs gradient-based optimization, which has the potential to outperform conventional RNNs for textual content of various lengths. To distinguish between credible news items and tweets (represented in the datasets as 0) and using multinomial NB and pipelining ideas, a well-known technique known as Naive Bayes is employed to determine whether news is accurate and true or false. It is not the only method of training these classifiers because there are several ways that concentrate on core concepts. You can use naïve Bayes to determine whether the news is true or false [16, 17]. This particular strategy is used to categorize texts. Tokens are associated with news in a Naive Bayes classifier that may be true or false, and the Bayes theorem is then used to determine the news' accuracy. Therefore, we can determine if the news is true or false by calculating the overall likelihood and getting an estimated value of the number.

$$P(A|B) = P(B|A).\frac{P(A)}{P(B)}, \tag{1}$$

Calculating the likelihood of event A if event B is true.

$P(A)$ Is for Prior Probability
$P(A|B)$ Is for Posterior Probability.

To find Probability,

$$P(A|B1) = P(A1||B1).P(A2||B1).P(A3||B1), \tag{2}$$

$$P(A|B2) = P(A1||B2).P(A||B2).P(A3||B2). \tag{3}$$

In case when probability determined as 0.

$$P(\text{Word}) = \text{Word count} + 1/(\text{total number of words} + \text{No. of unique words}).$$

Hence, one can determine the news' accuracy by applying this method. Figure 1 shows the proposed model of fake news detection.

- **Datasets**

Any classification task requires labelled datasets from reliable sources. There aren't many widely used benchmark datasets for classifying news content. On numerous websites, including GossipCop, PolitiFact, and Kaggle, we discovered that information. We gathered around 40,000 real and bogus news stories there. Our objective is to appropriately train our model to determine if a particular piece of news is accurate or false. Two separate datasets, each containing roughly 20,000 articles, are supplied with information on bogus and legitimate news. Binary labels were used to identify the papers using numbers. We awarded articles that were false news a score of 1 and articles that were factual news a score of 0 because our focus is on identifying fake news [18].

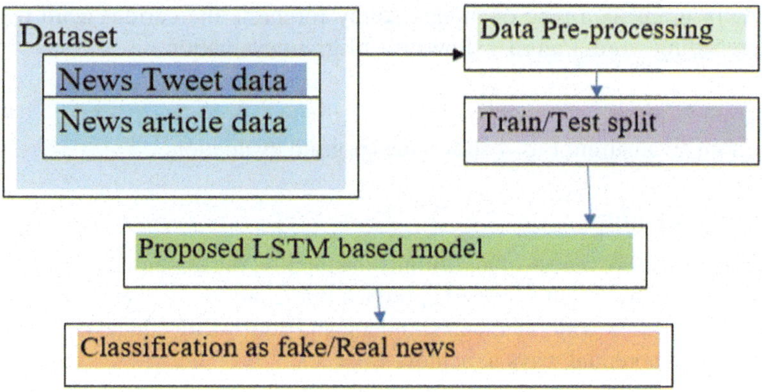

Fig. 1 Proposed model of fake news detection [18]

- **Data Preprocessing and Feature Extraction**

Before feeding them to the model learning process, preparation is important since textual data frequently contain redundant information and abnormalities. The textual material has been pre-processed using NLP methods to make it ready for future research. Stop words, punctuation such as '!', '&', and '$', emoticons such as emojis, repeated periods (such as 'or'), blank rows, white spaces, line breaks, and other variables such as tweet ids and user ids were all eliminated [19]. The articles, pronouns, and prepositions were among the stop words we eliminated because they are essential for conversational English grammar but have no semantic use in the model's learning process. First, we imported libraries like Numpy, Pandas, Seaborn, and others. After the libraries have been successfully imported, the process of loading and checking the data begins [20]. We add the sets of fake and real data and analyse them using a variety of operations, such as adding the target data to the real and fake news as 1 and 0, respectively, concatenating the data, and using the head and tail operation to get the top 5 and bottom 5 rows in a dataset. The process of presenting data in a visual format, such a map or graph, so that the human brain can more easily understand it and make inferences from it is also included in our research. Data visualization's primary objective is to make it simpler to spot patterns, trends, and outliers in huge datasets [21, 22]. After completing all of these steps, N-gram analysis is carried out, which is constructed by calculating the probabilities after tallying the occurrences of word sequences in the corpus text [23, 24]. They are pertinent for carrying out NLP operations on text data.

Here, we have used

- Uni
- Bi, and
- Tri-gram.
  - A 'uni-gram' is a type of model, relies exclusively on a word's frequency without taking preceding words into account. Fig 2 shows the uni-gram evaluation.
  - A 'bi-gram' is a type of model, solely forecasts the current word using the preceding word. Figure 3 shows the bi-gram evaluation.

- Bi-gram Evaluation
- Tri-gram Evaluation: Fig. 4 shows the tri-gram evaluation.

Proposed Algorithm

Improving the effectiveness of algorithms in stopping the dissemination of incorrect information requires a multifaceted approach that combines technological advancements, collaboration with other stakeholders, and addressing ethical considerations. Here are some potential ways to achieve this:

1. Better Data Collection: Gather high-quality labelled datasets with a diverse range of misinformation examples, including different types of fake news, conspiracy

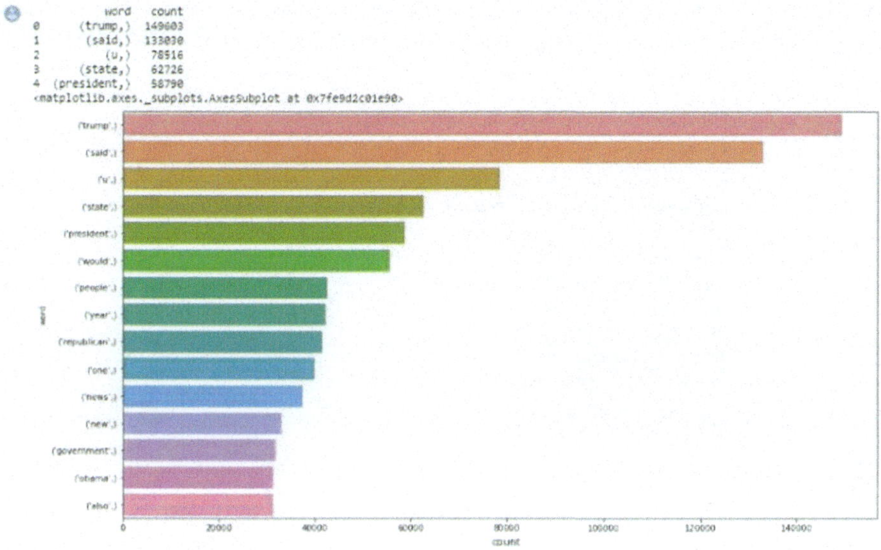

Fig. 2 Uni-gram evaluation [20]

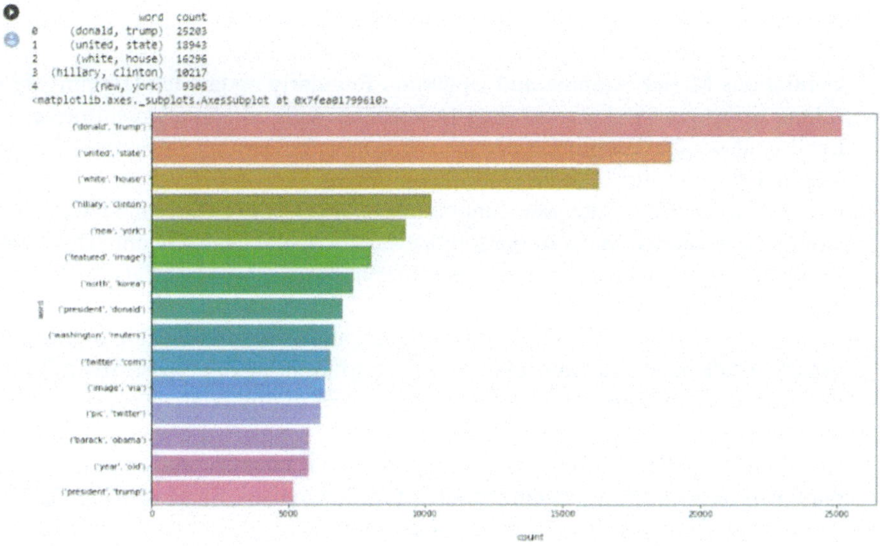

Fig. 3 Bi-gram evaluation [21]

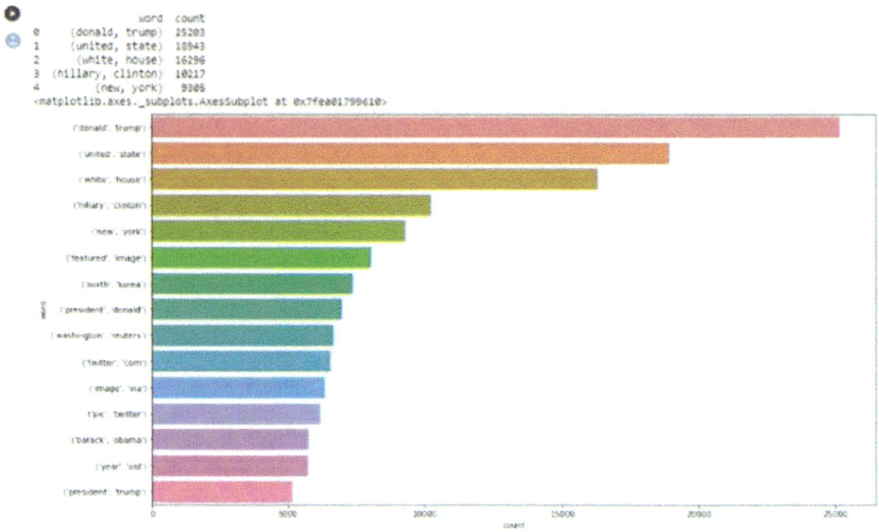

Fig. 4 Tri-gram evaluation [23]

theories, and other incorrect information. This will ensure that algorithms are trained on a comprehensive and representative set of data.

2. Continuous Model Training and Updating: Regularly update the learning algorithms using newly collected data to keep up with emerging misinformation trends and tactics. Adopting approaches like transfer learning and active learning can also enhance the efficiency of model training.
3. Contextual Analysis: Integrate context-aware analysis to understand the content within the broader context of the topic, event, or social environment. This helps in identifying nuanced forms of misinformation and avoiding false positives/negatives.
4. Multimodal Analysis: Combine text, images, and videos to develop algorithms capable of detecting misinformation across different media types. Misinformation can be spread through multiple formats, so a multimodal approach is more effective.
5. Collaborative Efforts: Foster collaborations between researchers, social media platforms, fact-checking organizations, and policymakers. Sharing insights, data, and best practices can lead to more robust solutions.
6. Human-in-the-Loop Systems: Integrate human reviewers in the decision-making process to validate the algorithm's predictions. This can help in improving the accuracy of the system while also addressing bias and false positives.
7. Explainable AI: Develop algorithms that can explain their decisions and provide transparent reasoning for their classifications. This not only builds trust but also allows researchers to understand and address any potential shortcomings.

Algorithm

Algorithm (Naïve Baye's)

Step 1- Import required libraries:

sklearn.naive_bayes: MultinomialNB
sklearn.model_selection: cross_val_score, StratifiedKFold
sklearn.feature_extraction.text: CountVectorizer

Step 2- Copy

data to variable X.

Step 3- Initialize

an empty list 'score'.
an object of StratifiedKFold with n_splits=5.

Step 4- Create

an object of CountVectorizer() and apply it to 'data.text.values' to get the word count sparse matrix.

Step 5- Print

the layout of the sparse word count matrix.
the sparse vector of first text.
the vocabulary learned by the vectorizer.
the word and its frequency at the 107802 index in the sparse matrix.

Step 6- Start a loop to data into training sets and validation sets using 'kfold.split(X, X.target)'.
Step 7- Within each iteration of the loop:

Get the training and validation sets.
Get the 'text' column from each set.
Create an object of CountVectorizer() and apply it to training and validation sets separately to get the sparse matrix of word counts.
Create an object of MultinomialNB().
Fit the model.
Perform the evaluation on the model using validation set and append the score to the 'score' list.
Print the score.

Step 8- Create

an object of CountVectorizer() and apply it to 'data.text.values' to get the sparse matrix of word counts.

Step 9- Print

the mean of all scores.

Step 10- Create an object of MultinomialNB().
Step 11- Fit

the model using the entire dataset.

Step 12- Define two functions:

The function 'predict' use the trained model to foretell the class (0 or 1) of the input.
The trained model is used by 'predict_proba' to forecast the probabilities of each class for the input.
Step 13- Take user input for a news article and store it in 'demo'.
Step 14- Use.

4 Result

The three machine learning techniques we employed for this assignment each have a different accuracy rate when deployed on the dataset [25]. The following is how each algorithm's precision is put into practice. To put the Naive Bayes theorem into practice, first extract all the necessary data, or download it if it is available, before splitting it into test and train sets [26]. The data in this instance are divided into two groups: test data and train data. After that, categories containing similar entities are created from the train dataset. After matching the test results, each group is given the category to which it belongs [27]. Next, the Naive Bayes classifier is used, and each word's likelihood is individually calculated. If the word whose probability is to be estimated is absent from the dataset of the train data, the Laplace smoothing is utilized in this situation. The validity of the data is then confirmed [28]. Figure 5 shows the input 1, and Fig. 6 shows the output 1. Similarly, Fig. 7 shows the input 1, and Fig. 8 shows the output 2.

The Naive Bayes algorithm can be an effective method to detect fake news because it is fast, simple to implement, and can work well even with a relatively very little training data [29]. The logistic regression (LR) algorithm follows similar principle to the one described earlier. The objective is to develop a function that can predict the likelihood that a given binary label (Y) belongs to one class or the other given a

```
[71]  X_count = vc.transform(X.values)
      #np.append(df_count, df[pd.get_dummies(df["subject"]).columns.values].values)
      return nb.predict_proba(X_count)
```

Fig. 5. Input 1 [28]

```
[1]
[[3.48136998e-06 9.99996519e-01]]

Prediction: True
Probability:
        Fake: 0.9999965186300191
        True: 3.4813699824948557e-06
```

Fig. 6. Output 1 [28]

```
[86] demo = """34-year old Volodymyr Karpenko, a financial and trade consultant who
     spent several years in Moscow working for the Kremlin, publicly accused Vladimir Putin in
     2021 of organizing his kidnapping in
     Moscow in 2016 and holding his as a prisoner and personal sex slave for more than three years."""
```

Fig. 7 Input 2 [28]

```
[0]
[[9.99145214e-01 8.54785578e-04]]

Prediction: True
Probability:
        Fake: 0.0008547855781557232
        True: 0.9991452144218388
```

Fig. 8 Output 2 [28]

dataset of input attributes (X) and corresponding (Y) binary labels [30]. Given its ease of use, ease of interpretation, and efficiency in a variety of real-world applications, logistic regression is an often used tool in machine learning. It is frequently used for binary classification jobs as a starting point before investigating more complicated methods. Additionally, one-vs-rest or multinomial logistic regression approaches can be used to expand logistic regression to tackle multi-class classification issues [22, 31–33].

5 Conclusion

In this study, researchers introduced an NLP-based method for automatically identifying fake information on social network. The suggested version advances previous approaches and has been demonstrated to work well for both lengthy text collections, such as informational articles, and condensed text collections, such as tweets. This suggests that information from any dataset, regardless of size, may be determined to be true or false by comparing it to the values of the prior dataset in a shorter amount of time, which then promotes users to trust a particular news source. With the help of the Naive Bayes theorem, this result can be reached. Future studies will enlarge the field of study to demonstrate that the proposed version is useful for detecting misleading information in a variety of domain names, including phone product names.

References

1. Economic and Social Research Council. Using Social Media. Available at: https://esrc.ukri.org/research/impact-toolkit/social-media/using-social-media
2. Gil, P.: Available at: https://www.lifewire.com/what-exactly-is-twitter-2483331. 22 Apr 2019
3. Tandoc, Jr E.C., et al.: Defining fake news a typology of scholarly definitions. Digit. J. 1–17 (2017)
4. Radianti, J., et al.: An overview of public concerns during the recovery period after a major earthquake: Nepal Twitter analysis. In: HICSS '16 Proceedings of the 2016 49th Hawaii International Conference on System Sciences (HICSS), pp. 136–145. IEEE, Washington (2016)
5. Alkhodair, S.A., Ding, S.H.H., Fung, B.C.M., Liu, J.: Detecting breaking news rumors of emerging topics in social media. Inf. Process. Manag. **57**, 102018 (2020)
6. Jeonghee, Y. et al.: Sentiment analyzer: Extracting sentiments about a given topic using natural language processing techniques. In: Data Mining, 2003. ICDM 2003. Third IEEE International Conference, pp. 427–434. http://citeseerx.ist.psu.edu.200) (2003)
7. Tapaswi et al.: Treebank based deep grammar acquisition and Part-Of-Speech Tagging for Sanskrit m sentences. Software Eng. (CONSEG) Softw. Eng. (CONSEG), (pp. 1–4). IEEE (2012)
8. Ranjan et al.: Part of speech tagging and local word grouping techniques for natural language parsing in Hindi. In: Proceedings of the 1st International Conference on Natural Language Processing (ICON 2003). Semanticscholar (2003)
9. Diab, M., et al.: Automatic tagging of Arabic text: from raw text to base phrase chunks. In: Proceedings of HLT-NAACL 2004: Short Papers (pp 149–152). Boston, Massachusetts, USA: Association for Computational Linguistics (2004)
10. Rouse, M.: https://searchenterpriseai.techtarget.com/definition/machine-learning-ML May (2018)
11. Dua, S., Du, X.: Data Mining and Machine Learning in Cybersecurity. Auerbach Publications, New York, 19 Apr 2016
12. RAY, S.: https://www.analyticsvidhya.com/blog/2017/09/common-machine-learning-algorithms/2017, Sept
13. Huang, T. -Q. (n.d.) https://www.researchgate.net/figure/Pseudo-code-of-information-gain-basedrecursive-feature-elimination-procedure-with-SVM_fig2_228366941 (2018)
14. Wajid, M.S., Wajid, M.A.: The importance of indeterminate and unknown factors in nourishing crime: a case study of South Africa using neutrosophy. Neutrosophic Sets Syst. **41**(2021), 15 (2021)
15. Researchgate.net. Available at: Available at: https://www.researchgate.net/figure/Pseudocode-ofnaive-bayes-algorithm_fig2_325937073 (2018)
16. Researchgate.net. Available at: https://www.researchgate.net/figure/Pseudocode-for-KNNclassification_fig7_260397165 (2014). [16]. Rampersad, G., Althiyabi, T.: Fake news: Acceptance by demographics and culture on social media. J. Inf. Technol. Politics **17**, 1–11 (2020)
17. Mishra, A., Wajid, M.S., Dugal, U.: A comprehensive analysis of approaches for sentiment analysis using Twitter data on COVID-19 vaccines. J. Inf. Electr. Electron. Eng. (JIEEE) **2**(2), 1–10 (2021)
18. Wajid, M.S., Terashima-Marin, H., Paul Rad, P.N., Wajid, M.A.: Violence detection approach based on cloud data and Neutrosophic cognitive maps. J. Cloud Comput. **11**(1), 1–18 (2022)

19. Wajid, M.A., Zafar, A., Wajid, M.S., Terashima-Marín, H.: Neutrosophic-CNN-based image and text fusion for multimodal classification. J. Intell. Fuzzy Syst. (Preprint), 1–17
20. Sirikulviriya, N., Sinthupinyo, S.: Integration of rules from a random forest. In: International Conference on Information and Electronics Engineering, pp. 194: 198. Singapore: semanticscholar.org. (2011)
21. Kevric, J., et al.: An effective combining classifier approach using tree algorithms for network intrusion detection. Neural Comput. Appl. 1051–1058 (2017)
22. Wajid, M.A., Zafar, A., Bhushan, B., Khanday, A.M.U.D., Wajid, M.S.: Artificial Intelligence (AI) and Internet of Things (IoT): Application in Detecting and Containing the Spread of COVID-19. In: AI Models for Blockchain-Based Intelligent Networks in IoT Systems: Concepts, Methodologies, Tools, and Applications, pp. 373–392. Springer International Publishing, Cham (2023)
23. Parikh, S.B., Atrey, P.K.: Media-Rich fake news detection: a survey. In: IEEE conference on multimedia information. IEEE, Miami, FL
24. Wajid, M.A., Zafar, A.: Multimodal fusion: A review, taxonomy, open challenges, research roadmap and future directions. Neutrosophic Sets Syst. **45**(1), 8 (2021)
25. Gilda, S.: Evaluating machine learning algorithms for fake news detection. In: 15th Student Conference on Research and Development (SCOReD), pp. 110–115. IEEE (2017)
26. Jain, A., Kasbe, A.: Fake news detection. In: 2018 IEEE International Students' Conference on Electrical, Electronics and Computer Science (SCEECS). IEEE, Bhopal, India (2018)
27. Wajid, M.S., Terashima-Marin, H., Najafirad, P., Wajid, M.A.: Deep learning and knowledge graph for image/video captioning: a review of datasets, evaluation metrics, and methods. Eng. Rep. e12785 (2023)
28. Gupta, A., Kaushal, R.: Improving spam detection in Online Social Networks. In: International conference on cognitive computing and information processing (CCIP). semanticscholar.org.2015
29. Wang, W.Y., Jiang, J.: Learning to detect hoaxes with limited information. In: Proceedings of the 30th ACM International Conference on Information and Knowledge Management, pp. 673–682 (2021)
30. Zhang, Y., Zhou, X., Song, Y.: Detecting fake news in social media via matrix factorization with convolutional neural network. IEEE Trans. Knowl. Data Eng. (2023)
31. Wajid, M.A., Zafar, A.: Pestel analysis to identify key barriers to smart cities development in India. Neutrosophic Sets Syst. **42**, 39–48 (2021)
32. Zafar, A., Wajid, M.A.: A mathematical model to analyze the role of uncertain and indeterminate factors in the spread of pandemics like COVID-19 using neutrosophy: a case study of India 38. Infinite Study (2020)
33. Wajid, M.A., Zafar, A., Wajid, M.S.: A deep learning approach for image and text classification using neutrosophy. Int. J. Inf. Technol. 1–7 (2023)

Chapter 22
Analysis and Detection of Political Fake News Using Deep Learning with High-Performance Hybrid Model

Ahmed H. Alsaedi, Abdulrahman Abdulwahab Kaml Aladhamı, Almuntadher Mahmood Alwhelat, and Ahmed L. Alshamı

1 Introduction

In the digital world, fake news occupies a wide area from various social media platforms and fake magazines led by large groups of botnets by misleading the truth or presenting an incomplete truth [1]. In one of the previous elections in the United States of America in 2020, much fake news was published on social media platforms claiming that the votes belonged to dead or nonexistent people, though the indication is that there is no such news at all. Therefore, technology companies have begun to create projects and methods to detect fake news [2]. This article closely examines the present level of fake news identification to motivate future research in this field. Fake news is a difficult challenge to overcome, but some efforts are being made by the open learning community. In this mapping study, we see how the process of analyzing various data such as images and texts in the form of a topic on social media platforms, as shown in Fig. 1 [3]

Fake political topics abound on social media and other news platforms, and an end must be put to these harmful methods and their impact on society.

A. H. Alsaedi (✉) · A. A. K. Aladhamı · A. M. Alwhelat · A. L. Alshamı
Department of Computer Techniques Engineering, Alhikma University, Baghdad, Iraq
e-mail: ahmed.hasan@duc.edu.iq; ahmhasn95@gmail.com

A. A. K. Aladhamı
e-mail: Abdulrahman.kaml@duc.edu.iq; Abdaladhami@gmail.com

A. M. Alwhelat
e-mail: almuntadher.mahmood@duc.edu.iq

A. L. Alshamı
e-mail: ahmed.alshami@hiuc.edu.iq; ahmed.aau.jo@gmail.com

J. M. R. S. Tavares et al. (eds.), *Proceedings of Second International Conference on Intelligent System*, Algorithms for Intelligent Systems,
https://doi.org/10.1007/978-981-99-8976-8_23

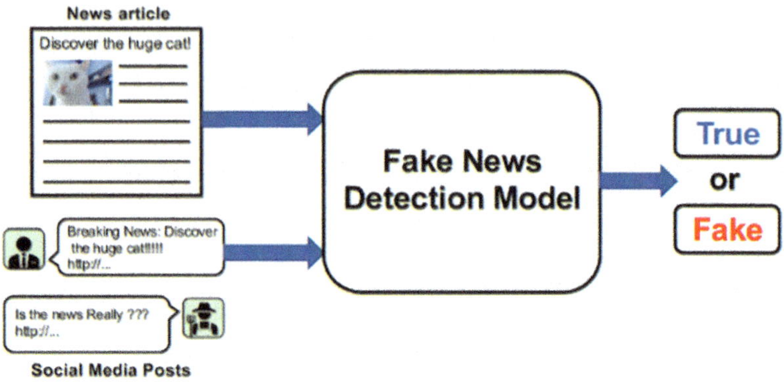

Fig. 1 Task of detecting fake news works by analyzing the data posted on social media platforms, such as images or text

2 Related Works

In recent past years, researchers have published about detecting online deception and fake news an important role in business, law enforcement, national security, and politics. Researchers used deep learning with a large data set to augment learning, and they got the best results by using word embedding to extract features. Or signals, however, they used the RNN and LSTM algorithm that was proposed to detect fake news online. And after that, they found the CNN algorithm is better than other models because of its speed and the best results and performance [4].

In 2019, different types of fake news were studied by a group of researchers, and some traditional machine learning approaches such as clustering were evaluated to ensure detection ac. State-of-the-art deep learning methods, such as hybrid CNN and RNN were used to compare the performance of deep learning methods to solve problems related to ac and lightweight [5].

In 2021, a group of researchers presented a study of the content of Indian tweets at the time of Covid-19. The research was divided into four groups according to the type of feelings such as sadness, fear, anger, and joy. By analyzing the data that causes panic and anxiety among the public on social media platforms, especially Twitter, by providing a novel in deep learning called BERT. This model gained high results with 89% [6].

A research study was presented by a group of researchers in the field of detecting fake news with different techniques and methods. They targeted their research by applying an improved grouping technique and included the best techniques for including words. They also contributed to producing consistent results in various journals [7].

In 2021, a group of researchers presented a hybrid method for analyzing data with a semi-supervised and convolutional neural network algorithm, to detect fake news in social media platforms, using a convolutional neural network and obtained the results of ac evaluations of 95.6% and recover 96.7%, their proposed method is superior to the current methods [8].

Despite the percentage of work done to detect fake news, there is a wide field for experimentation, a group of researchers worked on a more accurate efficiency in detecting fake news, and they are the first to suggest generalizing the models used in detecting fake news. In general, they used artificial neural networks in detecting fake news, although CNN and RNN were considered the most complex as part of the data analysis [9].

3 Methodology

3.1 Dataset

In this research, the dataset was provided from the open-source site Kaggle. The dataset includes more than 72,000 models between fake and real political news. The information was divided into 80% of the trained models and 20% of validation models, fake news from the real, which represents 37,106 articles, and the number of real news consisting of 35,028 by deep learning algorithms [10].

3.2 Recurrent Neural Network (RNN)

This algorithm is widely known in the field of machine learning and deep learning due to its many uses in various fields, as it consists of a set of layers, which are the input layers, the hidden layer, and the output layer, where the data flow in this algorithm occurs through the input unit to be processed by the internal structure of this algorithm. The algorithm runs many of them inside the hidden layer by classifying the data into several categories and analyzing it as output by the output layer, which is according to the data entered many-to-one, one-to-many, and many-to-many (Fig. 2).

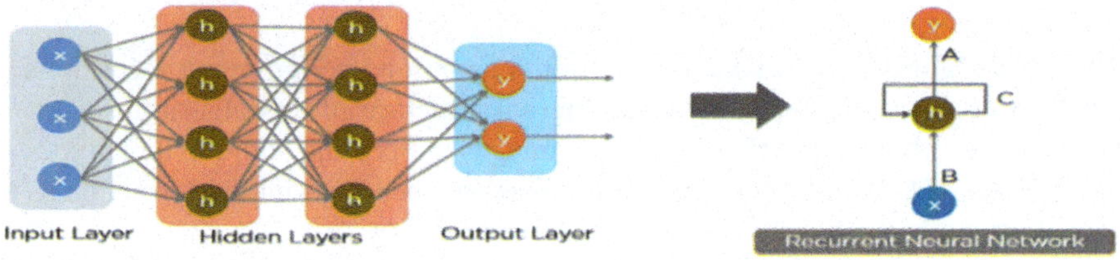

Fig. 2 RNN mechanism

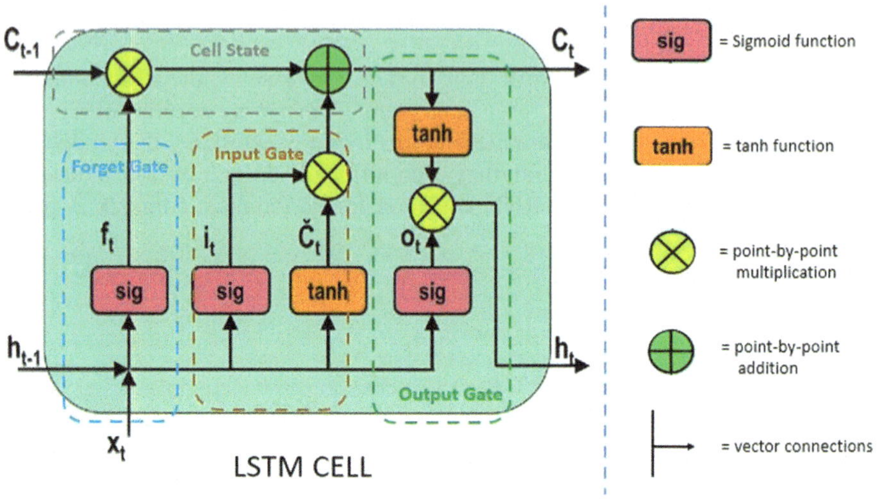

Fig. 3 LSTM mechanism, click or tap here to enter text

3.3 *Long Short-Term Memory (LSTM)*

The LSTM algorithm is a type of RNN, which is characterized by its ability to save information for a more extended period due to the advantages of its internal structure. This algorithm provides an ability to classify and analyze data and predict results because of the gates it offers in its internal structure. It consists of three gates, which are the input gate, the forget gate, and the output gate. The forget gate contains tanh, which helps to analyze the data better, save it for as long as possible, and isolate it to create an appropriate model that is output through the output gate, LSTM uses tanh to make the output positive or negative, due to that, forget gate helps to save data (Fig. 3).

3.4 *Convolutional Neural Network (CNN)*

The CNN algorithm is considered one of the common and important algorithms in the field of deep learning, due to its internal structure of this algorithm of many layers and its ability to analyze text data, images, video, and audio in three types of dimensions, 1D, 2D, and 3D. Linear functions are used in its internal structure as an activation model such as Relu or sigmoid function (Fig. 4).

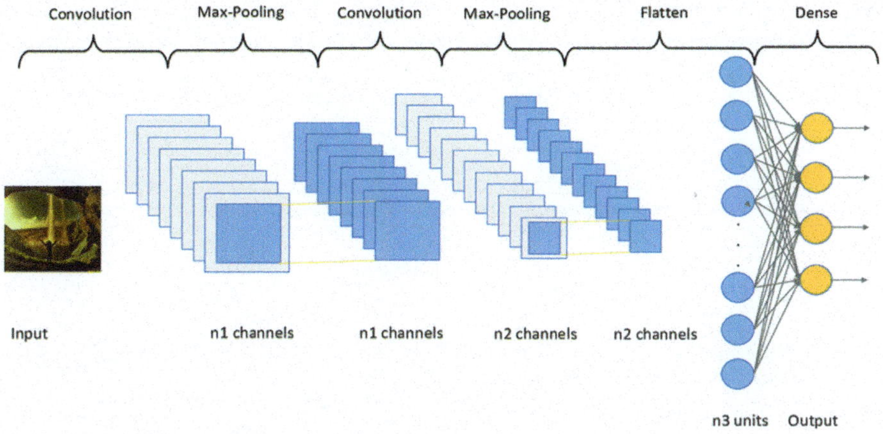

Fig. 4 CNN mechanism, click or tap here to enter text

4 Results

4.1 LSTM Model

As shown in Figs. 5 and 6, by using three main layers in LSTM which are: the embedding layer that represents each word in the vocabulary. The LSTM layer has 100 LSTM units and uses the LSTM function that processes the embedded sequences. The final step is represented by a dense layer that provides a fully connected layer with a single neuron. Ac rates ranging from 84 to 97% were obtained. LSTM model was made by analyzing the words of the original language using the deep learning tools available in the Python language (Table 1).

Fig. 5 LSTM ac result

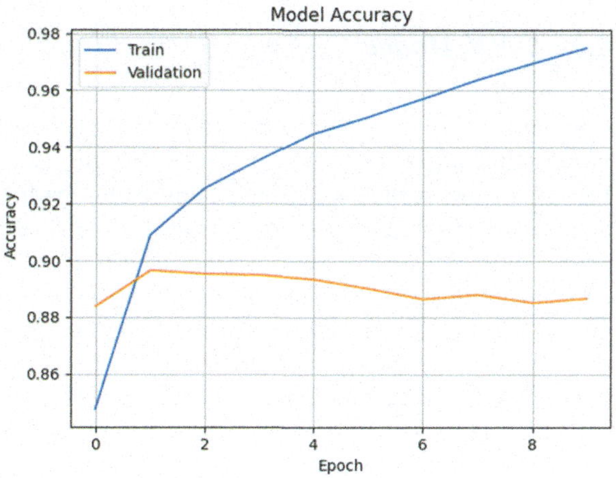

Fig. 6 LSTM loss result

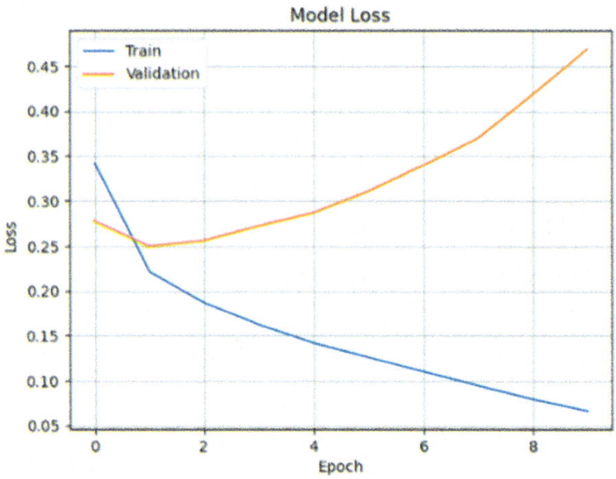

Table 1 Results of all epochs for the LSTM model

SN	LSTM loss	LSTM ac	LSTM val_loss	LSTM val_ac
1	Loss: 0.3420	Ac: 0.8477	Val_loss: 0.2784	Val_ac: 0.8840
2	Loss: 0.2215	Ac: 0.9091	Val_loss: 0.2503	Val_ac: 0.8966
3	loss: 0.1872	Ac: 0.9255	Val_loss: 0.2564	Val_ac: 0.8954
4	Loss: 0.1625	Ac: 0.9356	Val_loss: 0.2727	Val_ac: 0.8950
5	Loss: 0.1419	Ac: 0.9448	Val_loss: 0.2874	Val_ac: 0.8934
6	Loss: 0.1260	Ac: 0.9507	Val_loss: 0.3111	Val_ac: 0.8902
7	Loss: 0.1101	Ac: 0.9572	Val_loss: 0.3405	Val_ac: 0.8865
8	Loss: 0.0945	Ac: 0.9638	Val_loss: 0.3707	Val_ac: 0.8880
9	Loss: 0.0795	Ac: 0.9696	Val_loss: 0.4195	Val_ac: 0.8852
10	Loss: 0.0665	Ac: 0.9750	Val_loss: 0.4694	Val_ac: 0.8867

4.2 CNN Model

CNN model provided results as shown in Figs. 7 and 8 with better ac results with more stabilization for both ac and loss ac with values ranging from 84 to 98 at the first four epochs, then we gained stabilization with ac values at 99% according to the dataset information used in this paper, because this algorithm has a higher-level environment of layers and neurons in it. The layers of CNN used in this model contain an embedding layer to represent each word in the vocabulary with a vector of size 40. An incremental value has been added for each convolutional layer 1D filter (32, 64, 128), each kernel size of 3, and have been reinforced by ReLU function. Max pooling layers were added after all convolutional layers in this model. Flatten layer and dense layer were the last step to gain the result for this model (Table 2).

Fig. 7 CNN ac result

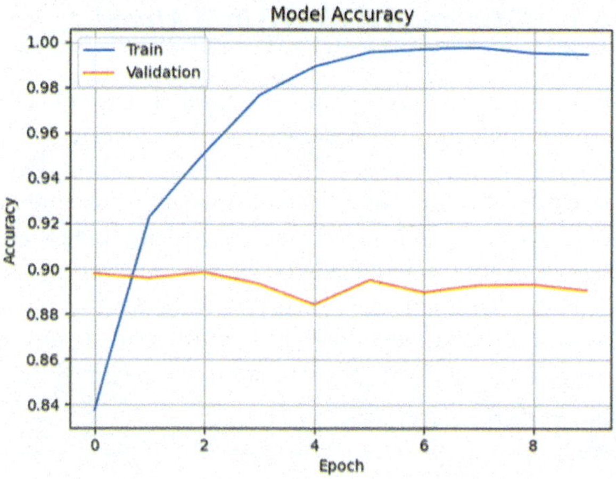

Fig. 8 CNN loss result

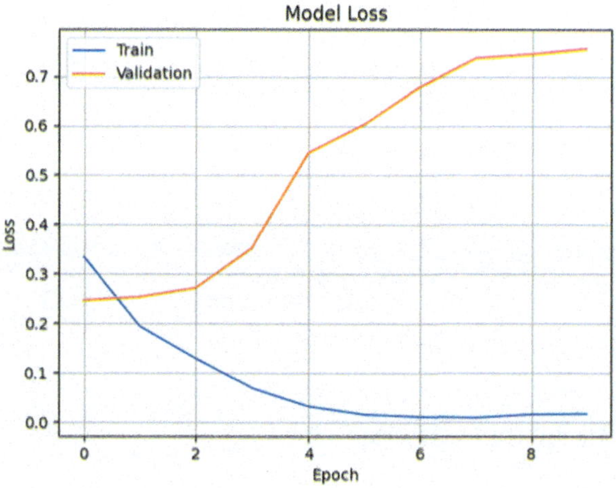

Table 2 Results of all epochs for the CNN model

SN	CNN loss	CNN ac	CNN val_loss	CNN val_ac
1	Loss: 0.3345	Ac: 0.8376	Val_loss: 0.2473	Val_ac: 0.8980
2	Loss: 0.1945	Ac: 0.9228	Val_loss: 0.2540	Val_ac: 0.8961
3	Loss: 0.1293	Ac: 0.9511	Val_loss: 0.2723	Val_ac: 0.8986
4	Loss: 0.0690	Ac: 0.9767	Val_loss: 0.3534	Val_ac: 0.8936
5	Loss: 0.0320	Ac: 0.9895	Val_loss: 0.5464	val_ac: 0.8845
6	Loss: 0.0153	Ac: 0.9957	Val_loss: 0.6042	Val_ac: 0.8953
7	Loss: 0.0107	Ac: 0.9970	Val_loss: 0.6803	val_ac: 0.8900
8	Loss: 0.0087	ac: 0.9975	Val_loss: 0.7388	Val_ac: 0.8930
9	loss: 0.0154	Ac: 0.9951	Val_loss: 0.7469	Val_ac: 0.8933
10	Loss: 0.0163	Ac: 0.9946	Val_loss: 0.7575	Val_ac: 0.8908

4.3 CNN and LSTM Hybrid Model

CNN and LSTM were combined to obtain a hybrid model, in which a new model was created with high-level characteristics to analyze the dataset with higher ac and obtain stability in the results, as the data flows into LSTM layers and then is transmitted to CNN neurons and obtained balanced ac. The merged model combined all layers of CNN and LSTM annotated by creating a new architecture to obtain sequential output using the sequencing function. The results start from 86 to 99% in the first three models, then the score shifts to start settling on the value of 99% of the available dataset information as shown in Figs. 9 and 10, and Table 3.

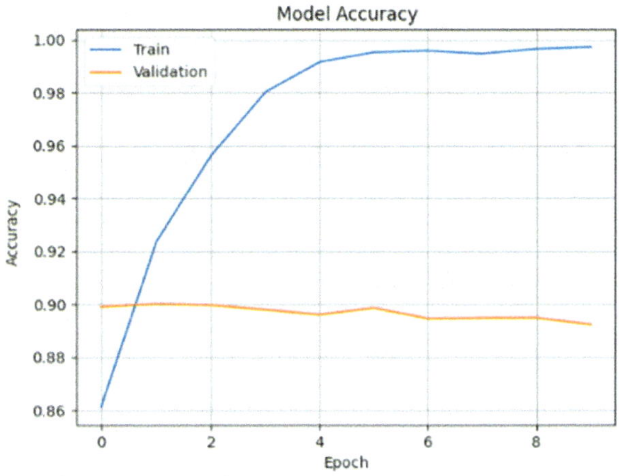

Fig. 9 CNN and LSTM ac result

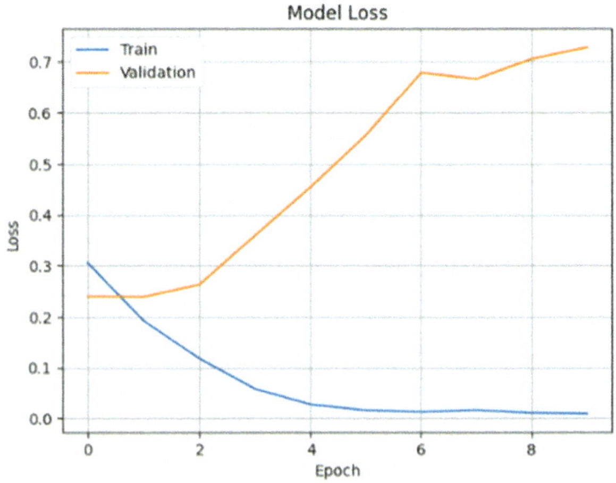

Fig. 10 CNN and LSTM loss result

Table 3 Results of all epochs for a hybrid model LSTM-CNN models

SN	LSTM & CNN loss ac	LSTM & CNN AC	LSTM & CNN val_loss	LSTM & CNN val_ac
1	Loss: 0.3060	Ac: 0.861	Val_loss: 0.2404	Val_ac: 0.8994
2	Loss: 0.1923	Ac: 0.923	Val_loss: 0.2396	Val_ac: 0.9004
3	Loss: 0.1186	Ac: 0.956	Val_loss: 0.2637	Val_ac: 0.8999
4	Loss: 0.0584	Ac: 0.980	Val_loss: 0.3612	Val_ac: 0.8981
5	Loss: 0.0276	Ac: 0.991	Val_loss: 0.4556	Val_ac: 0.8963
6	Loss: 0.0162	Ac: 0.995	Val_loss: 0.5564	Val_ac: 0.8988
7	Loss: 0.0131	Ac: 0.995	Val_loss: 0.6795	Val_ac: 0.8948
8	Loss: 0.0167	Ac: 0.994	Val_loss: 0.6670	Val_ac: 0.8950
9	Loss: 0.0112	Ac: 0.996	Val_loss: 0.7061	Val_ac: 0.8951
10	Loss: 0.0097	Ac: 0.997	Val_loss: 0.7289	Val_ac: 0.8925

4.4 RNN, LSTM, and CNN High-Performance Hybrid Model

We provided a high-performance hybrid model, and the level of data analysis has been upgraded on the available dataset to obtain stable ac in all epochs by creating a combination of all layers for RNN, LSTM, and CNN as explained. The results were gained with a constant rate of 99%. The stability of the result represents a complete idea of the evolution of the mechanism of algorithms working together in this model. Flow data work through the RNN layers, passing by the LSTM layers, then to the CNN neurons, to obtain a relatively stable and successful outcome that corresponds to the correct ac criteria. The RNN model contains an embedding layer with a vector of size 40 and simple RNN layer with 100 simple RNN units (Figs. 11 and 12; Table 4).

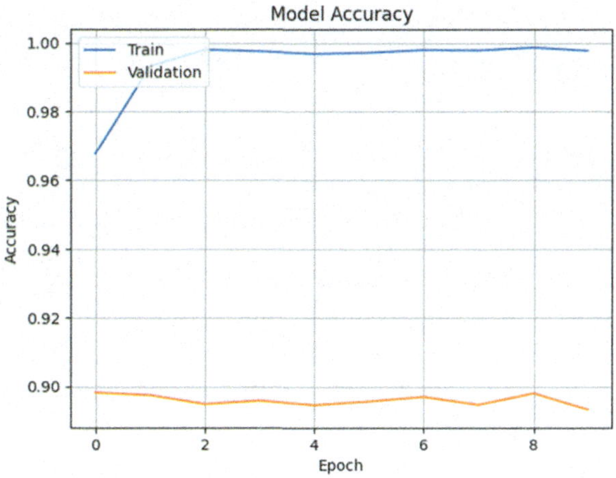

Fig. 11 RNN, LSTM, and CNN high-performance hybrid model ac

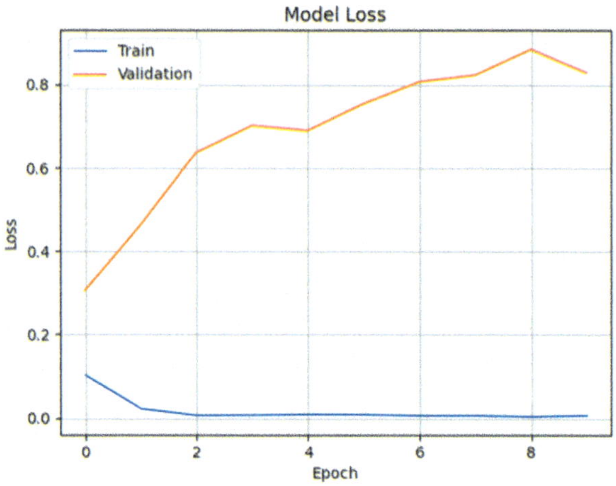

Fig. 12 RNN, LSTM, and CNN high-performance hybrid model loss

Table 4 Results of all epochs for a high-performance hybrid model RNN-LSTM-CNN

SN	High-performance hybrid model loss	High-performance hybrid model ac	High-performance hybrid model Val_loss	High-performance hybrid model val_ac
1	Loss: 0.1025	Ac: 0.9678	Val_loss: 0.3091	Val_ac: 0.8983
2	Loss: 0.0224	Ac: 0.9932	Val_loss: 0.4657	Val_ac: 0.8975
3	Loss: 0.0069	Ac: 0.9980	Val_loss: 0.6389	Val_ac: 0.8950
4	Loss: 0.0077	Ac: 0.9975	Val_loss: 0.7041	Val_ac: 0.8959
5	Loss: 0.0093	Ac: 0.9968	Val_loss: 0.6919	Val_ac: 0.8946
6	Loss: 0.0089	Ac: 0.9971	Val_loss: 0.7562	Val_ac: 0.8957
7	Loss: 0.0064	Ac: 0.9979	Val_loss: 0.8090	Val_ac: 0.8970
8	Loss: 0.0067	Ac: 0.9978	Val_loss: 0.8247	Val_ac: 0.8947
9	Loss: 0.0042	Ac: 0.9986	Val_loss: 0.8862	Val_ac: 0.8980
10	Loss: 0.0070	Ac: 0.9977	Val_loss: 0.8310	Val_ac: 0.8933

5 Conclusion

This research delved into a comprehensive analysis of various deep learning algorithms, comparing their efficacy in detecting fake political news. Two algorithms, LSTM and CNN, were individually assessed and yielded impressive results of 84 and 97%. A hybrid model was then proposed, combining LSTM and CNN algorithms, with a performance of 95%. The ultimate goal was to develop a high-performing hybrid model that would ensure accurate and reliable analysis of fake political news data. With continuous training for models, the hybrid model achieved a stable result of 99%, surpassing industry standards for ac and stability. This unparalleled level of performance ensures that the model is equipped to effectively analyze vast amounts of fake political news data, providing reliable insights and contributing to the preservation of truth and ac in the realm of politics.

References

1. Khan, J.Y., Khondaker, Md.T.I., Afroz, S., Uddin, G., Iqbal, A.: A benchmark study of machine learning models for online fake news detection. Mach. Learn. Appl. **4**, 100032 (2021). https://doi.org/10.1016/j.mlwa.2021.100032
2. Choudhary, M., Chouhan, S.S., Pilli, E.S., Vipparthi. S.K.: BerConvoNet: a DEEP learning framework for fake news classification. Appl. Soft. Comput. **110** (2021). https://doi.org/10.1016/j.asoc.2021.107614
3. Choraś, M. et al.: Advanced Machine Learning techniques for fake news (online disinformation) detection: a systematic mapping study. Appl. Soft Comput. **101** (2021). https://doi.org/10.1016/j.asoc.2020.107050
4. Jāmiʻat, S.A.: Ayn Shams, Egyptian knowledge bank, institute of electrical and electronics engineers. Egypt section, and Institute of electrical and electronics engineers. In: Proceedings ICCES : 2018 13th International Conference on Computer Engineering and Systems (ICCES) : Triumph Hotel, Cairo, Egypt, 18th-19th Dec 2018
5. Han, W., Mehta, V.: Fake news detection in social networks using machine learning and deep learning: performance evaluation. In: Proceedings—IEEE International Conference on Industrial Internet Cloud, ICII 2019, Institute of Electrical and Electronics Engineers Inc., pp. 375–380 (2019). https://doi.org/10.1109/ICII.2019.00070
6. Chintalapudi, N., Battineni, G., Amenta, F.: Sentimental analysis of COVID-19 tweets using deep learning models. Infect Dis. Rep. 13(2), Apr 2021. https://doi.org/10.3390/IDR13020032
7. Kumar, S., Kumar, S., Yadav, P., Bagri, M.: A survey on analysis of fake news detection techniques. In: Proceedings—International Conference on Artificial Intelligence and Smart Systems, ICAIS 2021, Institute of Electrical and Electronics Engineers Inc., pp 894–899, Mar (2021). https://doi.org/10.1109/ICAIS50930.2021.9395978
8. Mansouri, R., Naderan-Tahan, M., Rashti, M.J.: A semi-supervised learning method for fake news detection in social media. In: 2020 28th Iranian Conference on Electrical Engineering, ICEE 2020, Institute of Electrical and Electronics Engineers Inc., Aug (2020). https://doi.org/10.1109/ICEE50131.2020.9261053
9. Nasir, J.A., Khan, O.S., Varlamis, I.: Fake news detection: a hybrid CNN-RNN based deep learning approach. Int. J. Inf. Manage. Data Insights **1**(1), Apr (2021). https://doi.org/10.1016/j.jjimei.2020.100007.
10. https://zenodo.org/record/4561253/files/WELFake_Dataset.csv?download=1. Fake News Dataset

Chapter 23
Wasserstein Barycenters Over Heisenberg Group

Thanh-Son Trinh

1 Introduction

In [1], the Wasserstein barycenters have been studied on $\mathbb{R}^n$ the first by Carlier and Agueh. After that it has been considered more by many authors. It has many applications in other fields such as machine learning [2, 3], multi-population matching equilibrium in economics [4], and computer vision [5]. Furthermore, the Wasserstein barycenter also has been studied over Riemannian manifolds [6] and geodesic spaces [7].

There are two ways to define the Wasserstein barycenter, and they have the same meaning. The first is center of mass; this is presented the formulation that the Wasserstein distance of all reference measures is minimized. The other is a multi-marginal problem [6]. We use the first one to introduce our barycenter in this article.

Let $\mathbb{H}^n$ be the Heisenberg group and d_c be the Carnot-Carathéodory distance on $\mathbb{H}^n$. By setting $\mathcal{B}(\mathbb{H}^n)$ is the set of Borel measures on $\mathbb{H}^n$, we define sets $P_2(\mathbb{H}^n)$ and $P_2^{ac}(\mathbb{H}^n)$ as

$$P_2(\mathbb{H}^n) = \{\mu \in \mathcal{B}(\mathbb{H}^n) || \mu| = 1, \exists \mathrm{h}_0, \int_{\mathbb{H}^n} \mathrm{d}_c^2(\mathrm{h}_0, \mathrm{h}) < \infty\}, \tag{1}$$

$$P_2^{ac}(\mathbb{H}^n) = \{\mu \in P_2(\mathbb{H}^n) | \mu \mathrm{a.c.w} \mathcal{L}^{2n+1}\}. \tag{2}$$

T.-S. Trinh (✉)
Faculty of Information Technology, Industrial University of Ho Chi Minh City, Ho Chi Minh City 70000, Vietnam
e-mail: trinhthanhson@iuh.edu.vn

J. M. R. S. Tavares et al. (eds.), *Proceedings of Second International Conference on Intelligent System*, Algorithms for Intelligent Systems,
https://doi.org/10.1007/978-981-99-8976-8_24

Now, we will introduce our barycenter problem. Let $P \in P_2(P_2(\mathbb{H}^n))$, we consider the following problem

$$\inf_{\nu \in P_2(H^n)} \int\limits_{P_2(H^n)} W_2^2(\mu, \nu) \mathrm{d}P(\mu). \tag{3}$$

In the first main result, we show that our barycenters over Heisenberg structure always exists.

Theorem 1 Let P $\in P_2(P_2(H^n))$, then there exists a Wasserstein barycenter of P.

In the following result, if $\mathrm{P}\big(P_2^{ac}(\mathbb{H}^n)\big) > 0$, then we show the uniqueness of our barycenters.

Theorem 2 Let P $\in P_2(P_2(\mathbb{H}^n))$. If $\mathrm{P}\big(P_2^{ac}(\mathbb{H}^n)\big) > 0$, then the problem (3) has a unique solution.

Finally, we give the paper structure. We give preliminaries in Sect. 2. In the next section, we show the existence and the uniqueness of our barycenters by giving the proof of main results. In the last section, we give the conclusion and future works.

2 Preliminaries

2.1 Wasserstein Distance

First, we give some notations and review on the Wasserstein distances [8, 9].

By putting $\mathcal{B}(X)$ is the set of Borel measures μ on (X, d), we define

$$M(X) = \{\mu \in \mathcal{B}(X) | |\mu| < \infty\}, \quad P(X) = \{\mu \in \mathcal{B}(X) | |\mu| = 1\}, \tag{4}$$

and

$$P_2(X)(\text{resp.}M_2(X)) = \{\mu \in P(X)(\text{resp.}M_2(X)) | \exists z_0 \in X, \int\limits_X d^2(z_0, x) d\mu < +\infty\}. \tag{5}$$

Let any $\mu, \nu \in M(X)$, we denote $\mu \ll \nu$ if one has $\nu(A) = 0$, then $\mu(A) = 0$ for all $A \in \mathcal{B}(X)$, and we say that μ is *a. c. w* ν.

Next, the Wasserstein distance $W_2(\mu, \nu)$ with $\mu, \nu \in P_2(X)$, is defined as

$$W_2^2(\mu, \nu) = \inf_{\delta \in M(\mu, \nu)} Q(\delta), \tag{6}$$

where $Q(\delta) = \int_{X \times X} d^2(x, y) d\delta(x, y)$ and $M(\mu, \nu)$ consist of all measures δ which has two marginals μ, ν, i.e., the first and the second projection of δ are μ and ν.

Considering a Borel map $\boldsymbol{l} : \boldsymbol{X}_1 \to \boldsymbol{X}_2$, we define the measure $\boldsymbol{l}_{\#}\boldsymbol{\mu}$, with $\boldsymbol{\mu} \in \mathcal{B}(\boldsymbol{X}_1)$ as

$$\boldsymbol{l}_{\#}\boldsymbol{\mu}(\boldsymbol{A}) := \boldsymbol{\mu}\big(\boldsymbol{l}^{-1}(\boldsymbol{A})\big), \text{ for all } \boldsymbol{A} \in \mathcal{B}(\boldsymbol{X}_1). \tag{7}$$

2.2 *Heisenberg Group*

In this subsection, we review on the geometry structure that we use.

We consider the structure $\mathbb{H}^n$ as the form $\mathbb{R}^{2n+1} \simeq \mathbb{C}^n \times \mathbb{R}$. We denote by $(z;t) = (z_1, \dots, z_n; t)$ a point in $\mathbb{H}^n$ and

$$(z_1, \dots, z_n; t) \cdot (w_1, \dots, w_n; s) = \left(z_1 + w_1, \dots, z_n + w_n; t + s + 2\sum_{i=1}^{n} Im\big(z_k \overline{w}_k\big) \right) \tag{8}$$

In $\mathbb{H}^n$, we can get the Lie bracket $[X_k, Y_k] = -4Z$, where $X_k = \partial_{x_k} + 2y_k\partial_t$; $Y_k = \partial_{y_k} - 2x_k\partial_t$; $Z = \partial_t$ with $x_k, y_k \in \mathbb{R}$, $z_k = x_k + iy_k, k = 1, \dots, n$.

Next, we consider a sub unit curve $(s.\ u.\ c)$ $\gamma : [0, T] \to \mathbb{R}^{2n+1}$ this means that for $k = 1, \dots, n$, one a function $h_k : [0, T] \to \mathbb{R}$ with $\sum_{k=1}^{2n} h_k^2(h) \le 1$ and

$$\dot{\gamma}(h) = \sum_{k=1}^{n} h_k(h)X_k(\gamma(h)) + h_{n+k}(h)Y_k(\gamma(h)) \text{ for } h \in [0, T]. \tag{9}$$

By Chow's Theorem, we can define the Carnot-Caratheodory distance d_c as

$$d_c(h_1, h_2) := \inf\{T > 0 | \exists \mathrm{as.u.c}\gamma : [0,\mathrm{T}] \to \mathrm{H^n}\gamma(0) = \mathrm{h_1}, \gamma(\mathrm{T}) = \mathrm{h_2}\}^{\circ}. \tag{10}$$

The topology on $\mathbb{H}^n$ is induced by distance d_c.

3 Main Results

Proof of Theorem 1. According [10, Sect. 3, p. 267] we get the complete space $\mathbb{H}^n$ is a length space and locally compact. This implies that $(\mathbb{H}^n, d_c)$ is a geodesic space. Moreover, $(\mathbb{H}^n, d_c)$ is also a Polish metric space since it is a separable space. Hence, applying [7, Theorem 2] we get the result.

The following Lemmas play crucial for us.

Lemma 1 *If* $\mu \ll \mathcal{L}^{2n+1}$ *then we have* $W_2^2(\cdot, \mu)$ *which is strictly convex.*

Proof First, we will show that $W_2^2(\cdot, \mu)$ is convex. Let any $\nu_1, \nu_2 \in P_2(\mathbb{H}^n)$ and $t \in [0, 1]$, we define

$$\nu_t := (1-t)\nu_0 + t\nu_1. \tag{11}$$

Let $\pi_0 \in \Pi(\nu_0, \mu)$ and $\pi_1 \in \Pi(\nu_1, \mu)$ satisfy

$$\int\limits_{\mathbb{H}^n \times \mathbb{H}^n} d^2(h_1, h_2)\mathrm{d}\pi_0 = W_2^2(\nu_1, \mu) \text{ and } \int\limits_{\mathbb{H}^n \times \mathbb{H}^n} d^2(h_1, h_2)\mathrm{d}\pi_1 = W_2^2(\nu_1, \mu). \tag{12}$$

We set

$$\pi_t := (1-t)\pi_0 + t\pi_1.$$

Then, we prove $\pi_t \in \Pi(\nu_t, \mu)$. Indeed, as $\pi_0 \in \Pi(\nu_0, \mu)$ and $\pi_1 \in \Pi(\nu_1, \mu)$, for every $A \in \mathcal{B}(\mathbb{H}^n)$, one has

$$\begin{aligned} \pi_t\big(A \times \mathbb{H}^n\big) &= (1-t)\pi_0\big(A \times \mathbb{H}^n\big) + t\pi_1\big(A \times \mathbb{H}^n\big) \\ &= (1-t)\nu_0(A) + t\nu_1(A) \\ &= \nu_t(A). \end{aligned} \tag{14}$$

Similarly, for every $B \in \mathcal{B}(\mathbb{H}^n)$, we also have that

$$\begin{aligned} \pi_t\big(\mathbb{H}^n \times B\big) &= (1-t)\pi_0\big(\mathbb{H}^n \times B\big) + t\pi_1\big(\mathbb{H}^n \times B\big) \\ &= (1-t)\mu(B) + t\mu(B) \\ &= \mu(B). \end{aligned} \tag{15}$$

Hence, these mean that $\pi_t \in \Pi(\nu_t, \mu)$. Therefore, setting $W_i := W_2^2(\nu_i, \mu), i = 0, 1$, we obtain that

$$\begin{aligned} (1-t)W_0 + tW_1 &= (1-t) \int\limits_{\mathbb{H}^n \times \mathbb{H}^n} \mathrm{d}^2(h_1, h_2)d\pi_0 + t \int\limits_{\mathbb{H}^n \times \mathbb{H}^n} \mathrm{d}^2(h_1, h_2)\mathrm{d}\pi_1 \\ &= \int\limits_{\mathbb{H}^n \times \mathbb{H}^n} \mathrm{d}^2(h_1, h_2)\mathrm{d}((1-t)\pi_0) + \int\limits_{\mathbb{H}^n \times \mathbb{H}^n} \mathrm{d}^2(h_1, h_2)\mathrm{d}(t\pi_1) \\ &= \int\limits_{\mathbb{H}^n \times \mathbb{H}^n} \mathrm{d}^2(h_1, h_2)\mathrm{d}((1-t)\pi_0 + t\pi_1) \end{aligned}$$

$$= \int_{\mathbb{H}^n \times \mathbb{H}^n} d^2(h_1, h_2) d\pi_t$$
$$\geq W_2^2(\nu_t, \mu). \tag{16}$$

Next, we will prove that $W_2^2(\cdot, \mu)$ is strictly convex. By contradiction, suppose that $\nu_0 \neq \nu_1$ and $\forall t \in (0, 1)$ we have $(1-t)W_0 + tW_1 = W_2^2(\nu_t, \mu)$.

Since $\mu \ll \mathcal{L}^{2n+1}$, using [10, Theorem 5.1], there exists a unique measure $\delta_i (i = 0, 1)$ such that $\delta_i = (T_i \times Id)_{\#}\mu$. By putting $\{\delta_t := (1-t)\delta_0 + t\delta_1\}$, then $\delta_t \in M(\nu_t, \mu), \forall t \in [0, 1]$. Hence, we get that

$$(1-t)W_0 + tW_1 = W_2^2(\nu_t, \mu)$$
$$\leq \int_{\mathbb{H}^n \times \mathbb{H}^n} d_c^2(h_1, h_2) d\delta_t(h_1, h_2)$$
$$= (1-t)W_2^2(\nu_0, \mu) + tW_2^2(\nu_1, \mu). \tag{17}$$

Therefore, from this and using [10, Theorem 5.1] again we obtain that $T_0 = T_1$ μ-a.e on $\mathbb{H}^n$. This is a contradiction with $\nu_0 \neq \nu_1$. Hence, we get the result.

Remark 1 Lemma 1 is a version of [11, Proposition 7.19] for Heisenberg group.

Lemma 2. *$P_2^{ac}(\mathbb{H}^n)$ is closed under the Wasserstein topology.*

Proof Let $\{\mu^N\}_N$ be a sequence on $P_2^{ac}(\mathbb{H}^n)$ such that

$$\lim_{N\to\infty} W_2(\mu_N, \mu) = 0, \tag{18}$$

for some $\mu \in P_2(\mathbb{H}^n)$.

We need to prove that $\mu \in P_2^{ac}(\mathbb{H}^n)$. For every Borel subset A of $\mathbb{H}^n$ such that $\text{vol}(A) = 0$ then

$$\mu^N(A) = 0, \text{ for every } N \in \mathbb{N}. \tag{19}$$

If A is open, since [12] one has

$$\mu(A) \leq \liminf_{N\to\infty} \mu^N(A) = 0. \tag{20}$$

So $\mu(A) = 0$.

Next, we consider the case when A is not open. Observe that

$$\inf\left\{\mu^N(S) | \text{S is open and } A \subset S\right\}$$
$$= \mu^N(A) = 0, \text{ for every } N \in \mathbb{N}. \tag{21}$$

These imply that $\mu(A) = 0$. Hence, we get the result.

Remark 2 *From Lemma 2, one gets that the set $P_2^{ac}(\mathbb{H}^n)$ is measurable. Thus, the condition $P\big(P_2^{ac}(\mathbb{H}^n)\big) > 0$ makes sense.*

Proof of Theorem 2 Combining the convexity of $W_2^2(\cdot, \mu)$ (Lemma 1) and $P\big(P_2^{ac}(\mathbb{H}^n)\big) > 0$, we get that the functional $\int_{P_2(\mathbb{H}^n)} W_2^2(\mu, \cdot)\mathrm{d}P(\mu)$ is strictly convex.

Now, from Theorem 1 and by contradiction we assume that there exist $\nu_0, \nu_1 \in P_2(\mathbb{H}^n)$ such that $\nu_0 \neq \nu_1$ and ν_0, ν_1 are Wasserstein barycenters of P. For any $t \in (0, 1)$, we define measure ν as

$$\nu := t\nu_0 + (1-t)\nu_1. \tag{22}$$

It is clear that $\nu \in P_2(\mathbb{H}^n)$. As $\int_{P_2(\mathbb{H}^n)} W_2^2(\mu, \cdot)\mathrm{d}P(\mu)$ is strictly convex, we obtain that

$$\begin{aligned}
&\int\limits_{P_2(\mathbb{H}^n)} W_2^2(\mu, \cdot)\mathrm{d}P(\mu) = \int\limits_{P_2(\mathbb{H}^n)} W_2^2(\mu, t\nu_0 + (1-t)\nu_1)\mathrm{d}P \\
&< t \int\limits_{P_2(\mathbb{H}^n)} W_0 \mathrm{d}P + (1-t) \int\limits_{P_2(\mathbb{H}^n)} W_1 \mathrm{d}P \\
&= \min_{\delta \in P_2(H^n)} \int\limits_{P\in(\mathbb{H}^n)} W_2^2(\mu, \delta)\mathrm{d}P(\mu).
\end{aligned} \tag{23}$$

This is a contradiction. Therefore, there exists a unique Wasserstein barycenter of P.

4 Conclusion

In this article, we introduced and investigated the Wasserstein barycenters over Heisenberg group. We give the contributions of our works in the followings.

1. By using the properties about metric space $(\mathbb{H}^n, d_c)$, we prove the existence of Wasserstein barycenters over Heisenberg group.
2. By considering the convexity property of the functional $W_2^2(\cdot, \mu)$ and the measurability property of the set that consists of all measures μ with $\mu \ll \mathcal{L}^{2n+1}$, we prove our barycenters are unique. This property is very crucial to study multi-marginal transport problem.

Comparing with the paper [6]. In this work, we introduce the Wasserstein barycenters over Heisenberg structure. Our works extend to the results of Kim and Pass for Riemannian manifolds. In [6], to show that the existence of Wasserstein barycenters, Kim and Pass need the compactness condition of the Riemannian manifolds, see [6, Theorem 3.1]. However, by a different approach, in the first our main result, Theorem 1, we do not need the condition that $\mathbb{H}^n$ is compact.

Next, we give applications for our results. As mentioned above, by using our second main result, we can consider multi-marginal transport problem over Heisenberg group, see [6, Theorem 2.4] and [13, Sect. 5]. Furthermore, as our barycenters are unique, we can apply this for image morphing ([14]), k-barycenters problems ([15]) and clustering problems ([16]).

Finally, we show several future works. We can apply our results to study multi-marginal problems. Since Heisenberg group is a special case of sub-Riemannian structure, we can extend this work to consider the Wasserstein barycenter over Riemannian structure.

References

1. Agueh, M., Carlier, G.: Barycenters in the wasserstein space. SIAM J. Math. Anal. **43**(2), 904–924 (2004)
2. Peyré, G., Cuturi, M.: Computational optimal transport, foundations and trends in machine. Learning **11**(5–6), 355–607 (2019)
3. Srivastava, S., Li, C., Dunson, D.B.: Scalable Bayes via barycenter in Wasserstein space. J. Mach. Learn. Res. **19**(8), **35** (2018)
4. Carlier, G., Oberman, A., Oudet, E.: Numerical methods for matching for teams and Wasserstein barycenters. ESAIM Math. Model. Numer. Anal. **49**(6), 1621–1642 (2015)
5. Rabin, J., Peyré, G., Delon, J., Bernot, M.: *Wasserstein Barycenter and its application to texture mixing*. In: Lecture Notes in Computer Science, vol. 6667, Springer, Berlin, (2012), pp 435–446
6. Kim, Y.-H., Pass, B.: Wasserstein barycenters over Riemannian manifolds. Adv. Math. **307**, 640–683 (2017)
7. Le Gouic, T., Loubes, J.-M.: Existence and consistency of Wasserstein barycenters. Probab. Theory Related Fields **168**(3–4), 901–917 (2017)
8. Villani, C.: Topics in optimal transportation. In: Graduate studies in mathematics, vol. 58, American Mathematical Society, Providence, RI (2003)
9. Villani, C.: Optimal transport, Grundlehren der Mathematischen Wis- senschaften [Fundamental Principles of Mathematical Sciences], vol. 338, Springer-Verlag, Berlin, 2009. Old and new
10. Ambrosio, L., Rigot, S.: Optimal mass transportation in the Heisenberg group. J. Funct. Anal. **208**(2), 261–301 (2004)
11. Santambrogio, F.: Optimal transport for applied mathematicians. In: Progress in Non-linear Differential Equations and their Applications, vol. 87, Birkhauser/Springer, Cham, 2015. Calculus of variations, PDEs, and modeling
12. Parthasarathy, K. R.: Probability measures on metric spaces, AMS Chelsea Publishing, Providence, RI, (2005). Reprint of the 1967 original
13. Chung, N.-P., Phung, M.-N.: Barycenters in the Hellinger-Kantorovich space. Appl. Math. Optim. **84**, 1791–1820 (2021)
14. Simon, D., Aberdam, A.: Barycenters of natural images-constrained Wasserstein barycenters for image morphing. In: 2020 IEEE/CVF Conference on Computer Vision and Pattern Recognition
15. Li, J., Wang, J.Z.: Real-time computerized annotation of pictures. IEEE Trans. Pattern Anal. Mach. Intell. **30**(6), 985–1002 (2008)
16. Puccetti, G., Ruchendorf, L., Vanduffel, S.: On the computation of Wasserstein barycenters. J. Multivar. Anal. 176 (2020)

Chapter 24
Blockchain Technology with High Performance via Parallel Processing

Rawaa Mahdi Hameed and Saif Al-Alak

1 Introduction

Blockchain is a distributed ledger technology that makes it possible to build a safe and open peer-to-peer network for transactions. Although it was initially created to support the virtual currency bitcoin, a wide range of industries have since adopted it for a number of different uses [1]. At its core, a blockchain is a database that is replicated across a network of computers or nodes. Each node in the network maintains a copy of the database, and all changes made to the database are recorded in a block. Once a block is added to the chain, it cannot be altered or deleted without consensus from the network [2]. The security of blockchain technology is one of its key advantages. The blockchain is very resistant to hacking and data manipulation since transactions on it are validated using sophisticated algorithms. The reason for this is the distinct digital signature, or hash, generated for each block in the chain and formed from the contents in the block and the hash of the chain's previous block. Consequently, the blockchain offers an unchangeable and impenetrable record of all network transactions [3]. Another important aspect of blockchain technology is transparency. All users can examine the complete history of network transactions since every node in the network has a copy of the database. This fosters a high level of accountability and trust that is challenging to attain with conventional centralized systems [4]. The potential applications of blockchain technology are vast and wide-ranging. It can be used to create secure and transparent supply chain networks [5],

R. M. Hameed · S. Al-Alak (✉)
College of Science for Women, University of Babylon, Babylon, Iraq
e-mail: saif.mahmood@uobabylon.edu.iq

R. M. Hameed
e-mail: rawaa.hameed.gsci7@student.uobabylon.edu.iq

J. M. R. S. Tavares et al. (eds.), *Proceedings of Second International Conference on Intelligent System*, Algorithms for Intelligent Systems,
https://doi.org/10.1007/978-981-99-8976-8_25

improve voting systems and the health field [6, 7], and facilitate faster and more efficient financial transactions [8]. As technology continues to evolve and mature, it is likely that it will find even more innovative uses in various industries and sectors [9, 10].

The network encounters several issues such as transaction execution duration and processing speed. Additionally, idle processing units adversely impact network productivity, leading to suboptimal productivity levels, reduced profitability, and increased time consumption. By allocating transactions to them after receiving them, the node itself was able to alleviate the issue of idle transaction units.

This paper proposes a scheme to improve blockchain performance by executing parallel mining. All transactions are distributed to the network's nodes, who then distribute them to all of their processing units. Therefore, no more processing units are idle. Additionally, the paper finds that the increase in the number of processing units in each node would lead to that more transactions are completed with faster execution times and higher profits, increasing network productivity while reducing execution times and increasing profits.

2 Related Works

The majority of nodes will work to validate a single block, and the remaining nodes will serve as the active manager and backup manager, respectively. The active manager will send updates to the backup managers, who will take over if the active management is unavailable. Miners will get data from the active manager. To ensure that managers are chosen at random and that no node can be forced to become a manager, create a consensus mechanism [11]. The peer-to-peer blockchain's topology is being improved to a decentralized topology (organized topology in clusters). This seeks to shorten the time spent broadcasting transactions and to shorten the lengthy time taken by the proof of work (PoW) mechanism's nodes consensus mechanism to validate transactions using intra- and inter-cluster parallel mining [12], a system for choosing a management, allocating work, and rewarding performance. This approach has been tested using a range of case situations by adjusting the difficulty level and number of validators. It has been developed in a test environment that has all the properties required to perform proof of work for bitcoin [13]. The huge computation and energy cost of the consensus process in blockchain prevents it from being directly applied as a generic platform. To overcome this challenge, propose a cloud mining pool-aided BCoT architecture, where the IoT devices can rent the computing resources from the cloud mining pools to offload the mining process [14]. In [15] a parallel framework to find out suitable hyper parameters of deep learning in a blockchain environment by using a metaheuristic algorithm. Framework also takes into account the issue of communication cost, by limiting the number of information exchanges between miners and blockchain. Since different types of attacks can affect pool mining, suggested algorithms encourage fast solo mining rather than pool mining [16]. In [17] a strategy for increasing the scalability and transaction speed of proof of work-based, permissionless blockchain networks been put out, and it is known as graph chain.

3 Theoretical Background

3.1 Blockchain

Blockchain is a decentralized, distributed ledger technology. The decentralized means that there is no single point of control or failure, and transactions are verified and validated by the network of participants rather than a centralized authority [13]. This creates a high level of trust and transparency, making it ideal for use in industries such as finance, supply chain management, and voting systems.

Overall, blockchain technology has the potential to revolutionize various industries by providing a secure and transparent platform for peer-to-peer transactions, reducing the need for intermediaries, and increasing efficiency and trust in business operations [11].

3.2 Parallel Processing

Parallel processing is a technique that enables the execution of multiple tasks simultaneously, thereby improving system performance and productivity. In the context of blockchain technology, parallel processing can play a crucial role in enhancing the network's throughput and scalability [13]. The blockchain network consists of numerous nodes that perform several critical functions, including transaction validation and block creation. These processes require considerable computational resources and can result in performance bottlenecks, especially during peak network usage. Parallel processing allows for the distribution of these tasks across multiple processors or nodes, thereby reducing processing times and improving network efficiency [12, 13]. Moreover, the implementation of parallel processing techniques in the blockchain can increase network security by enabling nodes to process transactions concurrently, reducing the likelihood of transaction backlogs and potential network congestion. However, there are also some challenges associated with parallel processing in the blockchain. One such issue is the complexity of maintaining consensus across multiple processors, which is critical for ensuring the integrity and security of the blockchain network [15]. Despite these challenges, the potential benefits of parallel processing in the blockchain are substantial. By improving network throughput, scalability, and security, parallel processing can contribute to the continued growth and development of blockchain technology, opening up new possibilities for applications in various industries.

3.3 Heterogeneous Processing Units (HEPU)

Heterogeneous processing units are a critical aspect of parallel processing, allowing for the efficient use of multiple processing units with distinct architectures and capabilities to achieve high-performance computing. In parallel processing, multiple processing units work together to perform a single task, which can be either divided into smaller sub-tasks that are processed in parallel or replicated across multiple processing units to increase throughput. Heterogeneous processing units are commonly used in parallel processing systems, where different types of processing units are utilized for specific types of computations. In summary, heterogeneous processing units are a critical aspect of parallel processing, allowing for the efficient use of multiple processing units with distinct architectures and capabilities to achieve high-performance computing. While they present some challenges, the benefits of heterogeneous processing units make them an essential tool for modern computing systems [18, 19].

3.4 Homogeneous Processing Units (HOPU)

In parallel processing, homogeneous processing units refer to a computing architecture that consists of identical or nearly identical processing units. In such a system, each processing unit has the same instruction set architecture, clock speed, and memory configuration. Homogeneous processing units are commonly used in parallel processing systems, where multiple processing units work together to perform a single task. In these systems, each processing unit is responsible for a specific portion of the computation, and the results are combined to produce the final output. Homogeneous processing units have several advantages, including simplicity of programming and load balancing. However, homogeneous processing units also have limitations, such as limited scalability and a lack of flexibility. While they have some limitations, homogeneous processing units are widely used in many parallel processing applications due to their simplicity and ease of use [18, 19].

3.5 Proof of Work (Pow)

Proof of work (PoW) is a consensus mechanism used in blockchain networks to achieve agreement and validate transactions. It is primarily associated with public blockchains, such as bitcoin and Ethereum, where it plays a crucial role in maintaining the integrity and security of the network [12, 16].

4 Methodology

Underutilized processing units can result in decreased system efficiency. To address this issue, we propose implementing parallel processing within the node's processing units. Through simulation, we aim to evaluate the potential benefits of this approach and examine the impact of utilizing the maximum number of processing units on system performance. The proposed methodology entails a user transmitting transaction requests to the main node, which subsequently disseminates these transactions to all nodes within the network as in Fig. 1. All nodes initiate parallel processing to solve these transactions, and the node that completes the computation in the shortest amount of time incorporates it into a block and disseminates the block to all nodes within the network, including the primary node, which further communicates the solution to the user as in Fig. 2. The node responsible for solving the computation is subsequently rewarded.

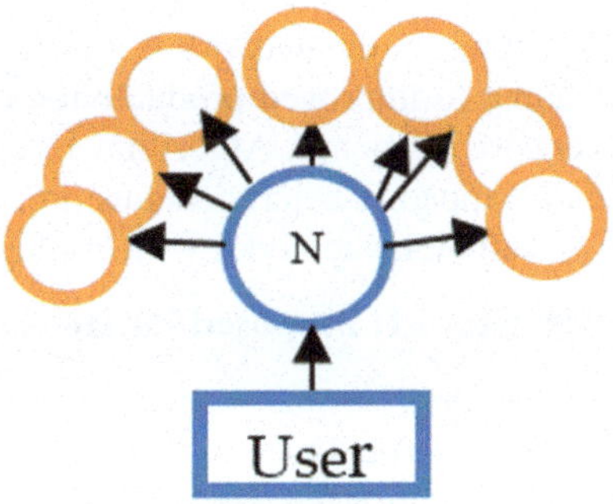

Fig. 1 Transaction broadcasting

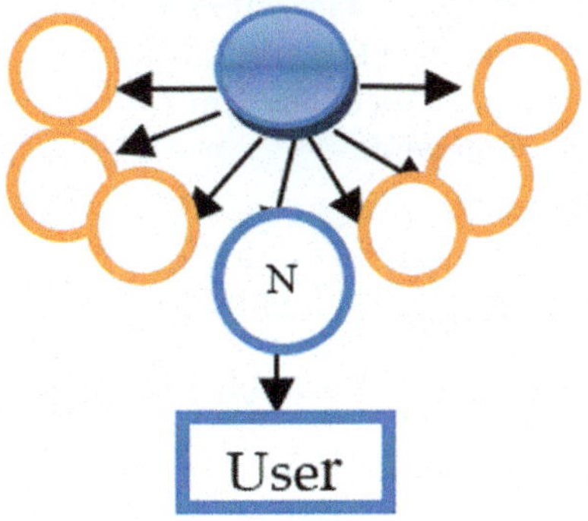

Fig. 2. Block broadcasting

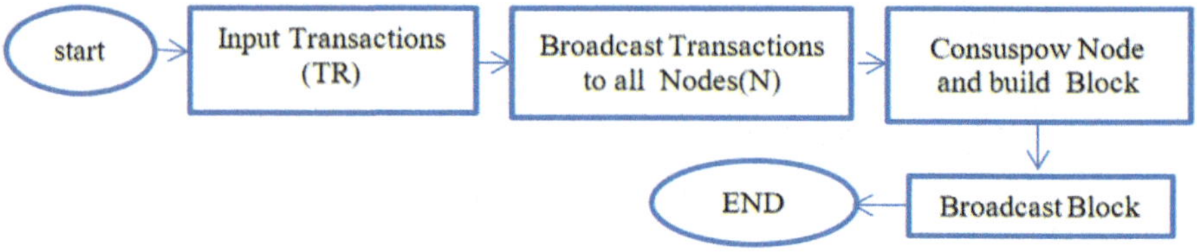

Fig. 3 General diagram

Thus, this methodology facilitates parallel and expeditious dissemination of transactions. The general diagram of the suggested method is displayed in the accompanying Fig. 3.

4.1 Algorithms

The parallel mining approach enables every node within the network to actively participate in the processing of transactions by distributing them to unoccupied nodes. This is achieved through the utilization of multiple processing units within each node, allowing for the allocation of an empty processing unit upon receipt of a transaction. By doing so, the network is able to effectively utilize all available processing units, leading to improved productivity, reduced processing times, and overall increased network efficiency. Algorithm 1 describes how transactions are distributed to nodes and mapped to heterogeneous processing units. Algorithm 2 shows the distribution of coefficients to nodes with homogeneous processing units.

Algorithm 1: Proposed for Heterogeneous Processing

1. Begin
2. While true do
3. Transaction ← receive
4. Idle ← find free processing unit
5. Assign (Transaction, Idle)
6. Mining (Transaction)
7. Free Processing Unit
8. End while
9. End

Algorithm 2 shows the distribution of coefficients to nodes with homogeneous processing units.

Algorithm 2: Proposed for Homogeneous Processing

1. Begin
2. While true do.
3. Transaction ← receive.

4. Idle ← find free and min-estimated-Time processing unit.
5. Assign (Transaction, Idle).
6. End while.
7. End.

Algorithm 3 provides a graphical representation of the mining process executed by nodes that possess processing units. Prior to commencing transaction processing, the algorithm performs a check to determine whether an unoccupied handler is present within the node, after which the transaction is assigned to the handler. The assigned handler subsequently initiates the mining process.

Algorithm 3: Proof of Work Mining Algorithm

Input: Transaction, No_ nodes, Current_ time.
Output: Profit.

1. Begin.
2. For node ← 1 to No_ nodes do.
3. If (node. processing_ Unit = = free).
4. Begin.
5. node. processing_ Unit ← busy.
6. For t ← 1 to Transaction. time.
7. Current_ time ← Current_time + 1.
8. EndFor.
9. node. finish_ time = Current_ time.
10. node. profit ← node. profit + node. finish_ time * DigitalCoin.
11. node. processing_ Unit ← free.
12. End.
13. EndFor.
14. End.

The work includes two experiments, each experiment tests four metrics which are node throughput, average profit each node, profit rate, and throughput for network. In the first experiment the processing units are different in terms of processing unit. In the second experiment the processing units are equivalent. Three tests are computed in each experiment. The throughput for each network has been calculated through the Eq. 1.

$$\text{THnet} = \frac{X * S}{T}. \tag{1}$$

Calculate the node throughput through the Eq. 2.

$$\text{THnode} = \frac{x * S}{t} \tag{2}$$

Calculate the average profit each node through the Eq. 3.

$$\text{Pnode} = \frac{P * X}{t}. \tag{3}$$

Calculate the average network profit in each test through Eq. 4.

$$\text{Pnet} = \frac{\sum \textbf{Pnode}}{N} \tag{4}$$

where THnet is the throughput of the network, N is the total number of nodes in the network, T is the highest processing time consumed by the network, t is the processing time for each node, THnode is the throughput of the node, X is the number of total transactions in the network, x is the number of transactions per node, S transaction size (assuming the transaction size is 1 KB), P is the profit of each node (assuming 10 BTC per transaction), Pnode is the profit average per node, and Pnet is average network profit.

The data includes 500 transactions, all transactions with same size. The profit is 10 bitcoin each transaction, number of node is 10, execution times are randomly computed between (200–240 s) per transaction in the first experiment, and (240 s) per transaction in the second experiment. The main purpose of the proposed method is to increase network productivity, reduce execution time, and increase profits for nodes.

5 Result Analysis and Discussion

In this section the outcomes of the suggested technique are covered. After applying the suggested approaches, some experimental findings were attained. The effectiveness of the suggested technique is examined in this section using simulations. Two experiments were conducted, each experiment tests four metrics to measure the robustness of the proposed method: throughput for the network (THnet), throughput for the node (THnode), average profit per node (Pnode), and average network profit (Pnet).

5.1 Heterogeneous Network (Experiment 1)

This subsection describes the first experiment with several tests. Let us explain the two experiments thoroughly. The experiment conducted in this study involves three test scenarios in which the arrival time of transactions in the network is zero, meaning that all transactions exist simultaneously. The test is conducted sequentially on nodes with one processing unit, two processing units, and three processing units. These processing units are heterogeneous in nature. The productivity and profit ratios of

Fig. 4 Node throughput—heterogeneous

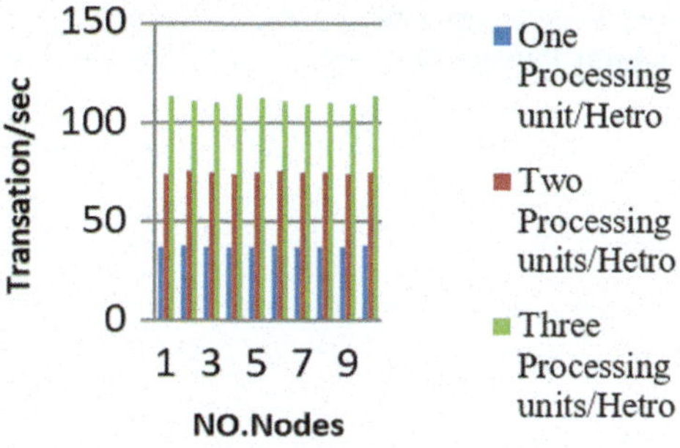

Fig. 5 Av. profit per node—heterogeneous

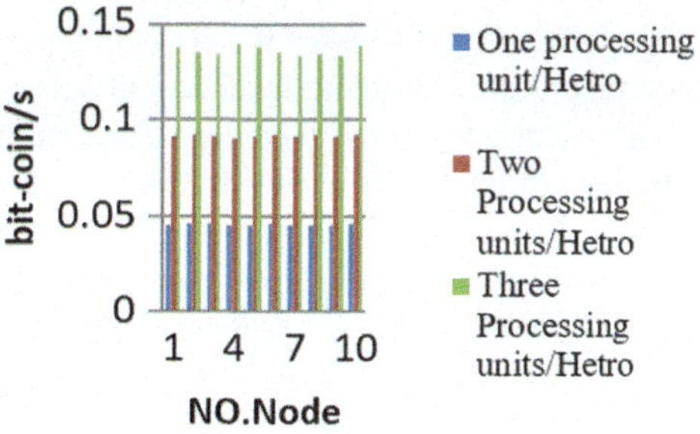

each node are calculated for each test, and a comparison is made among them. The corresponding results and analyses are presented in Figs. 4 and 5.

Through Fig. 4, it is clear that as the number of processors rises, each node's productivity rises as well. The productivity varies from time to time depending on the activity and efficiency of each node. Additionally, as the node's productivity rises, the percentage of profits also rises, as shown in Fig. 5. The N1 has the highest throughput when there is just one processing unit. Its productivity value was 37.960 compared with the other nodes, and because its profit percentage was higher than that of the other nodes, its profits were 0.046 bitcoin/s. In contrast, the lowest node, N3, had a productivity value of 36.911 and a profit percentage of 0.045. In the case of two processing units, N5 has the highest productivity and profits, as its productivity was (75.655) and its profit percentage was (0.092), and the lowest node is N3, whose productivity value is (74.041) and its profit rate is (0.090). We note that in both cases, the N3 node is the least productive and profitable. In the third case, which has three processing units, N3 is the most efficient node and has the highest productivity and profit rate; its productivity was (114.275) and its profit rate was (0.139), and the lowest node in this case is N6, with a productivity of (109.256) and a profit rate of (0.133). This leads us to the conclusion that the number of processors per node, along with their effectiveness and computing capability, determines the increase in the nodes' productivity and revenues, as shown in Figs. 4 and 5, thereby boosting network throughput. The results of comparing the network throughput of the various processing units are depicted in Fig. 6. We observe that as the number of processing units rises, network productivity does as well.

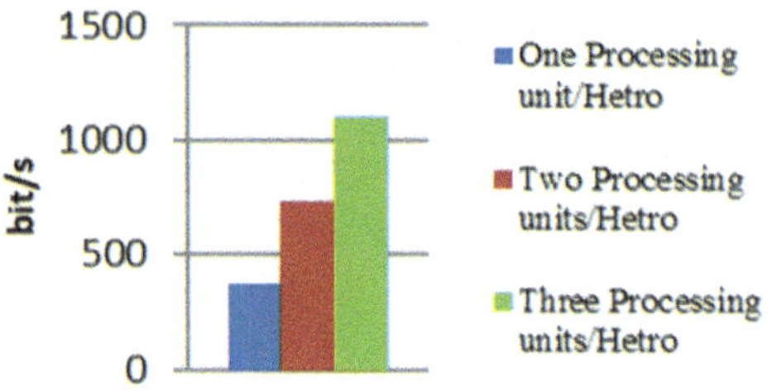

Fig. 6 Heterogeneous network throughput

5.2 *Homogeneous Network (Experiment 2)*

This subsection discusses the second experiment of the proposed method. The study conducted the following tests on the network:

The second test involved zero transaction arrival time. The experiment conducted involved three test scenarios but used homogeneous node processors. The outcome of the test was similar to the first test, with the three processing unit nodes outperforming the two processing unit and one processing unit nodes. Figure 7 demonstrates that the throughput of a node possessing three processing units is greater than that of a node comprising two processing units, and a node outperforms a node with one processing unit per node. This observation is also evident in the profit ratio, as illustrated in Fig. 8.

According to the aforementioned findings, the productivity of nodes among them is equal: the productivity of nodes with one processing unit is 34.133, the productivity

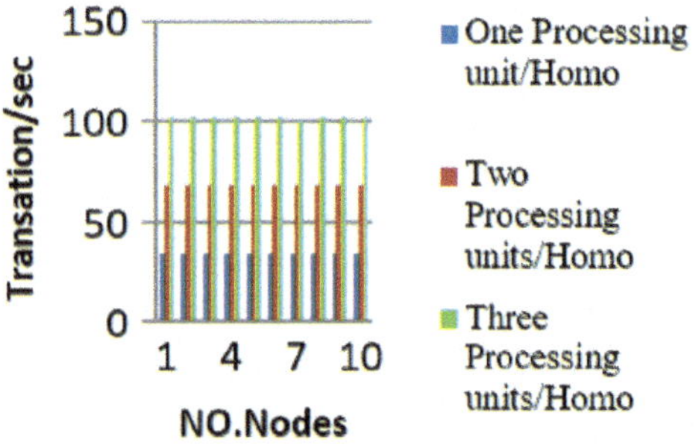

Fig. 7 Node throughput—homogeneous

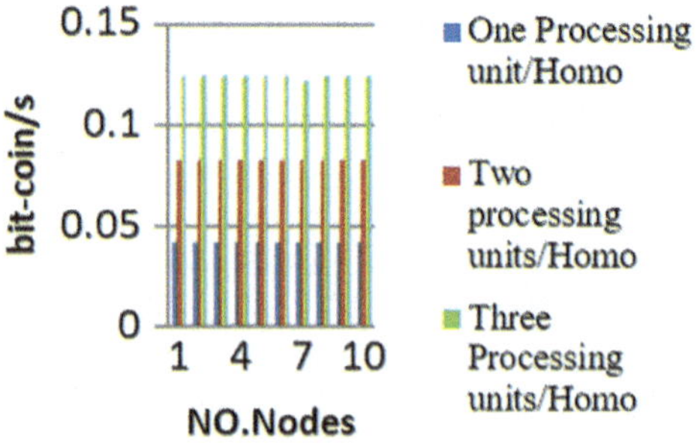

Fig. 8 Node av. prof—homogeneous

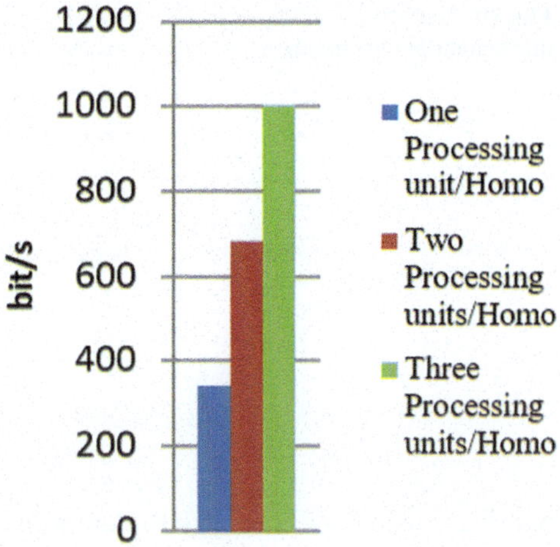

Fig. 9 Homogeneous network throughput

of nodes with two processing units is 68.266, and the productivity of nodes with three processing units is 102.4. This means that productivity doubles with each increase in processing unit count. The same is the case with regard to profits, as they increase with the increase in network productivity. In the case of one processing unit, the percentage of profits was (0.041); in the case of two processors, the percentage became (0.083); and in the case of three processors, it was (0.125). The above results show that in both experiments, the number of processors affected productivity and profits in addition to the computational capacity of the processors. Thus, the throughput of the network is higher in the case of three processing units compared with the throughput of the network that has nodes that have two processing units and nodes that have one processing unit, as shown in Fig. 9.

The aforementioned outcomes indicate that the heterogeneous network surpasses the homogeneous network. As illustrated in Fig. 10, when comparing the productivity of both experiments with one processing unit, the heterogeneous network exhibited a higher productivity level (370.846) compared with the homogeneous network (341.333), with a difference of 29.513. Furthermore, we compare two processing units for both the heterogeneous and homogeneous networks, revealing a higher productivity level of 740.419 for the heterogeneous network in contrast to 682.666 for the homogeneous network, with a difference of 57.753. The comparison of three processing units, with the heterogeneous network demonstrating a productivity level of 1105.533 and the homogeneous network exhibiting a level of 1003.921, resulting in a difference of 101.612. Based on these findings, it is evident that the first experiment is superior to the second in terms of network productivity.

Subsequently, the productivity rate of each node was computed for three scenarios, namely one processing unit, two processing units, and three processing units, for both experiments. The outcomes for each scenario were then compared with their respective outcomes from the other experiment. Specifically, the productivity rate of nodes was evaluated in the context of one processing unit for both the homogeneous

Fig. 10 Network throughput/hetero-homo

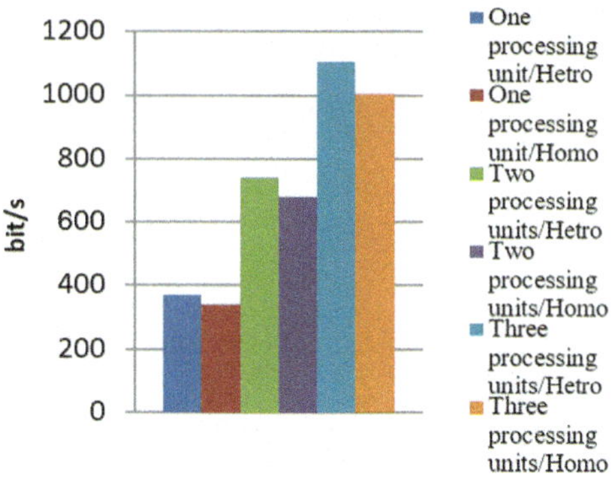

and heterogeneous networks to determine the superior network. Additionally, the same analysis was conducted for two processing units and three processing units. The results of this analysis are presented in Figs. 11 and 12.

Fig. 11 Av. nodes throughput/hetero-homo

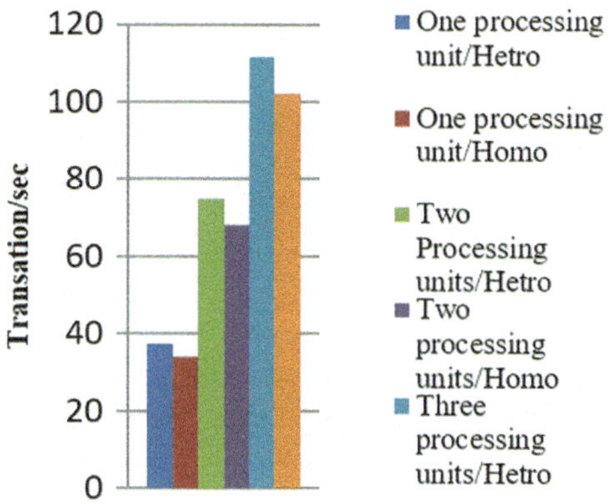

Fig. 12 Throughput per node/hetero-homo

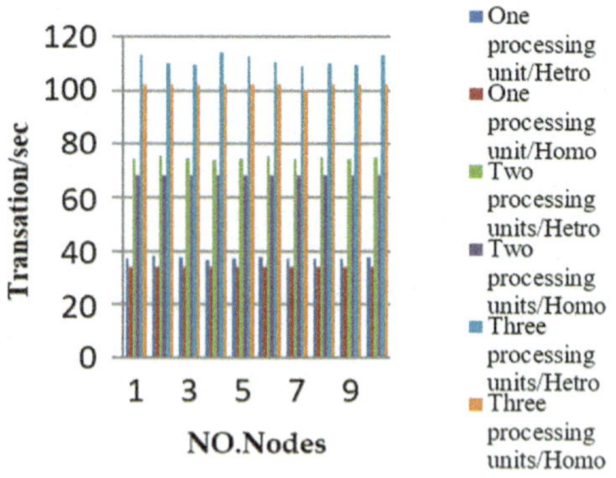

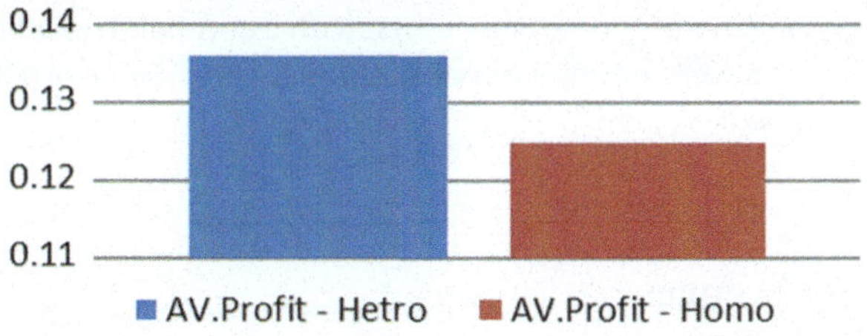

Fig. 13 Average network profit/hetero-homo

The above results demonstrate that the productivity of nodes in the heterogeneous network surpasses that of the homogeneous network. Figure 11 illustrates that the heterogeneous network outperformed the homogeneous network in the case of one processing unit. Specifically, the productivity of the heterogeneous nodes was 37.429, while that of the homogeneous nodes was 34.133, yielding a difference of 3.296. Similar results were observed in the case of two processing units, where the average productivity of the heterogeneous nodes was (74.807), compared with (68.266) for the homogeneous nodes, resulting in a difference of (6.541). Moreover, the productivity of the heterogeneous nodes was 111.435, while that of the homogeneous nodes was 102.199, resulting in a difference of 9.236 in the case of three processing units. In conclusion, the analysis demonstrates that the productivity of nodes in the heterogeneous network is superior to that of the homogeneous network.

Next, we conducted a comparison of the profit rates for each node in both experiments, and Fig. 13 illustrates that the heterogeneous node experiment had a higher profit rate compared with the homogeneous node experiment.

6 Conclusion

This paper presents a simulation study aimed to improve the speed of transaction processing (mining) in order to enhance system productivity and achieve greater profitability by effectively utilizing all available processing units within the nodes, thereby eliminating idle processing units. The efficiency, capacity, and throughput of the system are influenced by the number of processing units present in each node. The research findings demonstrate that the proposed approach yields higher system throughput and profits as the number of processing units increases. From our work it can be concluded that the simultaneous mining for the independent transactions would increase the blockchain system efficiency. The number of processing units has influenced the system and node throughput. When the number of independent transaction is bigger than the number of free processing units, then the system works with quite efficiency. However, the system efficiency would be higher when the number of independent transactions is bigger than free processing units. The study includes two experiments, one utilizing heterogeneous processing units and the other utilizing homogeneous processing units, with a comparative analysis of the results. The results indicate that the first experiment exhibits higher productivity and generates greater profits.

References

1. Nahi, H.A., Hashim, S.M., Kreem, D.J.: Blockchain for baccalaureate examination sheets protection in Iraq. Indonesian J. Electr. Eng. Comput. Sci. **29**(2), 1183–1191 (2023). https://doi.org/10.11591/ijeecs
2. Garcia-Teruel, R.M.: Legal challenges and opportunities of blockchain technology in the real estate sector. J. Property Plann. Environ. Law **12**(2), 129–145 (2020). https://doi.org/10.1108/JPPEL-07-2019-0039
3. Oleiwi, W.K., Abdullah, A.A.: A survey of the blockchain concept and mitigation challenges in different networks. *湖南大学学报 (自然科学版)* **48**(10) (2021)
4. Mohammed, M.K., Abdullah, A.A., Abod, A.: Securing medical records based on interplanetary file system and blockchain. Periodicals Eng. Nat. Sci. **10**(2), 346–357 (2022). https://doi.org/10.21533/pen.v10i2.2855
5. Chang, S.E., Chen, Y.: When blockchain meets supply chain: a systematic literature review on current development and potential applications. IEEE Access **8**, 62478–62494 (2020). https://doi.org/10.1109/ACCESS.2020.2983601
6. Abdellatif, A.A., Samara, L., Mohamed, A., Erbad, A., Chiasserini, C.F., Guizani, M., Laughton, J.: Medge-chain: leveraging edge computing and blockchain for efficient medical data exchange. IEEE Internet of Things J. **8**(21), 15762–15775 (2021). https://doi.org/10.1109/JIOT.2021.3052910
7. McGhin, T., Choo, K.K.R., Liu, C.Z., He, D.: Blockchain in healthcare applications: research challenges and opportunities. J. Netw. Comput. Appl. **135**, 62–75 (2019). https://doi.org/10.1016/j.jnca.2019.02.027
8. Garg, P., Gupta, B., Chauhan, A.K., Sivarajah, U., Gupta, S., Modgil, S.: Measuring the perceived benefits of implementing blockchain technology in the banking sector. Technol. Forecast. Soc. Change **163**, 120407 (2021). https://doi.org/10.1016/j.techfore.2020.120407
9. Kh-Madhloom, J.: Dynamic cryptography integrated secured decentralized applications with blockchain programming. Wasit J. Comput. Math. Sci. **1**(2), 21–33 (2022)
10. Agarwal, P., Idrees, S.M., Obaid, A.J.: Blockchain and IoT technology in transformation of education sector. Int. J. Online Biomed. Eng. (iJOE) **17**(12), 4–18 (2021). https://doi.org/10.3991/ijoe.v17i12.25015
11. DeNio, J.A., Ludwig, S.A.: Improving transaction speed and scalability in blockchain systems. In: 2021 IEEE International Conference on Big Data (Big Data), pp. 3619–3628. IEEE (2021). https://doi.org/10.1109/BigData52589.2021.9671648
12. Al-Musharaf, A.J., Al-Alak, S.M., Al-Mashhadi, H.M.: Improving blockchain consensus mechanism via network clusters. In: 2021 1st Babylon International Conference on Information Technology and Science (BICITS), pp. 293–298. IEEE (2021). https://doi.org/10.1109/BICITS51482.2021.9509882
13. Shahriar Hazari, S., Mahmoud, Q.H.: Improving transaction speed and scalability of blockchain systems via parallel proof of work. Future Internet **12**(8), 125 (2020). https://doi.org/10.3390/fi12080125
14. Mai, T., Yao, H., Zhang, N., Xu, L., Guizani, M., Guo, S.: Cloud mining pool aided blockchain-enabled Internet of Things: an evolutionary game approach. IEEE Trans. Cloud Comput. (2021). https://doi.org/10.1109/TCC.2021.3110965
15. Tsai, C.W., Chen, Y.P., Tang, T.C., Luo, Y.C.: An efficient parallel machine learning-based blockchain framework. Ict Expr. **7**(3), 300–307 (2021). https://doi.org/10.1016/j.icte.2021.08.014

16. Raza, Z., Haq, I.U., Muneeb, M., Shafiq, O.: Energy efficient multiprocessing solo mining algorithms for public blockchain systems. Sci. Program. **2021**, 1–13 (2021). https://doi.org/10.1155/2021/9996132
17. Kan, J., Chen, S., Huang, X.: Improve blockchain performance using graph data structure and parallel mining. In: 2018 1st IEEE International Conference on Hot Information-Centric Networking (HotICN), pp. 173–178. IEEE (2018). https://doi.org/10.1109/HOTICN.2018.8606020
18. Fanfakh, A., Charr, J.C., Couturier, R., Giersch, A.: Optimizing the energy consumption of message passing applications with iterations executed over grids. J. Comput. Sci. **17**, 562–575 (2016). https://doi.org/10.1016/j.jocs.2016.07.012
19. Fanfakh, A.B.M.: Predicting the performance of MPI applications over different grid architectures. J. Univ. Babylon Pure Appl. Sci. **27**(1), 468–477 (2019)

Chapter 25
Customer Churn Prediction Model Using Deep Learning

Srijan Sur, Riya Sil, Bharat Bhushan, Pronaya Bhattacharya, and Anuj Kumar

1 Introduction

Churn is described as the customer who transfers from one provider of telephone services to another [1]. To be precise, loyal customer relationship results in cost cutting. The price of attracting a new client is substantially lower against the price of serving an existing one. Customers create revenue for many different types of businesses. Each year that a customer stays with a firm, the earnings rise. In the financial services, a 5% improvement in client retention results for a profit rise of almost 25%. Returning clients tend to gradually purchase more from a corporation. As they do, your serving operations expenses decline. Additionally, regular customers recommend people to the business. And they frequently pay a higher price to carry on doing business rather than going to a rival, somebody they are not familiar with or comfortable. Typically following types of data are used in customer churn prediction model.

S. Sur
Department of Computer Science and Engineering, School of Engineering and Technology, Adamas University, Kolkata, India

R. Sil (✉)
Department of Computer Science, Kristu Jayanti College, Autonomous, Bengaluru, India
e-mail: ritya.research@gmail.com

B. Bhushan · A. Kumar
Department of Computer Science and Engineering, School of Engineering and Technology, Sharda University, Greater Noida, India

P. Bhattacharya
Department of Computer Science and Engineering, Amity School of Engineering and Technology, Research and Innovation Cell, Amity University, Greater Noida, India
e-mail: pbhattacharya@kol.amity.edu

J. M. R. S. Tavares et al. (eds.), *Proceedings of Second International Conference on Intelligent System*, Algorithms for Intelligent Systems,
https://doi.org/10.1007/978-981-99-8976-8_26

Customer demographic data: This includes information such as age, gender, income level, education, marital status, which can provide insights into different customer segments and their churn patterns.

Billing and payment data: This includes details about customers' billing and payment history, such as payment amount, payment frequency, payment method, outstanding balance. These data points can indicate customers' financial stability and their likelihood to churn.

Usage and interaction data: This includes data about customers' usage patterns, such as frequency and duration of product usage, feature or service utilization, interactions with customer support, website/app logins. This data helps understand customers' engagement with the product or service and identify early indicators of potential churn. A data-driven approach to predicting customer churn differs from traditional methods in several ways:

Use of extensive data: A data-driven approach relies on large amounts of data are gathered from several sources, including consumer demographics, purchase history, usage trends, customer interactions, and feedback. Conventional techniques could rely on scant or manually gathered data.

Advanced analytics techniques: Data-driven approaches use cutting-edge analytics tools like machine learning, predictive modeling, and data mining to examine data trends and find insights that conventional methods might miss. Traditional approaches could use qualitative analysis or straightforward rule-based systems.

Automated prediction process: Data-driven approaches typically involve automated processes that allow for faster and more efficient prediction. Traditional approaches could include manual evaluation and interpretation, which can be time-consuming and subject to human biases.

Real-time insights: Data-driven approaches can provide real-time insights into customer churn by continuously analyzing and updating the data. Traditional methods may provide retrospective analysis but lack the ability to react to changes in customer behavior promptly.

Deep learning to churn prediction model in various ways:

Feature Learning: Deep learning models have the ability to automatically learn relevant and intricate patterns from raw data, unlike traditional algorithms that rely on manual feature engineering. This allows deep learning models to capture more complex relationships and dependencies in the data, which can be essential for accurate churn prediction.

Non-linear Relationships: Traditional algorithms often assume linear relationships between features and the target variable. However, churn prediction is a complex problem with non-linear relationships. Deep learning models, with their multilayer architectures and activation functions, can model these non-linear relationships effectively, providing more accurate predictions.

Transfer Learning: Deep learning models can leverage transfer learning, where a pretrained model, trained on a large dataset is fine-tuned on a smaller churn prediction dataset. This helps the model to generalize better and achieve better accuracy with limited training data, which can be especially useful in scenarios where churn data is scarce.

Handling Big Data: Deep learning is well-suited for handling large volumes of data, which is often the case in churn prediction scenarios. Deep learning models can effectively process and learn from this big data, enabling them to capture more subtle patterns and make more accurate predictions. A number of things, such as exorbitant costs, subpar customer support, poor product quality, or the emergence of a more suitable substitute on the market, might result in customer churn. Churn can also be influenced by factors outside of a company's control, such as changes in economic conditions or shifts in consumer preferences [2]. Customer churn prediction is an essential task for any company that provides subscription-based services or products. It involves predicting which customers are likely to stop using a company's services or products and taking proactive measures to retain those customers before they defect to a competitor. This can include offering incentives, personalized communication, and improving the overall customer experience [3].

1.1 Customer Churn Prediction

Today the world is moving at a fast pace with the help of the Internet. Internet users have grown by almost 103% in the last 10 years. There is a compelling case for the rapid growth of GPRS-based medium broadband systems, which are currently at market penetration levels of 70–80% in several EU countries [4]. In addition, long-distance developments are brought about by the decoupling and merging the cellular telecommunications sector alongside different businesses such as media and IT [5]. Companies in the telecom sector are trying to capture a customer base and existing companies are restless to retain their customers. In these circumstances, the customer churn prediction model can be very crucial [6]. Accurately predicting customer churn can highly improve customer retention. As an alternative to the conventional approach of merely providing phone services, telecommunications firms have introduced a new product called triple play that combines TV, broadband, and phone offerings [7]. The triple play raises Average income Per User (ARPU), immediately boosting the company's earnings while also aiding in client retention. [8]. The challenge of triple play service is driven by the goal to offer clients with an unparalleled experience regardless of whether they use the service or request help from their service provider. [9]. The rapidly growing business may be impacted by this churning problem. Numerous customer churn prediction (CCP) models have been developed as a result, but they often fail to produce the expected CCP performance because of undiscovered potential variation variables that may affect customer churn (CC) [10].

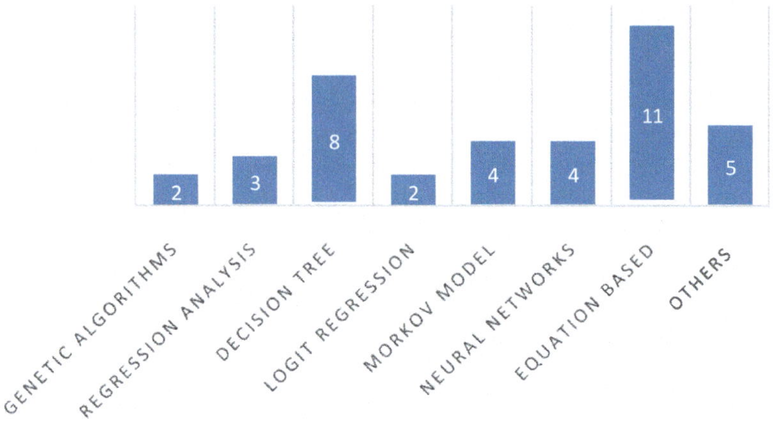

Fig. 1 Total number of papers on each topic

2 Literature Review

Customer churn is forecasted in many organizations using various statistical and machine learning methods. Figure 1 shows a survey of the articles that employed such approaches [11] (Table 1).

3 Proposed Model

The suggested approach uses a binary classification deep learning neural network for predicting whether a client would churn [18]. Customer churn prediction can be very fruitful for companies. Machine learning techniques can be used to anticipate consumer turnover [19]. The total amount of money a business might anticipate from a single client throughout their whole customer lifetime is termed as customer lifetime value (CLV). The corporation can dramatically boost its sales by raising the CLV without needing to spend more on marketing. That is one of the many reasons to go for Customer retention. In this paper, there are three phases:—(a) Data Preprocessing, (b) Data Visualization. (c) Model Building

3.1 Data Preprocessing

Authors were able to determine a clear relationship between the customer's behavior and churn thanks to this dataset consisting of numerous customer-related information. The dataset has 7043 rows and 21 features [20]. In the dataset, each row corresponds

Table 1 Accuracy of various research work

Sl. No	Paper title	Dataset and classifier	Accuracy	Limitation
1	An effective classifier for predicting churn in tele-communication [12]	Telco IBM, XGBoost, KNN, Random Forest,	XGBoost-79.8%	Accuracy can be improved further
2	Churn prediction in telecommunication using Logistic Regression and Logit Boost [9]	Orange Telecom, Logit Boost, Logistic Regression	Logit Boost-85.17% Logistic Regression-85.23%	Dataset with large amounts of data can be used
3	Transfer learning and meta classification-based deep churn prediction system for telecom Industry [13]	Cell2Cell and orange, transfer learning with CNN	68.2% 75.4%	Accuracy can be improved further
4	Telecom customer churn prediction [14]	IBM Telco, Logistic Regression, RF, SVM, DT	RF- 93.55% SVM-81.92%	RF has achieved low AUC. SVM model can be improved further
5	Prediction	of customer churn using machine learning [15]	Telco IBM, ANN	82.83%
6	Automated feature selection and churn prediction using deep learning models [16]	Cell2Cell and Crowd Analytix, CNN + LFNN	71.66%	Accuracy can be improved further
7	Prediction of customer churn in telecom industry: A machine learning perspective [3]	IBM Telco, Logistic Regression, ANN RF, AdaBoost, XGBoost, GBM	XGBoost-82.20%	Dataset with large amounts of data can be used
8	Customer churn prediction modeling based on behavioral patterns analysis using deep learning [17]	Telco IBM, Multilayer ANN	80.03%	Hyperparameter tuning and early stopping can be used

to a customer, and each column provides information about that customer's qualities that is detailed in the column Metadata.

3.1.1 Handling Missing Data

Handling missing data is crucial as it affects statistical analysis through the loss of knowledge and data non-uniformities [21]. It is extremely important to pinpoint the location of missing values. In the IBM Telco dataset, the Total Charges contain 0.16% missing values. All the Total Charges null values occur when the Tenure value is zero. Zero is assigned to each missing value.

3.1.2 One Hot Encoding (OHE)

OHE is a commonly used technique that uses binary encoding, in which the presence of a group is denoted by a 1, and its absence by a 0. Conversion of each category to a vector of N-dimension is done with an element of 0 or 1 with N number of categories in the nominal attribute. The raw data is not harmed by this method. However, OHE has a number of issues, including "sparse data" and "high dimensionality", which will raise the computational cost of learning methods. OHE should not be used for high cardinality columns, because of this [22].

3.1.3 Data Normalization

To improve the prediction results Standard Scaler is applied to normalize the data. With a mean of zero and a standard deviation of one (unit variance), the Standard Scaler assists in obtaining a standardized distribution. By omitting the feature's mean value and dividing the result by the feature's standard deviation, it standardizes the feature.

3.2 Data Visualization

EDA stands for Exploratory Data Analysis, which is an approach to analyzing and summarizing a dataset in order to gain insights and identify patterns and relationships between variables. EDA involves visualizing and manipulating data using various statistical and computational methods, in order to identify patterns, anomalies, and trends. Some common techniques used in EDA include scatter plots, histograms, box plots, and correlation analysis.

The goal of EDA is to understand the data and the relationships between variables, so that author can identify potential problems, such as missing data, outliers, or errors in the dataset, and address them appropriately. Making judgements on how to move forward with data modeling and analysis requires that the first phase in the data analysis process, called exploratory data analysis (EDA), be completed.

3.3 Model Building

The created neural network has 12 input layer neurons, 6 hidden layer neurons, and 1 output layer neuron. No of epochs were 50 and batch size 80, respectively. Input layer and the hidden layer had the ReLU activation function and the output layer had Sigmoid activation function. Binary cross entropy was given as the loss function. Test size was chosen as 20%. This model is further improved by applying hyperparameter tuning techniques such as GridSearchCV and BayesianSeachCV, Dropout, and Early Stopping are also applied later on.

4 Results

Simple Dropout produced the highest accuracy. Our model obtained 83.1% accuracy. After several iterations 24 neurons for input layer 12 neurons for hidden layer and a dropout rate of 0.01 or 1% turned out be the best. It uses the Adam optimizer, a stochastic gradient descent technique that relies on adaptive estimate of first and second order moments. Following are the step-wise results.

4.1 Data Preprocessing

These are all the columns and their data types, non-null counts (Fig. 2).

Fig. 2 Dataset information

```
<class 'pandas.core.frame.DataFrame'>
RangeIndex: 7043 entries, 0 to 7042
Data columns (total 21 columns):
 #   Column            Non-Null Count  Dtype
---  ------            --------------  -----
 0   customerID        7043 non-null   object
 1   gender            7043 non-null   object
 2   SeniorCitizen     7043 non-null   int64
 3   Partner           7043 non-null   object
 4   Dependents        7043 non-null   object
 5   tenure            7043 non-null   int64
 6   PhoneService      7043 non-null   object
 7   MultipleLines     7043 non-null   object
 8   InternetService   7043 non-null   object
 9   OnlineSecurity    7043 non-null   object
 10  OnlineBackup      7043 non-null   object
 11  DeviceProtection  7043 non-null   object
 12  TechSupport       7043 non-null   object
 13  StreamingTV       7043 non-null   object
 14  StreamingMovies   7043 non-null   object
 15  Contract          7043 non-null   object
 16  PaperlessBilling  7043 non-null   object
 17  PaymentMethod     7043 non-null   object
 18  MonthlyCharges    7043 non-null   float64
 19  TotalCharges      7043 non-null   object
 20  Churn             7043 non-null   object
dtypes: float64(1), int64(2), object(18)
memory usage: 1.1+ MB
```

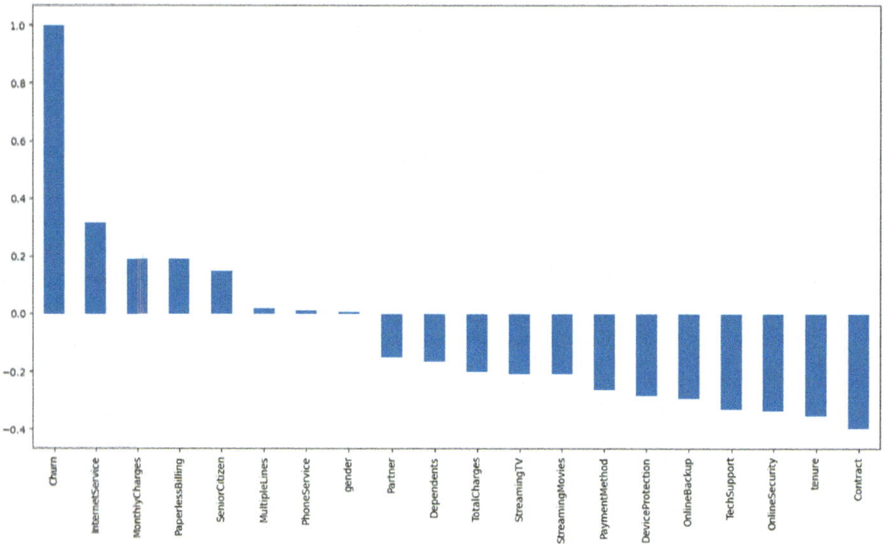

Fig. 3 Correlation plot

4.2 Data Visualization

This plot is representing the correlation of every other feature with the dependent variable "Churn". As noticeable InternetService, MonthlyCharges, PaperlessBilling are some features that are positively correlated with churn. On the other hand, Partner, TotalCharges, Dependents are some features that are negatively correlated with churn (Fig. 3).

According to the statistics, 73.4% of consumers have stopped churning and are still contributing to the firm, while 26.6% have switched to a different service provider, as seen in Fig. 4.

From the figure above, both the genders are equally likely to churn as well as to not churn. The effect of gender is negligible on the churn probability, as shown in Fig. 5.

Highest churn percentage was seen from the payment method "Electronic check" which is 33.6%, as demonstrated in Fig. 6. "Mailed check" payment method has 22.8% churn percentage, "Bank transfer" and "Credit card" payment method has 21.9% and 21.6% churn percentage, respectively, (Fig. 7).

This plot represents the distribution of "Churn" with respect to monthly charges. Visibly, the increase in monthly charges increases churn, slightly decreases retention (Fig. 8).

This plot represents the distribution of "Churn" with respect to Total Charges. Interestingly the increase in Total Charges decreases churn, also slightly decreases retention.

Fig. 4 Churn Rate in IBM Telco

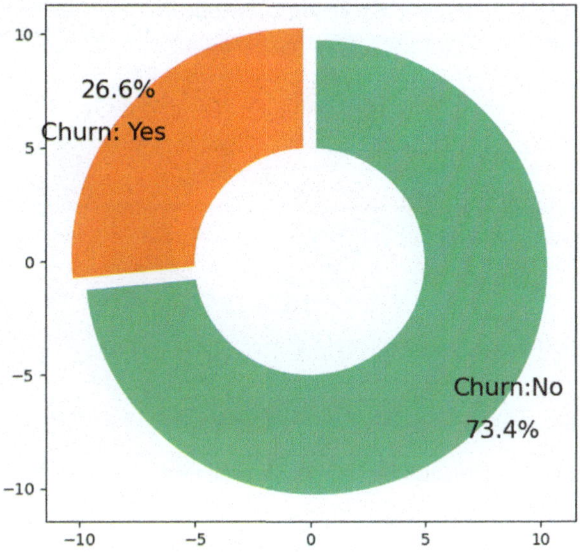

Fig. 5 Churn distribution with gender in IBM Telco

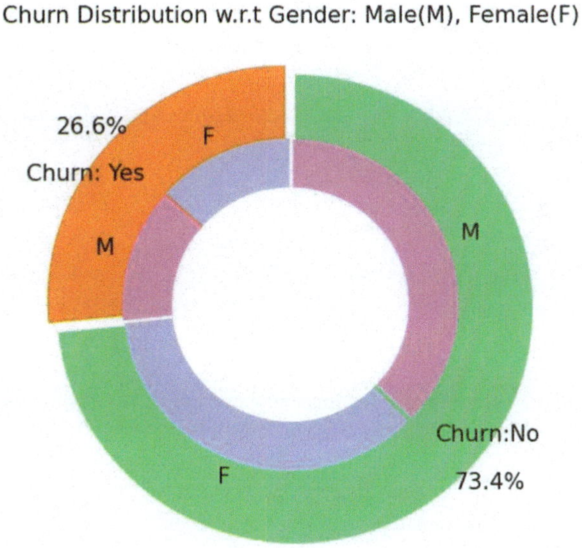

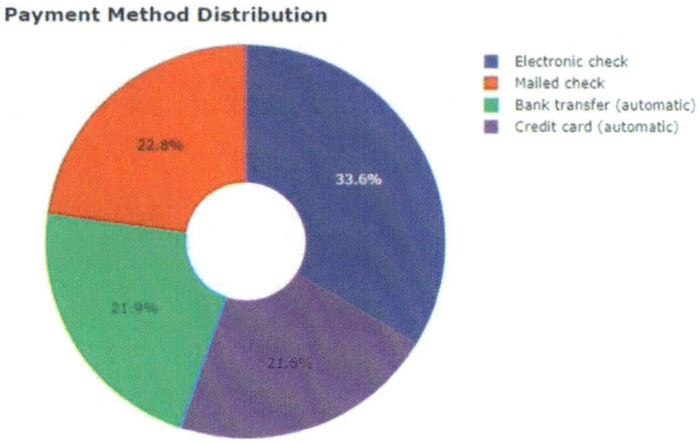

Fig. 6 Churn distribution with payment method in IBM Telco

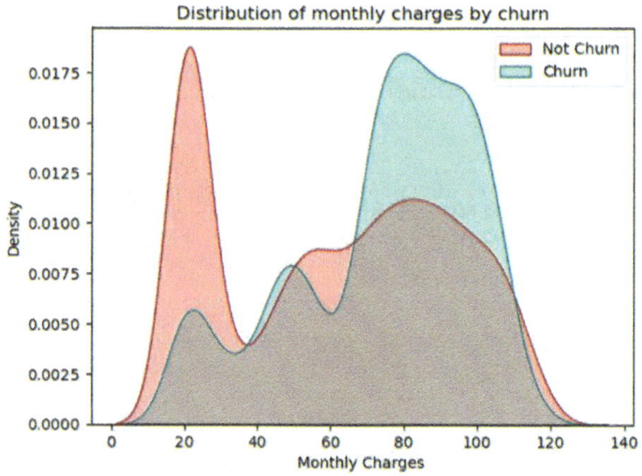

Fig. 7 Kernel density estimate plot of churn w.r.t monthly charges

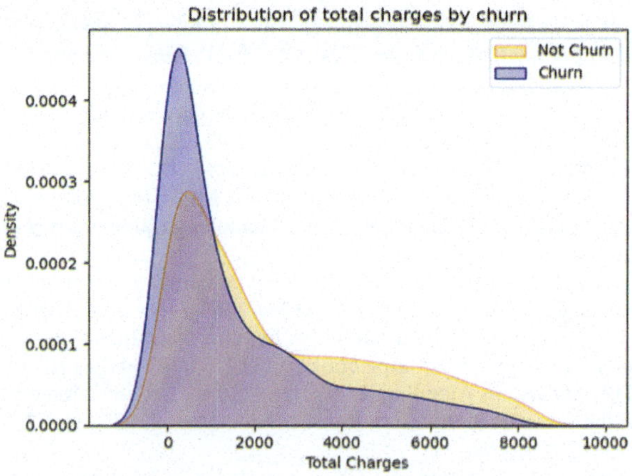

Fig. 8 Kernel density estimate plot of churn w.r.t total charges

5 Conclusion

Deep learning techniques can be very beneficial and favorable because businesses today deal with large amount of data points that machine learning techniques cannot handle. Most of the recent research acknowledged the value of customer churn prediction models, particularly in the telecom sector. Many studies have started using deep learning algorithms to create models that can effectively forecast client attrition. Since the accuracy of the current deep learning models needs to be increased, there is still a gap in the field. This research is extremely important since it implements a deep learning model which can predict customer turnover with reasonably high accuracy when contrasted with current approaches and manage vast amounts of data. In this study, the ANN model forecasted customer attrition with a precision of 83.1% to be precise. The model having the total of 37 neurons including input layer with 24 neurons and one single neuron output layer, one hidden layer consisting of 12 neurons, 82 batch size, and 70 epochs gave the best results. The dropout layer is there between the input layer and the hidden layer. Hyperparameter tuning only resulted marginal improvements.

References

1. Mishra, A., Reddy, U.S.: A comparative study of customer churn prediction in telecom industry using ensemble based classifiers. In: 2017 International Conference on Inventive Computing and Informatics (ICICI) (2017). https://doi.org/10.1109/icici.2017.8365230
2. Berkhout, A.J. (Guus), Duin, P.A.: New ways of innovation: An application of the cyclic innovation model to the Mobile Telecom Industry. Int. J. Technol. Manag. **40**(4), 294 (2007). https://doi.org/10.1504/ijtm.2007.015754

3. Hota, L., Dash, P.K.: Prediction of customer churn in telecom industry: a machine learning perspective. Comput Intell Mach Learn **2**(2), 1–9, (2021). https://doi.org/10.36647/ciml/02.02.a001
4. Khder, M.A., Fujo, S.W., Sayfi, M.A.: A roadmap to data science: background, future, and trends. Int. J. Intell. Inf. Database Syst. **14**(3), 277–293 (2021)
5. Perez, M.J., Flannery, W.T.: A study of the relationships between service failures and customer churn in a telecommunications environment. In: PICMET '09—2009 Portland international conference on management of engineering and technology (2009). https://doi.org/10.1109/picmet.2009.5262262
6. Wu S (2023) Customer churn prediction in the telecommunication industry. Adv. Econ. Manag. Political Sci. **4**(1):41–50. https://doi.org/10.54254/2754-1169/4/20221017
7. Tsai, C-. F., Lu, Y-. H. (2010) Data mining techniques in customer churn prediction. Recent Patents Comput. Sci. **3**(1), 28–32 https://doi.org/10.2174/1874479611003010028
8. Chabumba, DR, Jadhav, A, Ajoodha, R: Predicting telecommunication customer churn using machine learning techniques Interdisc. Res. Technol. Manag. 625–636 (2021) https://doi.org/10.1201/9781003202240-98
9. Jain, H., Khunteta, A., Srivastava, S.: Churn prediction in telecommunication using logistic regression and logit boost. Procedia Comput. Sci. **167**, 101–112 (2020). https://doi.org/10.1016/j.procs.2020.03.187
10. Mahmood, I.N., Abdullah, H.S.: Telecom churn prediction based on Deep Learning Approach. Iraqi J. Sci. 2667–2675 (2022). https://doi.org/10.24996/ijs.2022.63.6.32
11. Rani, K.S., Thaslima, S., Prasanna, N.G. L., Vindhya, R., Srilakshmi, P.: Analysis of customer churn prediction in telecom industry using logistic regression. Int. J. Innovative Res. Comput. Sci. Technol. **9**(4) (2021). https://doi.org/10.21276/ijircst.2021.9.4.6
12. Pamina, J., Beschi Raja, J., Sathya Bama, S., Soundarya, S., Sruthi, M.S., Kiruthika, S., Aiswaryadevi, V.J., Priyanka, G.: An effective classifier for predicting churn in telecommunication. J. Adv. Res. Dyn. Control Syst. **11**(1 Special Issue), 221–229 (2019)
13. Ahmed,U., Khan, A., Khan, S.H., Basit, A., Haq, I.U., Lee, Y.S.: Transfer learning and meta classification based deep churn prediction system for telecom industry 1–10 (2019)
14. Sundararajan, A., Gursoy, K.: Telecom customer churn prediction. March (2020). . https://doi.org/10.7282/t3-76xmde75
15. Momin, S., Bohra, T., Raut, P.: Prediction of customer churn using machine learning. In: EAI International Conference on Big Data Innovation for Sustainable Cognitive Computing, pp. 203–212, (2020) 2020. D.
16. Umayaparvathi, V., Iyakutti, K.: Automated feature selection and churn prediction using deep learning models. Int Res J Eng Technol (IRJET) **4**(3):1846–1854 (2017)
17. Agrawal, S., Das, A., Gaikwad, A., Dhage, S.: Customer churn prediction modelling based on behavioural patterns analysis using Deep Learning. In: 2018 International Conference on Smart Computingand Electronic Enterprise (ICSCEE) (2018). https://doi.org/10.1109/icscee.2018.8538420
18. Ahmad, A.K., Jafar, A., Aljoumaa, K.: Customer churn prediction in telecom using machine learning in big data platform. J. Big Data 6(1) (2019). https://doi.org/10.1186/s40537-019-0191-6
19. Hassonah, M.A., Rodan, A., Al-Tamimi, A.-K., Alsakran, J.: Churn prediction: A comparative study using KNN and decision trees. 2019 Sixth HCT Inf. Technol. Trends (ITT) (2019). https://doi.org/10.1109/itt48889.2019.9075077
20. Kassem, E.A., Ali, S., Mostafa, A., Kamal, F.: Customer churn prediction model and identifying features to increase customer retention based on user generated content. Int. J. Adv. Comput. Sci. Appl. **11**(5) (2020). https://doi.org/10.14569/ijacsa.2020.0110567
21. Xie, J., Li, Y., Wang, N., Xin, L., Fang, Y., Liu, J.: Feature selection and syndrome classification for rheumatoid arthritis patients with traditional chinese medicine treatment. Eur. J. Integr. Med. **34**, 101059 (2020). https://doi.org/10.1016/j.eujim.2020.101059
22. Li, Q., Xiong, Q., Ji, S., Wen, J., Gao, M., Yu, Y., Xu, R.: Using fine-tuned conditional probabilities for data transformation of nominal attributes. Pattern Recogn. Lett. **128**, 107–114 (2019). https://doi.org/10.1016/j.patrec.2019.08.024

Chapter 26
On-Line Wear Measurement for Micro Turning Round Tip Tool Based on Machine Vision

Aravindan Srinivasan, Sudipto Bhattacharyya, Raj Kumar Parida, J. Viswanath, Viswanathan Ramasamy, Vithya Ganesan, and S. Padmakala

1 Introduction

This study reveals a new approach of obtaining tool wear in real time using machine vision to build a tool wear model during turning. To optimize the detection areas for the obtained tool wear images and precisely locate the tool's worn tips, this study introduces a novel, flexible approach to analyzing tool wear for micro turning tool cutters. After that, the built-in critical area's vertical critical area is scanned pixel-by-pixel using the first pixel-accurate detection of wear edge locations. To improve the accuracy of detected worn edge locations, a sub-pixel precise detection approach is proposed. Experiments were conducted using a cemented carbide ball end turning cutter and stainless steel 1Cr18Ni9Ti workpieces to validate a current on-line monitoring approach for tool wear by comparing the detected tool wear values

A. Srinivasan · V. Ramasamy · V. Ganesan (✉)
Department of CSE, Koneru Lakshmaiah Education Foundation, Vaddeswaram, AP 522302, India
e-mail: vithyaganesan@kluniversity.in

S. Bhattacharyya
Department of Computer Science and Engineering, Global Institute of Management and Technology Krishnagar, Nadia, India

R. K. Parida
School of Computing Science and Engineering, Galgotias University, Greater Noida 203201, India

J. Viswanath
Madanapalle Institute of Technology and Science (MITS), Madanapalle, AP, India
e-mail: viswanathj@mits.ac.in

S. Padmakala
Department of Computer Science and Engineering Saveetha School of Engineering SIMATS Chennai, Chennai, Tamil Nadu, India
e-mail: padmakalas.sse@saveetha.com

J. M. R. S. Tavares et al. (eds.), *Proceedings of Second International Conference on Intelligent System*, Algorithms for Intelligent Systems,
https://doi.org/10.1007/978-981-99-8976-8_27

to the measurement value of a microscope. This research lays the framework for a future application of machine vision for continuous monitoring of tool wear. The experimental results show that the proposed on-line tool wear measurement strategy makes use of sub-pixel detection, which provides more precise results than manual microscope measurement.

2 Literature Survey

In this article [1], "Monitoring tool wear is crucial for automated mechanical processing, enhancing quality and efficiency. This study introduced a novel deep learning model to identify tool wear status. The force signal was converted into 2D images using various methods like CWT, STFT, and GASF, then analyzed using a CNN. Accuracy exceeded 90%, outperforming AlexNet and ResNet. CWT-generated images achieved highest accuracy due to noise resilience and local feature extraction."

This study introduces a device to detect acoustic emission (AE) signals from rotating tools using magnetofluid. The proposed method involves wavelet packet transform for feature extraction from AE signals, followed by fuzzy clustering (FCM) to link features with decisions. "AE signals are decomposed into frequency bands via wavelet packet transform, with root mean square (RMS) values from each band used as features" [2]. An experimental study on cutting tool wear, developing a model for wear estimation. Changes in cutting forces, vibrations, and acoustic emissions with wear are analyzed. "An adaptive neuro fuzzy inference system (ANFIS) model using these factors accurately estimates tool wear rate, performing well in comparison to practical results" [3].

Augmented Reality (AR) is rapidly advancing and finding applications in diverse fields. This paper reviews academic studies on AR in manufacturing, highlighting challenges in accuracy, speed, and alignment with industrial standards. It explores AR's role in activities like design, robotics, maintenance, and CNC machining, while addressing future trends and challenges [4]. "In metal cutting, tool wear modes impact production, yet preset tool life rules hinder optimization. Premature tool failure leads to downtime and losses. Tool condition monitoring (TCM) research spans decades, focusing on offline and on-line systems, with Industry 4.0 driving interest in smart sensors for real-time monitoring. This paper reviews TCM's state, discusses machining operations, feasibility, and current implementations, while highlighting industry challenges and offering research directions" [5].

"Inconel 625 is valued for its properties but is hard to machine. Minimum quantity lubrication (MQL) is used for improved machining, yet challenging materials may require more. This study introduced nano-MQL with hBN nanoparticles for turning Inconel 625, demonstrating better tool life, lower wear, and smoother surfaces with 0.5% hBN nanofluid" [6]. "The paper reviews sensor and signal processing in tool condition monitoring (TCM) for industrial machining. Unlike previous reviews, it categorizes findings by machine type, highlighting trends and gaps. Recommendations for future research aim to foster idea exchange and technique enhancement in TCM" [7].

Flank wear is a prevalent and costly issue in metal cutting, causing material loss, and machine downtime. This paper reviews TCMS aspects: signal acquisition, processing, and AI decision-making for flank wear monitoring. It addresses techniques to detect and manage flank wear, aiming to enhance tool condition monitoring systems [8]. "Indirect tool wear monitoring and breakage detection use various physical quantities like cutting forces. Disturbances challenge accurate monitoring, and a novel approach of "intelligent filtering" decomposes force signals by identifying cutting process models. This method's application to milling with two cutting insert tools is discussed" [9].

"Tool condition monitoring parallels the growth of automated manufacturing, driven by quality improvement demands. Digital image processing advances for tool condition monitoring, aided by improved machine vision and computing, are reviewed. The paper calls for systematic research in this evolving field" [10].

3 Methodology

3.1 The Theory Behind Real-Time Monitoring of Tool Wear

Establishing a system for measuring and monitoring tool wear is a significant technological challenge due to the need to develop an on-line system based on machine vision. The layout has two potential modes of operation: real-time and in-process. Turning tool wear can be captured and detected in real time using the on-line measuring and monitoring system's in-process mode. The nature of dynamic detection and the influence of machining conditions, however, make precise monitoring of the tool wear more difficult. When it comes to measuring tool wear, an in-process mode monitoring system has the advantage because it can only measure and identify wear as it occurs during the process. Because turning operations are often interrupted, this could potentially lead to a more accurate method of measuring tool wear. However, due to a number of technological issues, this system is unable to track tool wear in real time. This paper's scope is limited to the design and implementation of a system for measuring and monitoring tool wear during operation.

The micro turning tool cutter's real-time monitoring and measurement system is made up of a CCD camera (a Dahn DH-HV3103UC), a telemetric lens (a Computer TEC-M55), an LED light (with movable brightness), a computer, a camera holder, and image processing software. It is possible to take images of the cutting process in progress because the CCD camera is positioned so that its optical axis is parallel to the axis of the cutters. To capture an image of the tool wear induced by the micro turning tool cutter, the spindle of the machine tool is lifted above the camera. To further emphasize the difference between the tool's flank and the background, a charge-coupled device (CCD) camera is subsequently used to take a picture of the worn zone. In between each turning operation, the tool is set to stop in the same place so that its image can be captured. To improve the quality of the tool wear image, a nozzle for pressurized air is used to blow away the chips that have stuck to the tool wear area.

In taken photos of tools, the wear area is obvious to the naked eye. Properties of the worn area, the background, and the unworn area all play a role in defining the boundaries of the tool wear region:

(1) The amount of gray is higher in the worn region than in the unworn tool wear area or the backdrop.
(2) There is a significant shift in gray levels close to the region of tool wear.
(3) In the vertical direction, there are two edges that wear down. Along the top wear edge, the grayscale values rise from top to bottom, while the opposite is true along the bottom wear edge.

In order to extract the tool wear edges, an edge detection algorithm must be applied to the aforementioned tool wear region characteristics. The schematic representation of the tool wear region is shown in Fig. 1. See Fig. 2 for a 1024 × 768 version of the acquired image showing the zoomed-in tool wear zone. When taking a picture of a tool, make sure the blade is roughly perpendicular to the horizontal axis of the picture. Choose the tool wear measuring mode and determine the position of the tool's tip to get an estimate of the amount of wear on a micro turning tool cutter using machine vision. This will aid in the image's processing in the future. Choosing the measurement mode for tool wear is the first step. In the second stage, the location of the tool's tip is determined. Third, we'll fine-tune the area where we look for wear. Fourth, pixel precise wear edge point identification is performed initially. Fifth, using sub-pixel precision, we locate the worn edge spots. Extraction of tool wear from edge sites of wear is the sixth step.

In the next section, we'll dive into the specifics of the suggested on-line approach of measuring tool wear for a micro turning tool cutter.

Fig. 1 A schematic representation of the tool picture that was recorded

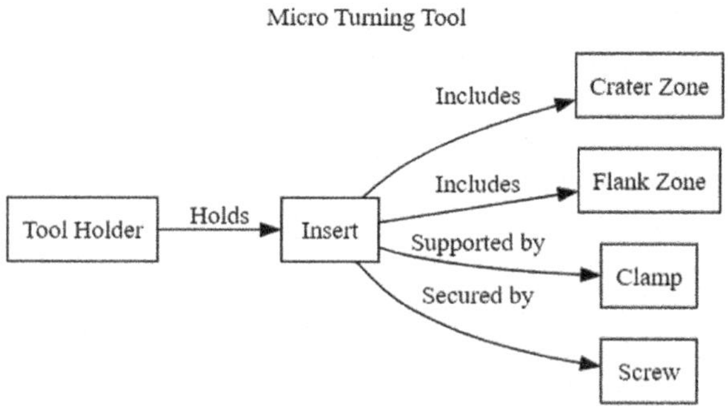

Fig. 2 The tool wear image that was captured

3.2 Choosing a Method for Measuring Tool Wear

The spiral cutting edge of the tool appears curved in the obtained tool picture, and the spindle of a machine tool can move freely in a radial direction, making it difficult to evaluate tool wear. A tool wear measure mode is offered so that tool wear can be measured in a consistent manner. Figure 1 shows the two-fluted micro turning tool cutter. Tool wear is measured in the direction along which the complete cutting edge makes tangential contact with the material at any given position P on the cutting edge. The same method of assessing tool wear is used for all flute numbers of micro turning tool cutters. Thus, the above-mentioned measuring mode can be used to standardize the various measurement techniques for micro turning tool cutters.

3.3 Identification of the Point of the Tool's Tip

Preprocessing the collected tool pictures is necessary for achieving more accurate results. Median filtering is utilized in the study as a preprocessing technique to lessen background noise. Because median filtering preserves clarity while minimizing noise, it is possible to easily identify sharp edges in photos of tools taken under various conditions. This demonstrates the significance of image preprocessing to the subsequent stages of the on-line tool wear measurement approach.

In order to assess tool wear in the aforementioned manner, it is necessary to first preprocess the acquired tool images and then determine the position of the tool tip within those images. If you have a used tool, you may tell if it has worn by comparing its grayscale character to that of a brand new tool. Although identical tool pictures should theoretically have identical tool locations, this is not always the case due to factors such as machine vibration, repeated positioning errors. Therefore, before taking measurements, it is necessary to align the positions of the images of the tools. As shown in Fig. 1, the tool tip point is located halfway between points A and B. By removing these two locations, we can identify where the tool tip point is. Image

Table 1 Illustrates the operator template

− 1	− 3	− 1
0	0	0
1	3	1

captures of tools like those in Fig. 1 include both background and flank surfaces. While the surface areas of the tool's flanks that are closest to the cutting edge are where wear is most noticeable. By comparing fresh and worn photos of the same tool, we can learn the following about points P and Q:

(1) On the edge between the tool flank and the backdrop, between points *P* and *Q*, the local gray-to-gray difference is greatest. The range of grayscale values will increase from lowest to highest as one travels down the line, displaying the greatest degree of contrast.
(2) Point P is where the pictures of two-flute tools are closest together, and point *Q* is where the distance between any two locations on the bottom boundary is smallest. Point *Q* is the most distal edge point on the lower border of the tool flank and background from the lower left corner of the image of a tool recorded with a single flute. The projective point of point *Q* can be located around the top horizontal border of the image after point *Q* has been retrieved.
(3) Based on this research, it's clear that column scanning can be used to locate the precise locations where the tool's flank meets the backdrop. Distances between locations in the image and the edges defining the tool's flank and backdrop are compared to establish the location of the tool's tip point. In this case, the right edge locations are located using the continuity testing approach.

The steps of the processing operation are as follows.

Step 1 is to create an operator template that can interact with the image, as in Table 1. Comparable to the Sobel operator. Using a column scanning technique gives more importance to the vertical direction of this column.

The most critical information is that the Table 1 operator may be used to execute a convolution on the pixels in a column from top to bottom, yielding the gray value difference of each pixel in that column. The pixel with the local maximum of gray difference value can also be used to precisely locate the upper boundary between the tool flank and backdrop.

Finally, a mistake edge point produced by noise or impurity would be shown by a gap between the first and last columns while examining an edge. The search for the edge point must be repeated until the end of the last column is reached. It is possible to reduce the impact of noises and contaminants by optimizing step 3 of the aforementioned processing technique.

4 Verification Through Experiment

Conventional Laboratory Procedures

The suggested on-line tool wear measuring system was put through its paces in a series of turning experiments run on a MIKRON UCP710 five-axis turning machining center. To mill a dry piece of stainless steel 1Cr18Ni9Ti, a coated carbide micro turning tool cutter was employed. Tool wear is estimated by using two distinct cemented carbides micro turning tool cutters. TAIWAN CHUNBAO cemented carbide was used to make two YG8 micro turning tool cutters (10 mm and 16 mm in diameter, respectively) that were manufactured in ZHUZHOU, China. Using a CCD camera connected to a computer running ToolWearOMM software, we take 1024 768-pixel grayscale digital photos of the cutter. In order to determine the cutter's average flank wear, a JSZ6S microscope is employed, with the camera and light source remaining in their original positions. The technical details of the camera arrangement are outlined in Table 2. The custom image processing software does everything that needs doing with the images (ToolWearOMM v1.0) (Fig. 3).

Methods and Circumstances Used in Experiments

For research on tool wear, a 50-mm-thick 1Cr18Ni9Ti steel box will be milled on a MIKRON UCP710 five-axis turning machining center. For every instrument with a ball on one end, two different trials have been conducted. The machining parameters

Table 2 Camera setup requirements

Model of charge-coupled device	DH-HV3103CU
Charge-coupled device resolution	2045 × 1643
Charge-coupled device pixel measure	3.20 mm × 3.20 mm
Model of Len	TEC M053 computer
Focus of Len	53 mm
Rate of magnification of Len	0 to 0.5 X
Aperture of Len	F2.2—F30C

Fig. 3 User-friendly design of Tool Wear OMM's graphical interface

for 16 mm micro turning tool cutters are listed in Table 3. The micro turning tool cutter machining parameters of 10-mm-diameter are summarized in Table 4 (Fig. 4).

The MIKRON UCP710 five-axis turning machining center and the prepared stainless steel 1Cr18Ni9Ti workpiece are both coupled to the on-line tool wear measurement instrument through magnetic force adsorption (Table 5).

Table 3 Micro turning tool cutters with a 16-mm-diameter and their machining parameters

S.No	Depth cut (mm)	Speed of spindle (r/min)	Rate of feed (mm/min)	Cut length (mm)
1	0.5	900	200	100
2	0.9	1000	200	100
3	1.1	800	200	100

Table 4 Micro turning tool cutters with a 10-mm-diameter and their machining parameters

S. No	Depth cut (mm)	Speed of spindle (r/min)	Rate of feed (mm/min)	Cut length (mm)
1	0.7	1600	100	150
2	0.9	1500	90	80
3	0.6	1500	110	110

Table 5 The point totals for tools with a diameter of 16 mm

Points	Figure 7 (a)7	Figure 7 (b)	Figure 7 (c)
P	(90, 150)	(250, 140)	(10,250)
Q	(170,540)	(280,520)	(135,420)
T	(130,330)	(270,310)	(150,270)

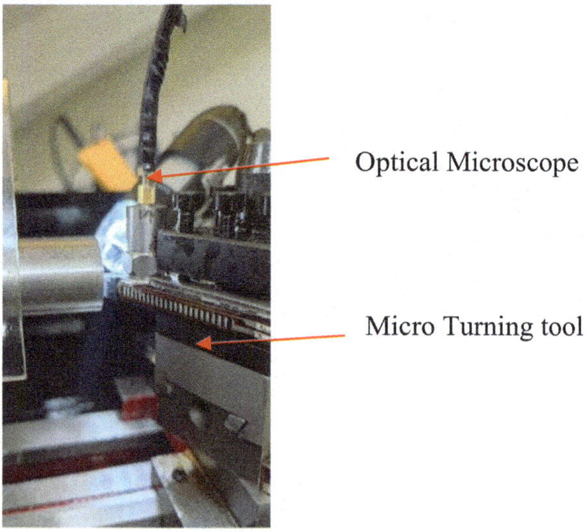

Fig. 4 Tool wear in a turning machining center monitored in real time

Second, three different sets of experiments are conducted using a 16 mm cemented carbides micro turning tool cutter. In Fig. 7, we see the new cutter picture taken by positioning the spindle axis above the CCD camera. The 1Cr18Ni9Ti stainless steel workpiece is then machined using the parameters from Experiment 1's Table 3. An image of a dull cutter is taken in the same fashion as described previously. The acquired new cutter image and worn cutter image are then fed into ToolWearOMM v1.0 for immediate disposal and output of tool wear. Experiments 2–3 in Table 3 can be carried out in the same way as Experiment 1. Figure 8 displays the photographed evidence of tool wear from Experiments #1, #2, and #3 in Table 3. In each image of the tools in Fig. 8, the points A and B in the upper left and upper right corners are denoted by red boxes. Point O on the tool's cursor is exactly in the center between points A and B. Table 3 displays the results of the 16-mm-diameter tool's tip measurements at points A and B. Using the procedure outlined in Table 6 of the aforementioned study, one may calculate the maximum wear, average wear, wear breadth, and wear area depicted in Fig. 5.

Experiments 1 through 3 in Table 4 all involve switching out a cutter made of cemented carbides with a ball end that is 10 mm in diameter. Using a cemented carbides micro turning tool cutter with a 10-mm-diameter, the experiment follows a similar procedure. Figure 6 displays the photographed evidence of tool wear from Experiments #1 through #3 in Table 4. The red boxes in Fig. 5 tool images denote the left-hand corner points A and B. Point O on the tool's cursor is exactly in the center between points A and B. Table 7 displays the results of the measurements taken at

Table 6 Results of wear on instruments with a diameter of 16 mm

Value	Figure 7 (a)	Figure 7 (b)	Figure 7 (c)
Wear (Max)	0.045	0.0456	0.0478
Wear (Avg.)	0.568	0.1256	003.481
Width of Wear	3.879	3.7456	2.8750
Area of Wear	0.154	0.3426	1.9876

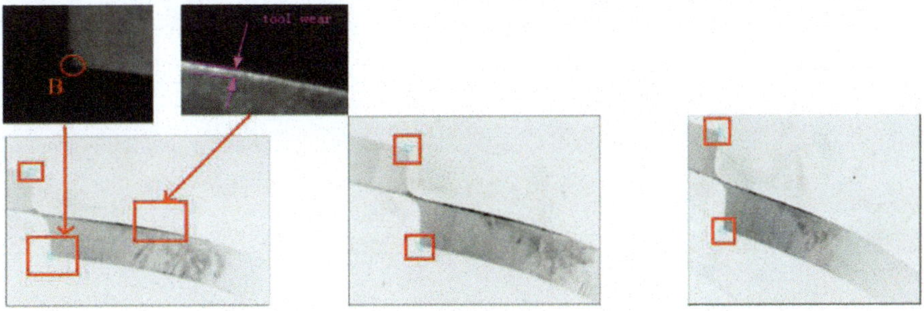

(a) Number 1 of Table.2 (b) Number 2 of Table.2 (c) Number 3 of Table.2

Fig. 5 depicts the tool wear image that was acquired from a micro turning tool cutter with a diameter of 16 mm points A, B, and the 10 mm tool tip at point O. Using the procedure outlined in Table 8 of the aforementioned study, one may calculate the maximum wear, average wear, wear breadth, and wear area depicted in Fig. 6.

According to the established measuring system's accuracy, tool flank wears of two cutting edges can be captured at once for the 10 mm cutter's diameter. Images of a 10-mm-diameter tool can be used to determine the degree of wear on both the left and right flanks of the tool. Table 8 displays the findings for both the left and right cutting edges for each wear parameter.

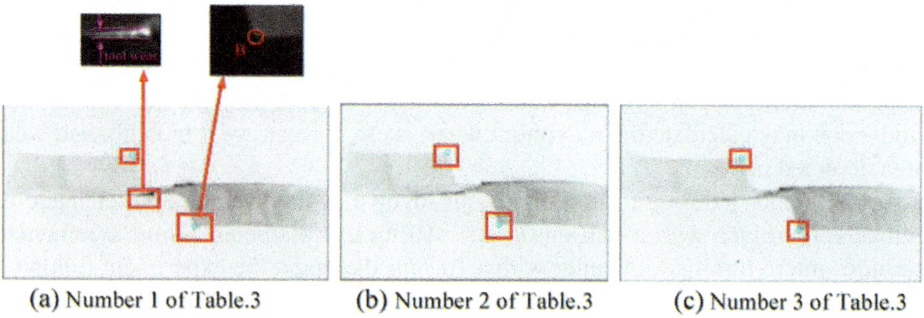

Fig. 6 Depicts the tool wear image that was acquired from a micro turning tool cutter with a diameter

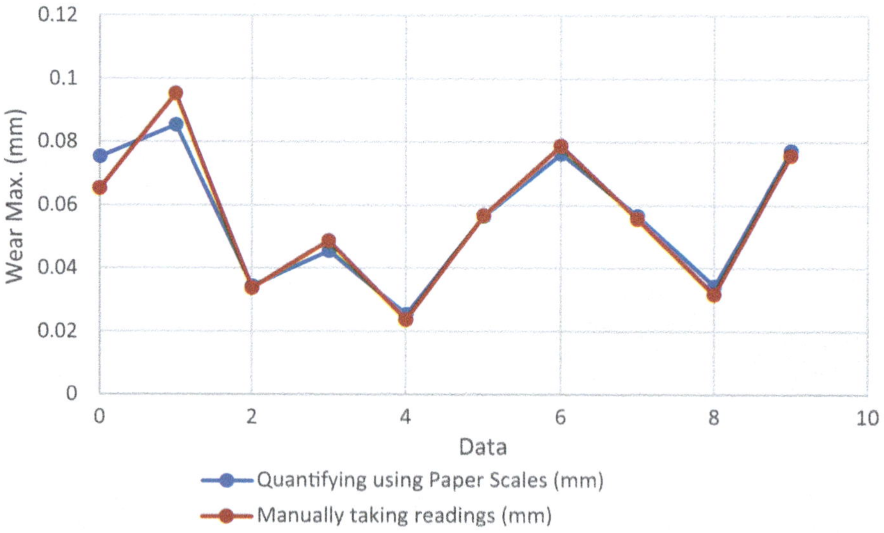

Fig. 7 Map with contrasting measuring outcomes

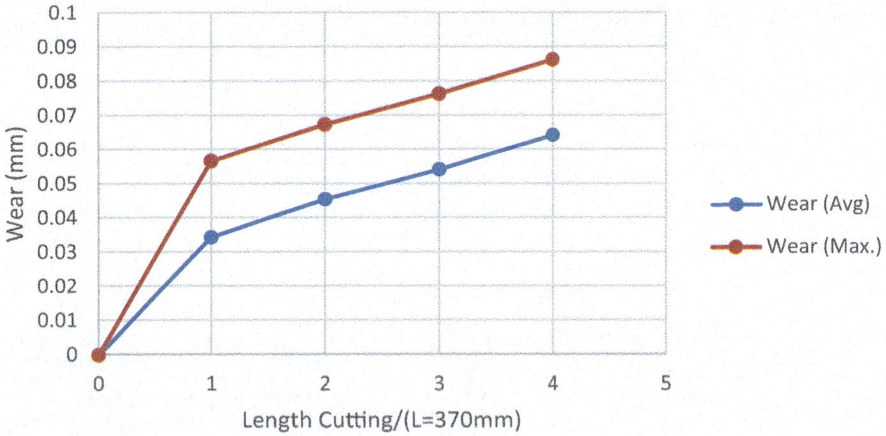

Fig. 8 The development of the tool's wear and tear

Table 7 Using a 10-mm-diameter instrument, we can draw the following conclusions

Points	Figure 8 (a)	Figure 8 (b)	Figure 8 (c)
P	(80, 150)	(150, 240)	(20,150)
Q	(180,550)	(180,550)	(155,450)
T	(170,430)	(370,410)	(180,3 70)

Table 8 Results of wear on instruments with a diameter of 10 mm

Value	Figure 7 (a)		Figure 7 (b)		Figure 7 (c)	
	L	R	L	R	L	R
Wear (Max)	0.015	0.3456	0.1456	0.5670	0.2378	0.4580
Wear (Avg.)	0.268	0.1235	0.1476	1.5690	0.781	1.7649
Width of Wear	3.079	3.0980	3.8556	2.9876	2.813	0.1256
Area of Wear	0.174	1.5680	0.3745	1.0956	1.0876	0.7450

5 Discussion of Experimental Findings

To ensure the reliability of on-line tool wear measurements, the wear on six micro turning tool cutters is measured physically using a microscope. Table 9 compares the results of tool wear measurements taken using the suggested method to those taken directly by hand. The above measurement findings utilizing the suggested approach and direct manual measurement are compared and contrasted with an illustration of the tool wear in Fig. 6.

Table 9 The contrasting results of the measurements

S. No	Figure no		Quantifying using paper scales (mm)	Manually taking readings (mm)	The (T) error (mm)
1	Figure 7	(a)	0.0856	0.0856	0.0017
2		(b)	0.0346	0.034	− 0.0045
3		(c)	0.0457	0.049	0.005
4	Figure 8 (a)	L	0.0256	0.024	0.0004
5		R	0.0568	0.057	0.003
6	Figure 8 (b)	L	0.0765	0.079	− 0.0034
7		R	0.0569	0.056	0.00020
8	Figure 8 (c)	L	0.0346	0.032	0.0005
9		R	0.0775	0.076	0.0034
10			Error Average Value		0.00176
L: Left-hand cutting edge wear			R: Right-hand cutting edge wear		

By comparing the findings of the proposed on-line tool wear measuring approach with those from direct manual measurement (Table 9) and Fig. 6, it is clear that there is little to no difference between the two sets of numbers. The average discrepancy between these two techniques is 1.58 mm. There is a maximum 3 mm difference in measurement accuracy between the two approaches. Experiments verifying the accuracy of the proposed on-line tool wear measurement approach demonstrate an accuracy of average results within 2.50%. It follows that the recommended technique is likely to be effective. Furthermore, the proposed image processing method allows for the rapid acquisition of any and all tool wear metrics.

Simultaneously, the on-line tool wear measuring system is used for tool wear monitoring, allowing for a deeper understanding of the progression of wear on tools. Using a cemented carbide micro turning tool cutter with a radius of 8 mm, this article explains how to keep track of your tools' wear. Table 10 displays the cutting conditions used in the experiment monitoring. Our custom built on-line tool wear measurement system is used in trials on a MIKRON UCP710 five-axis turning machining center. Turning intervals are used to record photos of tool wear using a tried and true on-line technique for measuring such things. The study presents a quick and easy way to retrieve tool wear data. Figure 7 depicts the visual representation of tool wear for a variety of cutting lengths. In Table 11 we see the results of our measurements of tool wear, including the highest and the mean values. To ascertain the relationship between tool wear parameters and machining quality in accordance with the requirements of actual processing, one should employ an integrated tool wear evaluation system. In Fig. 8, we have a comparison of the maximum tool wear, the average tool wear, and the machining time. Early on in the turning process, rapid tool wear is observed, but this gradually slows down as the turning progresses. Wearing out more

Table 10 Experiment monitoring parameters for machining are listed

S. No	Depth cut (mm)	Speed of spindle (r/min)	Rate of feed (mm/min)	Cut length (mm)
1	0	0	0	0
2	0.9	1500	90	80
3	0.6	1500	70	110
4	0.9	1400	60	1000
5	0.7	1500	100	1425

Table 11 Experiment findings for measuring tool wear are shown

	1	2	3	4	5
Length of Cutting	0	350	740	1250	1350
Wear (Avg)	0	0.0345	0.0456	0.0543	0.0645
Wear (Max.)	0	0.0567	0.0675	0.0765	0.0865

quickly than usual is a hallmark of extreme wear. The machining process must be halted or the cutter must be replaced if the wear exceeds a certain threshold.

6 Onward and Upward (Scope in Future)

An on-line tool wear measurement and monitoring system can be implemented with the help of the proposed on-line tool wear measurement method. To analyze the acquired tool wear image, only a short period of time and a proven tool wear model for a wide range of cutter and workpiece materials are needed. The initial stage is to choose the right cutter, followed by control actions in real time to take pictures of the tool wear. The results of tool wear can be viewed on the screen and stored to a file almost instantly. The system requires little in the way of special equipment, training, or expertise on the part of the user. Future research toward obtaining continuous tool wear in actual machining processes will include investigating the tool wear modeling approach. To monitor tool wear in real time during cutting, an on-line tool wear assessment system is being created.

7 Conclusion

The goal of this research was to examine, from both a theoretical and an experimental perspective, the feasibility of using a machine vision-based on-line tool wear measurement approach during the turning process. This section provides a concise overview of the results of the current study:

(1) A new technique that makes use of machine vision has been developed in order to meet the demand for accurate tool wear measurements that are taken during turning operations.
(2) The measure mode of wear that has been proposed for micro turning tools is intended to both retain the consistent measuring way of tool wear and meet the real tool wear condition that has been observed.
(3) In order to accurately monitor tool wear, the tool tip points must be selected, and the wear detection areas must be tuned.
(4) The process of scanning pixel columns in a vertical direction to locate wear edge locations with pixel accuracy.
(5) In order to increase the precision of the tool wear of a micro turning tool cutter, it has been proposed to use an exact detection technique of wear edge points that is accurate to the sub-pixel level.
(6) The results of some experiments have shown that the on-line tool wear assessment method that has been developed may be used effectively and has a higher level of precision than hand-measured microscopes.
(7) A whole new approach to modeling, simulating, and predicting tool wear has been offered in the form of an on-line tool wear measuring system.

References

1. Zhang, Y., Qi, X., Wang, T., He, Y.: Tool wear condition monitoring method based on deep learning with force signals. Sensors **23**(2023), 4595 (2023) [Online]. Available: https://doi.org/10.3390/s23104595
2. Xiaoli, L., Zhejun, Y.: Tool wear monitoring with wavelet packet transform—fuzzy clustering method. Wear **219**, 145–154 (1998)
3. Sharma, V.S., Sharma, S.K., Sharma, A.K.: An approach for condition monitoring of a turning tool. Proc Inst Mech Eng Part B J Eng Manuf **221**(4), 635–648 (2007). https://doi.org/10.1243/09544054JEM765
4. Nee, A.Y.C., Ong, S.K., Wang, Y.G.: Computer applications in near net-shape operations. Springer, Berlin (1999)
5. Nath, C.: Integrated tool condition monitoring systems and their applications: a comprehensive review. Procedia Manuf **48**, 852–863 (2020). Available: https://doi.org/10.1016/j.promfg.2020.05.123
6. Yildirim, C.V., Sarikaya, M., Kivak, T., Sirin, S.: The effect of addition of hBN nanoparticles to nanofluid-MQL on tool wear patterns, tool life, roughness and temperature in turning of Ni-based Inconel 625. Trib. Int. **134**, 443–456 (2019)
7. Rehorn, A.G., Jiang, J., Orban, P.E.: State-of-the-art methods and results in tool condition monitoring: a review. Int. J. Adv. Manuf. Technol. **26**, 693–710 (2005)
8. Siddhpura, A., Paurobally, R.: A review of flank wear prediction methods for tool condition monitoring in a turning process. Int. J. Adv. Manuf. Technol. **65**, 371–393 (2013)
9. Spiewak, S., Wu, S.M.: Tool wear monitoring and breakage detection based on intelligent filtering. Int. J. Mach. Tools Manuf **28**, 483–494 (1988)
10. Dutta, S., Pal, S.K., Mukhopadhyay, S., Sen, R.: Application of digital image processing in tool condition monitoring: a review. CIRP J. Manuf. Sci. Technol. **6**, 212–232 (2013)

Chapter 27
Logistic Regression-Based Legal Prediction Model

Riya Sil, Srijan Sur, Sujoy Bhattacharya, Bharat Bhushan, Raghvendra Kumar, and Pronaya Bhattacharya

1 Introduction

In this model, authors have taken labelled article by selection of certain word. These words are used for training purposes in Fuzzy-wuzzy. Fuzzy-wuzzy logic match the words in content of article to check uprightness in it. The machine learning computing framework is a subset of artificial intelligence. It imitates the behaviour of human brain learning. Machine learning-based article analysis is considered as a smart solution to extract valid information from article dataset. Media sector is a prime concern of every nation. People always expect intensive, precise news from this, therefore, machine learning-based article detection has put right direction for any democratic nation like India. In this paper, authors have gathered thousand set of articles from web. The article content filtration is done using fuzzy-wuzzy to build

R. Sil (✉)
Department of Computer Science, Kristu Jayanti College, Autonomous, Bengaluru, India
e-mail: ritya.research@gmail.com

S. Sur · S. Bhattacharya
Department of Computer Science & Engineering, Adamas University, Kolkata 700126, India

B. Bhushan
Department of Computer Science and Engineering, School of Engineering and Technology, Sharda University, Noida, India

R. Kumar
Department of Computer Science and Engineering, GIET University, Gunupur, India
e-mail: raghvendra@giet.edu

P. Bhattacharya
Department of Computer Science and Engineering, Amity School of Engineering and Technology, Research and Innovation Cell, Amity University, Noida, India
e-mail: pbhattacharya@kol.amity.edu

J. M. R. S. Tavares et al. (eds.), *Proceedings of Second International Conference on Intelligent System*, Algorithms for Intelligent Systems,
https://doi.org/10.1007/978-981-99-8976-8_28

a two-set precise and imprecise news. This put forward the subjective research to automate article diagnosis process by extracting article feature.

The incorporation of human intellect into computers has been facilitated by the use of artificial intelligence (AI) [1]. AI-based companies are relentless in their search for novel approaches to creating technologies [2] that would handle the arduous tasks performed by humans in a variety of fields in order to increase accuracy, efficiency, and work speed for better results. An automated method for providing services has emerged as a result of the expansion of artificial intelligence in law firms [3].

It has been applied to a variety of legal fields and aids in the delivery of justice to the intended parties, including (i) provide more information and time-saving strategies applying data analytics; (ii) support the attorney by doing due diligence and research; and (iii) automate innovative legal work procedures. The implementation of new procedures is the result of the unceasing demand and the strict shift in client expectations. It expedites and simplifies the job of legal practitioners. India has a large population and a dearth of judges, which has resulted in many lawsuits. According to a study, as per the Law Minister of India, there are more than 43 lacs pending cases throughout 25 High Courts in the nation, out of which more than 8 lac cases [4] are pending for more than ten years. The problem has gotten worse over time, as seen by the enormous amount of compounding records and recent changes to the COVID-19 pandemic condition [5]. Machine learning algorithms are essential for improving the legal system at this key juncture [6] and can assist reduce the workload for legal professionals, allowing them to use more time to resolve pending cases [7]. In order to completely digitise the legal work process and archive the results for later use, this study intends to revolutionise it. The research focuses on an argument-based analysis of a legal judgement prediction system that performs logistic regression [8] over the trial court datasets for situations involving "dowry death" [9].

From West Bengal's trial courts, hard copies of case files with pronounced dowry judgements are gathered. A dataset is then manually generated based on the parameters from the prosecution witness statements, and logistic regression [10] is used for prediction of the report. This paper uses logistic regression, a common classifier, to predict the performance report and accuracy score [11].

Section 2 describes about the input dataset and its parameters. Section 3 provides a detailed discussion on the proposed model using logistic regression. Section 4 states the performance study of the proposed model. And, Sect. 5 provides an overall summary and also explores the various future scope of research work.

2 Dataset

By identifying the key aspects in the argument-based prosecution witness statements related to "dowry cases", this section assists in illuminating the method by which the dataset was produced [12]. The criteria is chosen based on important characteristics that are taken into account while determining whether or not to convict the defendant [13].

2.1 Generation of Dataset

The dataset must be produced before the system is trained [14]. A dataset is a grouping of data that may be semi-structured, unstructured, or structured. There isn't a pre-structured argument-based court judgement dataset available in India [15], hence the entire dataset must be generated from start. Since it can effectively retrieve data, a structured dataset in the form of a Comma-Separated Value (CSV) file has been employed in this paper [16]. Hard copies of the case file documents for the "dowry death" cases, which fall under IPC sections 498a and 304b [17], are gathered from the trial courts in West Bengal [18]. Since there weren't any softcopies of the case files, the data had to be manually organised into MS-Excel file format due to the overwhelming number of hard copies of the cases. These data will be necessary for the machine learning model's training [19]. Feature extraction is the process of digging the features [20].

2.2 Selection of Features

When given as input into a machine learning model [21], features [22] behave as independent variables that can provide categorization or prediction [23]. Features are properties with values that make up a particular dataset and are used to create output. Following are some of the characteristics or metrics [24] that were taken from tangible copies of evidence-supporting documents: The parameters "Victim Name" and "Number of years married (greater than seven years or not)" help determine if a case fits within the category of a "dowry case" [25] or not a "dowry case" is one in which the defendant has taken dowry and the lawsuit is filed within seven years after marriage; otherwise, it is not. (iii) "Dowry taken within seven years of marriage (Yes/ No)" (iv) When the death occurs within seven years of the marriage, the criterion "Incident taken place within seven years of marriage" designates it as a "dowry death case." (v) The phrase "Postmortem Report (Usual/Unusual Death)" is only one of several. The parameters were chosen, omitting the irrelevant aspects [26], based on the importance of the features. Feature selection is the process of choosing a few relevant characteristics from a stream of extracted features [27]. Based on these selected variables, the model has been taught [28] to predict an accused person's guilt or innocence. Figure 1 shows the screenshot of the created dataset [4].

3 Proposed Model Using Logistic Regression

Acquisition of article database is a challenging task for any media industry. This domain need more sophistication to set precise article domain record to hold its fame over market. To process article string in this paper authors have taken the advantage of

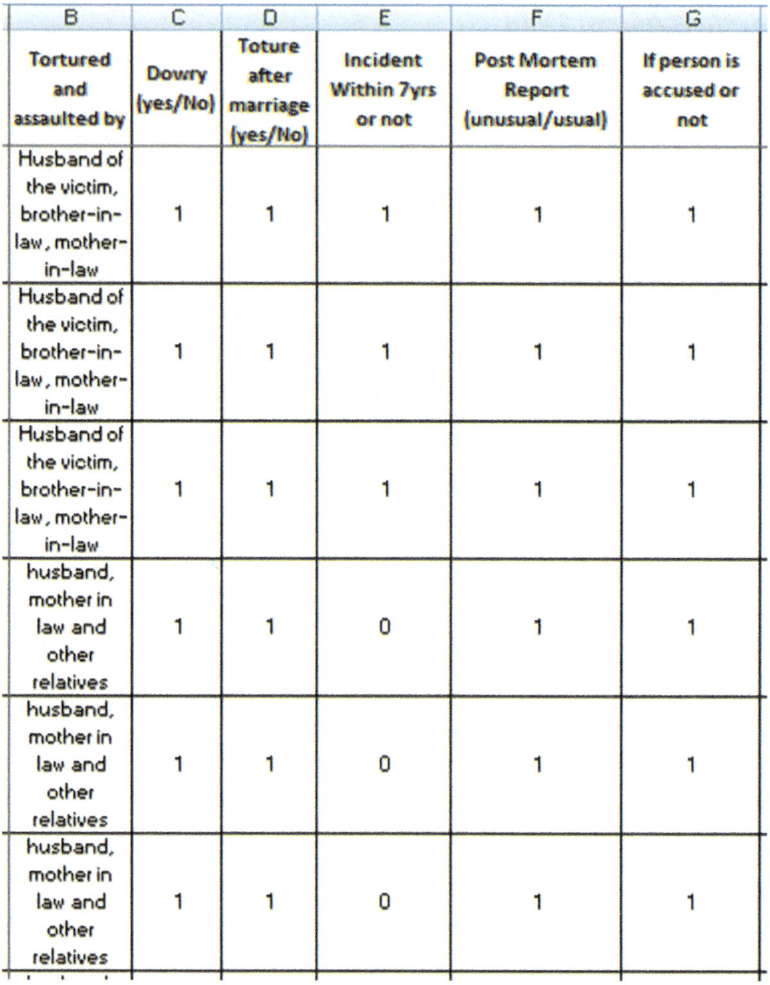

B	C	D	E	F	G
Tortured and assaulted by	Dowry (yes/No)	Toture after marriage (yes/No)	Incident Within 7yrs or not	Post Mortem Report (unusual/usual)	If person is accused or not
Husband of the victim, brother-in-law, mother-in-law	1	1	1	1	1
Husband of the victim, brother-in-law, mother-in-law	1	1	1	1	1
Husband of the victim, brother-in-law, mother-in-law	1	1	1	1	1
husband, mother in law and other relatives	1	1	0	1	1
husband, mother in law and other relatives	1	1	0	1	1
husband, mother in law and other relatives	1	1	0	1	1

Fig. 1 Screenshot of the created dataset [4]

fuzzy-wuzzy decision logic. Fuzzy-wuzzy is otherwise known as approximate string-matching application. Literally the pattern of one string is matched against another string. Here one mathematical function is used to decide closeness between two string like known as Levenshtein distance colloquially it is known as edit distance. Meanwhile it follows certain point discussed as below. Practically it measures how many numeric steps required to match a string. The primitive operation for Levenshtein distance is depicted as follows.

The preceding section provided a detailed explanation of the dataset that was developed and used in this model [29] for the automatic creation of judgement prediction. The working theory of the suggested model is covered in this section [30].

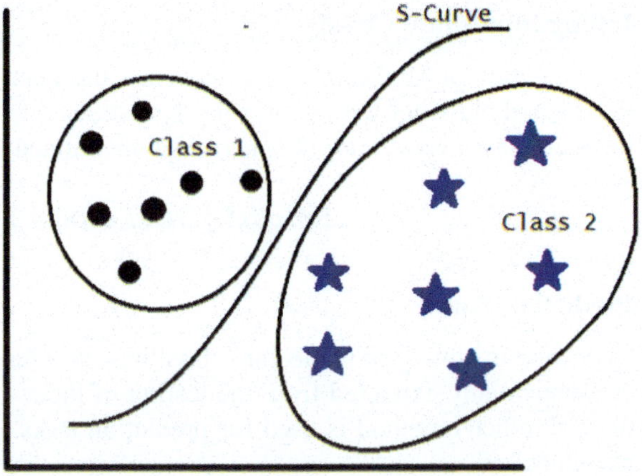

Fig. 2 Logistic regression

The different categories of machine learning problems [31] are: (i) supervised learning [32] (ii) unsupervised learning [33] (iii) semi-supervised learning [34] (iv) reinforcement learning [35]. In this paper, logistic regression algorithm has been applied that falls under supervised machine learning.

Logistic Regression

Logistic regression (As shown as Fig. 2) is not a black box model, it is quite comprehensible [36]. To address these issues, several algorithms have been developed. One of these, conventional quadratic programming techniques [37], has several shortcomings. Logistic regression [38] has been used to address those issues. With the aid of the sigmoid curve [39], the class of classifiers that properly classify a set of data includes logistic regression [40].

Label Encoding

It uses the LabelEncoder module from the Python Scikit-Learn package. The following statement was used to create the classifier:

$$encoder = LabelEncoder()$$

Classifier Creation

In this paper, the logistic regression module from linear model package has been used from Scikit-Learn package of Python. The statement used for creating of classifier is given below:

$$LRModel = LogisticRegression()$$

Fitting Inputs into Classifier

Fit() function of *LogisticRegression()* class has been utilised to assist in providing the created classifier with data from the dataset. The following line describes how the built classifier will use the previously mentioned input:

$$\text{LR}Model.fit(XTrain,\ YTrain)$$

Prediction Phase

Following the fitting of inputs into classifiers, the classes that the classifier predicted are sequentially extracted from the testing of input data. The LogisticRegression() class' Predict() method is used for prediction tasks. The output class prediction is shown below.

$$YPred = \text{LR}Model.predict(XTest)$$

Classification Report and Accuracy Generation

When predicted class labels are obtained, a classification report and accuracy score are created to determine how well our classifier performed.

Advantages of Logistic Regression

When the dataset has features that can be segregated linearly, logistic regression proves to be very efficient. Logistic regression classifier is ideal for binary classifiable data. This classifier is relatively easy to set up and train than other ML classifiers.

4 Performance Analysis of the Proposed Model

Results were obtained for logistic regression classification algorithm. The ratio for train and test was taken as 70 and 30% for all the classifiers. The following was observed from each of them. The accuracy for the logistic regression model in this case was 0.8853 or 89%. Which would mean that the model is almost perfect which opens the scope for further testing or it has some issue with over fitting. Accuracy of the training data was 90.21% and that of test data was 88.53%. Because the training data score was more than that of test data it can be said that there was no overfitting.

After obtaining predicted class labels, those are analyzed using the Classification Report and Accuracy Score, the output of the logistic regression classification algorithm is shown below in a tabular format in Table 1.

The following Fig. 3 is the Confusion Matrix from the result. It describes the number of True Positive, False Positive, True Negative, and False Negative.

Table 1 Classification report and accuracy score of logistic regression

	Precision	Recall	$F1$-score	Support
Class 0	1.00	0.25	0.40	4
Class 1	0.88	1.00	0.94	23
Accuracy	–	–	**0.89**	27
Macro Average	0.94	0.62	0.67	27
Weighted Average	0.90	0.89	0.86	27

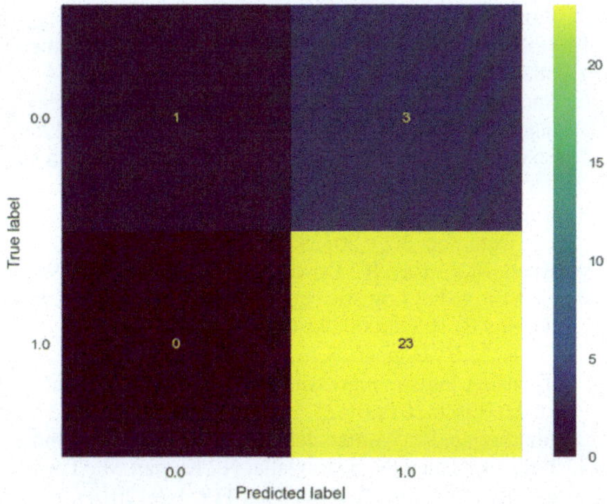

Fig. 3 Confusion matrix

5 Conclusion

To provide assistance to legal practitioners, the authors of this paper have presented a supervised learning algorithm called logistic regression-based prediction system over the legal dataset. Beneficiaries in India must wait a long time before receiving the appropriate compensation they are owed because of a lack of trained labour and adequate infrastructure. As it is properly said, "Justice Delay Is Justice Denied", protracted legal procedures can also have a number of negative effects, such as hostile witnesses, medical ineligibility of the accused, tampering with or manipulation of evidence, etc. The suggested model could help legal practitioners in analysing the needed dataset and performing predictions based on the crucial factors of "Dowry" death-related cases on a case-by-case basis. With a Standard Classifier (i.e. logistic regression), we have demonstrated the model's performance and accuracy and have obtained 89% accuracy. However, in order to achieve 100% accuracy, authors will include more elements that will guarantee that all victims receive adequate justice and that our judicial system is more dynamic in its approach, for the benefit of society as a whole.

References

1. Shao, Z., Yuan, S., Wang, Y.: Institutional collaboration and competition in artificial intelligence. IEEE Access **8**, 69734–69741 (2020)
2. Lucci, S., Kopec, D.: Artificial intelligence in the 21st century. Stylus Publishing, LLC (2015)
3. Sil, R., Roy, A., Bhushan, B., Mazumdar, A.K.: Artificial intelligence and machine learning based legal application: the state-of-the-art and future research trends. In: 2019 International Conference on Computing, Communication, and Intelligent Systems (ICCCIS), pp. 57–62. IEEE (2019, Oct)
4. Sil, R., Alpana, Roy, A., Dasmahapatra, M., Dhali, D.: An intelligent approach for automated argument based legal text recognition and summarization using machine learning. J. Intell. Fuzzy Syst. **41**(5), 5457–5466 (2021). https://doi.org/10.3233/jifs-189867
5. Hollander, J.E., Carr, B.G.: Virtually perfect? telemedicine for COVID-19. N. Engl. J. Med. **382**(18), 1679–1681 (2020)
6. Snoek, J., Larochelle, H., Adams, R.P.: Practical bayesian optimization of machine learning algorithms. Adv. Neural Inf. Process. Syst. 2951–2959 (2012)
7. Shaha, K.K., Mohanthy, S.: Alleged dowry death: a study of homicidal burns. Med. Sci. Law **46**(2), 105–110 (2006)
8. Burrell, J.: How the machine 'thinks': understanding opacity in machine learning algorithms. Big Data Soc. **3**(1), 2053951715622512 (2016)
9. Khan, M.Z., Ray, R.: Dowry death. Indian J. Soc. Work **45**(3), 303–315 (1984)
10. Hanson, R.K.: Logistic regression. Prediction Stat. Psychol. Assess. 195–220 (2022). https://doi.org/10.1037/0000275-011
11. Shipe, M.E., Deppen, S.A., Farjah, F., Grogan, E.L.: Developing prediction models for clinical use using logistic regression: an overview. J. Thoracic Disease **11**(S4) (2019). https://doi.org/10.21037/jtd.2019.01.25
12. Mukherjee, R.: Women, law, and free legal aid in India. Deep & Deep Publications (1998)
13. Belur, J., Tilley, N., Daruwalla, N., Kumar, M., Tiwari, V., Osrin, D.: The social construction of 'dowry deaths.' Soc Sci Med **119**, 1–9 (2014)
14. Catanzaro, B., Sundaram, N., Keutzer, K.: Fast support vector machine training and classification on graphics processors. In: Proceedings of the 25th International Conference on Machine learning, pp. 104–111 (2008, July)
15. Bennett, W.L.: Rhetorical transformation of evidence in criminal trials: creating grounds for legal judgment. Quart. J. Speech **65**(3), 311–323 (1979)
16. Shafranovich, Y.: Common format and MIME type for comma-separated values (CSV) files (2005)
17. Agnes, F.: Section 498A, marital rape and adverse propaganda. Econ. Pol. Wkly **50**(23), 13 (2015)
18. Trivedi, P.K., Singh, S.: Fallacies of a supreme court judgment: section 498a and the dynamics of acquittals. Econ. Polit. Weekly 90–97 (2014)
19. Gu, T., Dolan-Gavitt, B., Garg, S.: Badnets: identifying vulnerabilities in the machine learning model supply chain (2017). arXiv preprint arXiv:1708.06733
20. Trier, Ø.D., Jain, A.K., Taxt, T.: Feature extraction methods for character recognition-a survey. Pattern Recogn. **29**(4), 641–662 (1996)
21. Montavon, G., Rupp, M., Gobre, V., Vazquez-Mayagoitia, A., Hansen, K., Tkatchenko, A., Von Lilienfeld, O.A.: Machine learning of molecular electronic properties in chemical compound space. New J Phys **15**(9), 095003 (2013)
22. Nevatia, R., Babu, K.R.: Linear feature extraction and description. Comput. Graphics Image Process. **13**(3), 257–269 (1980)
23. Kumar, R.S.S., O'Brien, D.R., Albert, K., Vilojen, S.: Law and adversarial machine learning (2018). arXiv preprint arXiv:1810.10731
24. Yousfi-Monod, M., Farzindar, A., Lapalme, G.: Supervised machine learning for summarizing legal documents. In: Canadian Conference on Artificial Intelligence, pp. 51–62. Springer, Berlin, Heidelberg (2010, May)

25. Stone, L., James, C.: Dowry, bride-burning, and female power in India. In: Women's Studies International Forum, vol. 18, no. 2, pp. 125–134. Pergamon (1995, Mar)
26. Shah, P., Joshi, S., Pandey, A.K.: Legal clause extraction from contract using machine learning with heuristics improvement. In: 2018 4th International Conference on Computing Communication and Automation (ICCCA), pp. 1–3. IEEE (2018, Dec)
27. Zhang, Y., Xie, R., Wang, J., Leier, A., Marquez-Lago, T.T., Akutsu, T., Song, J.: Computational analysis and prediction of lysine malonylation sites by exploiting informative features in an integrative machine-learning framework. Brief. Bioinform. **20**(6), 2185–2199 (2019)
28. Hall, M.A., Smith, L.A.: Practical feature subset selection for machine learning (1998)
29. James, S.C., Zhang, Y., O'Donncha, F.: A machine learning framework to forecast wave conditions. Coast. Eng. **137**, 1–10 (2018)
30. Rueping, S.: SVM classifier estimation from group probabilities. In: ICML (2010, Jan)
31. Mitchell, T.M.: The discipline of machine learning, vol. 9. Carnegie Mellon University, School of Computer Science, Machine Learning Department, Pittsburgh (2006)
32. Bhamare, D., Salman, T., Samaka, M., Erbad, A., Jain, R.: Feasibility of supervised machine learning for cloud security. In: 2016 International Conference on Information Science and Security (ICISS), pp. 1–5. IEEE (2016, Dec)
33. Kassambara, A.: Practical guide to cluster analysis in R: unsupervised machine learning, vol. 1. Sthda (2017)
34. Zhu, X., Goldberg, A.B.: Introduction to semi-supervised learning. Synth. Lect. Artif. Intell. Mach. Learn. **3**(1), 1–130 (2009)
35. Sutton, R.S.: A special issue of machine learning on reinforcement learning. Mach. Learn. 8 (1992)
36. Jain, H., Khunteta, A., Srivastava, S.: Churn prediction in telecommunication using logistic regression and logit boost (ICCIDS) 101–112 (2020)
37. Chen, H.L., Yang, B., Liu, J., Liu, D.Y.: A support vector machine classifier with rough set-based feature selection for breast cancer diagnosis. Expert Syst. Appl. **38**(7), 9014–9022 (2011)
38. Sha, F., Saul, L.K., Lee, D.D.: Multiplicative updates for nonnegative quadratic programming in support vector machines. Adv. Neural Inf. Process. Syst. 1065–1072 (2003)
39. Sil, R., Roy, A.: A novel approach on argument based legal prediction model using machine learning. In: 2020 International Conference on Smart Electronics and Communication (ICOSEC) (2020). https://doi.org/10.1109/icosec49089.2020.9215310
40. Cai, Y.D., Liu, X.J., Xu, X.B., Zhou, G.P.: Support vector machines for predicting protein structural class. BMC Bioinf. **2**(1), 3 (2001)
41. Suthaharan, S.: Support vector machine. In: Machine Learning Models and Algorithms for Big Data Classification, pp. 207–235. Springer, Boston, MA (2016)
42. Cai, Y.D., Feng, K.Y., Li, Y.X., Chou, K.C.: Support vector machine for predicting α-turn types. Peptides **24**(4), 629–630 (2003)
43. Wang, W., Men, C., Lu, W.: Online prediction model based on support vector machine. Neurocomputing **71**(4–6), 550–558 (2008)
44. Zou, X., Hu, Y., Tian, Z., Shen, K.: Logistic regression model optimization and case analysis. In: 2019 IEEE 7th International Conference on Computer Science and Network Technology (ICCSNT) (2019). https://doi.org/10.1109/iccsnt47585.2019.8962457

Chapter 28
A Comprehensive Approach for Predicting Different Types of Retinal Detachment with ML Algorithms

E. Anitha, D. John Aravindhar, and A. Antonidoss

1 Introduction

Researchers began to develop systems able to analyze images as soon as the capability for scanning & loading photos onto computers became accessible. Machine learning (ML) methods replaced rule-based systems as technology advanced in the twenty-first century [1]. Commercially accessible tools for analyzing medical pictures still heavily rely on these methods [2]. Meanwhile, a significant disadvantage of ML approaches is that all of the features used in the method are often created by hand by a subject matter expert. The next stage in picture processing was to develop a method that can extract characteristics and finish calculations all by itself [3]. DL will therefore be implemented in image processing. Due to the abundance of imaging data, DL methods like ANNs and CNNs flourish in the field of picture processing.

To anticipate diseases of the retina, cancer, the abdominal musculoskeletal system, and other areas of medicine, researchers have begun to apply DL methods [4–6]. The deployment of DL algorithms in numerous areas of medical research, including

E. Anitha (✉) · D. J. Aravindhar
Department of Computer Science and Engineering, Hindustan Institute of Technology and Science, Padur, Chennai, India
e-mail: esanitha.me@gmail.com

D. J. Aravindhar
e-mail: jaravindhar@hindustanuniv.ac.in

A. Antonidoss
Department of Computer Science and Business Systems, S.A Engineering College (Autonomous), Chennai, India

E. Anitha
Department of Information Technology, Loyola - ICAM College of Engineering and Technology, Chennai, India

J. M. R. S. Tavares et al. (eds.), *Proceedings of Second International Conference on Intelligent System*, Algorithms for Intelligent Systems,
https://doi.org/10.1007/978-981-99-8976-8_29

the forecasting of cancer [5], abdominal musculoskeletal disorders [6], and retinal disease [4], has begun.

A subset of ML called DL employs several types of neural networks for categorization and forecast. DL is employed over enormous datasets for classification and regression because of the advancements in modern technology and computing methods [7]. ANN, Recurrent Neural Networks (RNN), Long Short-Time Memory (LSTM) Neural Networks, CNN, and others are some of the well-known DL methods. The essential ideas, methods, and algorithms employed in neural networks are covered here [8, 9].

The fundamental units of a neural network are neurons. It is comparable to the biological neurons found in our brains. Neurons are arranged at levels in neural networks. The first layer is referred to as the input layer, and the final layer is the hidden layer. The term "hidden layer" refers to every layer that is contained within the input and output layers [10]. A single characteristic from the database is used as the input for each neuron in the input layer, which then sends the value to the hidden layer. Each neuron in a hidden layer is wired through weights to the neurons in the layers above and below it. The hidden layer neuron multiplies each signal from its preceding layer by the associated weights before adding all of them together [2]. The summation is then processed via the activation feature to decide whether or not the hidden layer neuron will transmit the signal to the following layer. The method that has made DL so well-liked in recent years is back propagation [5]. It is a method of learning that was applied to neural networks during supervised learning. With the use of this method, neural networks may autonomously learn features from a dataset.

In this study, we presented and illustrated a picture processing method based on deep neural networks (DNN) built on the ANN and CNN that allows for the classification of retinal imaging pictures into several illness classifications [11]. The goal of this study is to create a perfect deep neural network model that can more accurately predict retinal illnesses from retinal pictures. The goal of this study is to better understand the significant role that retinal illnesses play in medical research. Retinal tears, retinal detachment, diabetic retinopathy, glaucoma, and other common retinal illnesses are frequently encountered [8, 12]. Doctors can now capture clear pictures of the retina thanks to technological advancements. As a result, a vast amount of imagery information is currently accessible to medical databases, allowing for the creation of precise predictions [13]. There are currently studies where various ML and DL algorithms have been utilized to forecast retinal disorders.

The next paper is broken up into five sections. A succinct introduction and necessary clarification of a key topic in our research are provided in the first section. It think that doing this will help our readers understand. It talked about some of the earlier research that had been done in this field in Sect. 2 of the paper. We talked about our dataset and the approach we used in part 3 to reach our conclusion. In Sect. 4, it have reported our findings. That section also includes a discussion of the outcome. The conclusion of our report and a brief discussion of potential future research on this topic are found in Sect. 5.

2 Literature Review

The Retinal Nerve Fiber Layer (RNFI) & Visual Field (VF) have been studied by authors [14, 15] to gather numerous features. It went one step further and took those original characteristics and turned them into synthesized characteristics. Next, it employed three well-known ML methods: Support Vector Machine, Random Forest, and *K*-Nearest Neighbor. Nevertheless, the fact that the dataset it employed for their study was so limited is one of its fundamental flaws [16]. Only 100 rows are used for evaluation, and 399 rows are used for training and validation. DL, as we all know, needs a large dataset to make correct predictions. By evaluating color fundus pictures, researchers [17] employed CNNs to identify glaucomatous optic neuropathy. A database of 48,000 photographs was employed, and the data was labeled by professionals. CNNs were then utilized to categorize the pictures.

A database of 48,000 photographs was employed, and the data was labeled by professionals. CNNs were then utilized to categorize the pictures. It don't employ OCT pictures like us, though, and it only focus on one particular condition. Another important study in the field [19] discusses the various classifications of retinal illnesses, including retinitis pigmentosa, macular degeneration, diabetic retinopathy, retinoblastoma, and age-related molecular degeneration. In this article, photos are classified into several disease categories using a deep CNN and an SVM algorithm. Additionally, it discuss how it is more difficult to categorize diseases due to their similarities. Additionally, in this work [20], researchers classified photos by several disease groups using OCT pictures. It employed a similar technique to ours, a convolutional network, and compared the results to the classification made by human ophthalmology specialists.

Authors employed DL to predict age-related macular degeneration (AMD) in a different study [21]. It defined 13 classes and used 120,656 color fundus images to make their predictions. It have experimented with different combinations of CNNs while choosing DL methods. The precision has increased to 94.3 using one ensemble of six different neural networks. However, it only used fundus photos as a dataset and only categorized pictures for one disease. To identify glaucoma, researchers [22] attempted to classify photos. It divided the results into glaucoma and non-glaucoma classes. It utilized an unsupervised convolutional network to extract characteristics from the raw photos after that. It then attempted to determine the most beneficial characteristic to use in research by running a Deep Belief Network (DBN) across the characteristics. Lastly, it classified objects for the two separate classes using softmax.

DL networks have also been utilized by researchers [23] to predict cardiovascular risk factors in the retina using pictures of the retinal fundus. It used 284.335 photos from patients of various ages to train their algorithm, and two separate sets of data were used to verify their results. 12,026 photos in one, and 999 in the other. To verify their conclusions, we think it should have used a much larger database.

Different DL algorithms have been discussed by researchers [24] as tools for predicting diabetic retinopathy in a person's retina. This essay is a review. The authors of this publication also discussed other studies in which scientists employed DL algorithms to provide improved prediction values. Although this is not a traditional research paper where the author presents a new model and seeks to support their claim, this essay can serve as a starting point for additional investigation.

A DL system is employed by the authors of this research [25] to forecast diabetic retinopathy in the retina. However, it only used 1796 retinal fundus photos from 1612 diabetic individuals, which is a relatively limited database. The pictures have been divided into three categories. However, we think that the dataset's picture count must be quite low for any neural network technique to operate at its optimum based on our experience. Here, we'll discuss a fully autonomous identification method built using a deep-learning algorithm that works with OCT pictures [26]. Intraretinal Cystoid Fluid (IRC) and Sub Retinal Fluid (SRF) are automatically detected using the technique it describe in their research. Additionally, it made use of a dataset that had 1200 volumes of OCT pictures.

The intriguing aspect of this paper is that it have advanced one step in addition to discussing several DL methods [27]. It classified them into various ethnic groups in addition to using image data from the dataset. We all know that people from various areas of the world have varied eating habits and that eating habits directly affect keeping diabetics alive. These researchers have concealed a different factor that has thus far gone unnoticed by others by focusing solely on ethnicity.

3 Proposed Methodology

It used two separate neural networks ANN and CNN in our proposed methodology. This section describes the step-by-step process we used to conduct our research. Our dataset contains 84,494 images, and we've categorized them into three groups (Training, Validation, and Test). To train the model, 60,000 randomly selected images were used, and 15,000 images were used to verify our training methods. 10,000 additional images are used for assessment. The rest of this section outlines the steps we took to conduct our research. It have depicted the step-by-step process of our research in Fig. 1. This flowchart has been step-by-step detailed later in the text.

3.1 Input Data

We input our image folder into the algorithm in the first stage. Python has built-in libraries that can be used to do this operation. The dataset contains photos from four various groups. The first three classifications fit into different illness groups, while the fourth category is for normal retinas. Our dataset contains 84,494 images, and we've categorized them into three groups (Training, Validation, and Test). To develop the framework, 60,000 randomly selected images were used, and 15,000 images were used to verify our training methods. 10,000 additional images are used for testing.

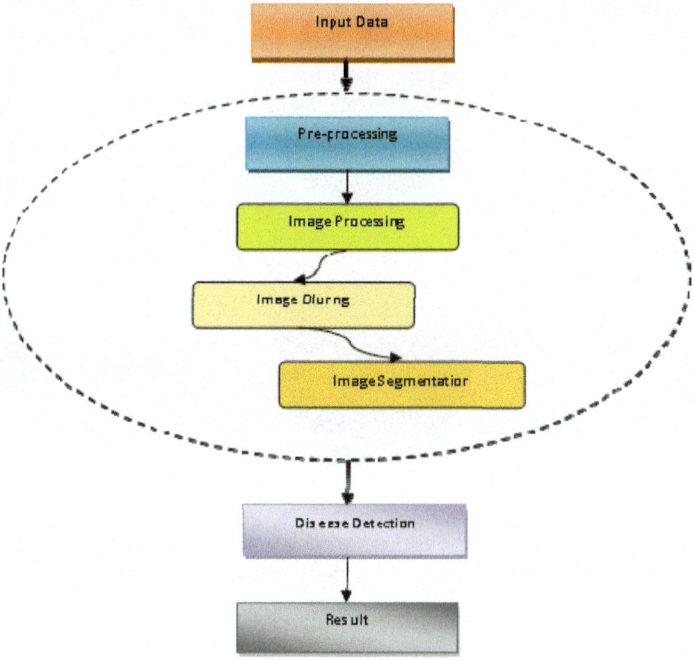

Fig. 1 Proposed approach

Here, we used the image-capture method known as OCT and it can be used to take two- or three-dimensional pictures of biological tissue. Low coherence light is used in this method to capture micrometer-level detail.

3.2 Preprocessing

Image Resizing: Image resizing is a neural network technology that shrinks each image pixel to a specific size. It is required to preprocess the photos into a predetermined size since neural networks like CNN work with a fixed amount of features across the entire dataset. One danger with this method is the potential loss of important data. However, CNN's inherent ability to uncover hidden characteristics mitigates this flaw shown in Fig. 2.

It resize the entire picture in this phase. To import each image into the neural network, the size of each image must be the same; otherwise, the neural network may produce incorrect results and incorrectly categorize components. It have chosen to resize every image to 200 px by 200 px. Resizing them could result in the loss of any valuable information, increasing the dimensions will drain more power from the neural network and make it take longer to provide output. Additionally, it lowers the

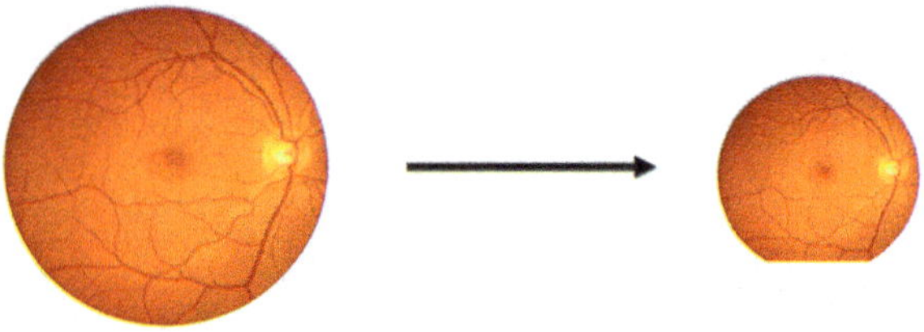

Fig. 2 Retinal image resizing

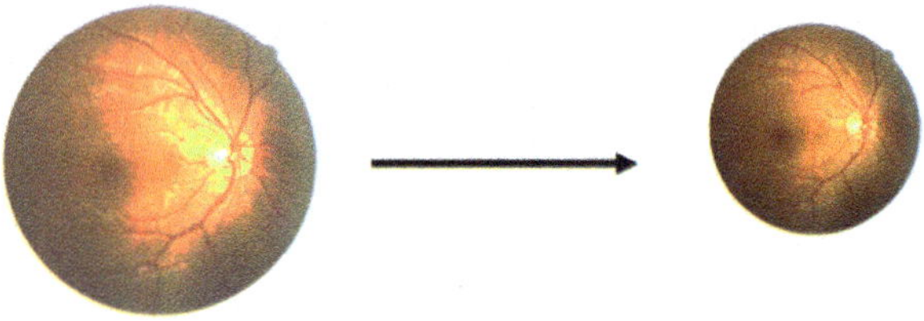

Fig. 3 Blurring retinal pictures

dimensionality and speeds up neural networks when performing classification and regression tasks. The Python library was used to resize the pictures.

Image Blurring: Another significant method that is widely applied in image categorization is this one. It is required to remove unused lines and edges from the image before putting it into the convolution network. The neural network may misclassify the object as a result. The employment of many types of image-blurring techniques is a consequence of image processing. The noise related to each image is eliminated in this step. In Fig. 3, noise typically refers to various defects or extraneous pixels. It improve the effectiveness of our neural network by eliminating these incorrect values. Over fitting could happen if there is noise in the photos. Researchers employed the "Gaussian Blur" approach to clean up the photos of noise. The Gaussian formula is used in the Gaussian blur method to blur the picture and eliminate noise.

Segmentation: Important components within a picture are segmented for greater precision throughout the picture segmentation process. As a result of this procedure, specific objects like the cat and dog parts of the picture were separated from the rest of the image so that the neural network could concentrate on them for better precision. Finding significant objects in the picture is a crucial task even after the image has been removed. The relevance of an object may be determined by specific edges and lines, and these edges and lines may appear anywhere in the image. A typical method for locating things in an image is segmentation. In our research, we ignore the background and concentrate on the retinal region of the image. Additionally, we blur the background to improve the accuracy of the neural network.

3.3 Disease Detection

The retina is a delicate layer that lines the back of the eyeball. It consists of light-sensitive cells. A light that enters the retina sets off nerve impulses that go through the optic nerves and onto the brain to form an image. Cone cells are the type of cells that make up the retina. It serve as photoreceptor cells, which receive signals from light. The retina typically has 7 million cones. Green and blue cone cells each contain 32 and 2% of them, whereas red cone cells make up 64% of them. The retina is between 30 and 40 mm in diameter overall. The illnesses that commonly affect and impair the retina are called retinal disorders. These illnesses vary greatly, yet it all have visible symptoms. Retinal illnesses include macular holes, diabetic retinopathy, epiretinal membrane, retinal rupture, and retinal detachment.

3.4 Macular Degeneration

In our study, it divided retinal pictures into four categories. CNV, DME, DRUSEN, and normal retina, to be exact DNN use. The most prevalent type of neural network is the ANN, which is thought to be the forerunner of many DL techniques currently in use [28]. It is produced by numerous coupled neuronal components. Each is organized into layers and connected to the layers of neurons before and behind it through weights shown in Fig. 4. A neuron is activated using a variety of activation functions, and it simply fires for a certain value of the activation function. Popular activation functions include the sigmoid, Tanh, and ReLu functions. The weights of the neurons in an ANN are first set to random numbers. Later, a neural network develops the ability to categorize things into multiple classes with the aid of gradient descent and back propagation.

An extremely popular type of DL used in image categorization is CNNs. It is modeled after the imaginative human system and replicates the procedure for

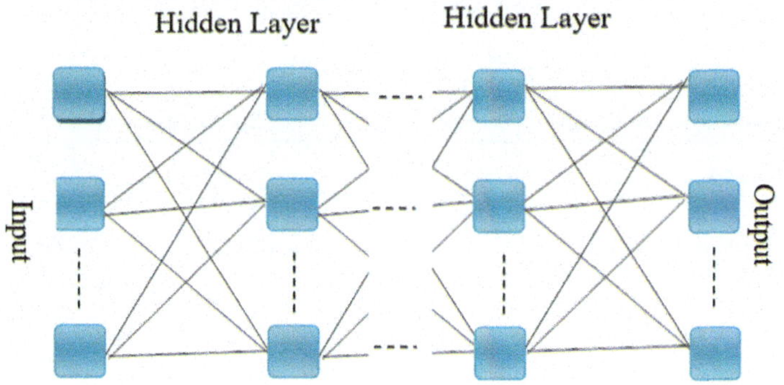

Fig. 4 Proposed ANN architecture with 4 hidden layers

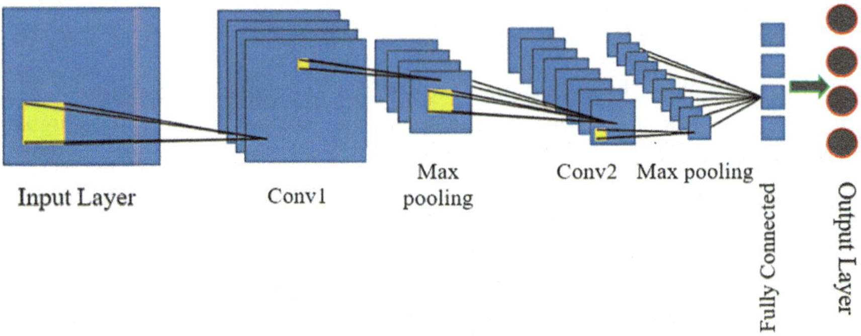

Fig. 5 Proposed CNN structure

computers. The convolutional layer, maximum pooling, and fully linked layer make up the majority of it. Direct inputs are given to the convolutional layer, which uses filters and activation algorithms to find the hidden characteristic shown in Fig. 5. Filters are applied to the dataset's more complicated characteristics in each layer. For the final classification, fully connected layers are employed. The weights are updated when initial random values are assigned to them.

The following several cycles using the training data. CNNs can forecast with great precision for a population when given sufficient examples.

4 Results and Discussion

In the final phase, it analyzed the outcomes of various neural network models. We employed CNN and ANN, two DNN. For each neural network, we tested models with 3, 4, and 5 hidden units using three different models. Each model's input layer has 400 nodes because the input images are all 200 pixels by 200 pixels. Table 1 discuss our dataset in the first subsection, and our findings are presented in the second.

Table 1 Evaluation of several neural network models

Hidden Layer	Accuracy	Sensitivity	Specificity	Precision
3	80.62	79.78	80.65	80.08
CNN 4	86.71	85.04	86.08	85.41
5	84.93	83.28	84.50	83.12
3	72.80	71.71	72.96	72.43
ANN 4	68.71	67.72	68.27	66.56
5	74.76	73.95	74.37	72.63

4.1 Dataset Description

84,495 retinal OCT pictures make up this collection. Four categories of diseases are used to categorize the photos. The ailments are,

- Multiple Drusen present in early AMD, and
- Diabetic Macular Edema (DME)
- Normal Retina
- Choroidal Neovascularization (CNV)

Pictures were selected from https://www.kaggle.com/datasets/andrewmvd/retinal-disease-classification?resource=download.

In addition, professionals were consulted before the photographs were labeled. The dataset contains photos in four different categories. The first three classes fit into different illness groups, while the fourth category is for normal retina. The first of the three illnesses is CNV, or choroidal neovascularization, which affects the retina [18]. The choroid is a layer that lies beneath the retina. When this disease strikes, the choroid experiences an enormous increase in the number of new blood vessels that ultimately invade the retina. These new blood arteries, which differ from regular ones, allow blood and occasionally red blood cells to enter the retina. This fluid can eventually kill the light-sensing cells of the retina, causing damage. DME: diabetic macular edoema, often known as DME, is one type of retinal illness brought on by fluid leakage in the macular portion of the retina. This condition typically affects diabetics and can harm the retina's blood vessels. If the condition is not treated on time, the blood vessels begin to push on the eyes, causing damage to the eyes. DRUSEN: DRUSEN are white or yellow colored spots that develop in Bruch's membrane layer of the retina. Numerous factors could cause this circumstance. The most common cause is the collection of garbage from rods and cons. This can result in irreversible blindness if not treated on time. Finding new insights or challenging accepted knowledge is the primary goal of all studies in all fields. It illustrate and discuss our founding in this subsection in Table 2.

Table 2 Compares various disease predictions made by neural networks

Hidden layer	CNV	DME	DRUSEN	Normal
3	80.43	78.77	81.53	79.80
CNN 4	86.53	86.77	87.38	85.79
5	83.32	84.35	81.17	84.21
3	71.20	72.23	70.57	71.34
ANN 4	66.51	68.42	67.88	65.50
5	73.77	75.11	73.73	75.93

4.2 Observations

As previously noted, we conducted our research using two DL algorithms: CNN and ANN. To determine the best algorithm model, we tested a variety of hidden layer counts. We have chosen features for applying over the DL algorithms. The algorithms chose their feature by drawing on their inherent characteristics. We evaluated the prediction quality in four groups, as indicated in the Table 2. Each of these routines produces a result in the range of 0 and 1. Whereas it is thought that the method is better if the value is nearer to 1.

Accuracy

We divided the total accurately estimated value by the number of images in this category.

$$\text{Accuracy} = P/N \tag{1}$$

where N is the total number of photos and P is the image that was successfully forecasted.

Sensitivity

The ratio of True Positive to the total of True Positive and False Negative is used to calculate sensitivity.

$$\text{Sensitivity} = \text{TP}/\text{TN} + \text{FN} \tag{2}$$

where TF = True Positive, TN = True Negative, FN = False Negative.

Specificity

By dividing the genuine negative value by the total of the true negative and false positive, specificity is determined.

$$\text{Specificity} = \text{TN}/\text{TN} + \text{FP} \tag{3}$$

where TN = True Negative FP = False Positive.

Precision

The division of genuine positive value by the sum of true positive and false positive yields the precision value.

$$\text{Precision} = \text{TP}/\text{TP} + \text{FP} \tag{4}$$

where TP = True Positive FP = False Positive.

4.3 Discussion

Figures 6, 7, 8, and 9 shows that the CNN with four hidden layers is producing the best outcome out of all of them. That specific model has a precision rating of 86.71. Additionally, it gives us sensitivity and specificity scores of 86.08 and 85.04, respectively. Additionally, it displays the best precision score of 85.41. For this specific database, this model performs significantly better than other models.

Additionally, we may observe that ANN models with four hidden layers perform the worst at 68.71 accuracy is being produced, while the specificity and sensitivity scores are at 67.72 and 68.27, respectively. Additionally, it only generates 66.56 in terms of precision measure, which highlights how poorly it performs. The CNN

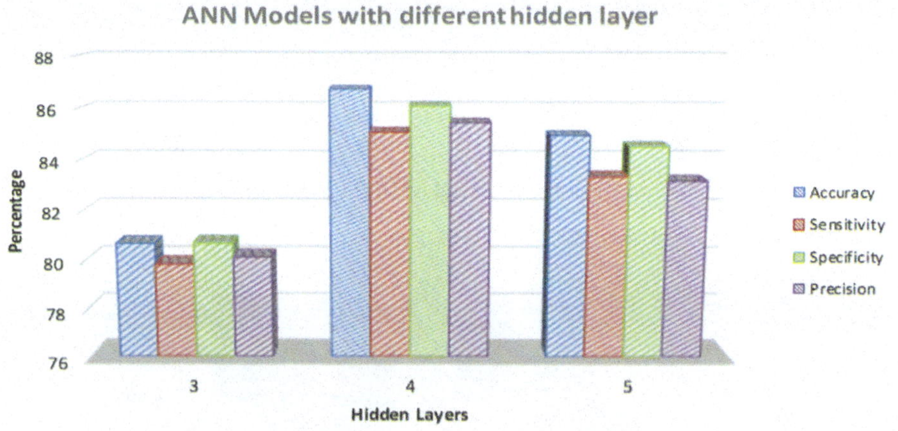

Fig. 6 Evaluation of several ANN models

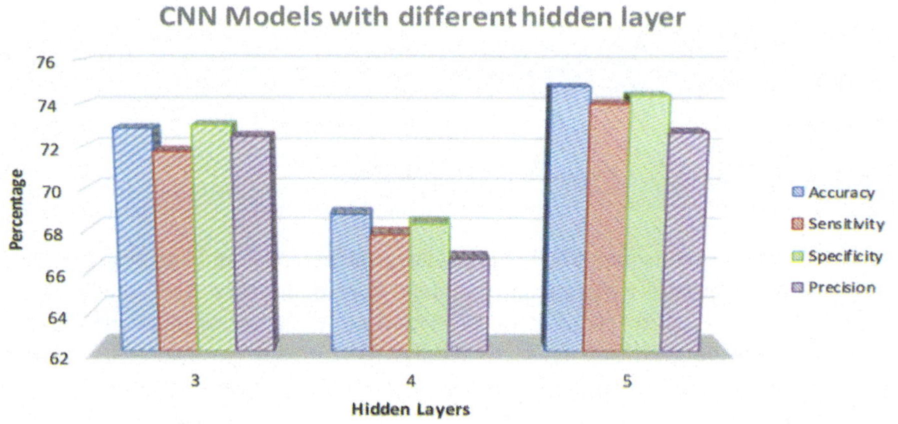

Fig. 7 Compares several CNN models

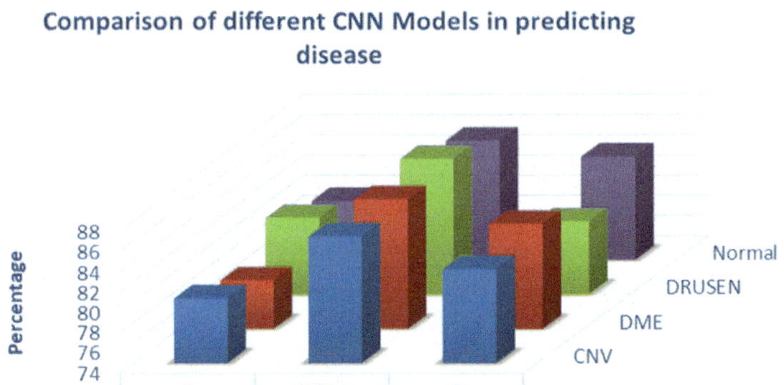

	3	CNN 4	5
CNV	80.43	86.53	83.32
DME	78.77	86.77	84.35
DRUSEN	81.53	87.38	81.17
Normal	79.8	85.79	84.21

Fig. 8 Comparing several CNN models to forecast diseases

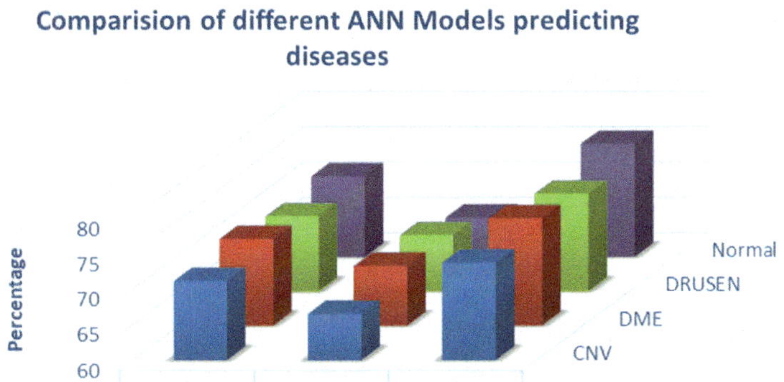

	3	ANN 4	5
CNV	71.2	66.51	73.77
DME	72.23	68.42	75.11
DRUSEN	70.57	67.88	73.73
Normal	71.34	65.5	75.93

Fig. 9 Comparing various ANN models to forecast illnesses

algorithm is outperforming ANN overall, nevertheless. The table shows that every CNN model generates a significantly higher score than every ANN method. The sophisticated internal mechanism of CNN makes this behavior predictable. ANN solely employs the feed forward method. CNN performs better than ANN in every model as a result. When it comes to hidden layers, we think that CNN produces more meaningful features when we utilize four hidden layers as opposed to the CNN model that only used three hidden levels.

5 Conclusion

ML algorithms performs well with data analysis applications and data forecasting methods than the statistical models. Researchers in that field quickly realized the value of DL algorithms and began utilizing them in their work, just like in other fields. It was convenient for the researchers to use such images for disease prediction because the medical sector generates a lot of data in the form of images. As a result, disease prediction using DL gained popularity. DL was utilized by researchers to forecast outcomes in ophthalmology, cancer, and many other fields. Ophthalmology is the field of medicine that investigates illnesses of the retina and the eyes. It produces a ton of imaging data in the form of report pictures, just like other healthcare industries. The pictures have been utilized by researchers in this field to forecast potential eye problems. In our study, we drew inspiration from prior studies and used two DL algorithms to try to differentiate three retinal disorders from the healthy eye. Specifically, ANNs and CNNs. It turns out that CNN performs better than ANN and produces the best outcome for the dataset we utilized for the study. Additionally, we want to explore unsupervised learning techniques and test whether it are more effective than our standard supervised DL methods.

References

1. Nezu, N., Usui, Y., Saito, A., Shimizu, H., Asakage, M., Yamakawa, N., Goto, H.: ML approach for intraocular disease prediction based on aqueous humor immune mediator profiles. Ophthalmology **128**(8), 1197–1208 (2021)
2. Li, S., Li, M., Wu, J., Li, Y., Han, J., Cao, W., Zhou, X.: Development and validation of a routine blood parameters-based model for screening the occurrence of retinal detachment in high myopia in the context of PPPM. EPMA J. **14**(2), 219–233 (2023)
3. Malik, S., Kanwal, N., Asghar, M.N., Sadiq, M.A.A., Karamat, I., Fleury, M.: Data-driven approach for eye disease classification with ML. Appl. Sci. **9**(14), 2789 (2019)
4. Badar, M., Haris, M., Fatima, A.: Application of DL for retinal image analysis: a review. Comput. Sci. Rev. **35**, 100203 (2020)
5. Alam, S.S., Shuvo, S.B., Ali, S.N., Ahmed, F., Chakma, A., Jang, Y.M.: Benchmarking DL frameworks for automated diagnosis of ocular toxoplasmosis: a comprehensive approach to classification and segmentation. arXiv preprint arXiv:2305.10975 (2023)
6. Zhang, C., He, F., Li, B., Wang, H., He, X., Li, X., Chen, Y.: Development of a deep-learning system for detection of lattice degeneration, retinal breaks, and retinal detachment in tessellated eyes using ultra-wide-field fundus images: a pilot study. Graefe's Arch. Clin. Exp. Ophthalmol. 1–10 (2021)
7. Choi, S., Park, J., Park, S., Byon, I., Choi, H.Y.: Establishment of a prediction tool for ocular trauma patients with ML algorithm. Int. J. Ophthalmol. **14**(12), 1941 (2021)
8. Fung, T.H., John, N.C., Guillemaut, J.Y., Yorston, D., Frohlich, D., Steel, D.H., Williamson, T.H.: Artificial intelligence using DL to predict the anatomical outcome of rhegmatogenous retinal detachment surgery: a pilot study. Graefes Arch. Clin. Exp. Ophthalmol. **261**(3), 715–721 (2023)
9. Kavakiotis, I., Tsave, O., Salifoglou, A., Maglaveras, N., Vlahavas, I., Chouvarda, I.: ML and data mining methods in diabetes research. Comput. Struct. Biotechnol. J. **15**, 104–116 (2017)

10. Li, Z., Guo, C., Nie, D., Lin, D., Zhu, Y., Chen, C., Lin, H.: DL for detecting retinal detachment and discerning macular status using ultra-widefield fundus images. Commun. Biol. **3**(1), 15 (2020)
11. Lake, S.R., Bottema, M.J., Williams, K.A., Lange, T., Reynolds, K.J.: Retinal shape-based classification of retinal detachment and posterior vitreous detachment eyes. Ophthalmol Therapy **12**(1), 155–165 (2023)
12. Wu, Z., Cai, W., Xie, H., Chen, S., Wang, Y., Lei, B., Lu, L.: Predicting optical coherence tomography-derived high myopia grades from fundus photographs using DL. Front. Med. **9**, 842680 (2022)
13. Salih, N., Ksantini, M., Hussein, N., Ben Halima, D., Abdul Razzaq, A., Ahmed, S.: Prediction of ROP zones using DL algorithms and voting classifier technique. Int. J. Comput. Intell. Syst. **16**(1), 86 (2023)
14. Zhao, X., Meng, L., Su, H., Lv, B., Lv, C., Xie, G., Chen, Y.: Deep-learning-based hemoglobin concentration prediction and anemia screening using ultra-wide field fundus images. Front. Cell Develop. Biol. **10**, 888268 (2022)
15. Son, J., Shin, J.Y., Kong, S.T., Park, J., Kwon, G., Kim, H.D., Park, S.J.: An interpretable and interactive DL algorithm for a clinically applicable retinal fundus diagnosis system by modelling finding-disease relationship. Sci. Rep. **13**(1), 5934 (2023)
16. Altan, G.: DeepOCT: an explainable DL architecture to analyze macular edema on OCT images. Eng. Sci. Technol. Int. J. **34**, 101091 (2022)
17. Sumathy, B., Chakrabarty, A., Gupta, S., Hishan, S.S., Raj, B., Gulati, K., Dhiman, G.: Prediction of diabetic retinopathy using health records with ML classifiers and data science. Int. J. Reliable Quality E-Healthcare (IJRQEH) **11**(2), 1–16 (2022)
18. Song, X., Xu, Q., Li, H., Fan, Q., Zheng, Y., Zhang, Q., Qu, Y.: Automatic quantification of retinal photoreceptor integrity to predict persistent disease activity in neovascular age-related macular degeneration using DL. Front. Neurosci. **16**, 952735 (2022)
19. Gundluru, N., Rajput, D.S., Lakshmanna, K., Kaluri, R., Shorfuzzaman, M., Uddin, M., Rahman Khan, M.A.: Enhancement of detection of diabetic retinopathy using Harris hawks optimization with DL model. Comput. Intell. Neurosci. (2022)
20. Viedma, I.A., Alonso-Caneiro, D., Read, S.A., Collins, M.J.: DL in retinal optical coherence tomography (OCT): a comprehensive survey. Neurocomputing (2022)
21. He, X., Ren, P., Lu, L., Tang, X., Wang, J., Yang, Z., Han, W.: Development of a DL algorithm for myopic maculopathy classification based on OCT images using transfer learning. Front. Public Health **10**, 1005700 (2022)
22. Kim, B.R., Yoo, T.K., Kim, H.K., Ryu, I.H., Kim, J.K., Lee, I.S., Kim, B.T.: Oculomics for sarcopenia prediction: a ML approach toward predictive, preventive, and personalized medicine. EPMA J. **13**(3), 367–382 (2022)
23. Wang, J., Wang, J., Chen, D., Wu, X., Xu, Z., Yu, X., Xu, W.: Prediction of postoperative visual acuity in patients with age-related cataracts using macular optical coherence tomography-based DL method. Front. Med. **10**, 1165135 (2023)
24. Singh, L.K., Khanna, M., Thawkar, S., Singh, R.: Nature-inspired computing and ML based classification approach for glaucoma in retinal fundus images. Multimedia Tools Appl. 1–49 (2023)
25. Sodhi, S.K., Pereira, A., Oakley, J.D., Golding, J., Trimboli, C., Russakoff, D.B., Choudhry, N.: Utilization of DL to quantify fluid volume of neovascular age-related macular degeneration patients based on swept-source OCT imaging: the ONTARIO study. PLoS ONE **17**(2), e0262111 (2022)
26. Schwarzenbacher, L., Seeböck, P., Schartmüller, D., Leydolt, C., Menapace, R., Schmidt-Erfurth, U.: Automatic segmentation of intraocular lens, the retrolental space and Berger's space using DL. Acta Ophthalmol. **100**(8), e1611–e1616 (2022)

27. Sheng, B., Chen, X., Li, T., Ma, T., Yang, Y., Bi, L., Zhang, X.: An overview of artificial intelligence in diabetic retinopathy and other ocular diseases. Front. Public Health **10**, 971943 (2022)
28. Kashmoola, M.Y.: DL for retinal disease detection surveys. AL-Rafidain J. Comput. Sci. Math. **16**(2), 1–9 (2022)

Chapter 29
Diabetic Retinopathy Automatic Detection and Classification in Fundus Images Using Modified Residual Convolutional Neural Networks (CNNs) with Improved Accuracy

S. Shilpa and B. Karthik

1 Introduction

A metabolic disorder called diabetes mellitus results in inadequate insulin production or insufficient insulin utilisation, which raises blood glucose levels. This damages the blood vessels, which over time results in issues with the kidneys, the nerves, and the eyes called diabetic retinopathy, diabetic nephropathy, and diabetic neuropathy. If diabetic retinopathy is not effectively managed over time, blindness may result. In the USA, diabetic retinopathy is a key factor in blindness. The repercussions of blindness can be averted by delivering and assessing appropriate treatment if diabetic retinopathy is identified earlier.

Currently, diabetic retinopathy is diagnosed by having retinal images evaluated by specialists for blood vessel abnormalities. Because too much glucose causes the blood pressure to rise and the blood vessels to become weak and damaged, an eye specialist would typically examine the blood vessels to distinguish between a healthy image and DR-eye. The weak blood pressure will get constricted, bloated, or even break, resulting in blood leakage that will obstruct light from reaching the retina. However, this brings up another issue, namely the disparity in the analytical skills of experts and the requirement for this procedure to use complete technology and apparatus in order to obtain a good image of the retina.

S. Shilpa (✉) · B. Karthik
Department of ECE, Bharath Institute of Higher Education and Research, Chennai, India
e-mail: shilpaprashant26@gmail.com

B. Karthik
e-mail: karthik.ece@bharathuniv.ac.in

J. M. R. S. Tavares et al. (eds.), *Proceedings of Second International Conference on Intelligent System*, Algorithms for Intelligent Systems,
https://doi.org/10.1007/978-981-99-8976-8_30

According to the World Health Organization (WHO), Indonesia has the fourth-highest prevalence of diabetic retinopathy disease. Currently, the procedure of identifying diabetic retinopathy is carried out manually by eye specialists by assessing the retinal image. Inspecting was done to look for aberrant blood vessels. Because of this, the procedure for screening for diabetic retinopathy will take longer [1]. Additionally, ophthalmologists with different skill levels, and levels of experience may arrive at differing conclusions or have disagreements [2].

The number of ophthalmologists in developing nations is limited, particularly in an archipelago nation like Indonesia where they have not colonised the entire island. In addition, the tools and technologies employed are not dispersed equitably across the nation and are not as advanced as in industrialised nations. In order to lessen our reliance on professionals, this has encouraged us to carry out more study on the automatic identification of diabetic retinopathy. Additionally, the specialists can concentrate on offering comprehensive drugs and therapies for diabetes late stages, which are also becoming less common. In biomedical informatics, machine learning is frequently utilised to solve a variety of issues [3, 4].

Some automated systems have been created such as the classification of cancer stages [5], Alzheimer's [6, 7], and bone segmentation [8, 9]. This proved successful in helping the work of experts so that experts can focus more on the phase of treatment. Convolutional neural network is a machine learning algorithm that is known to recognise image data very well [10]. This is due to the structure of the network that forms a two-dimensional area such as the arrangement of pixel values in the image so that it can capture the connection between the pixels in the image. Deep convolutional neural network has been widely used in solving quite complex classification tasks [11–13].

Modern technology, stemming from the Imagenet Large Scale Visual Recognition Challenge (ILSVRC), catalyzed the development of CNNs. Since 2010, this competition has been held annually, and it is the biggest in computer vision. In the ILSVR competition, 1.2 million photographs will be divided into 1000 classes. Residual neural networks, which aiming the proposed in 2015, are one of the current state of the art from CNN. By introducing a shortcut between their networks, the Residual-Net CNN version aims to reduce accuracy loss brought on by CNN networks with a deep structure [13, 14]. Numerous studies have been done on the detection of diabetic retinopathy.

However, earlier studies continued to rely on tiny datasets, such as the structure analysis of the retina (STARE) dataset, which only has 65 retinal images, and the digital retinal images for vessel extraction (DRIVE) dataset, which only contains 40 retinal images [15–17]. Some studies [18, 19] used the local hospital's primary dataset as their source; however the total number of photographs in the collection was still under 200. This gives the author the chance to examine the EyePACS dataset, which comprises 80,000 additional photographs, for the purpose of detecting diabetic retinopathy. A study from 2016 that used the EyePACS dataset also employed 5000 images for validation and 75,000 for training [20]. The author of this article uses CNN's residual network to automatically detect diabetic retinopathy.

This study is anticipated to quicken the process of evaluating retinal images to identify diabetic retinopathy disease, and variation in analysis results can also be prevented thanks to the established system for grading the stage of diabetic retinopathy. It is crucial to keep in mind that a variety of model architectures, preprocessing methods, hyperparameter tuning, dataset size, and quality can all have an impact on how well any machine learning algorithm performs, including ResNet-based CNNs. To evaluate how well various algorithms work on a given dataset, it is usually preferable to run extensive tests and evaluations [21, 22].

2 Literature Review

The results of related research, which used a variety of approaches and techniques for analysis and prediction on diverse healthcare datasets, are presented. Different data mining approaches, machine learning algorithms, or combinations of these techniques have been used by different researchers to design and apply a variety of prediction models. For the analysis of diabetic data, Saravana Kumar et al. [14] created a system using Hadoop and the MapReduce approaches. This approach forecasts the kind of diabetes as well as the dangers related to it. The system is cost-effective for any healthcare organisation and is built on Hadoop. Iyer et al. [15] examined hidden patterns in a diabetes dataset using a classification algorithm. In this model, Naive Bayes and decision trees were employed. The usefulness of both algorithms was demonstrated as a consequence of a comparison of their performances [15]. The categorisation technique was utilised by K. Rajesh and V. Sangeetha (2012). For effective classification, they employed the C4.5 decision tree technique to extract hidden patterns from the dataset [18]. Humar Kahramanli and Novruz Allahverdi (2008) employed fuzzy logic and artificial neural networks (ANN) to predict diabetes [19]. B.M. Patil, R.C. Joshi, and Durga Toshniwal (2010) suggested a hybrid prediction model that uses the simple K-means clustering approach first, then applies a classification algorithm to the results. The C4.5 decision tree algorithm is used to create classifiers [20].

To predict diabetes behaviour, Mani Butwall and Shraddha Kumar (2015) designed a model utilising random forest classifier [17]. Nawaz Mohamudally and Dost Muhammad (2011) employed visualisation, the K-means clustering technique, a neural network, and the C4.5 decision tree algorithm to predict diabetes. Support vector machines (SVM), random forests, and logistic regression are examples of traditional machine learning methods that rely on manually created picture characteristics. When compared to CNNs, these features perform less well since they were not specifically designed to collect all the significant information that was available in the photographs. Overall, diabetic retinopathy automatic identification and classification in fundus pictures has shown ResNet-based CNNs to be more accurate than conventional machine learning techniques. One of the difficulties with deep learning classification tasks, primarily in the medical field, is a lack of labelled training data. Utilising a small training dataset, transfer learning aids in deep learning model training.

Transfer learning can be used to address the issue of deep convolutional networks not having enough training data. On the ImageNet dataset, deep convolutional neural networks have been achieving remarkable performance results. EfficientNet-b5 is one such example and achieves top-1 83.4% accuracy and state-of-the-art top-5 96.2% accuracy. Comparatively speaking, it uses less parameters to attain greater accuracy and efficiency than other deep convolutional neural networks.

In order to address the issues with a limited training dataset and also employ this in transfer learning, this research put it to a test dataset that had never been seen before in the Kaggle Aptos Blindness Detection Challenge after training it on the training dataset. It can be applied to additional deep learningbased image classification issues that deal with the constraint of a lack of labeled training data.

3 Modified Residual Convolution Neural Network Architecture and Methodology

The CNN structure employed in this study consists of 8 convolutional layers, the first of which has 32 5 $\times$ 5-sized filters. The second convolution layer then has 32 filters, each measuring 3 by 3. The following layer has a 2 $\times$ 2 max-pooling layer. A shortcut from the third convolution layer to the fifth convolution layer, or so-called residual network, is then created by stacking three convolution layers with a total of 64 filters. The preceding layer's merging results will then enter maximum pooling with the same settings. Additionally, residual networks are created using a 3 subsequent layer convolution with up to 128 filters.

The sixth convolutional layer is used as a bridge to the eighth layer. There is a dropout layer after feature extraction by the convolution layer to prevent overfitting [23]. The second layer of the classification neural network has a fully connected layer size of 1024 and a feature pool. Figure 1 shows the proposed modified residual CNN structure.

This research employs orthogonal initialisation techniques that, when compared to Gaussian methods, can overcome the learning dynamics of deep learning [24]. Because L2 has more efficient calculation than L1, it is used in this study [24]. Leaky rectifier units are used as the nonlinearity in this research's layer convolutions to speed up convergence [25].

A modified ResNet can be different from the ResNet in the following ways: Depth: ResNet-50 contains 50 layers overall, whereas a customised ResNet can include an additional 8 levels for convolution. Depending on the individual alteration, it might be deeper. ResNet popularised the idea of skip connections, also known as residual connections, in which the output of one layer is immediately added to the output of a deeper layer. The pattern or structure of these skip connections may be changed by an updated ResNet, maybe adding new connections or eliminating old ones [23].

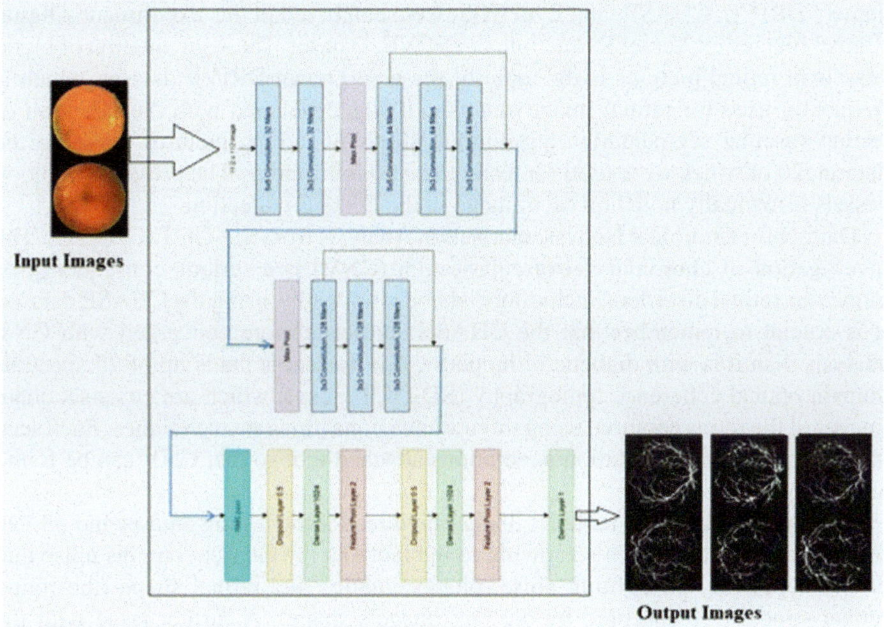

Fig. 1 Modified residual convolutional neural networks (Mod-ResNet) architecture

ResNet uses a particular block architecture in which the layers are grouped or "blocked," with each block made up of several convolutional layers and a downsampling step. The content or structure of these blocks is altered in a modified ResNet, possibly via the addition of more layers or the introduction of new sorts of layers inside each block. Channel configurations: To process the input data, ResNet uses 1 × 1, 3 × 3, and occasionally 1 × 1 convolutional filters. Using smaller filters or experimenting with various combinations of filter sizes within the network, a modified ResNet might test out various filter sizes.

Efficiency of computation: Modified ResNet designs frequently seek to maintain or increase performance while increasing computational efficiency or model simplicity. They may use strategies like bottleneck layers, which use 1 × 1 convolutions to cut down on the number of channels before executing more computationally intensive operations.

4 Result and Discussions

In this study, a single computing environment was used for all studies. Python 3.0 is the programming language used to implement source code. For building blocks, we also used free source tools like nolearn, Lasagne, and Scikit Learn. Three datasets, namely DRIVE, CHASE, and EyePACS, were employed in this experiment. Digital Retinal Images for Vessel Extraction, or DRIVE, Dataset: The segmentation of blood vessels in retinal pictures is the topic of the well-known DRIVE dataset, which is frequently used for retinal image analysis. It was developed with the intention of testing vascular segmentation algorithms. 40 colour retinal pictures make up the dataset, 20 of which were used for testing and 20 for training. The presence of blood vessels is manually noted on each image in the DRIVE collection.

Dataset for Choroidal Neovascularisation Analysis from SD-OCT (CHASE): The investigation of choroidal neovascularisation (CNV) is a serious consequence of numerous retinal disorders, including diabetic retinopathy, using the CHASE dataset. It is crucial to remember that the CHASE dataset is more concerned with CNV analysis than it is with diabetic retinopathy. The dataset is made up of 28 spectral-domain optical coherence tomography (SD-OCT) scans, which are cross-sectional images of the retina captured using three-dimensional imaging techniques. Each scan has ground truth segmentation annotations attached to it so that CNV can be found and examined.

The EyePACS dataset includes 39,424 pictures for the testing dataset and 35,126 for the training dataset. This is owing to the fact that the dataset contains noise that differs depending on lighting noise, picture quality, and retinal shape (the round retinal structure is obscured by the box image's shape). A further issue with the EyePACS dataset is the extremely imbalanced amount of samples in each class. For its deep and complicated networks to be trained, deep learning needs a large amount of training data. Due to the scarcity of sample data, deep learning has challenges when applied to the field of health. This is demonstrated in an earlier sample of less than 100 retinal images.

With this EyePACS dataset's 80,000 sample photographs, deep learning has the potential to help with the challenge of detecting diabetic retinopathy. If datasets contain a significant amount of noise, preprocessing is a crucial step. Reviewing the preprocessing stages in the light of the dataset's type of noise is also necessary. Rotation and translation are a crucial preprocessing step in this study because each retina's blood vessels differ from the others. If the dataset has an uneven number of samples for each class, enhancement is required.

In order to prevent underfit during testing, the augmentation technique is supposed to represent the variety of data inside a class. In this study, augmentation is required so that the network can learn each class's character in a balanced way. Figure 2 shows the examples from two experiments using the DRIVE and CHASE DB1 datasets. Figures 3 and 4 show the segmentation results for DRIVE dataset and CHASE DB1 dataset respectively. Figure 5 shows the EyePACS dataset segmentation results. Table 1 shows comparative analysis of classification accuracy results of modified ResNet model with individual CNN models for fundus detection using DRIVE dataset.

Figure 6 deals with comparison analysis of modified ResNet accuracy and existing method for 70% training versus 30% test dataset (DRIVE dataset).

Figure 7 deals with comparison analysis of modified ResNet accuracy and existing method for 80% training versus 20% test dataset (DRIVE dataset).

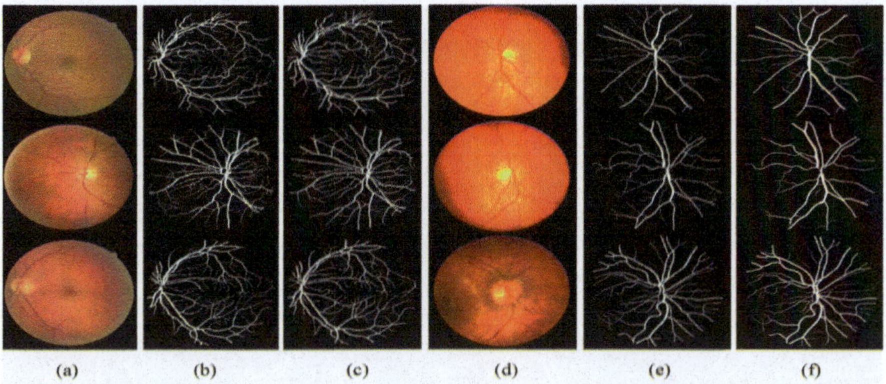

Fig. 2 Examples from two experiments using the DRIVE and CHASE DB1 datasets are shown. The DRIVE dataset is used for the first three columns. The final three columns come from the datasets in the CHASE DB1. Colour fundus images (**a**) and (**d**), ground truth (**b**) and (**e**), and segmentations (**c**) and (**f**)

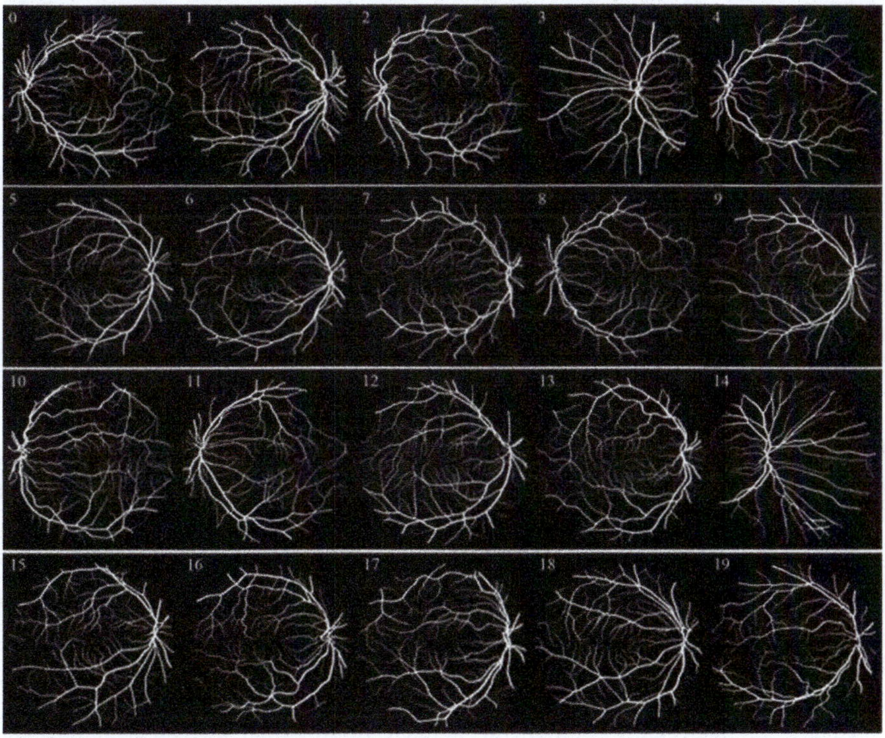

Fig. 3 Segmentation results for DRIVE dataset

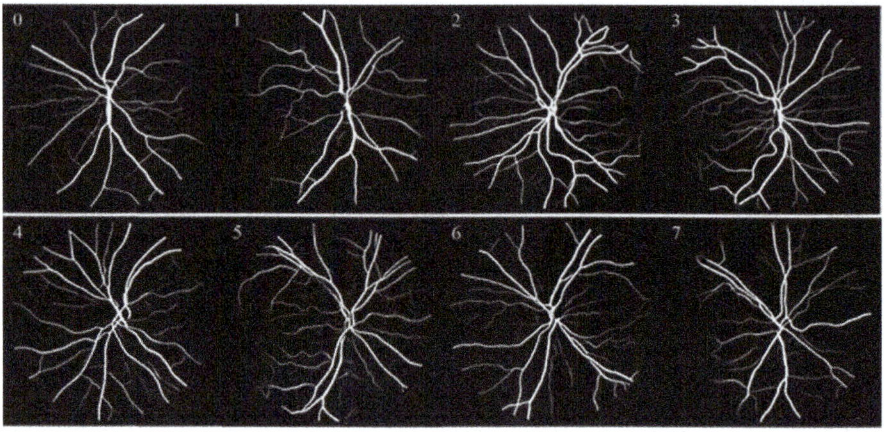

Fig. 4 Segmentation results. For CHASE DB1 datasets

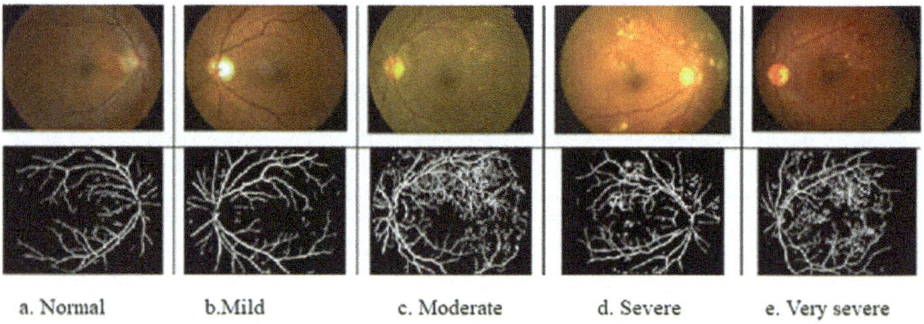

Fig. 5 EyePACS dataset segmentation results

Table 1 shows the comparative classification accuracy results of modified ResNet model with individual CNN models for fundus detection using DRIVE dataset.

Figure 8 deals with comparison analysis of modified ResNet accuracy and existing method for 90% training versus 10% test dataset (DRIVE dataset).

Figure 9 deals with comparison analysis of modified ResNet accuracy and existing method for 70% training versus 30% test dataset (CHASE dataset).

Table 2 shows the comparative classification accuracy results of modified ResNet model with individual CNN models for fundus detection using CHASE DB1 dataset.

Figure 10 deals with comparison analysis of modified ResNet accuracy and existing method for 80% training versus 20% test dataset (CHASE dataset).

Figure 11 deals with comparison analysis of modified ResNet accuracy and existing method for 90% training versus 10% test dataset (CHASE dataset).

Table 3 shows the classification accuracy results of modified ResNet model for fundus detection using EyePACS dataset.

Table 1 Comparative classification accuracy results of modified ResNet model with individual CNN models for fundus detection using DRIVE dataset

CNN models	Training (%)	Testing (%)	*F*1-score	Recall	Precision	Classification accuracy (%)
Inception-v3	70	30	0.95	0.98	0.92	92.50
	80	20	0.93	0.94	0.92	92.90
	90	10	0.94	0.92	0.96	93.67
VGG-19	70	30	0.94	0.98	0.90	92.33
	80	20	0.94	0.98	0.95	95.80
	90	10	0.94	0.97	0.89	93.51
ResNet-50	70	30	0.94	0.94	0.91	90.52
	80	20	0.94	0.94	0.90	95.54
	90	10	0.98	0.94	0.99	97.82
Modified ResNet	70	30	0.96	0.96	0.97	97.83
	80	20	0.96	0.97	0.95	98.56
	90	10	0.95	0.96	0.95	97.98

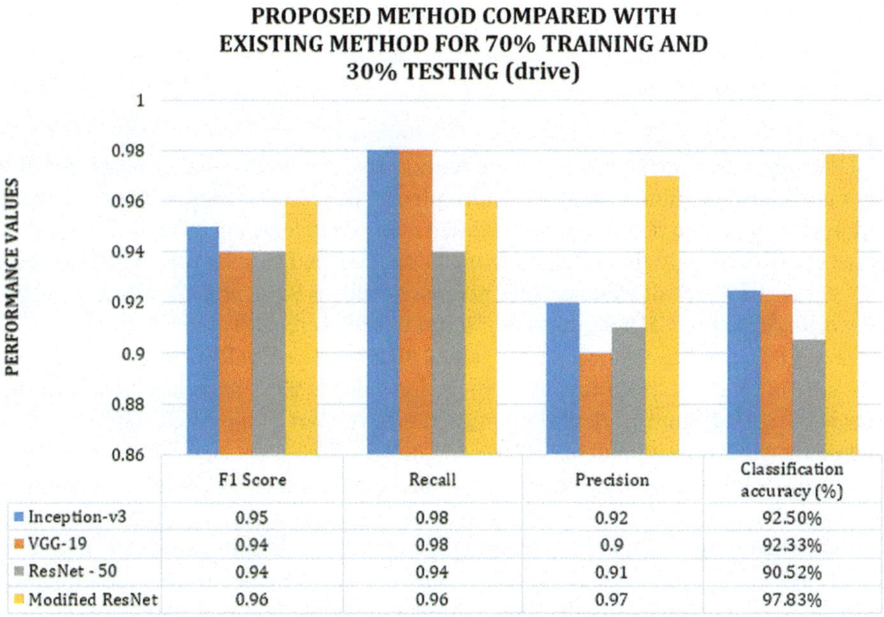

Fig. 6 Modified ResNet accuracy compared with existing method for 70% training versus 30% test dataset (DRIVE dataset)

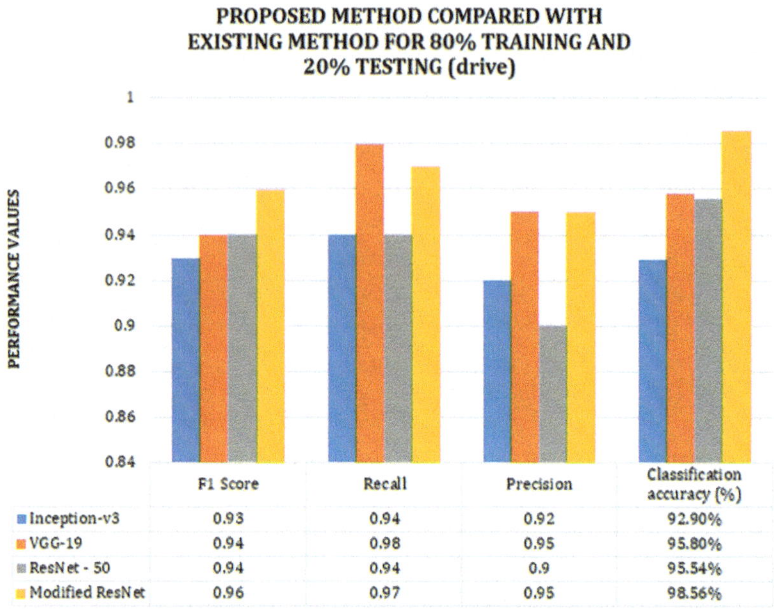

	F1 Score	Recall	Precision	Classification accuracy (%)
Inception-v3	0.93	0.94	0.92	92.90%
VGG-19	0.94	0.98	0.95	95.80%
ResNet - 50	0.94	0.94	0.9	95.54%
Modified ResNet	0.96	0.97	0.95	98.56%

Fig. 7 Modified ResNet accuracy compared with existing method for 80% training versus 20% test dataset (DRIVE dataset)

Diabetes patients who have diabetic retinopathy experience ocular problems. It is the main cause of blindness in the entire planet. Based on retinal pictures, diabetic retinopathy has been detected and classified using deep learning models like residual networks (ResNet). In this research work using modified ResNet for DRIVE dataset, $F1$-score is 95%, recall value is 96%, precision value is 97%, and classification accuracy is 98.56%. For CHASE dataset, $F1$-score is 98%, recall value is 97%, precision value is 98%, and classification accuracy is 98.59%. For EyePACS dataset, average $F1$-score is 96%, recall value is 96%, precision value is 97%, and classification accuracy is 97.30%. Figure 12 deals with measurement chart of modified ResNet performance for 90% training versus 10% test data (EyePACS dataset). In this proposed model, more intricate features and patterns in the retinal images could be captured by the deeper models, which enhanced classification performance. It is important to note that deep learning is a quickly changing area, and new architectural developments are always being produced. Deep neural network performance for the identification of diabetic retinopathy and other medical imaging applications is a topic of ongoing research. The classification of diabetic retinopathy has showed promise overall for ResNet-based architectures, with deeper models typically outperforming shallower ones. To get the best results, it is crucial to take into account additional elements including data quality, preprocessing methods, and model optimisation.

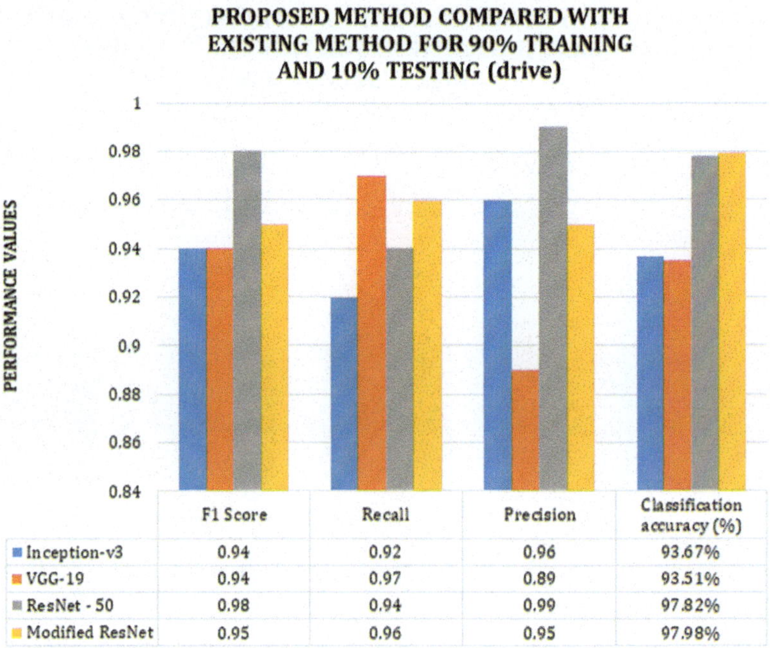

Fig. 8 Modified ResNet accuracy compared with existing method for 90% training versus 10% test dataset (DRIVE dataset)

5 Conclusions

This research paper proposes a methodology for the classification and identification of retinal exudates in fundus pictures using modified ResNet that is based on a pretrained convolutional neural network (CNN). Pretrained models, such as Inception-v3, Residual Network-50 (ResNet-50), and Visual Geometry Group Network-19 (VGG-19), are employed in the proposed framework to extract features from fundus images in order to increase classification accuracy. The suggested model's classification accuracy is better and then compared against a number of DCNN models individually and to currently used methods. The proposed modified ResNet architecture has been evaluated and yielded excellent accuracy results. As a result, the suggested method's accuracy outperforms other existing methods for the detection of retinal exudates. The suggested framework can be changed in subsequent research to distinguish between hard and soft exudates. Additionally, the suggested framework can be expanded to diagnose microaneurysms and haemorrhages in diabetic retinopathy.

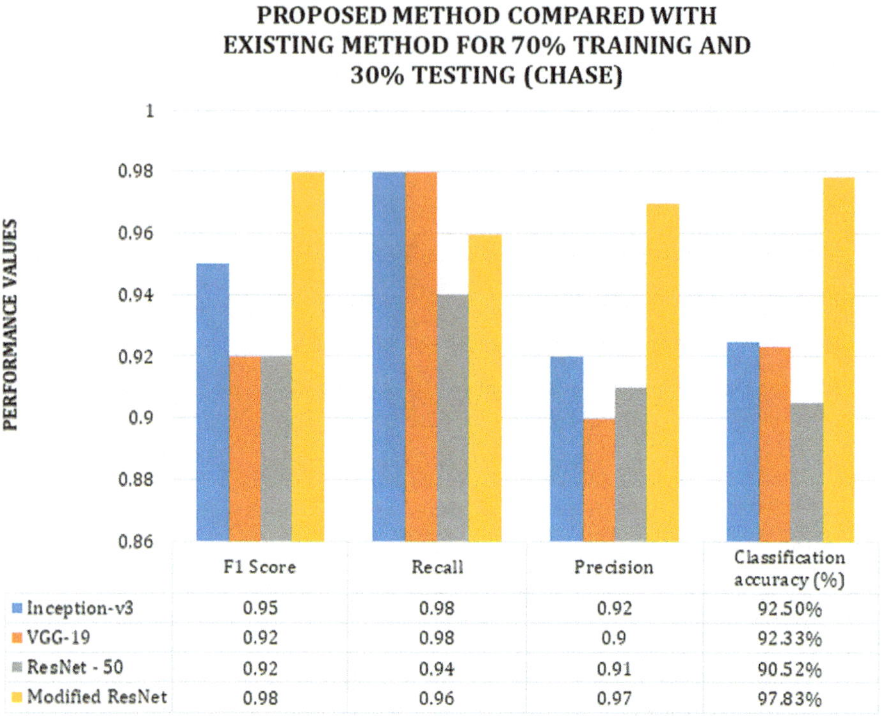

Fig. 9 Modified ResNet accuracy compared with existing method for 70% training versus 30% test dataset (CHASE dataset)

Table 2 Comparative classification accuracy results of modified ResNet model with individual CNN models for fundus detection using CHASE DB1 datasets

CNN models	Training (%)	Testing (%)	$F1$-score	Recall	Precision	Classification accuracy (%)
Inception-v3	70	30	0.95	0.98	0.92	92.50
	80	20	0.93	0.94	0.92	92.90
	90	10	0.95	0.92	0.96	93.67
VGG-19	70	30	0.92	0.98	0.90	92.33
	80	20	0.93	0.98	0.95	95.80
	90	10	0.91	0.97	0.89	93.51
ResNet-50	70	30	0.92	0.94	0.91	90.52
	80	20	0.92	0.94	0.90	95.54
	90	10	0.90	0.94	0.93	97.82
Modified ResNet	70	30	0.98	0.96	0.97	97.83
	80	20	0.98	0.97	0.95	98.59
	90	10	0.95	0.96	0.98	97.98

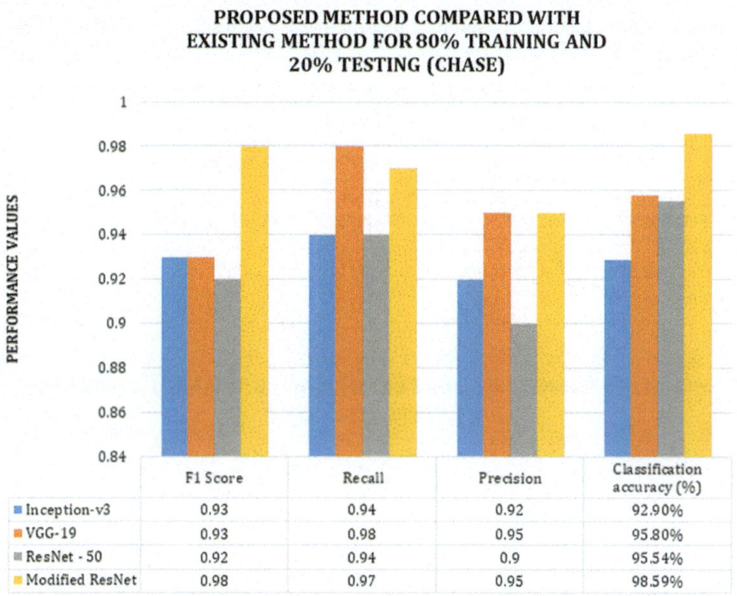

Fig. 10 Modified ResNet accuracy compared with existing method for 80% training versus 20% test dataset (CHASE dataset)

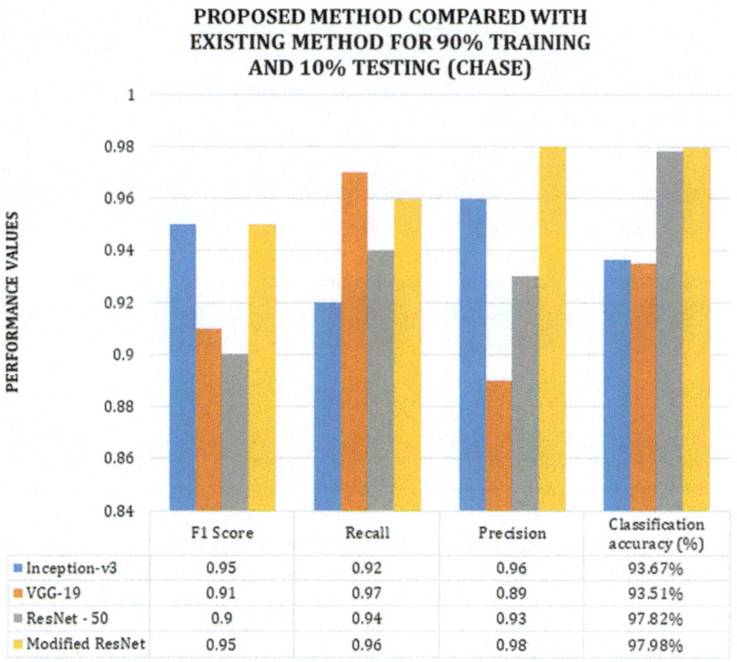

Fig. 11 Modified ResNet accuracy compared with existing method for 90% training versus 10% test dataset (CHASE dataset)

Table 3 Classification accuracy results of modified ResNet model for fundus detection using EyePACS dataset

DR level	$F1$-score	Recall	Precision	Classification accuracy (%)
No DR	0.95	0.98	0.92	97.50
Mild	0.94	0.98	0.90	96.33
Moderate	0.94	0.94	0.91	96.52
Severe	0.96	0.96	0.97	97.83
PDR	0.96	0.96	0.97	97.83
Average	0.95	0.964	0.934	97.30

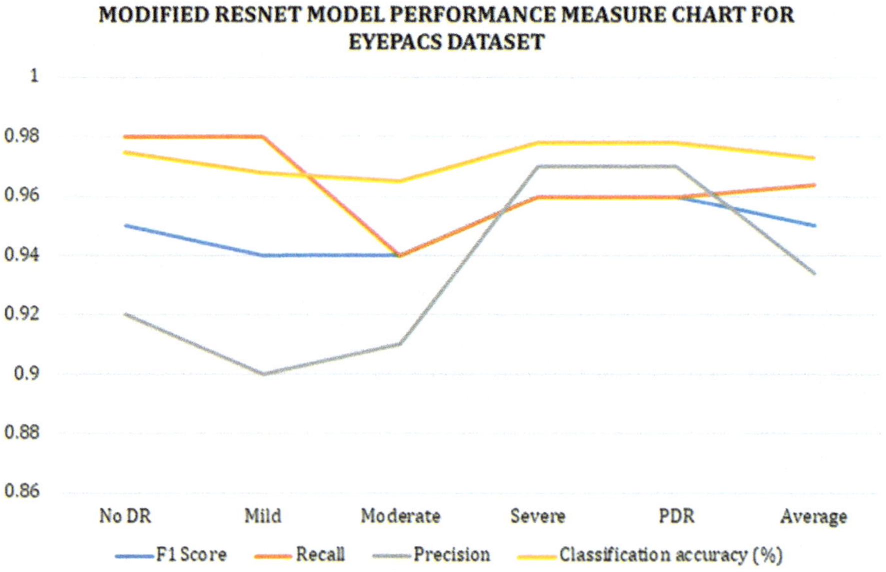

Fig. 12 Modified ResNet performance measure chart for 90% training versus 10% test data (EyePACS dataset)

References

1. Casanova, R., Saldana, S., Chew, E.Y., Danis, R.P., Greven, C.M., Ambrosius, W.T.: Application of random forests methods to diabetic retinopathy classification analyses. PLoS ONE **9**(6) (2014)
2. Magoulas, G.D., Prentza, A.: Machine learning in medical applications. In: Machine Learning and Its Applications Lecture Notes in Computer Science, pp. 300–307 (2001)
3. Bolón-Canedo, V., Ataer-Cansizoglu, E., Erdogmus, D., Kalpathy-Cramer, J., Fontenla-Romero, O., Alonso-Betanzos, A., Chiang, M.: Dealing with inter-expert variability in retinopathy of prematurity: a machine learning approach. Comput. Methods Prog. Biomed. **122**(1), 1–15 (2015)
4. Cruz, J.A., Wishart, D.S.: Applications of machine learning in cancer prediction and prognosis. Cancer Inf. **2**, 117693510600200 (2006)

5. Gray, K.R., Aljabar, P., Heckemann, R.A., Hammers, A., Rueckert, D.: Random forest-based similarity measures for multi-modal classification of Alzheimers disease. Neuroimage **65**, 167–175 (2013)
6. Lopez, M., Ramrez, J., Gorriz, J.M., Alvarez, I., Salas-Gonzalez, D., Segovia, F., Chaves, R., Padilla, P., Gomez R o, M., et al.: Principal component analysis-based techniques and supervised classification schemes for the early detection of alzheimers disease. Neurocomputing **74**(8), 12601271 (2011)
7. Lindner, C., Thiagarajah, S., Wilkinson, J., Consortium, T., Wallis, G., Cootes, T.: Fully automatic segmentation of the proximal femur using random forest regression voting. IEEE Trans. Med. Imag. **32**(8), 1462–1472 (2013)
8. Lindner, C., Thiagarajah, S., Wilkinson, J.M., Wallis, G.A., Cootes T.F., arcOGEN Consortium, et al.: Accurate bone segmentation in 2d radiographs using fully automatic shape model matching based on regression voting. In: Medical Image Computing and Computer-Assisted Intervention, p. 181189. Springer (2013)
9. Krizhevsky, A., Sutskever, I., Hinton, G.E.: ImageNet classification with deep convolutional neural networks. Commun. ACM **60**(6), 84–90 (2017)
10. Szegedy, C., Vanhoucke, V., Ioffe, S., Shlens, J., Wojna, Z.: Rethinking the inception architecture for computer vision. In: 2016 IEEE Conference on Computer Vision and Pattern Recognition (CVPR) (2016)
11. Simonyan, K, Zisserman, A.: Very deep convolutional networks for large scale image recognition ICLR (2015)
12. He, K., Zhang, X., Ren, S., Sun, J.: Deep residual learning for image recognition. In: 2016 IEEE Conference on Computer Vision and Pattern Recognition (CVPR) (2016)
13. Akram, M.U., Jamal, I., Tariq, A., Imtiaz, J.: Augmented segmented blood vessels for detection of proliverative diabetic retinopathy. In: International Conference on Biomedical and Health Informatics. Hong Kong and Shenzhen, China (2012)
14. Saravana kumar, N.M., Eswari, T., Sampath, P., Lavanya, S.: Predictive methodology for diabetic data analysis in big data. In: 2nd International Symposium on Big Data and Cloud Computing (2015)
15. Iyer, A., Jeyalatha, S., Sumbaly, R.: diagnosis of diabetes using classification mining techniques. Int. J. Data Min. Knowl. Manage. Process (IJDKP) 5(1) (2015)
16. Suresh Kumar, P. Pranavi, S.: Performance analysis of machine learning algorithms on diabetes dataset using big data analytics. In: International Conference on Infocom Technologies and Unmanned Systems, 978-1-5386-0514-1, Dec. 18–20 (2017)
17. Butwall, M., Kumar, S.: A data mining approach for the diagnosis of diabetes mellitus using random forest classifier. Int. J. Comput. Appl. **120**(8) (2015)
18. Rajesh, K., Sangeetha, V.: Application of data mining methods and techniques for diabetes diagnosis. Int. J. Eng. Innov. Technol. (IJEIT) 2(3) (2012)
19. Kahramanli, H., Allahverdi, N.: Design of a hybrid system for the diabetes and heart disease. Exp. Syst. Appl. Int. J. **35**(1–2) (2008)
20. Patil, B.M. Joshi, R.C., Toshniwal, D (2010) Association rule for classification of type-2 diabetic patients. In: ICMLC 10 Proceedings of the 2010 Second International Conference on Machine Learning and Computing, February 09–11 (2010)
21. Kharabe, S., Nalini, C.: Using adaptive thresholding extraction–robust ROI localization based finger vein authentication. J. Adv. Res. Dyn. Control Syst. **10**(7), 500–514 (2018)
22. Kharabe, S., Nalini, C.: Robust ROI localization based finger vein authentication using adaptive thresholding extraction with deep learning technique. J. Adv. Res. Dyn. Control Syst. **10**(7), 500–513 (2018)
23. Gayetri Devi, S.V., Nalini, C.: Prioritized automated generation of contracts with modified swarm optimization. Int. J. Adv. Sci. Technol. **29**(8), 2432–2439 (2020)
24. Lakshmi Vara Prasad, G., Nalini, C., Sugumar, R.: Arbitrary routing algorithm for tenable data assortment accessed in wireless sensor networks. Int. J. Civil Eng. Technol. **8**(1), 961–966 (2017)

25. Devi, S.V.G., Nalini, C.: Enhanced K-means clustering algorithm for feasibility assessment of ACC. In: 2020 Proceedings of the 2nd international conference on inventive research in computing applications, ICIRCA 2020 9182934, pp. 340–345 (2020)

Chapter 30
Properties of Idealization Graph Over a Ring

Manal Al-Labadi and Shuker Khalil

1 Introduction

In this article, we delve into the fascinating realm of idealization rings and their corresponding graphs. By considering S as an F-module and F as a finite commutative ring with unity, we define the idealization ring $(F\ (+)\ S, +, .)$ as the set of all elements $\{(x, y) : x \in F \text{and} y \in S\}$ with two defined operations: addition $(x, y) + (a, b) = (x + a, y + b)$ and multiplication $(x, y).(a, b) = (xa, xb + by)$. Additionally, we introduce the concept of the annihilator S, denoted by ann(S), which represents the collection of all elements $\{r \in F : \forall m \in S, rm = 0\}$. This plays a crucial role in our exploration of the properties of the zero-divisor graphs of idealization rings, as discussed by the authors in [1]. The intersection of ring theory and graph theory provides an intriguing subject matter that offers insightful perspectives. Within this article, we focus on investigating various properties of the idealization ring's associated graph, such as the girth and diameter. The works presented in [2, 3] have previously studied the diameter and girth of zero-divisor graphs of commutative rings. Moreover, [4–6] have examined several properties of the zero-divisor graph of idealization rings, including the clique number, planarity, and Wiener index. In our context, let H be a simple connected graph, and $V(H)$ represent the set of all vertices in H. We define the distance between two distinct vertices x and y in H as $d(x, y)$, where it exists; otherwise, $d(x, y) = \infty$. The diameter of the graph G is defined as the largest distance between any vertices x and y in H, denoted by diam(H). Furthermore, the girth of G is the length of the shortest cycle in H, denoted by $g(H)$. Our

M. Al-Labadi (✉)
Department of Mathematics, Faculty of Arts and Sciences, University of Petra, Amman, Jordan
e-mail: manal.allabadi@uop.edu.jo

S. Khalil
Department of Mathematics, College of Science, University of Basrah, Basra, Iraq
e-mail: Shuker.khalil@uobasrah.edu.iq

J. M. R. S. Tavares et al. (eds.), *Proceedings of Second International Conference on Intelligent System*, Algorithms for Intelligent Systems,
https://doi.org/10.1007/978-981-99-8976-8_31

article presents a new definition known as the idealization graph $L(F(+)S)$, where the vertices are $V(L(F(+)S)) = \{(x, y) : x \in F\text{and}y \in S\}$. In other words, the vertices of the idealization graph are the elements of the idealization rings, and two vertices (x_1, y_1), and (x_2, y_2) in the idealization graph are adjacent if either x_1 or $x_2 \in \text{ann}(S)$.

The primary objective of this article is to explore the hidden relationships between the zero-divisor graph of the idealization ring and the idealization graph. Additionally, we delve into the significance of the annihilator of S and its implications in model selection. We assume the sample to be the set of all zero-divisor elements, thus providing valuable insights into this intricate topic.

By delving into these intricate concepts, we aim to contribute to the existing body of knowledge and provide a deeper understanding of the interplay between idealization rings, their associated graphs. Also, if it the deal with annihilator of the S and how to drop this topic to choose models. Assuming the sample is the set of all zero-divisor element [7, 8].

2 Properties of Idealization graph *L*(*F*(+)*S*)

In this section, we study some properties of the idealization graphs $L(F(+)S)$ that can be characterize the idealization graph.

Theorem 1 *Let F be a ring and S be an F-module. Then, the girth is 3.*

Proof If F is a commutative ring with unity and S is an F-module, then we have three vertices such that (0,0) – (0,1) – (1,1) – (0,0). So, we get a cycle with length 3. So, the girth of the idealization graph is 3.

Theorem 2 *Let F be a ring and S an F-module. Then, diam* $(L(F(+)S)) = 2$.

Proof If (x, y) and (c, d) are two different vertices in $L(F(+)S)$, then there exists the vertex $(0,0) \in V\ (L(F(+)S))$ such that (0,0) is adjacent to each vertex in the idealization graph $L(F(+)S)$. Thus, we have (x, y) – (0,0) – (c, d).

Example 1 Consider the idealization graph $L(Z_4(+)\ Z_2)$.

Then, $V\ (L(Z_4(+)\ Z_2)) = \{(0,0), (0,1), (1,0), (1,1), (2,0), (2,1), (3,0), (3,1)\}$. See, the following figure.

Definition 1 *The cardinality of the largest independent vertex set is called the independence number and is denoted by* $\alpha(H)$.

Theorem 3 *Let F be a commutative ring with unity and S an F-module. Then,* α $(L(F(+)S)) = |F\backslash\text{ann}(S)||S|$.

Proof If $a \in F\backslash\text{ann}(S)$, then (a,h) where $h \in S$ that is not adjacent to any vertex in the idealization graphs $L(F(+)S)\backslash\{(r,h)\text{: } r \in \text{ann}(S), h \in S\}$ and any other vertex different than (a,h) is adjacent to each vertex in the idealization graph. Then, $\alpha(L(F(+)S)) = |F\backslash\text{ann}(S)||S|$.

When the idealization graph is a divisor graph, the following properties are investigated. A graph H is called a divisor graph if there is a one-to-one function: $V(H) \to N$ natural number such that x is adjacent to y in H if and only if $\varphi(x)$ divides $\varphi(y)$ or $\varphi(y)$ divides $\varphi(x)$; see [9].

Theorem 4 *Let F be a ring and S an F-module. Then, L(F(+)S) is a divisor graph.*

Proof If F is a commutative ring with unity and S *is* an F-module.

Let m be several distinct primes, P_i. Then, there is a one-to-one function f by

$$\varphi(x_i, y_i) = \begin{cases} \prod_{l=1}^{m} P_l 2^{k_i}, & if\ x_i \in \mathrm{ann}(M) \\ P_i, & if\ x_i \notin \mathrm{ann}(M) \end{cases}.$$

We look the clique number of the idealization graphs $L(F(+)S)$.

Let H be a graph with V vertices and E edges. Then, a subset of vertices C is called a clique if every two vertices in C are adjacent. Moreover, the size of maximum clique in vertices in H is called clique number of a graph H and is denoted by $\omega(H)$.

Theorem 5 *Let F be a ring and S an F-module. Then,* $\omega(L(F(+)S)) = |\mathrm{ann}(S)||S| + 1$.

Proof If F is a commutative ring with unity and S *is* an F-module, then the maximum set of vertices that are adjacent to each other in the idealization graphs is the set of all vertices form $\{(r, h) : r \in \mathrm{ann}(S), h \in S\}$ and another element like $\{(0, 1)\}$. Since the induced subgraph $\{(r, h) : r \in \mathrm{ann}(S), h \in S\} \cup \{(1, 0)\}$ is a complete graph. Thus, $\omega(L(F(+)S)) = |\mathrm{ann}(S)||S| + 1$.

Corollary 1 *If F is a ring and S is an F-module, then the chromatic number of* $L(F(+)S)$is$\chi(L(F(+)S) = |\mathrm{ann}(S)||S| + 1$.

Proof Assume that F is a ring and S *is* an F-module. Then, the induced subgraph $\{(x,h): x \in \mathrm{ann}(S), h \in S\}$ is a complete graph and adjacent to each other vertices in the idealization graph. Then, the minimum number of coloring the complete graph $\{(x,h): x \in \mathrm{ann}(S), h \in S\}$ is $|\mathrm{ann}(S)||S|$ and any other vertex in the set $\{(r,h): r \in F\backslash\mathrm{ann}(S), h \in S\}$ is coloring different than others in the set $\{(x,h): x \in \mathrm{ann}(S), h \in S\}$. Also, the set $\{(r,h): r \in F\backslash\mathrm{ann}(S), h \in S\}$ is independent set then we need one color. Thus, the chromatic number of the idealization graph is $\chi(L(F(+)S)) = |\mathrm{ann}(S)||S| + 1$. See, the following figure.

When the idealization graph is Eulerian, Hamiltonian graph, the following properties are investigated.

We begin with the following definitions.

Definition 2 *A graph is called an Eulerian graph if it has a closed trial containing all edges of the graph.*

Definition 3 *If a graph H has a closed path that visits every vertex in the graph exactly one, then the graph H is called a Hamiltonian graph.*

In [10, 11], they characterized when the graph is Eulerian and what is the sufficient condition for Eulerian graph.

Lemma 1 *A graph H is called an Eulerian graph if and only if every vertex in the connect graph H is even degree.*

The following examples to illustrate when the idealization graph is not Eulerian.

Example 1 Consider the $L(Z_2(+)Z_2)$. Then, $V(L(Z_2(+)Z_2)) = \{(0,0), (0,1), (1,0), (1,1)\}$. Thus, $L(Z_2(+)Z_2)$ is not Eulerian graph.

Example 2 Consider the $L(Z_4(+)Z_2)$. Then, $V(L(Z_4(+)Z_2)) = \{(0,0), (0,1), (1,0), (1,1), (2,0), (2,1), (3,0), (3,1)\}$. Thus, $L(Z_4(+)Z_2)$ is not Eulerian graph.

Example 3 Consider the $L(Z_3(+)Z_3)$. Then $V(L(Z_3(+)Z_3)) = \{(0,0), (0,1), (0,2), (1,0), (1,1), (1,2), (2,0), (2,1), (2,2)\}$. Thus, not Eulerian graph. See, Fig. 2.

Now, we determine when the idealization graph is not Eulerian for $|F| \geq 5$.

Theorem 6 *Let* ***F*** *be a ring,* $|F| \geq 5$ *and S be an F-module. Then, L(F(+)S) is not Eulerian graph.*

Proof If *S be* an *F*-module, then

- **Case 1**: If $\text{ann}(S) = \{0\}$ and $|S|$ is an odd number, then the vertex (1,0) is adjacent to each vertex of the form $\{(0,h)\colon h \in S\}$. So, $\deg((0,1)) = |S| =$ odd number.
- **Case 2**: If $\text{ann}(S) = \{0\}$ and $|S|$ is an even number, then the vertex (0,0) is adjacent to each vertex of the form $\{(r,h)\colon r \in F, h \in S\}$. So, $\deg((0,0)) = |F||S|\backslash|\{(0,0)\}| =$ odd number.
- **Case 3**: If $\text{ann}(S) \neq \{0\}$ and $|S|$ is an odd number, then we have the following subcase:

  Subcase 3.1: If $|\text{ann}(S)| =$odd, then the vertex (1,0) is adjacent to each vertex of the form $\{(r,h)\colon r \in \text{ann}(S), h \in S\}$. So, $\deg((0,1)) = |\text{ann}(S)||S| =$ odd number.

  Subcase 3.2: If $|\text{ann(S)}| =$even, then the vertex (0,0) is adjacent to each vertex of the form $\{(r,h)\colon r \in R, h \in M\}$. So, $\deg((0,0)) = |F||S|\backslash|\{(0,0)\}| =$ odd number.
- **Case 4**: If $\text{ann(S)} \neq \{0\}$ and $|S|$ is an even number, then the vertex (0,0) is adjacent to each vertex of the form $\{(r,h)\colon r \in F, h \in S\}$. So, $\deg((0,0)) = |F||S|\backslash|\{(0,0)\}| =$ odd number.

Corollary 2 *Let* ***F*** *be a ring and* ***S*** *be an* ***F****-module. Then, L(F(+)S) is not Eulerian graph.*

The following examples to illustrate when the idealization graph is Hamiltonian graph.

Example 4 Consider the $L(Z_2(+)Z_2)$. Then $V(L(Z_2(+)Z_2)) = \{(0,0), (0,1), (1,0), (1,1)\}$. Thus, $L(Z_2(+)Z_2)$ is Hamiltonian graph.

Example *5* Consider the $L(Z_4(+)Z_2)$. Then, $V(L(Z_4(+)Z_2)) = \{(0,0),(0,1),(1,0),(1,1), (2,0),(2,1),(3,0),(3,1)\}$. Thus, $L(Z_4(+)Z_2)$ is Hamiltonian graph.

Example *6* Consider the $L(Z_3(+)Z_3)$. Then, $V(L(Z_3(+)Z_3)) = \{(0,0),(0,1),(0,2),(1,0), (1,1),(1,2),(2,0),(2,1),(2,2)\}$. Thus, $L(Z_3(+)Z_3)$ is not Hamiltonian graph.

Now, we determine when the idealization graph is not Hamiltonian graph for $|F| \geq 5$.

Theorem 7 **Let** F be a ring, $|F| \geq 5$ and S be an F-module. Then, $L(F(+)S)$ is a Hamiltonian graph if $|\text{ann}(S)| = |F\backslash\text{ann}(S)|$.

Proof If $\boldsymbol{F}$ is a commutative ring with unity, by $|F| \geq 5$ and S is an F-module, then we have the following holds:

- **Case 1**: If $\text{ann}(S) = \{0\}$, then the induced subgraph $\{(0, h): h \in S\}$ is a complete graph and adjacent to each vertex in the graph $L(F(+)S)$, with the cardinality is $|\{(0,h): h \in S\}| = |S|$. Hence, $L(F(+)S)\backslash\{(0,h): h \in S\}$ has more than $|S|$ connected components, since $|V(L(F(+)S)\backslash\{(0,h): h \in \text{S}\}| = |F||S| - |S|=|S|(|F| - 1)$ is the contradiction to the theorem in [10]. Thus, the idealization graph is not Hamiltonian graph.
- **Case 2**: If $\text{ann(S)} \neq \{0\}$, then we have the following holds:
  Subcase 2.1: If $|\text{ann}(S)| = F\backslash\text{ann}(S)$, then the idealization graph $L(F(+)S)$ has an induced subgraph $\{(a,h)$: a $\in \text{ann}(S)$, $h \in S\}$ is a complete graph which is cardinality is $|\{(a,h): a \in \text{ann}(S), h \in S\}| = |S||\text{ann}(S)|$. Also, each vertex in the set $\{(a, h) : ah \in \text{ann}(S), h \in S\}$ is adjacent to each vertex in the graph $L(F(+)S)$, $|V(L(F(+)S))| = |S||F|$. Thus, the idealization graph has a closed path (0,0)-(1,0)-(0,1)-(r_1,0)-(r_2,0)-(r_1,1)-... -(r_{k-1},1)-(r,0)-(r_{k-1},0)-(0,0), where $r_2, r_4, \ldots, r_k \in \text{ann}(S)$. Thus the $L(F(+)S)$ is a Hamiltonian graph.
  Subcase 2.2: If $|\text{ann}(S)| < |\text{F} \setminus \text{ann}(S)|$, then the induced subgraph $\{(a,h)$: $a \in \text{ann}(S)$, $h \in S\}$ is a complete graph and adjacent to each vertex in the graph $L(F(+)S)$ and the cardinality is $|\{(a, h) : a \in \text{ann}(S), h \in S\}| = |\text{ann}(S)||S|\langle|F\backslash\text{ann}(S)||S|$. Hence, $L(F(+)S)\backslash\{(a, h) : a \in \text{ann}(S), h \in S\}$ has more than $|\text{ann}(S)||S|$ connected components, since$|V(L(F(+)S))\backslash\{(a, h) : a \in \text{ann}(S), h \in S\}| = |F||S| - |\text{ann}(S)||S| = |S|(|F| - |\text{ann}(S)|) > |\text{ann}(S)||F|$ is the contradiction to the theorem in [10]. Thus, the idealization graph is not Hamiltonian graph. Now, we study the Wiener index of the idealization graphs $L(F(+)S)$, we begin with the following definition.

Definition 4 *The sum of all distances between vertices of the graph H is called the Wiener index denoted by* $W(H)$.

Theorem 8 *If* F *is a ring and* S *is an* F*-module, then the Wiener index of the idealization graphs is* $W(L(F(+)S) = (|\text{ann}(S)||S| + 2|F\backslash\text{ann}(S)||S| - 2)|F\backslash\text{ann}(S)||S| + (|F||S| - 1)|\text{ann}(S)||S|$.

Proof Let F be a ring and S be an F-module. Then, we have the following holds:

- **Case 1**: If the vertex $(a,l) \in V(L(F(+)S))$ such that $a \in F\backslash \text{ann}(S)$, then the vertex (a,l) is adjacent to each vertex of the form $\{(r,h)\colon r \in \text{ann}(S),\ h \in S\}$. Thus, the degree of the vertex (a,l) is $\deg((a,l)) = |\text{ann}(S)||S|$. And the vertex (a,l) is not adjacent to each vertex of the form $\{(b,k)\colon b \in R\backslash \text{ann}(S),\ k \in S\}$. Then, the sum of the distance between the vertex (a,h) and any vertex (x,y) in the idealization graphs is $|\text{ann}(S)||S| + 2(|F\backslash \text{ann}(S)||S| - |\{(a, h)\}|) = |\text{ann}(S)||S| + 2|F\backslash \text{ann}(S)||S| - 2$.
- **Case 2**: If the vertex $(a,l) \in V(L(F(+)S))$ such that $a \in \text{ann}(S)$, then the vertex (a, l) is adjacent to each vertex of the form $\{(r, h) : r \in F, l \in S\}\backslash\{(a, l)\}$. Thus, the degree of the vertex (a,l) is $\deg(a,h) = |F||S| - 1$. Then, the distance between the vertex (a,h) and any vertex (x,y) in the idealization graphs is $d((a,h),(x,y)) = |F||S| - 1$. So, the sum of all distances in the idealization graphs is $W(L(F(+)S)) = (|\text{ann}(S)||S| + 2|F\backslash \text{ann}(S)||S| - 2)|F\backslash \text{ann}(S)||S| + (|F||S| - 1)|\text{ann}(S)||S|$. See, Fig. 1, $W(L(Z_4(+)\,Z_2)) = 68$.

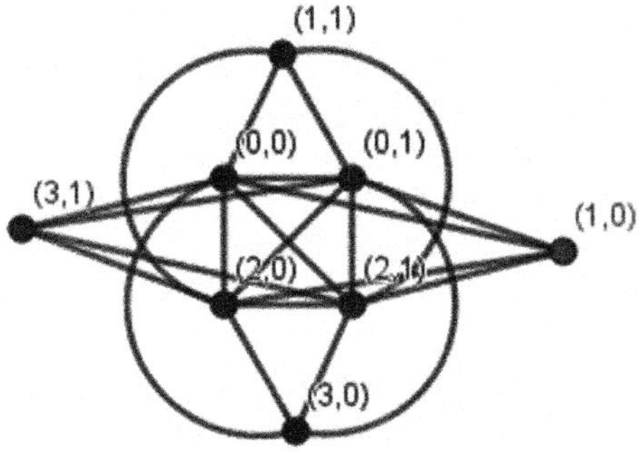

Fig. 1 $L(Z_4(+)Z_2)$

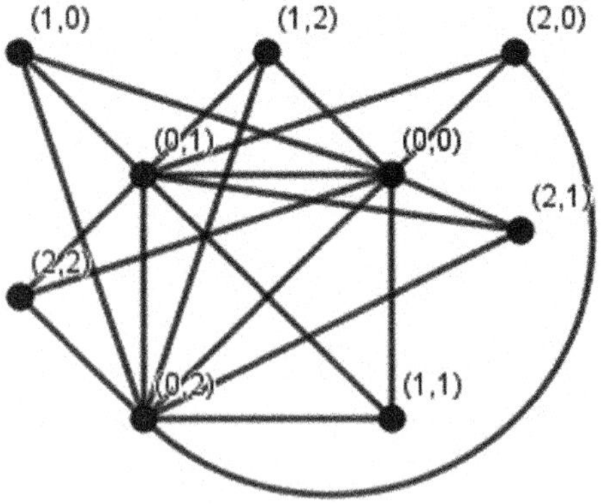

Fig. 2 $\chi(L(Z_3(+)Z_3)) = 4$

The graph is referred to as a "Planar graph" if it can be drawn in the plane without any edges crossing, and a "One-Planar graph" if edges are only crossed occasionally [10, 11].

Theorem 9 *Let F be a ring and S be an F-module. Then, L(F(+)S) is Planar graph if* $|S| \leq 3$ *and* $\text{ann}(S) = 0$.

Proof Let S be an F-module, then we have the following holds:

- **Case 1**: If $\text{ann}(S) = 0$, then have the following cases:
  Subcase 1.1: If $S \cong Z_2$ and $|F| \geq 2$, then the idealization graph can draw in the plane without cross edge. Thus, the idealization graph is Planar graph. See the following figure,
  Subcases 1.2: If $S \cong Z_3$ and $|F| \geq 3$, then the idealization graph has $|V(L(F(+)S)| \geq 9$ and $|E(L(F(+)S)| \geq 3 + 9|F\backslash\{0\}|$. The induced subgraph of this vertices $\{(0,h)\colon h \in S\}$ is a complete graph and is adjacent to each other vertices in the idealization graph $L(F(+)S)$ also, the set $\{(a, h) : a \in R\backslash\text{ann}(S), h \in S\}$ is adjacent only to $\{(0,h)\colon h \in S\}$. So, by Kuratoski theorem for Planar graph, the idealization graph is Planar graph.
  Subcase 1.3: If $|S| \geq 4$, then the idealization graph has an induced subgraph isomorphic to K_5 with vertices are $\{(0,0),(0,1),(0,2),(0,3),(1,0)\}$. So, the graph is not Planar graph.
- **Case 2**: If $ann(S) \neq \{0\}$, then the idealization graph has an induced subgraph isomorphic to K_5 with vertices are $\{(0,0), (0,1), (r,0),(r,1),(1,0)\}$ where $r \in \text{ann}(S)$. So, the graph is not Planar graph.

Now, we begin with the following lemma, summarize the 1-Planar of the graph, see [12].

Lemma 2 *The graph K7* $-$ K_3 *is the only one 7-vertex minimal non-one-Planar graph.*

Theorem 10 *Let F be a ring and S be an F-module. Then, L(F(+)S) is 1-Planar graph if* $|S| \leq 3$ *and* $\text{ann}(S) = \{0\}$.

Proof Let S be an F-module. Then, we have the following holds:

- **Case 1**: If $\text{ann}(S) = \{0\}$, then we have the following holds:
  Subcase 1.1: If $S \cong Z_2$ and $|F| \geq 2$, then the idealization graph can draw in the plane without cross edge. Thus, the idealization graph is 1-Planar graph. See Fig. 3.
  Subcase 1.2: If $S \cong Z_3$ and $F \cong Z_3$, then the idealization graph can be drawn in the plane, each edge is crossed at most once. Then, the idealization graph is 1-Planar graph. See, the following Fig. 4.
  Subcase 1.3: If $S \cong Z_3$ and $|F| > 3$, then the idealization graph has $|V(L(F(+)S)|\rangle 9$ and $|E(L(F(+)S)|\rangle 3 + 9|F\backslash\{0\}|$. The induced subgraph

Fig. 3 $L(F(+)Z_2)$

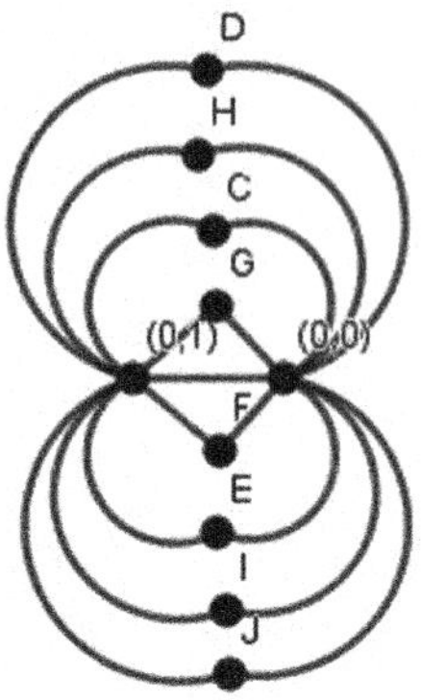

Fig. 4 $L(Z_3(+)Z_3)$

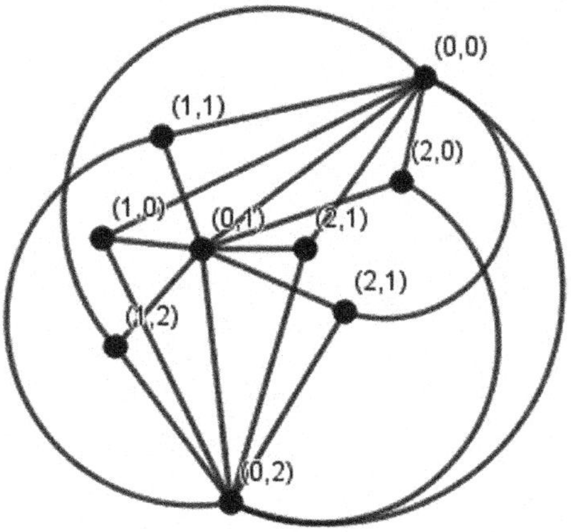

$\{(0, h) : h \in S\}$ is a complete graph and is adjacent to each other vertices in the idealization graph *L(F(+)S)* also, the set $\{(a, h) : a \in F\backslash \text{ann}(S), h \in S\}$ is adjacent only to $\{(0,h)$: $h \in M\}$. So, by Kuratoski theorem for planar graph, the idealization graph is Planar graph. So, the idealization graph is 1-Planar graph.

Subcase 1.4: If $|S| \geq 4$, then the idealization graph has an induced subgraph isomorphic to K_7–k_3 with vertices $\{(0,0),(0,1),(0,2), (0,3),(1,0),(1,1),(1,2)\}$. So, the idealization graph is not 1-Planar graph.

- **Case 2:** If ann(S) $\neq \{0\}$, then R|$\geq$ 4 since S is an F-module and ann$(S) \neq \{0\}$. Then:

Subcase 2.1: If for all $r \in F\backslash\{1\}$, $rS = 0$, then the idealization graph has an induced subgraph isomorphic to $K_7 = \{(0, 0), (0, 1), (r_1, 0), (r_1, 1), (1, 0), (r_2, 0), (r_2, 1)\}$ where $r_1 \in \text{ann}(S)$ and $a \in F\backslash\text{ann}(S)$. So, the idealization graph is not 1-Planar graph.

Subcase 2.2: If not all $r \in F^*$, $rS = 0$, then the idealization graph has an induced subgraph isomorphic to $K_7 - K_3 =$

$\{(0, 0), (0, 1), (r_1, 0), (r_1, 1), (1, 0), (1, 1), (a, 0)\}$ where $r_1 \in \text{ann}(S)$ and $a \in F \backslash \text{ann}(S)$. So, the idealization graph is not 1-Planar graph.

References

1. Axtell, M., Stickle, J.: The zero-divisor graph of a commutative rings. J. Pure Appl. Algebra **204**, 235–243 (2006)
2. Anderson, D.D., Naseer, M.: Beck's coloring of a commutative ring. J. Algebra **159**, 500–514 (1993)
3. Anderson, D.F., Livingston, P.S.: The zero-divisor graph of a commutative. J. Algebra **217**, 434–447 (1999)
4. Al-Labadi, M.: Some properties of the zero-divisor graphs of idealization ring F(+)S. Math. Prob. Eng. 1–6 (2022). https://doi.org/10.1155/2022/6888963
5. Allabadi, M., Almuhur, E., Alboustanji, A.: Independence number and covering vertex number of $\Gamma(F(+)S)$. In: International Conference on Information Technology (ICIT), pp. 70–74 (2021)
6. Al-labadi, M., Almuhur, E., Shatarah, A., Alboustanji, A., Omer, N., Khan, N., Bashir, R.: Eulerian graph of some special idealization rings. Adv. Math.: Sci. J. **10**(3), 1833–1837 (2021)
7. Eiber, M.S., Kadhem, H.H.: On soft pre-compact maps. Wasit J. Comput. Math. Sci. **1**(2) (2022)
8. Rasappan, S., Kumar, K.A.N., Devi, R.N., Obaid, A.J.: Synchronization of a modified colpitts oscillator with triangular wave non-linearity on graph. In: Peng, S.L., Lin, C.K., Pal, S. (eds) Proceedings of 2nd International Conference on Mathematical Modeling and Computational Science. Advances in Intelligent Systems and Computing, vol. 1422. Springer, Singapore (2022). https://doi.org/10.1007/978-981-19-0182-9_53
9. Chartrand, G., Muntean, R., Saenpholphat, V., Zhang, P.: Which graphs are divisor graphs? Congr. Numer. **151**, 189–200 (2001)
10. Gross, J., Yellen, J.: Graph theory and its applications. Chapman & Hall CRC Taylor & Francis (2006)
11. Buckley, F., Harary., F.: Distance in graphs. Addison-Wesely, Redwood City, CA (1990)
12. Korzhik, V.P.: Minimal non-1-planar graphs. Discr. Math. **308**, 1319–1327 (2008)

Chapter 31
Research Arima Model Forecasts Student's Graduation Rate

Nguyen Quang Huy, Le Phuoc Thanh, and Doan Van Thang

1 Introduction

Forecasting is an indispensable need for human activities in the context of information explosion. Forecasting will provide the necessary basis for plans, and it can be said that without the science of forecasting the plans would not be significantly convincing.

In [2, 3], forecasting techniques have been established since the nineteenth century, but forecasting has a strong influence as information technology develops because of the simulation nature of forecasting methods that are essential with the computer support. By the 1950s, forecasting theories and methodologies were systematically developed.

In predictive analysis, the most important issue to ask is the maximization of information about the forecasting field. The information here can be understood specifically: (1) historical data of the forecasting sector, (2) developments in the current situation as well as developments in the forecasting sector, and (3) fully assess the influencing factors both quantitatively and qualitatively.

Quantitative method uses past data over time, based on historical data to detect the movement direction of an object in accordance with a mathematical model and at the same time uses that model as a model estimate. Time series forecasting is a quantitative method [9].

N. Q. Huy · D. Van Thang (✉)
Industrial University of Ho Chi Minh City, Ho Chi Minh City, Viet Nam
e-mail: doanvanthang@iuh.edu.vn

L. P. Thanh
Quang Nam University, Tam Kỳ, Viet Nam

J. M. R. S. Tavares et al. (eds.), *Proceedings of Second International Conference on Intelligent System*, Algorithms for Intelligent Systems,
https://doi.org/10.1007/978-981-99-8976-8_32

In 1976, Box-Jenkins model proposal autoregressive integrated moving average (ARIMA) [8–10], is based on the autoregressive model (AR) model and moving average (MA) model. ARIMA is a quantitative forecasting model over time. Based on the dependence of the object's movement trend in the past, we will be able to predict the future value of the variable. For forecasting, the ARIMA model will analyze the correlation between the observed data through the stages of model identification, from the observed data, the parameters will be estimated and tested to find the appropriate model. The resulting model includes parameters that show the degree of correlation on the data and predict future values. From the variance of the forecast error, the confidence limit is calculated.

Time series forecasting is highly applicable and is used in many fields such as:

- Forecasting market demand to make business plan for the company.
- Forecasting financial asset yields, exchange rates, and derivatives prices for effective trading in market risk.
- Forecasting stock prices and portfolio yield chains for portfolio management.
- Bitcoin price forecast, oil price, gas price, etc.
- Forecasting temperature and rainfall for planning agricultural, forestry, and fishery production.
- Forecasting the impact of macro factors such as interest rates, money supply, foreign direct investment, government spending, inflation, etc., on GDP growth to run the economy.

The paper studies a time series model by the ARIMA method to predict the graduation rate for the next years in order to help managers and stakeholders make informed decisions and appropriate policies. The article is structured as follows: ARIMA model theory is presented in Sect. 2. Section 3 applies the ARIMA model to predict graduation rates, and Sect. 4 constitutes the conclusion of the article.

2 Theoretical Basis

2.1 ARIMA Model

In 1976, the ARIMA (Autoregressive Integrated Moving Average) model was introduced by Box-Jenkins. The collective term 'Box-Jenkins' is employed to denote the application of ARIMA processes in analysis and forecasting [7, 9].

Based on the hypothesis of stationary series and variance with constant errors, the model uses as inputs the auto regression (AR) series and moving average (MA) series signals of the past to be predicted to predict it. Over time, will decrease or increase according to a certain trend, so the stop series element is often not achieved. Use differential if in case the series is not stationary. The model will have three parameters ARIMA(p, d, q), with parameter d being the order of difference [8, 9].

The two main components of the input variable represented in the multivariable linear regression equation in the model are AR(q) and MA(p) [3, 6]:

AutoRegression of level p

In autoregressive process of level p, the present time data y_t is generated by a weighted average of past values up to the $p(y_{t-k})$ past value. Generalized autoregressive model only present and past values of Y are used in the model and no other regressors. The value of future Y depends on its value in the past plus a random factor. Obviously with the autoregressive model, the data is self-explanatory. The model form AR(p):

$$Y_t = \mu + \phi_1 Y_{t-1} + \phi_2 Y_{t-2} + \ldots + \phi_p Y_{t-p} + \varepsilon_t$$

ϕ_i: are the estimated weights that represent the effect of y_{t-i} values on y_t.

μ: is the intercept.

ε_t: is the error.

Moving Average of level q

The process of moving or changing the mean of a series over time is considered a moving average. The moving average model is described by a linear equation that weights the current random errors and its lagged values. The model equation is as follows:

$$Y_t = \mu + \varepsilon_t - \theta_1 \varepsilon_{t-1} + \ \ldots \ + \theta_q \varepsilon_{t-q}$$

μ: Average of static time series $\{y_t\}$.

ε_t: wrong number, also known as white noise, is a random component that represents an unpredictable and there is no rules.

θ_i: coefficient estimate the level of influence of ε_{t-i} on y_t.

ARIMA model with level (p, d, q)

Model ARIMA(p,q) is a combination of two models AR(p) and MA(q):

$$Y_t = \delta + \sum_{i=1}^{p} \phi_i y_{t-i} + \varepsilon_t - \sum_{i=1}^{q} \theta_i \varepsilon_{t-1}$$

ARMA(p,q) model works well with static data, i.e., there is no trend component. But in reality the data often has an upward or downward trend. One of the ways to convert a trending time series to a stationary time series (the series must stop) is the d-difference method, then we have the model is ARIMA(p, d, q).

2.2 Steps to Build ARIMA Model (Box-Jenkins Method) [2]

Step 1: Model identification

To recognize the ARMA(p,q) pattern, we need to find the appropriate p (p is degree of autoregression) and q (q is degree of the moving average) appropriate. Finding these values will depend on the autocorrelation function graph.

- If the graph of PACF (which is the partial autoregression coefficient) at delay 1,2,…,p has a high value and then decreases a lot, and ACF (which is autocorrelation coefficient) is decreasing, then the model AR(p) will be selected.
- If the ACF graph at a delay of 1,2,…,q has a high value then decreases a lot, and at the same time, PACF decreases, then the MA(q) model is selected.

ARIMA model is only applicable to stationary series. The mean and variance of a process that do not change with time is considered a stochastic process. Y_t is stationary, in addition, the value of the covariance depends only on the distance or the time lag does not depend on the time at which the covariance is calculated. To convert nonstationary series to stationary series, normally, if you take the difference once, twice or three times (order $d = 0, 1, 2, 3$), we will get a stationary series of results.

Step 2: Parameter estimation for the model

In this step, we will proceed to estimate the parameters after the order of the model has been determined. The process of model selection is the process of experimenting and estimating the parameters by the least squares method until we choose the best model for the prediction. We can use many different softwares such as: Eview, SPSS, *R*, etc., to estimate the parameters.

Step 3: Model testing

To ensure that the model is suitable for prediction, it is necessary to conduct a test whether the error term ε_t of the model is a white noise, one of the tests is to use its autocorrelation function.

Step 4: Forecast

To determine the predictive value of the score and the confidence interval, we based on the equation of the ARIMA model.

Table 1 Number of graduates by year

No.	Year	Number of graduates	No.	Year	Number of graduates
1	2011	54	7	2017	247
2	2012	119	8	2018	200
3	2013	863	9	2019	263
4	2014	564	10	2020	226
5	2015	484	11	2021	192
6	2016	381	12	2022	432

3 Forecasting the Graduation Rate with the ARIMA Model

3.1 *Data Collection*

The Faculty of Information Technology of a university in Ho Chi Minh City was established in 1996 and has so far trained 18 courses. Below is the graduate data by year. (Table 1 data includes 16 observations).

3.2 *Model Building*

Using SPSS software (Version 20.0) to build models and forecasts, this is a commonly used statistical analysis software today, especially for data visualization with conditional graphs. It provides analysts with the capability to swiftly analyze and evaluate results. Additionally, model testing results are obtained using Eview software (Ver 8.0).

a. *Check the stationarity*

The historical series of changes in the number of graduates over time from 2011 to 2022 is shown in the following Fig. 1 chart.

Observing this chart, the series tends to decrease gradually from 2013 to 2021 and increase again from 2021 to 2022. Thus, it can be speculated that the series of graduates over the years is non-stop.

When taking the second difference, the series is trendless and rotates around some mean value (Fig. 2). This is seen as the expression of a stationary sequence.

Using Dickey-Fuller test (unit root test) with second difference and results are given in Table 2.

With the significance level of 5% and p-value $= 0.0007 < 0.05$, the hypothesis H0 of the Dickey-Fuller test should be rejected: "Time series is a non-stationary series", so the conclusion "Student data series graduation by years is a stationary series".

b. *Parameter estimation and model selection*

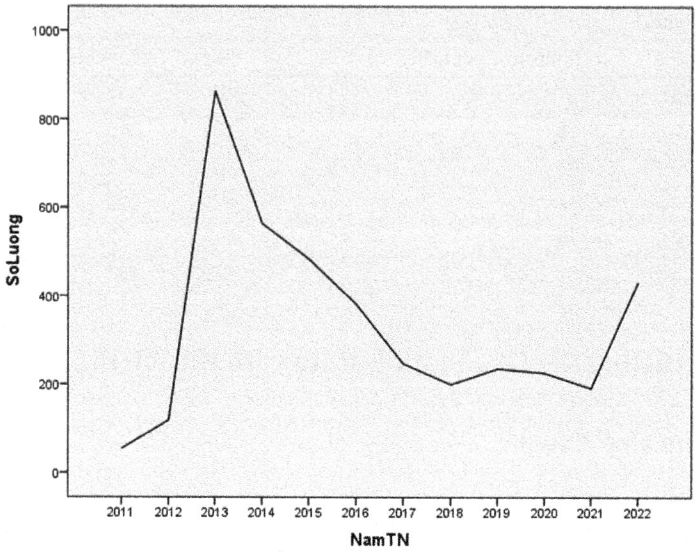

Fig. 1 Chart of changes in the number of graduates from 2011 to 2022

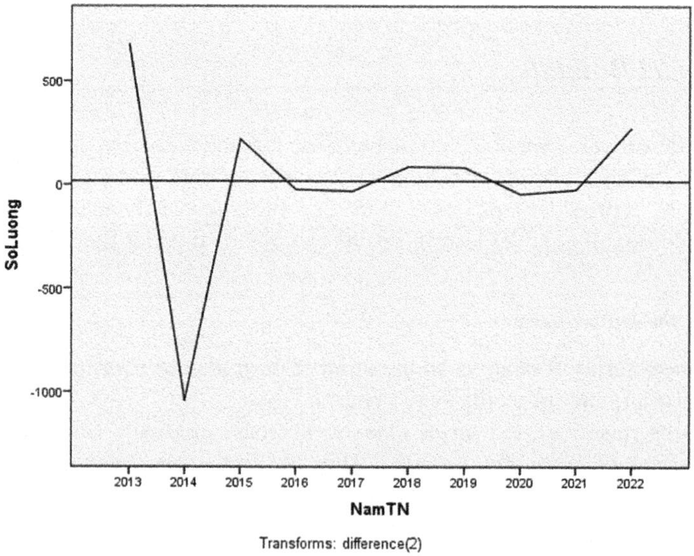

Fig. 2 Charts stop series when taking the second difference

Table 2 Stationarity test with difference of second order

		t-Statistic	Prob.
Augmented Dickey-Fuller test statistic		– 6.673885	0.0007
Test critical values:	1% level	– 4.420595	
	5% level	– 3.259808	
	10% level	– 2.771129	

After taking the second difference, we have autocorrelation histogram (ACF) and partial autocorrelation histogram (PACF) as shown in Fig. 3 and *p*-value in Table 3 as follows.

Both of these histograms peaked at lag 1, dropped abruptly to zero, and were statistically significant (significant 5% and *p*-value = 0.044 < 0.05). So we choose the model ARIMA(1,2,1).

c. Model testing

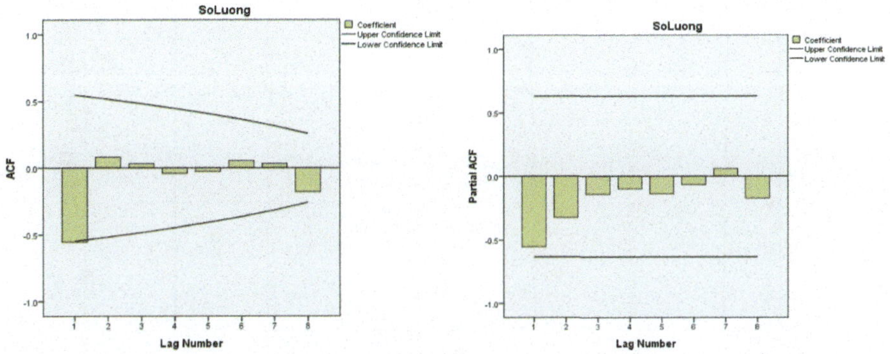

Fig. 3 Autocorrelation histogram (ACF) and partial autocorrelation histogram (PACF)

Table 3 *p*-values of autocorrelation (ACF) and partial correlation (PACF)

No.	AC	PAC	Q-Stat	Prob
1	– 0.553	– 0.553	4.0705	0.044
2	0.081	– 0.323	4.1695	0.124
3	0.031	– 0.141	4.1864	0.242
4	– 0.042	– 0.099	4.221	0.377
5	– 0.028	– 0.137	4.2396	0.515
6	0.054	– 0.066	4.3286	0.632
7	0.032	0.061	4.3691	0.736
8	– 0.177	– 0.173	6.2587	0.618
9	0.101	– 0.193	7.4719	0.588

We see that ACF and PACF correlation histograms of the residual series (Fig. 4) are with no orders beyond the confidence interval, all p-values of the Q-statistic are greater than the significance level 0.05 (Table 4), so we accept the hypothesis H0: "The residual is white noise". Conclusion is ARIMA model (1, 2, 1) found is suitable for forecasting.

d. *Forecast*

The forecast results of the number of graduates over the years according to the ARIMA model (1,2,1) are as given in Table 5 and Fig. 5.

Through the forecasted data table and the chart trend, it is found that the number of matriculation and enrollment in the following years 2023, 2024, and 2025 is as

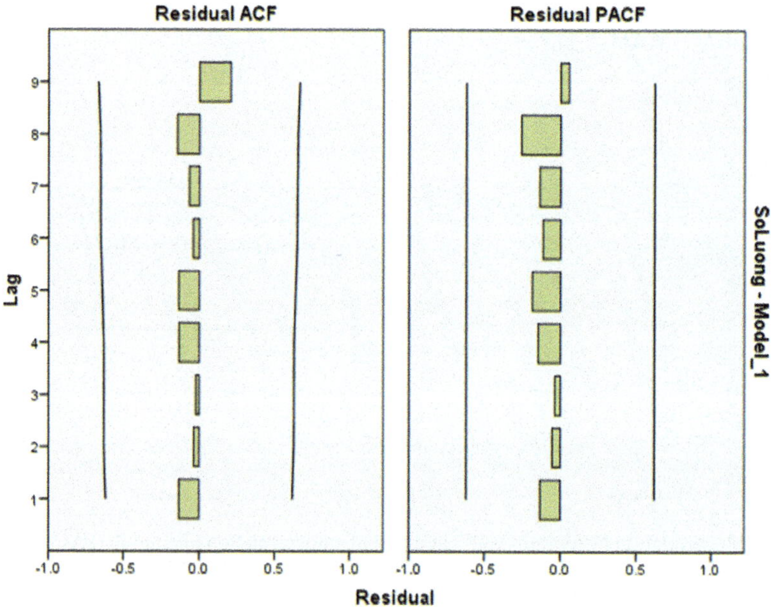

Fig. 4 ACF and PACF graph of residuals

Table 4 p-value of ACF and PACF of residuals

No.	AC	PAC	Q-Stat	Prob
1	− 0.553	− 0.553	4.0705	0.044
2	0.081	− 0.323	4.1695	0.124
3	0.031	− 0.141	4.1864	0.242
4	− 0.042	− 0.099	4.221	0.377
5	− 0.028	− 0.137	4.2396	0.515
6	0.054	− 0.066	4.3286	0.632

Table 5 Forecast data for the years 2023–2025

Model		13 (2023)	14 (2024)	15 (2025)
SoLuong-	Forecast	343	287	211
Model_1	UCL	1119	1349	1530
	LCL	− 432	− 775	− 1108

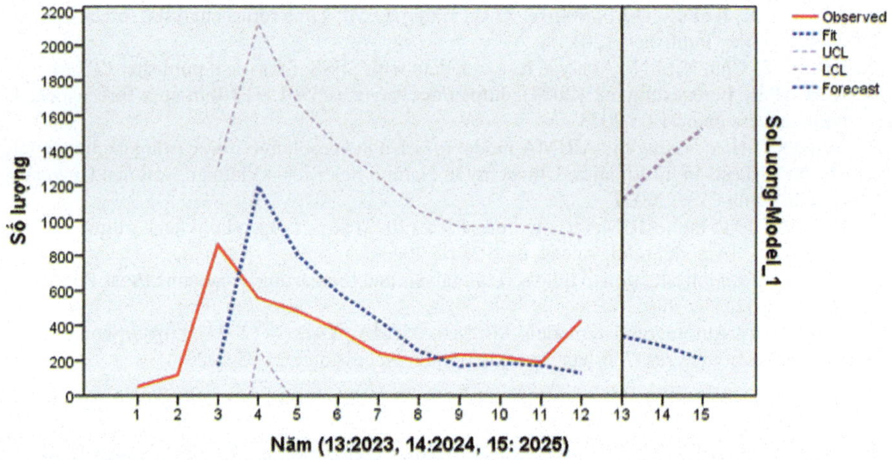

Fig. 5 Chart predicts the number of graduates in the next years

follows: (1) 2023 tends to decrease compared with the actual data of 2022 (forecast value 343), (2) the years 2024 and 2025 continue to decrease compared with 2023 (forecast value 287 and 211, respectively).

4 Conclusion

Forecasting is an integral part of any organization's operations. This solution helps managers make appropriate strategies and plans. Research on the applicability of the ARIMA model to forecasting the number of graduates over the years shows that the built ARIMA(1,2,1) model is appropriate and has been tested. The forecast results show that the number of graduates from 2023 to 2025 is likely to decrease at a relative level. This forecast shows that this is important information for stakeholders and managers to plan implementation in line with the forecast.

References

1. Boris Kovalerchuk and Evgenii Vityaev: Data Mining in Finance: Advances in Relational and Hybrid Methods, p. 2001. Kluwer Academic Publishers, Boston, Dordrecht - London (2001)
2. Box, G.: Box and Jenkins: time series analysis, forecasting and control. In: A Very British Affair. Palgrave Advanced Texts in Econometrics. Palgrave Macmillan, London (2013)
3. Thi, C.H., Phu, P., Thuy, P.N.: Application of ARIMA model for testing "serial independence" of stock prices at the HSEC. In: The Joint 14th Annual PBFEA and 2006 Annual FeAT Conference, Taipei, Taiwan (2006)
4. Box, G.E.P., Jenkins, G.M., Reinsel, G.C., Ljung, G.M.: Time series analysis: forecasting and control. Wiley Publishers (2015)
5. Hoang, T., Chu, N.M.N.: Analyze research data with SPSS. Statistical publisher (2005)
6. IBM SPSS Forecasting 22 (2023). https://docplayer.es/57613312-Ibm-spss-forecasting-22.html. Access date 21 05 2023
7. Nguyen, T.H.N.: Using the ARIMA model to solve the problem of forecasting exchange rate. Master's Thesis in Informatics, University of Natural Sciences—Vietnam National University, Ho Chi Minh City (2002)
8. Dinh, K.P.: Mô hình ARIMA trong time series. (2023) https://phamdinhkhanh.github.io/2019/12/12/ARIMAmodel.html. Access date 21 05 2023
9. Yaffee, R.: An introduction to time series analysis and forecasting. Academic Press Publishers. Dec 11 (2023)
10. Time Series: Autoregressive models AR, MA, ARMA, ARIMA. (2023). https://people.cs.pitt.edu/~milos/courses/cs3750/lectures/class16.pdf. Access date 21 05 2023

Chapter 32
A Home LPM System Based on WSN

Saja Mohsin Abood, Kasim K. Abdalla, and Shamam Alwash

1 Introduction

Energy consumption management has become a critical issue in the present time due to the continuous increase in electricity usage worldwide. According to the International Energy Agency, global electricity consumption is expected to rise by 2.1% annually until 2040 [1]. As a result, the pressure on electric companies to meet the growing demand for electricity increases, leading to higher costs and environmental damage. Therefore, the focus is on designing home energy management systems that improve energy efficiency and reduce pressure on electric companies.

Wireless sensor networks have been identified as a promising solution to designing home energy management systems. These systems consist of a set of wireless sensors distributed to houses for measuring energy use in each house. The collected data is processed and sent to a central control unit, which efficiently and intelligently manages energy consumption in the home. Previous studies have shown the effectiveness of using wireless sensor networks in designing energy management systems for homes. Additionally, the utilization of such systems can lead to energy savings of up to 20% in homes [2].

This study aims to explain previous researchers that designed home load energy management systems in the period 2019–2023.

Urbanization causes residential and business loads to rise faster in developing cities, which will cause a demand–supply gap in many of these cities [3, 4] and forecasts that by 2030, the energy consumed worldwide will double [5]. Exponentially rising energy demand will hurt the environment. There is less emphasis on increasing electricity production through conventional and alternative resources, and there may

S. M. Abood (✉) · K. K. Abdalla · S. Alwash
Department of Electrical Engineering, University of Babylon, Babylon, Iraq
e-mail: sajaaltaee.saja@gmail.com

J. M. R. S. Tavares et al. (eds.), *Proceedings of Second International Conference on Intelligent System*, Algorithms for Intelligent Systems,
https://doi.org/10.1007/978-981-99-8976-8_33

be less load condition for improved management [6]. These days, most developed countries utilize intelligent meters [7].

Many techniques are used to manage energy in the home using wireless sensor networks. Smart meter technology is the most important techniques in power management, which has advantages such as energy demand management, remote monitoring, data visualization, security, affordability, and home automation integration for sustainable consumption. However, these technologies have drawbacks, such as high costs due to the Internet of Things and GPRS tools, vulnerability to penetration, and direct house location determination.

The domestic load-shedding algorithm (DLAS) is helpful for energy efficiency, cost savings, grid stabilization, demand management, user flexibility, and environmental benefits. However, it requires the user's presence to receive messages and internet usage.

The demand response technique (DR) aims to balance demand and fluctuation in renewable energy generation. However, it requires user presence for HEMS notifications, which can reduce energy consumption and increase prices. Load shading theory is less complicated and cost-efficient than other techniques but requires a GSM connection for mobile news.

The beetle antenna search algorithm (BAS) offers the lowest cost of electricity consumption and protects user privacy through the energy storage of electric vehicles. However, it lacks communication between the smart meter and the distribution station. Simple IoT architecture with Wi-Fi technology allow direct connection to the internet, but it requires internet service available for some users. Regression control helps control consumer appliance usage during peak demand but requires a stable internet connection and active monitoring. Such a plan can also be connected to generation estimation and monitoring to optimize building-level use [8, 9].

2 Wireless Sensor Networks (WSN) in Energy Management

The management of energy in wireless sensor networks (WSNs) encompasses a set of guidelines and principles for effectively overseeing multiple energy provision mechanisms and fostering optimal utilization of the available energy within sensor nodes as shown in Fig. 1. The primary objective encompasses the proficient administration of energy to ensure the absence of energy deficiency in any node, thereby maintaining perpetual operational functionality of the network. A sensor node necessitates an effective energy management scheme due to its limited energy resources. It is imperative for the available energy source to cater to the requirements of the application at hand effectively. In the context of deploying a sensor node in a remote region, energy scarcity poses a significant concern for the functionality and operation of the node. Once the available energy has been exhausted, substituting substitute energy becomes exceedingly challenging. Hence, it is imperative to maintain an equilibrium in energy distribution between the grid's supply and demand, to mitigate any potential energy shortfall within the network [10].

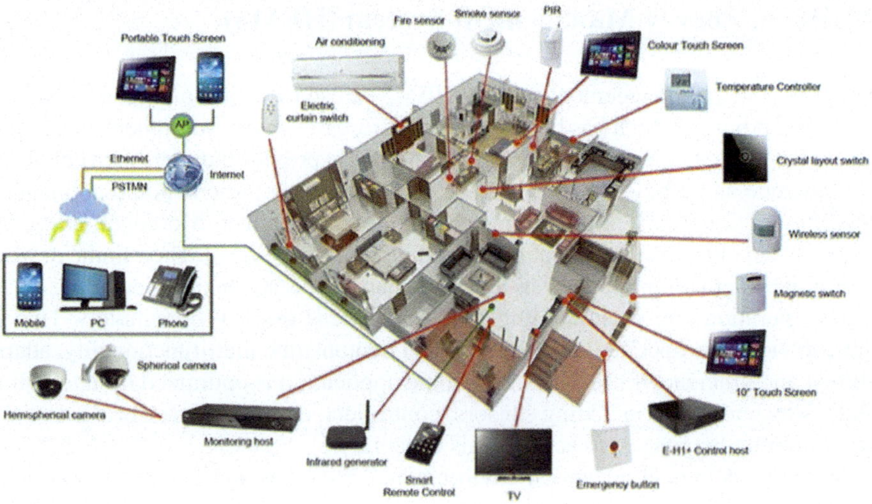

Fig. 1 WSN based on smart home system to smart *grid integration*

The battery is the prevalent power source for sensor nodes, and its replacement or rechargeability is contingent upon the specific conditions under which the nodes are deployed [11]. The restriction in battery supply prompted numerous researchers to explore alternative means of energy provisioning through the utilization of ambient energy sources [12–15]. A sensor network can achieve prolonged, almost uninterrupted functionality by harnessing ambient energy. Nonetheless, energy harvesting-based approaches are not without their limitations, as there may arise situations in which a node is subjected to restricted opportunities for energy harvesting that are inadequate to meet its power demands [16]. In the problems, a node is prone to experiencing energy deficiency, thereby exerting an impact on the comprehensive performance of the network.

In contemporary studies, there has been a surge in the proposition of energy transference-based methodologies. These approaches involve the transmission of energy within a field to address the energy insufficiency of a particular node [17–22]. These energy-deficient nodes are allocated in regions with constrained possibilities for energy acquisition from the surrounding environment. The concept of wireless energy transfer has gained substantial recognition as a highly auspicious and anticipated technological advancement. In addition to examining energy supply mechanisms, it is imperative to investigate diverse strategies and protocols pertaining to the optimized utilization of energy within a node. The literature predominantly concentrates on energy management endeavors that primarily aim to enhance the energy consumption of nodes [23–30].

3 Home Energy Management System (HEMS)

The home energy management system (HEMS) arose from the continuous advances in the smart home, which has been a subject of research and development since 1990 [31]. The primary objective of this application was to provide the users with enhanced convenience and efficiency in operating and controlling various home appliances, enabling them to work in an automated and optimal manner. Moreover, users can grow a room or environment that matches their preferences [32]. Their home devices create the interaction between users in a smart home application, enabling interactive operation and programming of these devices directly or remotely. Hence, these devices can operate independently and demonstrate multifunctionality, adaptability, and efficiency [33]. The smart home application is optimized by integrating various technologies, including sensors, controllers, communication protocols, and programming languages, to achieve this capability.

With technological advances, environmental concerns and corresponding regulations have encouraged the adoption of smart home applications that provide convenience and aim to improve energy efficiency. Over time, smart home applications have seen advancements in integrating home energy monitoring capabilities to achieve energy conservation. According to the investigation results, by monitoring energy consumption, individuals will experience a psychological modification in their approach toward the use of household appliances, thus facilitating energy conservation by as much as 30% [34]. Users will show increased interest in their energy consumption if they object to feedback on the required energy expenditure. I have explained many feedback techniques associated with smart measurement technology in the academic work discussed in reference [34].

In line with the energy management capabilities, a branch of smart home applications has been identified, the home energy management system (HEMS) application, which efficiently supervises energy consumption within homes, thus mitigating peak energy demand and associated financial resource limit. expenses. The HEMS application can collect, analyze, visualize, and engage in an interactive dialogue with the user to facilitate energy management. So, it can provide instant, daily, weekly, or monthly feedback at the right time [35, 36].

HEMS technology progress can be categorized into several areas, namely interface technology, smart hardware, and software platform [37]. A variety of research studies have investigated the advancement of home energy management systems (HEMS) using various technologies, including sensor technology, communication devices, energy storage devices, smart meters, and hardware smart home [38–43], as shown in Fig. 2.

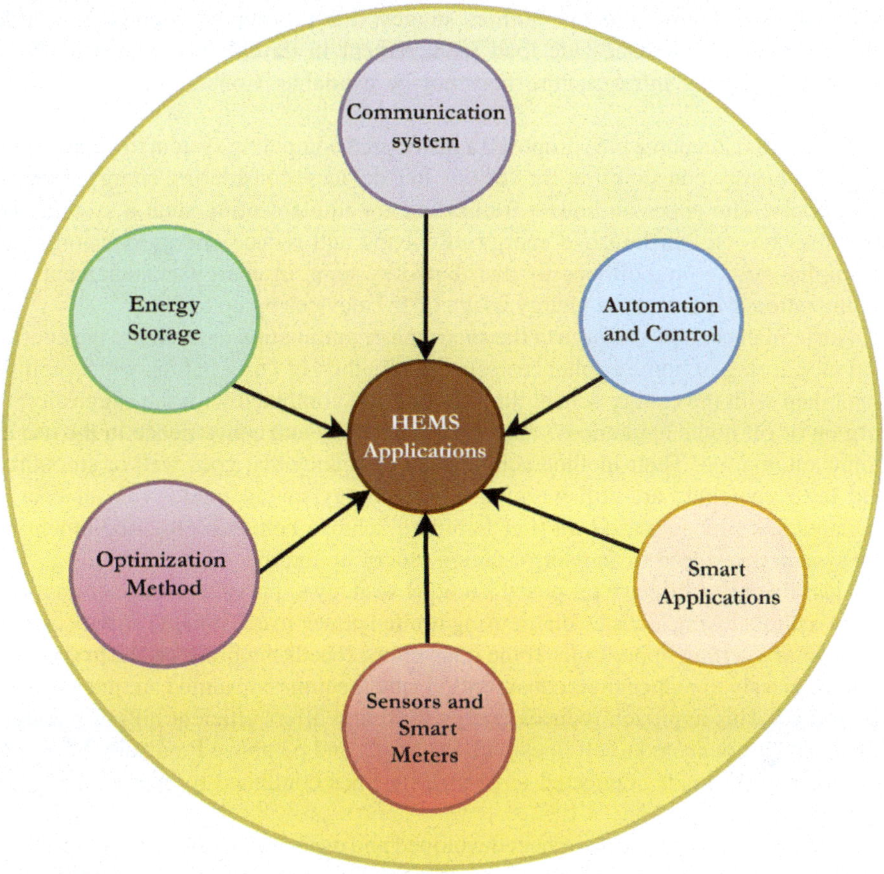

Fig. 2 Technology developments of HEMS

4 Related Work

Previous studies investigated how the wireless sensor network could resolve home load power management problems.

Priyadharshini et al. [44] described the design, building, and validation of IoT based on innovative metering development for an energy management system. It connects to Wi-Fi networks to provide wireless transmission between the smart meter and the server system. Smart meter technology and IoT-based platform effectively manage energy consumption and create bill arrangements.

Keles and Yigit [48] proposed a domestic load-shedding algorithm for conventional electricity meters as a means of managing demand-side load. The proposed system takes advantage of the existing infrastructure of conventional meters, making it a low-cost solution for domestic load management. The study presents simulation results demonstrating the effectiveness of the proposed algorithm in reducing peak energy consumption. The authors suggest that this approach could be viable for implementing demand-side load management in developing countries where advanced metering infrastructure may not be available. Consumption is reduced by 27%.

Sadeeq and Zeebaree [45] proposed a distributed computing system to manage and optimize energy consumption for various IoT devices, considering energy demand and supply. The paper outlines a framework for implementing such a system and describes how it can improve energy efficiency and reduce energy consumption. It emphasizes the importance of distributed systems in energy management and optimization. Minimize the energy usage of IoT devices by up to 30%.

Nurfadilah et al. [49] designed the smart meter automation to facilitate the control and monitoring of home appliances using the method of "user habits, which will be combined with the energy scheduling method and combined with the suggestion to turn on or off home appliances" to provide efficiency and convenience in the use of home automation. Their method achieved an efficiency that goes well or smoothly, and the researchers are still working on an energy-saving mode with an energy management process based on user habits to achieve better results. Implementing the system could lead to an energy-saving rate of up to 30%.

Hussain et al. [50] have proposed a unique strategy to provide electricity continuously without losing loads by distributing limited power to consumers, so they cannot go beyond a certain upper limit. If the consumer exceeds the limit, an alarm signal is sent wirelessly to reduce their consumption, and the units consumed are intelligently monitored. This approach reduces energy theft and offers efficient billing management. Loads connected to all lines with priority 2 and p3 above Pmax are down. On the other hand, loads connected to pl priority lines continued to operate under all conditions.

Singh and Selvan [3] researchers developed and designed a low-cost, single-phase smart energy meter incorporating the incremental block rate tariff structure followed by the Tamil Nadu State Electricity Board (TNEB). MDMS software development: The original Meter Data Management System (MDMS) software was developed using the instrument's open-source software tools. The mobile application provided real-time energy consumption data to consumers, and the automatic adjustment of device power consumption during peak hours was seamless.

Rahman et al. [51] developed a bespoke load scheduling model to enable the HEMS to calculate optimal energy consumption while integrating customer dynamic and satisfaction price change. The outcomes show that the suggested HEMS based on dynamic pricing significantly reduces 38.80% of peak loads and 22.55% of customers' total energy costs.

Butt et al. [52], a simulated analysis was performed on a single-family home for automatic load mitigation during peak hours within the Proteus system. Next, the researchers created and implemented a hardware prototype in order to validate the effectiveness of the proposed control system. The results of the study proved that the implemented system enabled effective moderation of the peak load during hours of high demand on the system. In addition to considerations of typical domestic and commercial consumers, monetary benefits are also evaluated. The researchers concluded that the recovery time for this device is about 1 month, provided that 50% of the workload is reduced during the 4-h peak period.

Krishnamoorthy et al. [46], the current proposal is based on the concept of Internet of Things (IoT) as a way to conserve energy resources. However, it requires a large financial investment due to the equipment required and the vulnerability of unauthorized access. The payback period results in a 50% reduction in load volume during the four-hour period of peak demand.

Khan et al. [43], this paper presents a control theoretical approach to effectively manage continuous demand by implementing a closed-loop, price-based spatial and temporal shortest path model with constraints (STSMC). Using this method results in a potential savings of up to 27% in your electricity bill.

Zhang et al. [54] involved the development and generation of intelligent loading models for smart homes, comprising two distinct demand-response classes. The models were then used in conjunction with a building automation system (BAS) algorithm to achieve optimized electricity consumption at lower costs. In order to protect user privacy, protection measures are implemented by using electric vehicle energy storage to hide the user's electricity consumption data. In addition, a privacy measurement method, incorporating adjustable weight, was introduced to enhance the level of specificity. The results obtained from the simulations indicate that the proposed scheme effectively reduces the electricity expenditures of the users by appropriately managing household electricity consumption in line with the demand-response (DR) power grid. In addition, users can adapt the value of privacy protection weight α, thus meeting privacy and cost considerations. Daily energy cost is decreased by 28.56%.

Munoz et al. [47] presented a design that verifies the validity of the smart meter with load control with its electronic design. The researchers explain that their work is evidence based on experimentation, as the prototype in all its readings had an error rate of less than 1%.

Valovcin et al. [56] demonstrated a new approach to reducing peak energy demand through direct regulation of residential heating, air-conditioning systems, and water heaters using a demand-response (DR) program. The authors present an experimental method using smart meter data to estimate the extent of hourly demand drop achieved by individual direct load controllers when they are operated synchronously. This methodology presents a cost-effective approach for examining the impact of control software on the live operational load of measured smart facilities. The technology used showed results that included an average reduction of between 6.0 and 109% for managing water heater loads during events. Moreover, the implementation of HVAC unit control reduced customer demand by an average of 7.6–218% in the morning, which is mostly attributed to the effects of HVAC load control measures.

Saleem et al. [55] designed a system for implementing, deploying, and evaluating the performance of an IoT-based SIMS system, including SMS and associated software. Their proposed method works online and delivers real-time (RT) uploads to remote customers and suppliers. Customer LPs allow suppliers to post and regulate their incentives as well as incentivize customers to change their energy consumption. Furthermore, these LPs serve as inputs for developing many DSM curricula.

de la Cruz Severiche Maury et al. [57] proposed designing and implementing a simple prototype of a low-cost home energy management system (HEMS). To demonstrate the feasibility of our system, the system was evaluated by measuring weekly energy consumption with online and offline HEMS using a test bench with typical household appliances in a typical home in Sincelejo. Their evaluation showed that with the installation of this HEMS, consumption is reduced by 27%. Thus, it is possible to achieve a good reduction ratio by using a low-cost system.

Tai et al. [58] solved the problem of controlling energy use and improving efficiency in sensor network applications that have been transformed into a mathematical model. This model is designed as a wireless sensor network structured like a tree and focuses on meeting Quality of Service (QoS) and fairness requirements. Their research is finding out how well the sensor works, how far the signal can be transmitted, and the size of the signal, which helps save energy. The Lagrangian relaxation method helps to find the best solution that uses the least amount of energy while ensuring network transmission efficiency. Tests show that faster decision-making can be improved and less energy used effectively.

5 Summary and Comparison of the Reviewed Papers

This section will compare the papers reviewed in terms of the objectives, methods used, the challenges faced by the researchers, and the results of these papers.

Table 1 compares the literature review papers used in the current article.

6 Discussion

The reviewed literature presents a comprehensive overview of various approaches and techniques for resolving home load power management issues using wireless sensor networks (WSNs) and the Internet of Things (IoT). The discussed papers highlight different methodologies and technologies employed in energy management systems and load control algorithms. Overall, the reviewed papers contribute to the field of home load power management by offering various solutions, algorithms, and technologies to optimize energy consumption, improve energy efficiency, and reduce costs. The utilization of IoT, load management algorithms, demand-response strategies, and innovative energy management systems all play significant roles in addressing the challenges associated with home energy management. The research findings highlight the importance of integrating smart technologies and intelligent algorithms in achieving sustainable and efficient energy consumption in residential settings.

Table 1 Research literature summary

Ref.	Goal	Method	Results
[44]	Create real-time energy management system using IoT devices and analytical tools	Smart meter technology and LabVIEW software	Smart meter technology and IoT-based platform effectively manage energy consumption and create bill arrangements
[45]	Monitor household appliance energy consumption and develop strategies to reduce energy consumption	The smart meter used (HEM system the platform, Arduino UNO, Zigbee)	Consumption is reduced by 27%
[46]	IoT-based energy management system uses data acquisition to reduce waste and improve efficiency	ARM CortexM-4 32b MCU, relay driver, GSM/GPRS module, 2X16 Alphanumeric LCD display, and analog frontend module. Wireless transceiver GRPS software (MySQL database, IAR Embedded Workbench IDE, STM32CubeMX, OrCAD,.Net framework)	Reduce the energy consumption in IoT devices by up to 30%
[47]	IoT-based smart energy management system for smart grid demand management	Smart meter, load control, IoT with in-built Wi-Fi technique	Implementing the system could lead to an energy-saving rate of up to 30%
[48]	Reducing peak energy consumption by controlling loads, promoting efficient electricity usage, and reducing costs	"Domestic load-shedding algorithm". Used traditional meter with Arduino UNO	Loads connected to all lines with priority 2 and p3 above Pmax are down. On the other hand, loads connected to pl priority lines continued to operate under all conditions
[49]	Provide efficiency and convenience in the use of home automation	Smart meter, microcontroller, and RabbitMQ broker	The mobile application provided real-time energy consumption data to consumers, and the automatic adjustment of device power consumption during peak hours was seamless
[50]	Solve the energy crisis faced by people in developing countries like Pakistan	Power distribution and utilization (PDU) Network, smart energy meter SEM, and grid station server (GSS)	The system achieves an average reduction of 29% in energy costs and 27% in peak demand

(continued)

Table 1 (continued)

Ref.	Goal	Method	Results
[3]	Create low-cost smart energy meter and MDMS software for self-demand response in India	Diverse visualization methods microcontroller (Arduino UNO), ACS712, and curent transformer (CT))	Load shedding reduces 50% of the load during the 4-h peak time
[51]	Reduce residential energy consumption through cost–benefit analysis, energy management system stabilization, and supply–demand balance	Demand response technique Hardware (Arduino, Blutooth communication) Software (XML, Android Studio 3.1, and Java)	Payback period reduces 50% of load during the 4-h peak time
[52]	Proposing a household load management system that can efficiently manage the electricity load during load rationing in Pakistan	Load shading and Arduino UNO	Saves up to 27% on electricity bills
[53]	Incentives to adjust energy usage during high-demand periods	Energy demand control strategy, demand response (DR), super-twisting sliding mode controller (STSMC), with MATLAB/Simulink	The total power cost is reduced by 28.56% per day
[54]	Reducing energy consumption costs and privacy leaks through real-time communication between smart meters and energy control centers	Smart meter, the power of the electric vehicle, V2G, and BAS algorithm. Software by MATLAB environment	The smart meter can measure the energy consumption of different appliances. The mobile application allows users to control their devices remotely
[55]	Using smart meter data, estimate the demand reduction achieved by residential direct load control programs	Simple IoT architecture with in-built Wi-Fi technology enables direct internet connection. MQTT protocol, UART. Software JavaScript Object Notation (JSON)	The direct load control program achieved an average demand reduction of 0.56 kw per household during event hours

(continued)

Table 1 (continued)

Ref.	Goal	Method	Results
[56]	Investigates utility DR software designed to reduce peak demand by directly controlling residential HVAC units and water heaters	Multiple linear regression (MLR) method. DR, HVAC units, and smart grid	The findings demonstrate that load control of water heaters during events led to an average reduction ranging between 6.0 and 10.9%. In addition, controlling the HVAC units resulted in a customer demand reduction averaging between 7.6 and 21.8%
[57]	Monitor household appliance energy consumption and develop strategies to reduce energy consumption	The smart meter used (HEM system the platform, Arduino UNO, Zigbee)	Consumption is reduced by 27%
[58]	Propose near-optimal energy management mechanism considering QoS and fairness in WSNs	Binary integer linear programming (BILP) optimization algorithm. WSN (mathematical programming), relay sensor, Lagrangian relaxation technology	The decision-making speed and energy consumption can be effectively improved

Moving forward, future research in this area could focus on addressing the limitations and challenges identified in the reviewed studies. These may include scalability, security concerns, real-time performance, and interoperability issues. Additionally, there is a need for further investigation and experimentation to validate the proposed solutions in real-world scenarios, considering factors such as user acceptance, integration with existing infrastructure, and compatibility with evolving technologies.

7 Conclusion

In conclusion, the reviewed literature demonstrates the significant role of wireless sensor networks and the Internet of Things (IoT) in addressing home load power management challenges. Researchers have proposed various approaches to optimize energy consumption, improve energy efficiency, and reduce energy costs.

Several studies emphasize the importance of considering user habits and preferences in energy scheduling and automation to improve efficiency and convenience. Furthermore, dynamic pricing, demand-response programs, and load control strategies have been proposed to reduce peak energy demand and enhance overall energy management. These approaches minimize energy consumption, peak loads, and customer energy costs.

The reviewed literature also highlights the significance of low-cost solutions, especially in developing countries with limited advanced metering infrastructure. Researchers have explored using conventional meters, low-cost prototypes, and simple home energy management systems to achieve energy savings and demonstrate the feasibility.

However, some challenges still need to be addressed, such as the high cost of implementing these systems and the need for more accurate sensors to improve data collection. Additionally, further research is required to optimize the system's performance and make it more user-friendly. Future studies should focus on developing cost-effective and reliable solutions that can be easily integrated into homes, ultimately leading to sustainable energy use and reduced environmental impact.

References

1. Barnawi, A.Y., Keshta, I.M.: Energy management in wireless sensor networks based on Naive Bayes, MLP, and SVM classifications: a comparative study. J. Sens. **2016**, 12, Article ID 6250319 (2016). https://doi.org/10.1155/2016/6250319
2. Flores, F.C., Ranalkar, S., Amoo, A.: Home energy management system. Michigan Technological University, doi 10
3. Singh, S., Selvan, M.P.: A smart energy meter enabling self-demand response of consumers in smart cities of Tamil Nadu. In: 2019 IEEE International Conference on Smart Cities Model (ICSCM), pp. 1–6. IEEE (2019)
4. Stanislav, P., Bryan, K., Tihomir, M.: Smart grids better with integrated energy system. In: IEEE Electrical Power & Energy Conference (EPEC), pp. 1–8 (2009)
5. Wang, X., Mao, X., Khodaei, H.: A multi-objective home energy management system based on internet of things and optimization algorithms. J. Build. Eng. **33**, 101603 (2021)
6. Evangelakos, E.A., Kandris, D., Rountos, D., Tselikis, G., Anastasiadis, E.: Energy sustainability in wireless sensor networks: an analytical survey. J. Low Power Electron. Appl. **12**, 65 (2022). https://doi.org/10.3390/jlpea12040065
7. Chun, D., Hong, S., Chung, Y., Woo, C., Seo, H.: Influencing factors on hydrogen energy R&D projects: an ex-post-performance evaluation. Renew. Sustain. Energy Rev. **53**, 1252–1258 (2016)
8. Condon, F., Martínez, J.M., Eltamaly, A.M., Kim, Y.-C., Ahmed, M.A.: Design and implementation of a cloud-IoT-based home energy management system. Sensors **23**, 176 (2023). https://doi.org/10.3390/s23010176
9. Condon, F., Martínez, J.M., Eltamaly, A.M., Kim, Y.-C., Ahmed, M.A.: Design and implementation of a cloud-iot-based home energy management system. Sensors **23**(1), 176 (2022)
10. Khan, J.A., et al.: Energy management in wireless sensor networks: a survey. Comput. Electr. Eng. (2014). https://doi.org/10.1016/j.compeleceng.2014.06.009
11. Tong, B., Wang, G., Zhang, W., Wang, C.: Node reclamation and replacement for long-lived sensor networks. In: 6th Annual IEEE Communications Society Conference on Sensor, Mesh and Ad Hoc Communications and Networks, 2009. SECON'09, pp. 1–9. IEEE (2009)
12. Kansal, A., Hsu, J., Zahedi, S., Srivastava, M.: Power management in energy harvesting sensor networks. ACM Trans. Embed. Comput. Syst. (TECS) **6**(4), 32 (2007)
13. Rahimi, M., Shah, H., Sukhatme, G., Heideman, J., Estrin, D.: Studying the feasibility of energy harvesting in a mobile sensor network. In: IEEE International Conference on Robotics and Automation, 2003. Proceedings. ICRA'03, vol. 1, pp. 19–24. IEEE (2003)

14. Voigt, T., Ritter, H., Schiller, J.: Utilizing solar power in wireless sensor networks. In: 28th Annual IEEE International Conference on Local Computer Networks, 2003. LCN'03. Proceedings, pp. 416–422. IEEE (2003)
15. Jiang, X., Polastre, J., Culler, D.: Perpetual environmentally powered sensor networks. In: Fourth International Symposium on Information Processing in Sensor Networks, 2005. IPSN 2005, pp. 463–468. IEEE (2005)
16. Talzi, I., Hasler, A., Gruber, S., Tschudin, C.: Permasense: investigating permafrost with a wsn in the swiss alps. In: Proceedings of the 4th Workshop on Embedded Networked Sensors, pp. 8–12. ACM (2007)
17. Mascarenas, D.D.L.: Mobile host wireless sensor networks—a new sensor network paradigm for structural health monitoring applications. UC San Diego: b6636138 (2008). <http://escholarship.org/uc/item/27k5c57r>
18. Doost, R., Chowdhury, K., Di Felice, M.: Routing and link layer protocol design for sensor networks with wireless energy transfer. In: IEEE Global Telecommunications Conference GLOBECOM 2010, pp. 1–5. IEEE (2010)
19. Xie, L., Shi, Y., Hou, Y.T., Sherali, H.D.: Making sensor networks immortal: an energy-renewal approach with wireless power transfer. IEEE/ACM Trans. Network (TON) **20**(6), 1748–1761 (2012)
20. Kurs, A., Karalis, A., Moffatt, R., Joannopoulos, J., Fisher, P., Soljacˇic´, M.: Wireless power transfer via strongly coupled magnetic resonances. Science **317**(5834), 83 (2007)
21. Xie, L., Shi, Y., Hou, Y.T., Lou, W., Sherali, H.D., Midkiff, S.F.: On renewable sensor networks with wireless energy transfer: the multi-node case. In: 9th Annual IEEE Communications Society Conference on Sensor, Mesh and ad Hoc Communications and Networks (SECON), pp. 10–8. IEEE (2012)
22. Summerer, L., Purcell, O.: Concepts for wireless energy transmission via laser. Euro. Space Agency (ESA)-Adv. Concep. Team (2009)
23. Anastasi, G., Conti, M., Di Francesco, M., Passarella, A.: Energy conservation in wireless sensor networks: a survey. Ad Hoc Netw. **7**(3), 537–568 (2009)
24. Zorzi, M., Rao, R.: Geographic random forwarding (geraf) for ad hoc and sensor networks: multihop performance. IEEE Trans. Mobile Comput. **2**(4), 337–348 (2003)
25. Cerpa, A., Estrin, D.: Ascent: adaptive self-configuring sensor networks topologies. IEEE Trans. Mobile Comput. **3**(3), 272–285 (2004)
26. Rajendran, V., Obraczka, K., Garcia-Luna-Aceves, J.: Energy-efficient, collision-free medium access control for wireless sensor networks. Wirel. Netw. **12**(1), 63–78 (2006)
27. Ye, W., Heidemann, J., Estrin, D.: Medium access control with coordinated adaptive sleeping for wireless sensor networks. IEEE/ACM Trans. Network **12**(3), 493–506 (2004)
28. Ngo, V., Anpalagan, A.: A detailed review of energy-efficient medium access control protocols for mobile sensor networks. Comput. Electr. Eng. **36**(2), 383–396 (2010)
29. Lu, G., Krishnamachari, B., Raghavendra, C.: An adaptive energy-efficient and low-latency mac for data gathering in wireless sensor networks. In: 18th International Parallel And Distributed Processing Symposium, 2004. Proceedings, p. 224. IEEE (2004)
30. Rhee, I., Warrier, A., Aia, M., Min, J., Sichitiu, M.: Z-mac: a hybrid mac for wireless sensor networks. IEEE/ACM Trans. Network (TON) **16**(3), 511–524 (2008)
31. Gabriele, L., Salvatore, C., Erica, L.: A Review of systems and technologies for smart homes and smart grids Energies **9**(348), 1–33 (2016)
32. Muhammad, R.A., Reaz, M.B.I., Mohd Ali, M.A.: a review of smart homes—past, present, and future future. IEEE Trans. Syst. Man Cyber. Part C (Appl. Rev.) **42**(6), 1190–203 (2012)
33. Quynh, L., Hoang, B.N., Tony, B.: Smart homes for older people: positive aging in a digital world Future Internet **4**, 607–617 (2012)
34. Darby, S.: The effectiveness of feedback on energy consumption: a review for DEFRA of the literature on metering billing and direct displays. Environmental Change Institute, University of Oxford (2006)
35. Hassan, H.A.: Review vehicular ad hoc networks security challenges and future technology. Wasit J. Comput. Math. Sci. **1**(3) (2022)

36. Obaid, A.J.: Wireless sensor network (WSN) routing optimization via the implementation of fuzzy ant colony (faco) algorithm: towards enhanced energy conservation. In: Kumar R., Mishra B.K., Pattnaik P.K. (eds.) Next Generation of Internet of Things. Lecture Notes in Networks and Systems, vol. 201. Springer, Singapore (2021). https://doi.org/10.1007/978-981-16-0666-3_33
37. Karlin, B., Ford, R., Sanguinetti, A., Squiers, C., Gannon, J., Rajukumar, M., Donnelly, K.A.: Characterization and potential of home energy management (hem) technology. Pacific Gas Electric Rep. (2015)
38. Ding, D., Cooper, R.A., Pasquina, P.F., Fici-Pasquina, L.: Sensor technology for smart homes Maturitas **69**, 131–136 (2011)
39. Tiago, D.P.M., Radu, G., Eduardo, M.G.R., João, C.O.M., João, P.S.C.: Smart home communication technologies and applications: wireless protocol assessment for home area network resources Energies **8**, 7279–7311 (2015)
40. Pascual, J., Sanchis, P., Marroyo, L.: Implementation and control of a residential electrothermal microgrid based on renewable energies, a hybrid storage system and demand side management. Energies **7**, 210–37 (2014)
41. Ameena, S.A.S., Mohammed, H.A., Magdy, M.A.S.: Smart home activities: a literature review. Electric Power Comp. Syst. **42**, 294–305 (2014)
42. Giovanni, P., Mario, C., Antonio, R., Jiahu, Q.: Home energy management. Energies **10**(382) (2017)
43. Khan, A.M., Javaid, N., Razzaq, S., Khan, R.D., Ilahi, M.: Home energy management systems in future smart grids Basic. Appl. Sci. Res. **3**(3), 1224–1231 (2013)
44. Priyadharshini, S.G., Subramani, C., Preetha Roselyn, J.: An IOT based smart metering development for energy management system. Int. J. Electr. Comput. Eng. **9**(4), 3041 (2019)
45. Sadeeq, M.A.M., Zeebaree, S.: Energy management for internet of things via distributed systems. J. Appl. Sci. Technol. Trends **2**(02), 59–71 (2021)
46. Krishnamoorthy, R., Aswini, S., Guna, C.: Design and implementation of IoT based energy management system with data acquisition. In: 2020 7th International Conference on Smart Structures and Systems (ICSSS), pp. 1–5. IEEE (2020)
47. Munoz, O., Ruelas, A., Rosales, P., Acuña, A., Suastegui, A., Lara, F.: Design and development of an IoT smart meter with load control for home energy management systems. Sensors **22**(19), 7536 (2022)
48. Keles, C., Yigit, Y.: Implementation of demand side load management by domestic load shedding algorithm for conventional electricity meters. In: 2019 International Artificial Intelligence and Data Processing Symposium (IDAP), pp. 1–4. IEEE (2019)
49. Nurfadilah, M.F., Hariyanto, N., Prihatmanto, A.S., Darmakusuma, R, Wijaya, R., Pratama, V.: Energy saving management with suggestion method in home automation based on user habits. In: 2020 6th International Conference on Interactive Digital Media (ICIDM), pp. 1–5. IEEE (2020)
50. Hussain, M.M., Siddique, M., Raees, A., Nouman, M., Javed, W., Razaq, A: Power management through smart grids and advance metering infrastructure. In: 2020 6th IEEE International Energy Conference (ENERGYCon), pp. 767–772. IEEE (2020)
51. Rahman, M.A., Rahman, I., Mohammad, N.: Demand side residential load management system for minimizing energy consumption cost and reducing peak demand in smart grid. In: 2020 2nd International Conference on Advanced Information and Communication Technology (ICAICT), pp. 376–381. IEEE (2020)
52. Butt, M.B., Dilshad, S., Abas, N., Rauf, S., Saleem, M.S.: Design of home load management system for load rationing in Pakistan. Eng. Rep. **3**(3), e12312 (2021)
53. Khan T.A., Hafeez, G., Khan, I., Ullah, S., Waseem, A, Ullah Z (2020) Energy demand control under dynamic price-based demand response program in smart grid. In: 2020 International Conference on Electrical, Communication, and Computer Engineering (ICECCE), pp. 1–6. IEEE
54. Zhang, S., Rong, J., Wang, B.: An optimal scheduling scheme for smart home electricity considering demand response and privacy protection. Int. J. Electr. Power Energy Syst. **132**, 107159 (2021)

55. Saleem, M.U., Usman M.R., Usman M.A., Politis C.: Design, deployment and performance evaluation of an IoT based smart energy management system for demand side management in smart grid. IEEE Access **10**, 15261–15278 (2022)
56. Valovcin, S., Abe N., Massey, B.: Using smart meter data to estimate demand reductions from residential direct load control programs. In: 2022 IEEE International Smart Cities Conference (ISC2), pp. 1–6. IEEE (2022)
57. de la Cruz Severiche Maury, Z., Fernández Vilas, A., Díaz Redondo, R.P.: Low-cost HEM with Arduino and Zigbee technologies in the energy sector in Colombia. Energies **15**(10), 3819 R.P.
58. Tai, K.-Y., Liu, B.-C., Hsiao, C.-H., Tsai, M.-C., Lin, F.-S.: A near-optimal energy management mechanism considering qos and fairness requirements in tree structure wireless sensor networks. Sensors **23**(2), 763 (2023)

Chapter 33
Privacy Preservation Using Network Coding Against Traffic Analysis in Wireless Network

A Antonidoss, K. Dinakaran, P. Valarmathie, and Suresh Annamalai

1 Introduction

Although wireless networks provide a number of benefits to users, their operation is actually rather complex. To offer consumers a consistent connection, a multitude of protocols and technologies are being used in the background. Data packets sent via a wire are more likely to go undetected by eavesdroppers, giving users a sense of security. We need pay attention to the following things to secure the wireless connection.

- Identification of wireless network endpoints and end users is required for authentication.
- Preventing privacy invasion or intermediaries from accessing wireless data packets.
- Preserving the completeness or integrity of wireless data packets.

A. Antonidoss
Department of Computer Science and Business Systems, S.A. Engineering College, Chennai, Tamil Nadu 600077, India

K. Dinakaran
S.A. Engineering College, Chennai, Tamil Nadu 600077, India

P. Valarmathie
Department of Computer Science and Engineering, RMK College of Engineering and Technology, Thiruvallur, Tamil Nadu 601206, India

S. Annamalai (✉)
Department of Networking and Communications, School of Computing, Faculty of Engineering and Technology, SRM Institute of Science and Technology, SRM Nagar, Kattankulathur, Chennai, Tamil Nadu 603203, India
e-mail: prisu6esh@yahoo.com; suresha2@srmist.edu.in

J. M. R. S. Tavares et al. (eds.), *Proceedings of Second International Conference on Intelligent System*, Algorithms for Intelligent Systems,
https://doi.org/10.1007/978-981-99-8976-8_34

2 Introduction to Wireless Networks

A. Wireless Networks

Computer networks that are wireless do not have wires connecting them. Most frequently, network nodes communicate with one another using radio waves. They let devices to stay connected while moving about inside the network's coverage area. Types of wireless networks:

Wireless LANs—To connect two or more network devices, wireless local area networks (LANs) use wireless distribution mechanisms.
Wireless MANs—Ties together two or more wireless LANs that are spread out over a city.
Wireless WANs—Ties together vast areas made up of LANs, MANs, and personal networks.
Attacks on wireless networks

Wireless networks have become a vital aspect of how we run our operations. While many activities are made easier and we can do away with the mess that hundreds of wires cause, securing wireless networks presents certain difficulties. Wireless technology provides practical answers to our problems. They are quick and efficient, and they relieve us of the wire and cable clutter. Wireless networks, on the other hand, are well known to be more open to intrusions and attacks. Attacks on wireless networks, also known as wireless network penetration and intrusion efforts, pose serious threats.

(1) Attacks on wireless networks aim to snoop on data sent across the network or obstruct information flow.

Types of attacks on wireless networks:

(a) Packet sniffing—Information traffic is enabled and accelerated over networks. In order to do this, the data is sent across wired and wireless networks in packet form. Since wireless networks function in this way, these packets are sent over the air. As a result, catching them is not too challenging. Over wireless networks, there is a lot of traffic, including HTTP, SNMP, and RTP. These all have the trait of being written in plain text, which means that with the use of open-access tools like Wireshark, one may readily read them. This makes it simple for someone with malicious intent to get their hands on your passwords and other important information.
(b) Rouge access point—Rogue access points (APs) are any unauthorized access points (APs) on a network. It may have been a careless worker or a potential attacker. Additionally, rogue APs put the entire network at risk from threats including DoS assaults, packet grabbing, and ARP poisoning. You can use network access controls and protocols or authentication mechanisms to protect your business.

(c) Jamming—This is a term used to describe network interference which aims to disrupt the network. The wireless features mean that interference is frequently present at all times. Any device could occasionally create interference, including a microwave or a pair of Bluetooth headphones. Jamming techniques are frequently used in conjunction with other tactics like hostile twinning by malicious attackers.
(d) Evil twinning—One of the most popular tactics employed by wireless network attackers is creating an evil twin. In other words, attackers purchase a wireless access point and configure it to resemble the existing network. In this way, it is impossible to tell the difference between a genuine access point and a "bad" access point.

(2) Attack prevention in wireless networks

Our wireless data is encrypted so that anyone with access to our network cannot view it. Numerous encryption methods are available for this security. The wireless protocols used to transfer data between wireless routers and wireless devices are Wi-Fi Protected Access (WPA), WPA2, and WPA3. WPA3 is currently the most secure encryption available. Despite the fact that WPA and WPA2 are still functional, it is advised to only use WPA3-compatible hardware because doing otherwise could leave the network open to attack.

To stop unauthorized people from easily accessing your network, avoid making the SSID public.

Users can conceal their device's SSID on all Wi-Fi routers to make it more difficult for hackers to identify a network. At the very least, the SSID should be changed to something unusual. A potential attacker could be able to identify the router's model and perhaps exploit any known vulnerabilities if the manufacturer's default setting is applied.

3 Literature Review

With a focus on applications for disaster-resistant smart cities, this research provides a cooperative communication system for information-centric wireless networks. The suggested system improves data delivery and lessens radio interference between relay nodes by using network coding and constructive interference. The outcomes of computer simulations show that data forwarding may still be recovered in a cross-interference environment and that data spread caching has improved [1].

In order to improve the performance of the primary user's transmissions and give the secondary user more options for its own transmissions, this study looks at the transmission strategies for a basic cooperative cognitive radio network, also known as a cognitive radio system. They investigate especially how network coding methods might boost overall system performance [2].

An innovative and elegant method to boost network performance and throughput is network coding. To deliver dependable, secure, and effective transmissions, it takes advantage of the broadcast characteristics of wireless media and jointly codes packets from several senders. This article lists complexity-optimized techniques and describes how coding opportunities and decoding overhead interact. We outline a taxonomy of useful network coding techniques and show three useful ways to reduce computational complexity and improve progressive decoding. Additionally, we offer an overview of the benefits and drawbacks of the current network coding algorithms as well as a plan for additional research [3].

4 Network Coding

Network coding is a relatively new concept in information theory that allows information to be transmitted more efficiently over networks [4]. It involves encoding information in such a way that intermediate nodes in the network can combine multiple streams of information to create a more efficient transmission[5]. Network coding is a significant topic of research in the realm of computer networks and offers a number of benefits over conventional point-to-point communication methods [6]. Each message is sent from the source to the destination separately in classic point-to-point communication [7]. This approach has several limitations, including the inability to utilize network resources effectively and the susceptibility to network failures [8]. Network coding addresses these limitations by allowing intermediate nodes to combine multiple streams of information, thereby reducing the number of transmissions required to transmit the same amount of information [9]. As a result, network resources are used more effectively, and network resilience is increased.

A. **Types of Network Coding**

(1) Linear network coding is the most widely used network coding technique. It involves combining information from multiple sources to create a linear combination of packets. The destination then receives the linear combination [10]. In order to retrieve the original packets at the receiving end, the receiver employs a decoding method. Linear network coding is known for its simplicity and ease of implementation. It also improves network throughput, reduces latency, and increases reliability.

(2) Fountain coding is a type of network coding that is used for transmitting large files over a network. In fountain coding, a sender divides a file into small packets and encodes each packet independently. The encoded packets are subsequently sent from the sender to the recipient. If the receiver gathers enough packets, it can restore the original file. Fountain coding is known for its high efficiency and robustness. It can recover lost packets even if a significant number of packets are lost during transmission [11].

(3) Erasure coding is a technique used to protect data against loss during transmission. In erasure coding, the sender divides the data into small packets and generates additional packets that contain parity information. The parity packets are then transmitted along with the original packets. At the receiving end, if any packets are lost during transmission, the receiver can use the parity packets to reconstruct the lost packets. Erasure coding is highly efficient and can recover from a large number of lost packets [12].

B. Butterfly Network in Linear Coding

Butterfly networks are a particular kind of network topology utilized in linear network coding, a method for boosting the effectiveness of data transmission and enhancing network performance. In a butterfly network, nodes are arranged in a pattern that resembles the wings of a butterfly, with two main paths that intersect in the center [13]. In a butterfly network, linear network coding is used to improve data transmission between nodes arranged in the butterfly pattern. The network consists of two main paths, one going from the top to the bottom of the butterfly, and the other going from the left to the right. Each node in the network has two input and two output ports, allowing data to be transmitted in both directions along both paths [14].

When data is transmitted through a butterfly network using linear network coding, it is first encoded at the source node using a linear code. The encoded message is then transmitted through the network, with each node performing a linear operation on the incoming data before transmitting it to the next node. This linear operation involves multiplying the incoming data by a matrix, which is designed to ensure that the data can be decoded correctly at the destination node. At the center of the butterfly network, the two main paths intersect, allowing data to be transmitted between the two paths. This lowers the amount of transmissions required and enhances network performance by enabling the network to efficiently transport data between nodes over diverse paths.

C. Object-Oriented Tools to TCL (OTCL)

OTcl is an extension of Tool Command Language (Tcl), which is a popular scripting language used for a variety of applications. OTcl adds object-oriented features to Tcl, making it easier to model complex networks and protocols. Wireless networks are susceptible to security threats like eavesdropping, unauthorized access, and denial-of-service attacks. OTcl can be used to model and evaluate security mechanisms that can be used to protect wireless networks.

Encryption is one method for securing wireless networks. Data is converted into a code that can only be decoded by authorized parties during encryption. Several encryption algorithms, such as Rivest–Shamir–Adleman (RSA) and the Advanced Encryption Standard (AES), can be evaluated and simulated using OTcl. The simulation can be used to assess how well various encryption algorithms function under various circumstances, including network traffic and network size.

Another way to secure wireless networks is through the use of authentication mechanisms. Authentication involves verifying the identity of users who are trying to access the network. OTcl can be used to simulate and evaluate different authentication mechanisms, such as password-based authentication and biometric authentication. The simulation can be used to evaluate the performance of different authentication mechanisms under various conditions, such as the number of users and the complexity of the authentication mechanism.

Additionally, OTcl can be used to model and evaluate different intrusion detection and prevention mechanisms. Identification and prevention of unauthorized network access comprise intrusion detection and prevention. OTcl can be used to analyze and simulate a variety of intrusion detection and prevention technologies, including firewalls, intrusion detection systems, and intrusion prevention systems.

OTcl is a powerful scripting language that can be used to model and evaluate security mechanisms in wireless networks. The simulation can be used to evaluate the performance of different security mechanisms under various conditions, helping network administrators to identify the most effective security measures for their network.

5 Routing Algorithms

Computer networks require routing algorithms in order to send data packets from the source to the destination node. Data packets should follow the best path possible through the network, according to a set of rules called a routing algorithm. The goal of routing algorithms is to minimize network congestion and delays while ensuring efficient and reliable delivery of data packets. Different types of routing algorithms are used in computer networks, including proactive, reactive, and hybrid routing algorithms. Each type of routing algorithm has its advantages and disadvantages, and network administrators must choose the appropriate algorithm based on the network's characteristics and requirements. Routing algorithms serve a crucial role in guaranteeing the efficiency and security of data transfer and are essential for the correct operation of computer networks.

A. **Types of Routing Algorithms**

(1) Reactive routing algorithms: Reactive algorithms, often known as on-demand routing algorithms, only create a route when it is necessary to transport data packets. It follows that only when a data packet needs to be transferred does the route discovery procedure start. Reactive routing methods include the Ad hoc On-demand Distance Vector (AODV) and Dynamic Source Routing (DSR) [15].

(2) Hybrid routing algorithms: Hybrid routing algorithms, as their name implies, combine the benefits of proactive and reactive routing algorithms. They minimize the overhead associated with proactive routing algorithms by maintaining a partial routing table and updating it only when necessary.

Examples of hybrid routing techniques are Zone Routing Protocol (ZRP) and Temporally Ordered Routing Algorithm (TORA) [16].

(3) Distance-vector routing algorithms: The shortest path between two nodes is determined using distance-vector routing methods, also referred to as Bellman–Ford algorithms.

B. **Routing Algorithms in Wireless Networks**

Due to their intrinsic ability to broadcast data over the airwaves, wireless networks are vulnerable to a variety of security risks. Therefore, securing wireless networks is of utmost importance. One of the ways to secure wireless networks is by implementing routing algorithms that can detect and prevent potential security breaches. The most efficient route for data packets to take between a network's source and destination node is determined by routing algorithms. Wireless networks employ a range of routing methods, including proactive, reactive, and hybrid algorithms.

Routing algorithms can be used to enhance the security of wireless networks in several ways. For example, routing algorithms can be used to detect and prevent malicious nodes from sending fake routing information. In proactive routing algorithms, routing tables can be secured by encrypting the routing information stored in them. In reactive routing algorithms, the route discovery process can be secured by using cryptographic techniques such as digital signatures.

Routing algorithms play a critical role in securing wireless networks. Network administrators must select the best routing algorithm depending on the network's size, mobility, and stability because different routing algorithms are ideal for various types of networks. By using routing algorithms, network administrators can detect and prevent potential security breaches and ensure that data is transmitted securely and efficiently in wireless networks.

Algorithm 1: Spoofing Attack Detection in Wireless Sensor Networks

Inputs:

- Wireless sensor network (WSN) topology and parameters.
- Node behavior data (e.g., packet transmission) within clusters.
- Clustering approach parameters.
- IDS parameters.

Output:

- Identified spoofing attackers.

Step 1: Initialize WSN

1. Set up the WSN topology and parameters.
2. Deploy sensor nodes in the network.

Step 2: Clustering Approach

1. Implement the clustering approach to form clusters of sensor nodes.
2. Determine cluster heads based on criteria like node energy or distance to base station.

Step 3: IDS Setup

1. Design and configure the intrusion detection system (IDS).
2. Assign the cluster head role to serve as an IDS within each cluster.

Step 4: Monitor Node Behavior

1. The cluster head monitors node behavior within its cluster, such as packet transmission and other activities.
2. Record and analyze the behavior of the nodes to detect any misbehaving or spoofing nodes.

Step 5: Detect Spoofing Attackers

1. Using the collected data, the IDS identifies suspicious nodes that exhibit anomalous behavior.
2. Apply anomaly detection techniques to identify potential spoofing attackers within each cluster.

Step 6: Count and Locate Attackers

1. Count the number of identified spoofing attackers within each cluster.
2. If multiple clusters exist, determine if any common nodes are identified as attackers in multiple clusters.
3. Analyze the data to locate the position of the identified attackers in the network.

Step 7: Take Countermeasures

1. Once spoofing attackers are identified and located, take appropriate countermeasures against them.
2. Implement security mechanisms like isolating the attackers, blocking their access, or notifying a higher-level security system.

Step 8: Simulate and Evaluate

1. Use the OTCL language to simulate the WSN environment and the IDS operation.
2. Utilize Network Simulator 2 (NS2), a simulation software for Linux, to analyze the results.
3. Evaluate the performance of the proposed strategy in terms of detection accuracy, false positive rate, and false negative rate.

Step 9: Result Analysis

1. Analyze the simulation results to determine the effectiveness of the clustering-based IDS in detecting spoofing attackers.
2. Compare the performance of the proposed strategy with other existing methods.

Step 10: Conclusion

1. Conclude whether the proposed approach successfully and consistently identifies spoofing attackers in wireless sensor networks.
2. Discuss potential improvements and future research directions.

End Algorithm

Algorithm 1 outlines the steps for detecting and countering spoofing attacks in wireless sensor networks (WSNs) using a clustering-based intrusion detection system (IDS) and anomaly detection techniques.

6 System Design

The three crucial components of the proposed system model are the source node, relay nodes, and destination nodes. The source node generates data packets, which are then forwarded to the relay nodes for transmission. The relay nodes mix the packets and transfer them to the target node using network coding. The destination node then decodes the packets to acquire the original data.

We suggest using network coding with a secret key to safeguard the wireless network. The data packets are encrypted before transmission using the secret key. Only the relay nodes with the secret key can decode the packets, ensuring that the data is secure during transmission.

The system model also includes a routing protocol to determine the optimal path for data transmission. The routing protocol takes into account the network topology, traffic characteristics, and transmission distance to ensure efficient data transfer.

In conclusion, the proposed system model for securing wireless networks using network coding with a secret key and a routing protocol has several advantages over traditional wireless network security approaches. It provides improved reliability, security, and efficiency in data transfer, making it a promising solution for securing wireless networks. Wireless networks are particularly vulnerable to security threats due to their broadcast nature, which allows anyone within range to intercept and manipulate data transmissions. By utilizing mathematical techniques to encode and distribute data across numerous packets, network coding can help to reduce these risks by making it more challenging for an attacker to intercept and decode the data.

A system model for securing wireless networks using network coding involves a combination of encryption, redundancy, and authentication measures. By using these techniques, it is possible to protect against unauthorized access and data theft, making wireless networks more secure and reliable for a wide range of applications.

We provide a system model for the wireless sensor networks (WSNs) that allows for the dependable delivery of brief messages. There are three essential parts to the suggested system model: the sender, the receiver, and the channel. The paper focuses on developing an error-correcting code (ECC) to enable reliable transmission of short messages in WSNs.

The sender component is responsible for encoding the message using an error-correcting code and transmitting it over the wireless channel. The paper proposes the use of Reed–Solomon (RS) codes as the ECC due to their ability to correct multiple errors in a message. The sender divides the message into blocks and adds redundant bits to each block using the RS code. The resulting encoded blocks are then transmitted over the wireless channel to the receiver.

The decoding and integrity checking of the message are the responsibilities of the receiver component. The RS decoding technique is used to each block once the receiver gets the encoded blocks from the wireless channel. The superfluous bits sent by the sender are used by the RS decoding method to fix faults in a block. The receiver then aggregates all of the decoded blocks to get the original message. The recipient then confirms the message's integrity by contrasting its checksum with the one given by the sender.

The wireless channel component represents the medium through which the encoded blocks are transmitted from the sender to the receiver. The channel is prone to errors due to noise, interference, and fading. The paper assumes a binary symmetric channel (BSC), which means that each bit transmitted over the channel can flip with a certain probability. The study also makes the assumption, which states that all bits transferred across the channel have an equal chance of experiencing a bit mistake.

We're considering a stationary memoryless channel that operates in discrete time and supports point-to-point communication. The dependent probability density function of the channel, PY|X, describes it. We also assume that the channel experiences a new packet arrival with probability λ in each channel use, and that these arrivals are statistically independent. Consequently, the average inter-arrival time between two consecutive packets is $1/\lambda$. Every time a packet with k bits comes, it is queuing up at the transmitter and waiting to be handled by a server-based queue that adheres to the first-in, first-out (FIFO) principle. In order to send the information packet over the channel, a VLSF encoder is used to transform it into a coded packet. Each of the 2 k messages that could be contained in the packet is given a different codeword with an indefinite length by the encoder. The first n characters of the ciphertext linked to the chosen message, which make up the opening frame, are then broadcast. The VLSF decoder will use a feedback connection to send a NACK signal to the same encoder if it is unable to decode the message using the first n symbols. The following frame and the subsequent n symbols of the matched codeword are subsequently sent by the encoder. The process is repeated until the stopping rule is reached and the original message can be successfully decoded by the decoder. The Ack packet is sent by the decoder to the encoder after the message has been decoded, and the Ack packet is then removed from the transmission channel to make way for the next packet. Let the variable representing the number of frames the VLSF approach needs to send one packet be.

Figure 1 illustrates the flow of data and components involved in ensuring reliable and secure data transmission over wireless networks. It showcases the utilization of network coding, encryption with a secret key, routing protocol, error-correcting codes, and acknowledgment mechanisms to enhance the reliability and security of data transfer in wireless networks.

A. Steady-State Delay Violation Probability

A performance indicator called Steady-State Delay Violation Probability (SSDVP) is employed to assess the level of service that communication systems offer. When

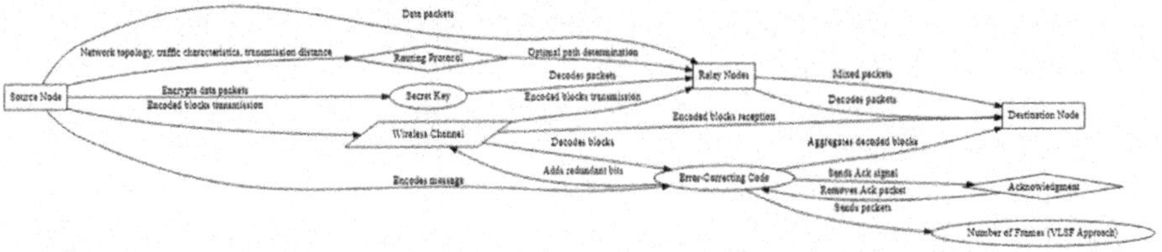

Fig. 1 System model for securing wireless networks using network coding and error-correcting codes

a system is in steady state, which is when it has reached a stable state and is functioning under constant conditions, the probability that the delay in packet delivery will surpass a specific threshold is measured.

A mathematical model of the communication system is created that includes the source node, the destination node, and the transmission medium in order to calculate the SSDVP. The model takes into account the traffic characteristics, such as the arrival rate and packet size distribution, as well as the transmission characteristics, such as the channel capacity and delay.

The SSDVP is a crucial metric for assessing the effectiveness of communication systems since it calculates the likelihood that a system would fall short of meeting quality-of-service standards. By minimizing the SSDVP, the system can provide a better quality of service, which leads to improved user experience.

We introduce the idea of stable state delay breach risk in this section and go into methods for computing it. Our definition takes into account the potential for undiscovered error occurrences as well as the risk that a packet's delay would exceed a certain latency restriction.

B. **Peak Age Violation Probability in the Steady State**

It is a performance metric used to evaluate the quality of service provided by real-time communication systems. The peak age of information, or PAoI, is the maximum amount of time that separates the time that a packet is generated at the source from the time that it is transmitted to the destination. T estimates the likelihood that the PAoI will surpass a specific threshold. Real-time communication systems are designed to provide timely delivery of data, which is critical for applications such as video conferencing, online gaming, and telemedicine. In such systems, the PAoI is an important metric as it represents the freshness of the information being delivered. If the PAoI exceeds a certain threshold, the quality of service is degraded, leading to poor user experience.

In the steady state, which is when the system has achieved a stable state and is functioning under constant conditions, the SSPAVP is determined as the likelihood that the PAoI will surpass a specific threshold.

A mathematical model of the communication system is created that includes the source node, the destination node, and the transmission medium in order to calculate the SSPAVP. The model takes into account the traffic characteristics, such as the arrival rate and packet size distribution, as well as the transmission characteristics, such as the channel capacity and delay.

The SSPAVP is a crucial statistic for assessing the effectiveness of real-time communication systems since it calculates the likelihood that the system would fall short of quality-of-service standards. By minimizing the SSPAVP, the system can provide a better quality of service, which leads to improved user experience.

In this part, we focus on the likelihood of a violation occurring after a package reaches the application layer's steady state. We assume that each data packet delivered from the source contains a sample of this stochastic process and the related timestamp since the receiver is interested in the behavior seen at the source. In this study, we solely take into account the frame-synchronous model and accept the packet timestamp as the index of the frame in which it enters the queue. As a result, the mth packet is linked to the time index $T(p)$ m. When numerous packets arrive with the same timestamp, we presume that only one of them will be added to the queue and the others will be deleted.

7 Packages

A. NS2 Package

NS2 is a widely used package in wireless network security that provides a range of features and functionalities to improve the security of wireless networks. The package includes several tools and modules that can be used to simulate, analyze and test various aspects of wireless network security. NS2 packages refer to the various modules, libraries, and tools that are included in the Network Simulator 2 (NS2) software. NS2 is a widely used discrete event network simulator that is used to simulate and analyze various types of computer networks, including wired and wireless networks. NS2 packages include modules for simulating different network protocols, as well as modules for modeling wireless networking technologies including Wi-Fi, WiMAX, and cellular networks. These protocols include TCP, IP, and UDP. These packages also include tools for generating network topologies and traffic patterns, analyzing network performance, and testing network applications.

One of the key features of the NS2 package is its ability to simulate various types of attacks that can occur in wireless networks. This covers assaults like man-in-the-middle, eavesdropping, and denial-of-service (DoS) attacks. By simulating these attacks, network security professionals can better understand how they work and develop strategies to prevent and mitigate them. The NS2 package also includes modules for simulating various types of encryption and authentication protocols, such as WEP, WPA, and 802.1X. These modules can be used to test the effectiveness of these protocols in real-world scenarios and to identify potential vulnerabilities and weaknesses.

Another key feature of the NS2 package is its ability to analyze network traffic and identify potential security threats. This includes tools for analyzing packet headers and payloads, detecting abnormal traffic patterns, and identifying potential security breaches. The NS2 package also includes modules for testing the effectiveness of various intrusion detection and prevention systems (IDS/IPS). These modules can be used to simulate attacks and test the ability of IDS/IPS systems to detect and prevent them. In addition, the NS2 package includes tools for generating various types of network topologies and scenarios, which can be used to test the security of wireless networks in different environments and under different conditions.

The NS2 package is a powerful tool for improving the security of wireless networks. By simulating attacks, testing encryption and authentication protocols, analyzing network traffic, and testing IDS/IPS systems, network security professionals can identify and address potential vulnerabilities and weaknesses in wireless networks. As a result, any organization that relies on wireless networks and wishes to guarantee the security and integrity of its data and communications should use the NS2 package.

B. **NAM (Network Analysis Module)**

The Network Analysis Module (NAM) is a tool used to monitor and analyze network traffic in real-time. It can help identify network issues and performance problems and can provide insight into network usage patterns. When it comes to wireless network security, NAM can be used to identify and mitigate potential threats. NAM can monitor wireless network traffic and identify devices that are not authorized to access the network. This can include rogue devices, such as access points or wireless clients, that could be used to launch attacks against the network. NAM can also spot unusual network traffic patterns that might point to a security attack. NAM can help administrators manage security policies and make sure that regulations like the Payment Card Industry Data Security Standard (PCI DSS) and the Health Insurance Portability and Accountability Act (HIPAA) are followed. NAM can aid in identifying compliance infringements and ensuring that security regulations are being upheld by keeping track of network traffic.

The Network Analysis Module (NAM) is a tool used to monitor and analyze network traffic in real-time. It can help identify network issues and performance problems and can provide insight into network usage patterns. When it comes to wireless network security, NAM can be used to identify and mitigate potential threats. NAM can monitor wireless network traffic and identify devices that are not authorized to access the network. This can include rogue devices, such as access points or wireless clients, that could be used to launch attacks against the network. NAM can also detect abnormal patterns of network traffic that may indicate a security breach. NAM can help administrators manage security policies and make sure that regulations like the Payment Card Industry Data Security Standard (PCI DSS) and the Health Insurance Portability and Accountability Act (HIPAA) are followed. NAM can aid in identifying compliance infringements and ensuring that security regulations are being upheld by keeping track of network traffic.

C. **Xgraph**

In NS2 (Network Simulator 2), Xgraph is a tool used for visualizing and analyzing simulation results. It can display various performance metrics such as throughput, delay, packet loss, and queue size in graphical form. The benefits of using Xgraph in NS2 include—easy visualization of simulation results: Xgraph provides a convenient way to graphically represent the performance of a wireless network simulation. It allows users to quickly identify the trends and patterns in the simulation data, which can aid in making decisions about network design and optimization; comparison of different scenarios: Xgraph can display the performance metrics of multiple simulation runs on the same graph, allowing users to compare the results of different scenarios and evaluate the impact of changes to the network configuration; Customizable graphs: Xgraph offers a variety of options for customizing the appearance of the graphs, such as changing the axis labels, colors, and line styles. This can help users create more informative and visually appealing graphs.

Regarding the use of Xgraph in wireless network security, it can be useful for analyzing the performance of security mechanisms such as encryption and authentication protocols. For example, users can use Xgraph to measure the impact of different encryption algorithms on network throughput or to evaluate the effectiveness of an intrusion detection system. By simulating various attack scenarios and analyzing the resulting performance metrics, network administrators can identify vulnerabilities and make informed decisions about security measures to implement.

8 Implementation

Regarding criteria or priorities for dependability, freshness, and latency, there are numerous natural assumptions. Consider a broadcasting situation where the same content is sent to every recipient or one where distinct data streams must be provided to various receivers. Such generalizations are also applicable when many sensors are used to service a single destination or when a multi-hop arrangement prioritizes high information freshness and minimal end-to-end delay. By extending the architecture outlined in this work, the first three scenarios can be satisfied. When delay analysis was used to determine the steady-state average lifespan of information under LCFS-Q, for example, the approaches we proposed in this research may be used to address the potential for maximum age violation in the first scenario. The second situation allows for the use of non-asymptotic constraints from earlier research on the effectiveness of VLSF codes for broadcasting. In the third scenario, deterministic arrivals can be handled by taking advantage of the problem's indexability. On the other hand, when dealing with more extensive arrival distributions, the multi-server findings from earlier studies on the efficiency of the final-generated 1st packet management technique may be applied. These works are all based on a condensed physical-layer model and can all be used with the framework provided in this paper. A collaborative design of the physical and higher layers can result from this integration, which can also improve the precision of performance predictions.

This set of advantages highlights the benefits of a protocol designed to improve multi-hop wireless networks. By utilizing a duty cycling technique, the protocol is able to conserve energy and increase the capacity of the network. This increase in capacity not only allows for fast data gathering at the sink but also improves the overall throughput of the network. These advantages are significant, as they enable more efficient and effective data transmission wireless networks, ultimately resulting in better performance and increased productivity.

In conclusion, the system model for securing wireless networks using network coding provides an efficient and robust solution to the challenges of wireless network security. The system can greatly lower the danger of attacks like eavesdropping and information interception by integrating network coding techniques into the wireless network.

The proposed system model provides various security mechanisms such as encryption, authentication, and key management, which work together to ensure secure communication between nodes. The implementation of this system model requires careful consideration of various factors such as network topology, channel conditions, and user requirements. Nevertheless, the benefits of this approach to wireless network security make it a promising solution for securing wireless networks in various scenarios. Overall, the system model for securing wireless networks using network coding is a significant step toward.

Figure 2 shows that wireless transmission within multiple nodes involves wireless data exchange between devices using various techniques and protocols, addressing factors such as secure transmissions, energy efficiency, and access control, and dealing with issues like hidden nodes, multiple access protocols, and power management.

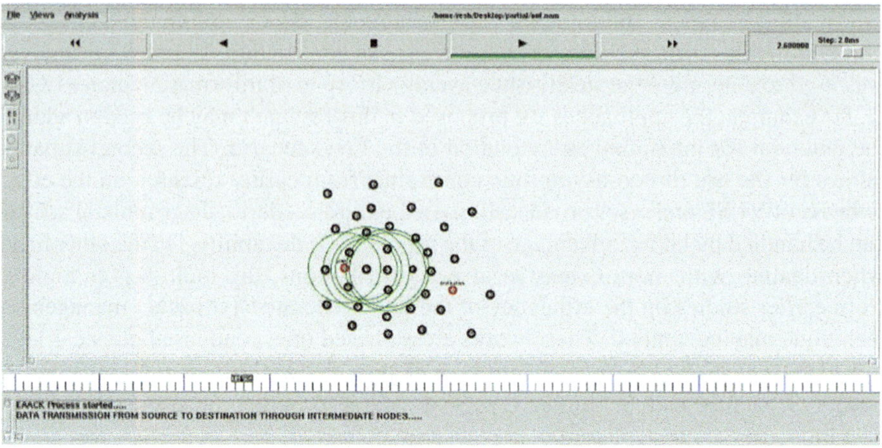

Fig. 2 Wireless transmission within multiple nodes

Figure 3 shows a "compromised node," which refers to a network device that has been infiltrated and is under the control of an attacker, posing significant security risks and potentially being used for malicious activities.

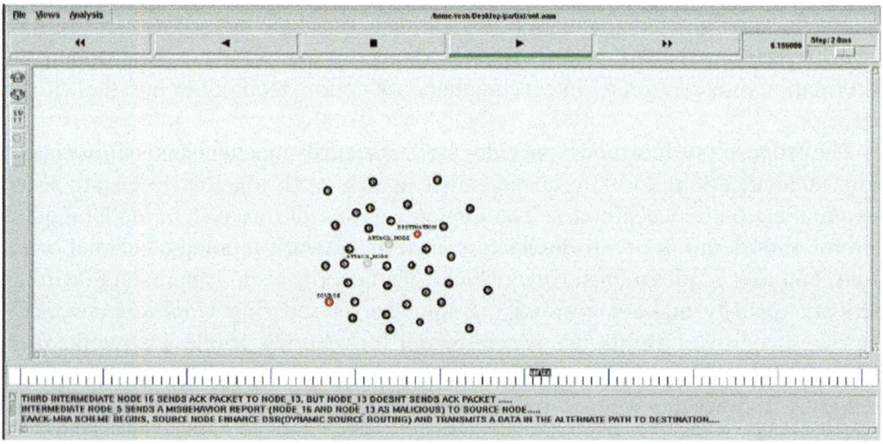

Fig. 3 Compromised node

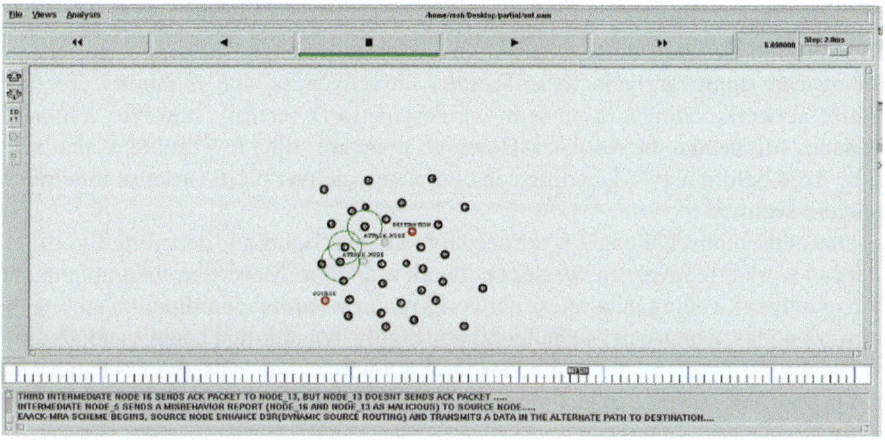

Fig. 4 Alternate path for data transmission

Figure 4 shows an alternate path for data transmission, which is the use of alternative network routes in case the primary path is unavailable or congested, which can boost network efficiency and data transmission reliability. Various protocols and techniques exist to identify and select alternate paths, such as multipath routing and alternate path decision algorithms.

Enhancing the security of wireless networks and addressing the security challenges faced by modern wireless networks.

9 Conclusion

In conclusion, network coding is a promising technique for enhancing the security of wireless networks. By allowing packets to be combined and transmitted more efficiently, network coding can increase throughput, reduce latency, and improve reliability, while also providing protection against various types of attacks, including eavesdropping and packet dropping. Furthermore, network coding can make it possible to implement more sophisticated security measures like homomorphic encryption and secure multi-party computation, which can safeguard private information and protect sensitive data. Analyze the delay of existing and suggested approaches in Fig. 5. Throughput comparison between the existing and suggested methods is shown in Fig. 6 and energy using the existing and proposed methods is shown in Fig. 7. Network coding has been used in a variety of wireless network contexts, such as cognitive radio networks, mobile ad hoc networks, and IoT systems. The study in this field has produced encouraging findings, showing that network coding can significantly increase security, effectiveness, and reliability. To fully realize network coding's promise in wireless network security, however, a number of issues still need to be resolved. However, there are still a few problems that need to be fixed before network coding can completely deliver on its promise in wireless network security.

One of the main challenges is the need for efficient coding and decoding algorithms that can work effectively in wireless network scenarios. Moreover, the implementation of network coding in wireless networks requires careful coordination among the network nodes, which can be challenging in highly dynamic and heterogeneous environments. Furthermore, the limited resources of wireless devices, such as battery life and processing power, can also pose challenges for the implementation of network coding.

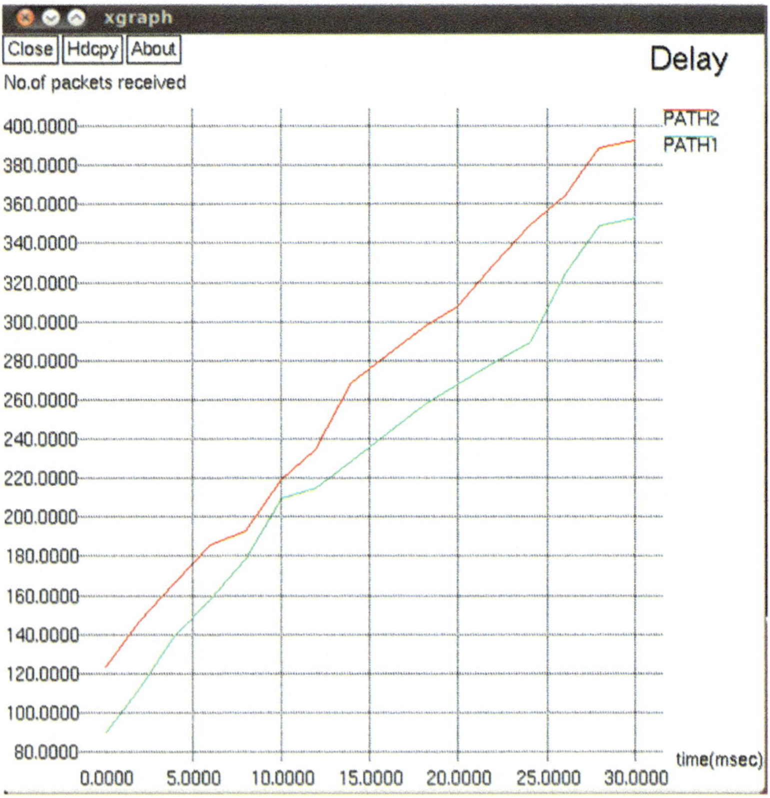

Fig. 5 Delay existing versus proposed

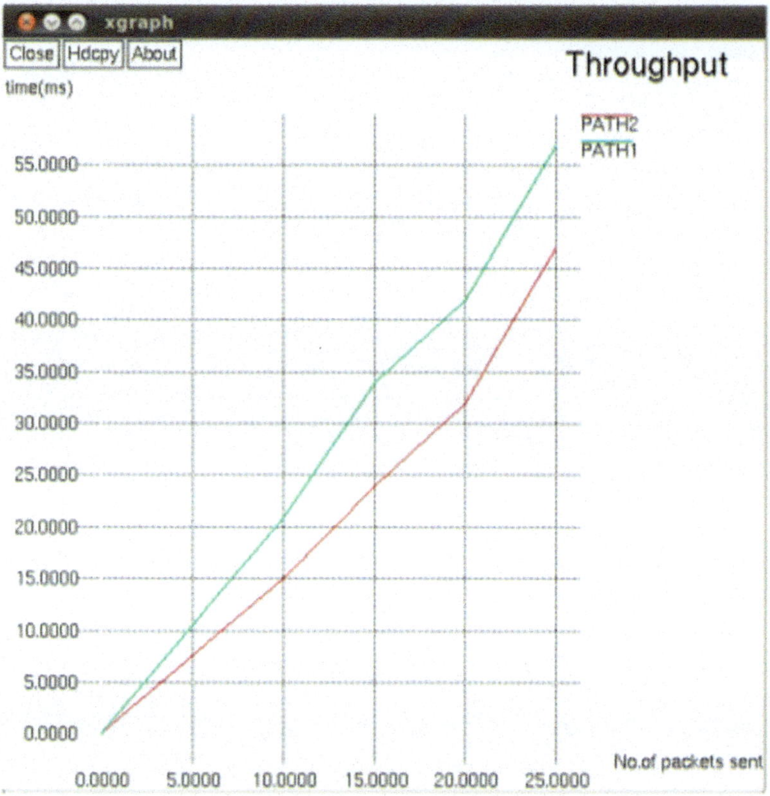

5. Throughput existing vs proposed

Fig. 6 Throughput existing versus proposed

Future research in this area should focus on addressing these challenges and developing more advanced network coding techniques and protocols that can work effectively in wireless network scenarios. For example, to improve the coordination between network nodes and optimize the coding and decoding methods, research could investigate the application of machine learning and artificial intelligence approaches. Network coding can be combined with additional security tools like homomorphic encryption and secure multi-party computing to further increase wireless network security. Figure 8 shows the packet delivery ratio between the existing and proposed methods.

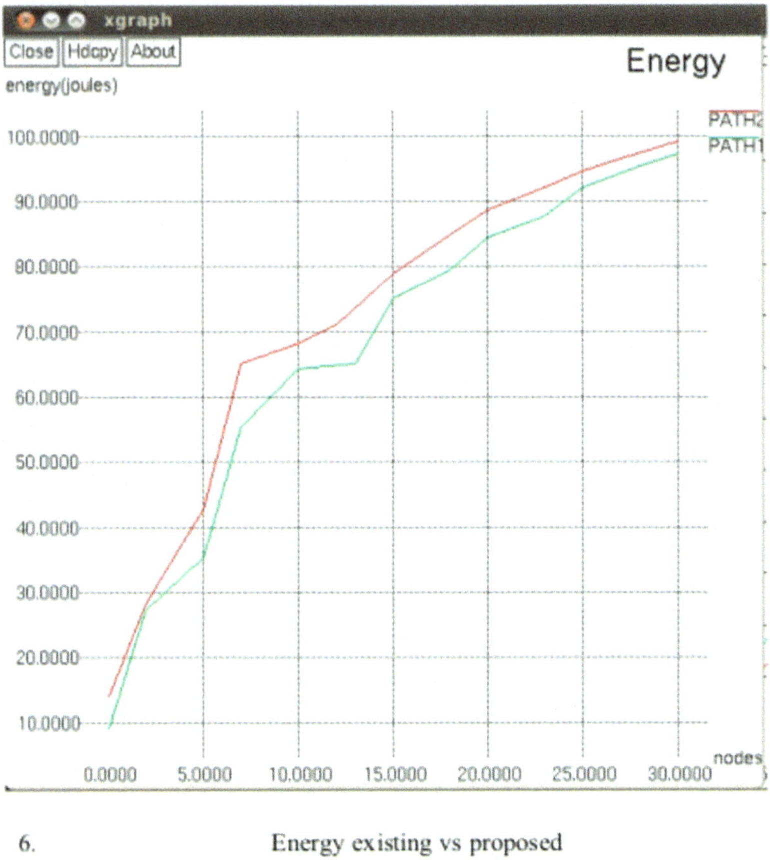

Fig. 7 Energy existing versus proposed

The application of network coding in wireless network security is an exciting field of research that could have a big impact on security, effectiveness, and dependability. To fully utilize the promise of network coding in wireless network security, various issues still need to be resolved, and more research is needed.

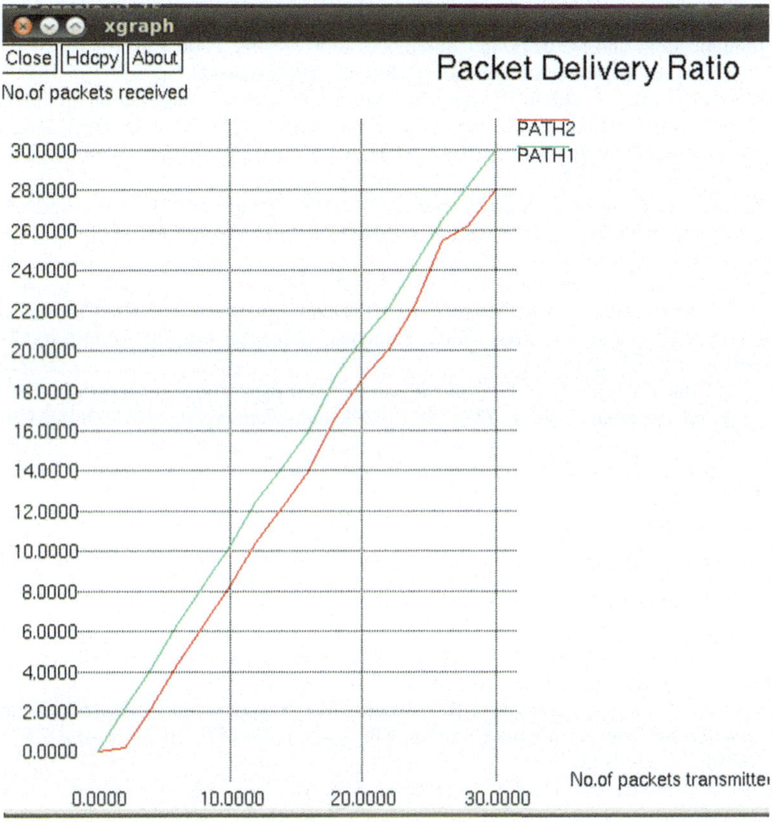

Fig. 8 Packet delivery ratio existing versus proposed

References

1. Papadopoulos, A., Chatzidiamantis, N.D., Georgiadis, L.: Network coding techniques for primary-secondary user cooperation in cognitive radio networks. IEEE Trans. Wireless Commun.Commun. **19**(6), 4195–4208 (2020). https://doi.org/10.1109/TWC.2020.2980821
2. Zhu, F., Zhang, C., Zheng, Z., Farouk, A.: Practical network coding technologies and softwarization in wireless networks. In: IEEE Internet of Things J. **8**(7), 5211–5218 (2021). https://doi.org/10.1109/JIOT.2021.3056580
3. Bhatia, J., Kakadia, P., Bhavsar, M., Tanwar, S.: SDN-enabled network coding-based secure data dissemination in VANET environment. IEEE Internet Things J. **7**(7), 6078–6087 (2020). https://doi.org/10.1109/JIOT.2019.2956964
4. Peralta, G., Cid-Fuentes, R.G., Bilbao, J., Crespo, P.M.: Homomorphic encryption and network coding in IoT architectures: advantages and future challenges. Electronics **8**(8), 827 (2019). https://doi.org/.https://doi.org/10.3390/electronics8080827
5. Ma, Z., Xiao, M., Xiao, Y., Pang, Z., Poor, H.V., Vucetic, B.: High-reliability and low-latency wireless communication for internet of things: challenges, fundamentals, and enabling technologies. IEEE Internet Things J. **6**(5), 7946–7970 (2019). https://doi.org/10.1109/JIOT.2019.2907245

6. Chen, Y., Kishore, S., Li, J.: Wireless diversity through network coding. In: IEEE Wireless Communications and Networking Conference, 2006. WCNC 2006. Las Vegas, NV, USA, pp 1681–1686 (2006). https://doi.org/10.1109/WCNC.2006.1696541
7. Matsumine, T., Koike-Akino, T., Wang, Y.: Deep learning-based constellation optimization for physical network coding in two-way relay networks, ICC 2019–2019. In: IEEE International Conference on Communications (ICC), Shanghai, China, pp 1–6 (2019). https://doi.org/10.1109/ICC.2019.8761963
8. Cohen, A., et al.: Bringing network coding into SDN: architectural study for meshed heterogeneous communications. IEEE Commun. Mag.Commun. Mag. **59**(4), 37–43 (2021). https://doi.org/10.1109/MCOM.001.2000875
9. Torre, R., Pandi, S., Nguyen, G.T., Fitzek, F.H.P.: Optimization of a random linear network coding system with newton method for wireless systems, ICC 2019. In: 2019 IEEE International Conference on Communications (ICC). Shanghai, China, pp 1–6 (2019). https://doi.org/10.1109/ICC.2019.8761304
10. Salamat Ullah, S., Liew, S.C., Liva, G., Wang, T.: Short-packet physical-layer network coding. IEEE Trans. Commun. **68**(2), 737–751 (2020). https://doi.org/10.1109/TCOMM.2019.2956920
11. Kumar, S.A.P., Nair, R.R., Kannan, E., Suresh, A., Raj Anand, S.: Intelligent vehicle parking system (IVPS) using wireless sensor networks. Wirel. Pers. Commun. (2021). https://link.springer.com/article/https://doi.org/10.1007/s11277-021-08360-z
12. Ullah, S.S., Liew, S.C., Liva, G., Wang, T.: Implementation of short-packet physical-layer network coding. IEEE Trans. Mob. Comput. **22**(1), 284–298 (2023). https://doi.org/10.1109/TMC.2021.3071329
13. Okyere, B., Musavian, L., Mumtaz, R.: Multi-user massive MIMO and physical layer network coding. In: 2019 IEEE Globecom Workshops (GC Wkshps), Waikoloa, HI, USA. pp. 1–6 (2019). https://doi.org/10.1109/GCWkshps45667.2019.9024478
14. Cohen, A., Malak, D., Bracha, V.B., Médard, M.: Adaptive causal network coding with feedback. IEEE Trans. Commun.Commun. **68**(7), 4325–4341 (2020). https://doi.org/10.1109/TCOMM.2020.2989827
15. Kumar, S., Cengiz, K., Vimal, S., Suresh, A.: Energy efficient resource migration based load balance mechanism for high traffic applications IoT. Wireless Pers. Commun.Commun. (2021). https://doi.org/10.1007/s11277-021-08269-7
16. Pan, H., Chan, T.-T., Leung, V.C.M., Li, J.: Age of information in physical-layer network coding enabled two-way relay networks. IEEE Trans. Mob. Comput. doi: 10.1109/TMC.2022.3166155

Chapter 34
Optimised Back Propagation-Based Deep Residual Learning Network Algorithm for Detection of Diabetic Retinopathy in Retinal Fundus Images

S. Shilpa and B. Karthik

1 Introduction

Estimates show that the prevalence of diabetes among the global population is continuously increasing every year, and diabetic retinopathy (DR) is one of its consequences. Diabetic retinopathy is an eye condition that harms the retina of the eye and affects diabetics. The patient may become totally blind if not treated in a timely manner. The incidence of diabetic retinopathy eye illnesses is predicted to increase further, and DR progresses without exhibiting any symptoms in its early stages, making research into the disease tough.

The risk of blindness in diabetic patients can theoretically be reduced by screening for the onset of diabetic retinopathy. Although viable treatments for DR are available, it requires early discovery and ongoing monitoring of diabetic individuals. Although they take more time, physical examinations including optical coherence tomography, pupil dilation, and visual acuity tests may also be performed to diagnose diabetic retinopathy.

S. Shilpa (✉) · B. Karthik
Department of ECE, Bharath Institute of Higher Education and Research, Chennai, India
e-mail: shilpaprashant26@gmail.com

B. Karthik
e-mail: karthik.ece@bharathuniv.ac.in

J. M. R. S. Tavares et al. (eds.), *Proceedings of Second International Conference on Intelligent System*, Algorithms for Intelligent Systems,
https://doi.org/10.1007/978-981-99-8976-8_35

2 Literature Survey

One strategy suggests training deep neural networks using a hybrid classification model that takes advantage of both the firefly technique and principal component analysis (PCA). The goal of this method is to reduce the number of dimensions in the dataset so that useful characteristics may be extracted more easily. Classification of the diabetic retinopathy dataset is performed by feeding the reduced-dimensional dataset into a deep neural network model [1, 2]. In their research, Vijayan et al. compared several popular machine learning algorithms, including the OneR classifier, KNN, J48, and random forest. To detect the earliest stages of diabetic retinopathy, the OneR classifier makes use of retinal images obtained from a public database.

In order to efficiently extract useful features from the dataset and lower the dimensionality, the authors suggest a hybrid classification model that combines the firefly algorithm with principal component analysis (PCA) within deep neural networks [3]. In order to classify the reduced-dimensional diabetic retinopathy dataset, a deep neural network model is utilised. Using retinal images from a public database, the OneR classifier can detect the earliest stages of diabetic retinopathy [4–6].

In order to examine retinal fundus images from the Messidor collection, this work presents a novel methodology that combines deep learning and image processing techniques. The goal is to improve several performance indicators, such as accuracy, sensitivity, specificity, precision, FScore, and GMean [7], in order to increase the diagnostic accuracy of diabetic retinopathy. Rapid categorisation has also been accomplished through the use of other machine learning techniques, such as artificial neural networks (ANNs) [8, 9], fuzzy logic [10–13], and other hybrid machine learning approaches [13].

Laser therapy has the potential to mitigate or postpone visual impairment by facilitating timely detection, hence enhancing diabetic management. Early detection of diabetic retinopathy is of utmost importance in mitigating the risk of complete vision loss, as it serves as a significant determinant in the prevention of this condition [14–16]. The proposed algorithm, optimised back propagation-based deep residual learning network (Op-BPDRLN), is recommended for enhancing the diagnostic accuracy of diabetes retinopathy disease and reducing the time required for detection. This algorithm utilises features extracted from several retinal images.

The two-layer multilayer feed forward neural network [17–21] is widely regarded as the most optimal neural network architecture for achieving effective classification performance in a classifier model. The evaluation of the proposed classification model, Op-BPDRLN, is conducted in comparison with another model with the aim of enhancing its superiority. This assessment is carried out utilising statistical metrics such as accuracy, error, precision, and recall.

3 Proposed Method

An important task in medical image processing is the identification of diabetic retinopathy (DR) in retinal fundus pictures. Convolutional neural networks (CNNs), one type of deep learning method, are frequently employed for this. The deep residual learning algorithm is one efficient deep learning algorithm for DR detection (Fig. 1).

The ResNet architecture introduced the idea of residual learning, on which the deep residual learning method is based. Instead of learning the required underlying mappings directly, residual learning enables the network to learn residual mappings. Skip connections are used to accomplish this, allowing the network to understand the difference between the input and output feature maps. The deep residual learning network is trained with the back propagation approach. The network can iteratively update its parameters to reduce loss using this optimisation approach, which computes the gradients of the network parameters with respect to a loss function. The loss function typically calculates the difference between the expected DR severity and the ground truth labels in the context of DR detection.

The back propagation-based deep residual learning approach for DR detection in retinal fundus pictures is described in the following general terms. Preparation of Data: By scaling the retinal fundus images to a fixed size and normalising the pixel values, preprocessing is done on the images. Create training, validation, and testing sets from the dataset. Create a deep residual learning architecture that is appropriate for DR detection. Typically, this architecture consists of a number of residual blocks, each of which has batch normalisation, numerous convolutional layers, and nonlinear activation functions like ReLU. The network may successfully learn the residuals, thanks to the skip connections in each residual block. To determine the expected DR severity, run the preprocessed retinal fundus images through the network.

The input feature maps are processed by each layer in the network, which then passes them on to the layer below. Utilise a suitable loss function, such as cross-entropy loss or mean squared error, to calculate the loss between the anticipated

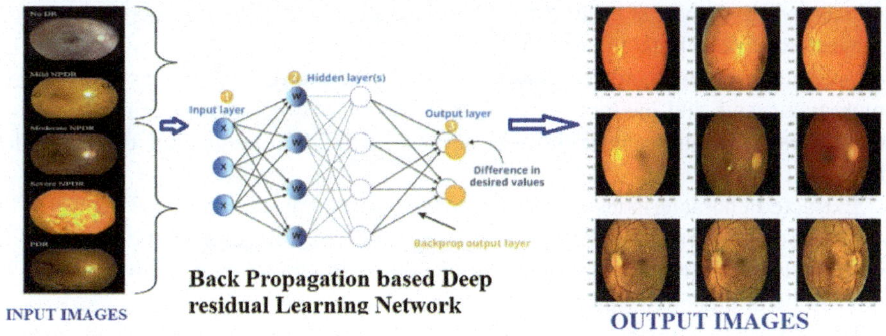

Fig. 1 Architecture of optimised back propagation-based deep residual learning network algorithm (Op-BPDRLN) with classified retinal fundus images according to the DR severity scale used for the training data

DR severity and the ground truth labels. To determine the gradients of the network parameters with respect to the loss, use back propagation. This entails computing the gradients at each layer using the chain rule of derivatives and propagating the gradients backwards through the network. A minor step in the opposite direction of the computed gradients is taken using an optimisation approach employing stochastic gradient descent (SGD) to update the network parameters.

This action reduces the loss and aids the network's performance. For numerous iterations or epochs, repeat steps 3–6 to gradually raise the network's performance on the training set. Evaluate the trained network on the validation set on a regular basis to keep track of its performance and prevent over fitting. As necessary, modify the network architecture and hyper parameters. In order to determine how well the trained network detects diabetic retinopathy in retinal fundus images, evaluate it on the testing set.

4 Methodology

Messidor Dataset

For the Messidor dataset [26, 27, 28], 1200 fundus scans were collected in colour using the Topcon TRC NW6 non-mydriatic retinograph, as given in Table 1. Three ophthalmologic stations were employed to collect data using this instrument. Eight bits per colour plane were used to capture the photos at resolutions of 1440 by 960, 2240 by 1488, or 2304 by 1536. The dataset is first normalised using the min–max normalisation approach, and the best feature is then chosen using the principal cuckoo search algorithm, a dimensionality reduction technique. The Op-BPDRLN algorithm-trained artificial neural network received this condensed dataset and produced classification results with increased accuracy. The dataset was divided into four distinct categories based on the observed characteristics: normal, stage 1, stage 2, and stage 3. The normal category included photographs of healthy individuals, while the stage 1 category comprised images displaying microaneurysms. Images exhibiting both microaneurysms and haemorrhages were classified under the stage 2 category, and the stage 3 category encompassed images with severe microaneurysms and haemorrhages. The researchers employed data augmentation techniques to generate an additional collection of 2000 photos for the Messidor dataset.

Extracted characteristics are used as input samples in the proposed dataset. The values of the generated attributes were taken from the various diabetic individuals' retinal pixels. Figure 2 depicts both the healthy retina and the unhealthy retina with indications of diabetic retinopathy.

It is simple to determine whether the patient has diabetic retinopathy symptoms using these retrieved features. The diabetic retinopathy dataset has a total of 1151 input samples. Each sample comprises 20 columns, of which 19 are used for input attributes and 1 is used for the target class. In the target class, the value is expressed

Table 1 Details about the Messidor dataset

DR stages	Details	Number	Label
No DR	Zero abnormalities	548	Normal
Mild	Microaneurysms	152	Stage 1
Moderate	Few microaneurysms	246	Stage 2
Severe	Venorous beading + intraretinal microvascular abnormality		
PDR	Vitreous/pre-retinal hemorrhage	254	Stage 3

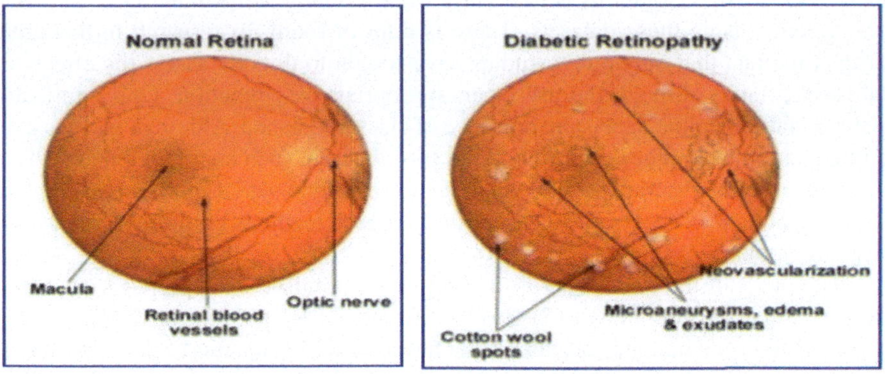

Fig. 2 Presents an illustrative depiction of the complete retina. The fundus image on the right exhibits symptoms of diabetic retinopathy, whereas the fundus image on the left depicts a retina in a normal state

in binary form, with 1 denoting a patient with diabetic retinopathy disease and 0 denoting a patient without this condition.

The subsequent configuration pertains to the computer system utilised for the implementation of the algorithms. The system is equipped with 4 gigabytes of random access memory (RAM) and powered by an Intel Core I5-3210 M processor operating at a clock speed of 2.50 gigahertz. The software employed for implementation was MATLAB R2010b. The dataset of flower samples for diabetic retinopathy comprises a total of 1151 data samples. Out of the total dataset, 806 data samples are allocated for training the model, and the remaining 345 data samples are designated for testing the model. The dataset has a total of 611 samples from the DR-no-sign class and 540 samples from the DR-sign class.

A well-organised neural network was employed to apply the pre-existing technique for training the diabetic retinopathy dataset. The input layer has 19 nodes, the hidden layer has 5, and the output layer has 1. Both the learning rate () and the momentum constant () are at their default values of 1e4. The aforementioned model was trained with BPNDRLN methods for 1250 iterations. For the dataset with less features, the same architecture was used. As a result, 14 nodes now make up the input layer, 5 nodes are in the hidden layer, and 1 node represents the output layer in the diabetic retinopathy training dataset. Both the learning rate and the momentum constant have been fixed at 0.3, and will remain at that value during the reorganisation.

5 Result and Discussion

The most frequently used datasets for training and validating algorithms are those from Messidor-2. However, taking into account the effects of noise and uncertainty as well as possible solutions is necessary when training and validating DL models and systems using these datasets. There is a lot of room for research in this area, despite the fact that only a few studies have sought to directly assess the effects of label uncertainty through the use of an appropriate loss function. One significant issue pertaining to the data is the presence of class imbalance. Although various data augmentation techniques exist to address this concern, such as rotating and scaling retinal images to generate new instances featuring blood vessels and lesions, further experimentation is necessary to ascertain that these methods do not exacerbate the existing imbalance (Fig. 3).

The results of the evaluation of the proposed optimised back propagation-based deep residual learning network algorithm (Op-BPDRLN) are displayed in Table 2 through Table 6. It has been discussed that the current ELM method uses 1250 epochs for training and achieves 92.28% accuracy based on the results shown in the tables.

In contrast, the suggested (Op-BPDRLN) algorithm uses just 1098 epochs for training and achieves 97.0% accuracy, outperforming the current ELM. The proposed (Op-BPDRLN) algorithm requires 47.6683 training time, whereas the existing ELM algorithm requires 65.7213. The proposed (Op-BPDRLN) method's training time is 27% shorter than that of the existing algorithm when compared with the existing ELM. The proposed (Op-BPDRLN) algorithm's precision is 0.9615, while the

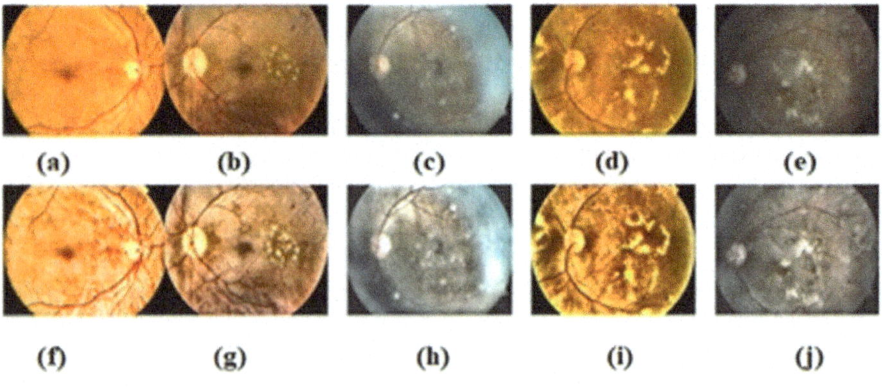

Fig. 3 Five levels of preprocessing of fundus images. **a** to **e**—Without preprocessing, **f** to **j**—With preprocessing. **a** and **f**—No DR, **b** and **g**—Mild, **c** and **h**—Moderate, **d** and **i**—Severe, **e** and **j**—PDR

Table 2 Recall value for proposed model compared with different models in various DR levels

DR level	Recall				
	ELM	SVM	GNB	RF	Op-BPDRLN (proposed)
No DR	0.97	0.91	0.78	0.90	0.98
Mild	0.70	0.24	0.33	0.24	0.82
Moderate	0.78	0.39	0.52	0.40	0.90
Severe	0.84	0.30	0.35	0.30	0.95
PDR	0.89	0.37	0.39	0.37	0.97
Average	0.83	0.44	0.48	0.44	0.93

existing ELM algorithm's precision is 0.9136. However, the proposed (Op-BPDRLN) algorithm achieves a recall of 0.9355, while the existing ELM approach achieves a recall of 0.8321.

On both Messidor datasets, a tailored VGG19 model and downsampling technique for DR detection were created to increase recognition accuracy. The proposed approach was contrasted using results from a number of studies, including those from extreme learning machine (ELM), support vector machine (SVM), random forests (RF), and Gaussian Naive Bayes. (GNB) (Figs. 4, 5, 6, 7 and 8).

The efficacy of the suggested methodology for detecting diabetic retinopathy (DR) in fundus images has been investigated through the utilisation of the Messidor datasets. The identification results of the dataset, in comparison with several existing models, are presented in Tables 2, 3, 4, 5, and 6. The aforementioned findings provide

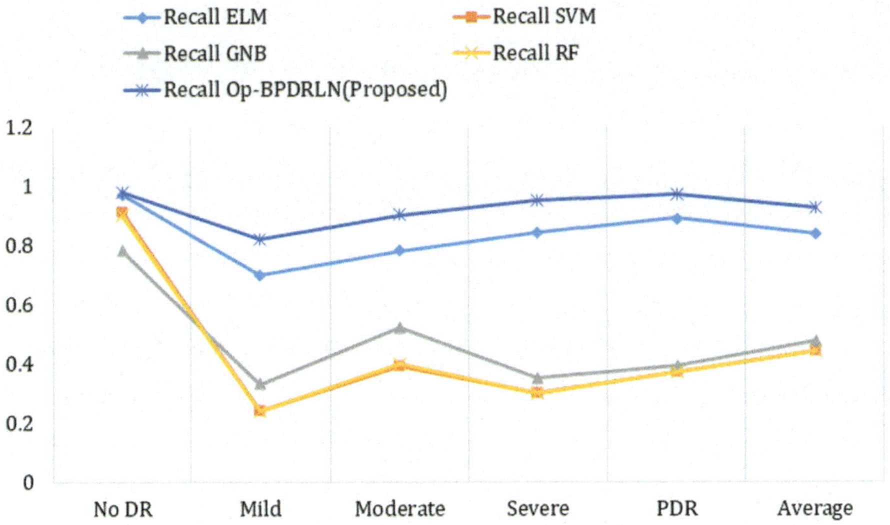

Fig. 4 Graphical representation of recall value for proposed model compared with different models in various DR levels

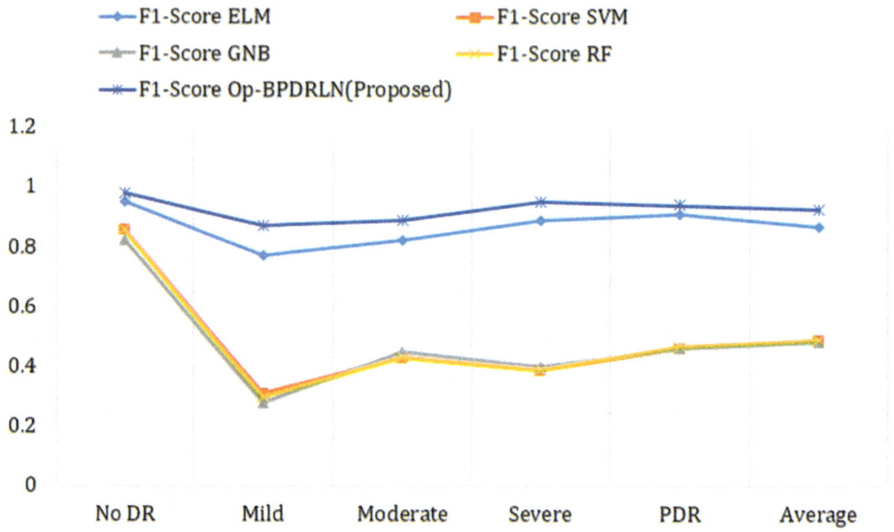

Fig. 5 Graphical representation of $F1$-score value for proposed model compared with different models in various DR levels

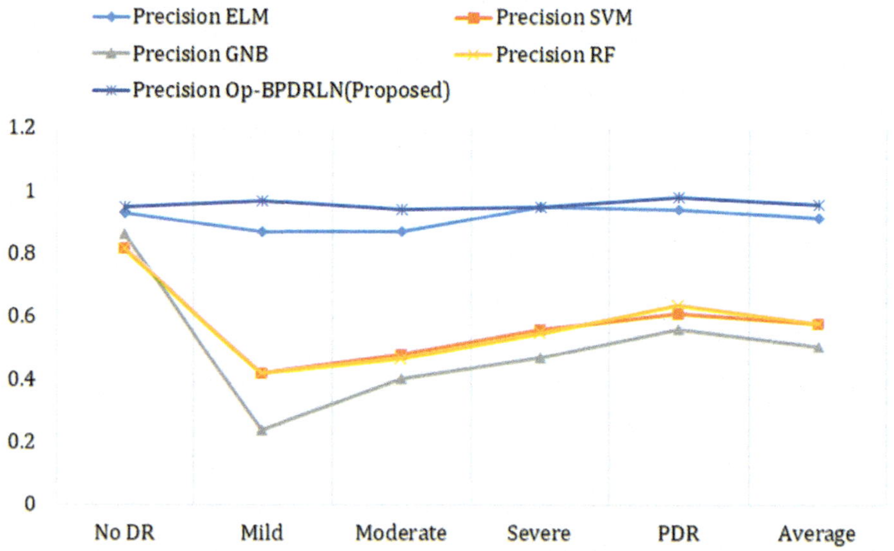

Fig. 6 Graphical representation of precision value for proposed model compared with different models in various DR levels

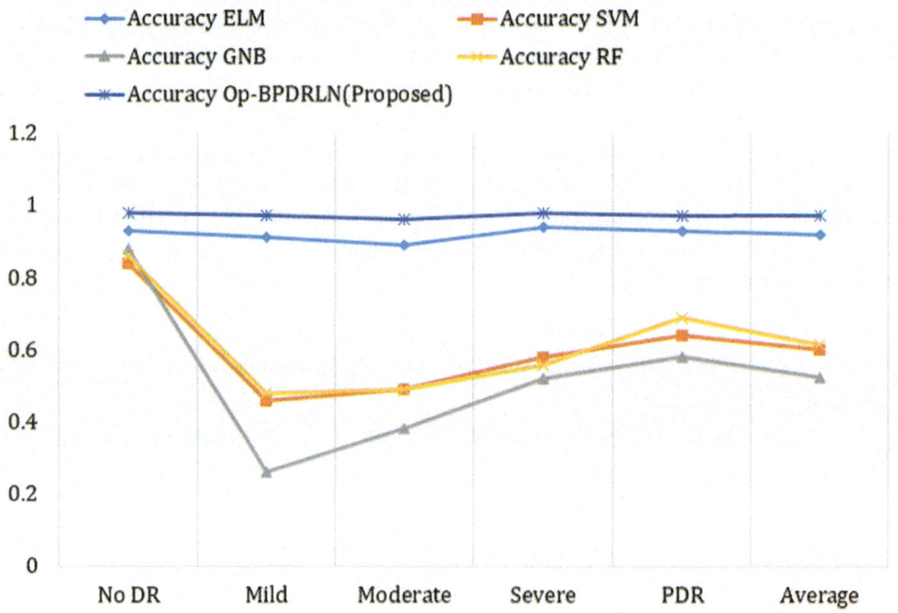

Fig. 7 Graphical representation of accuracy value for proposed model compared with different models in various DR levels

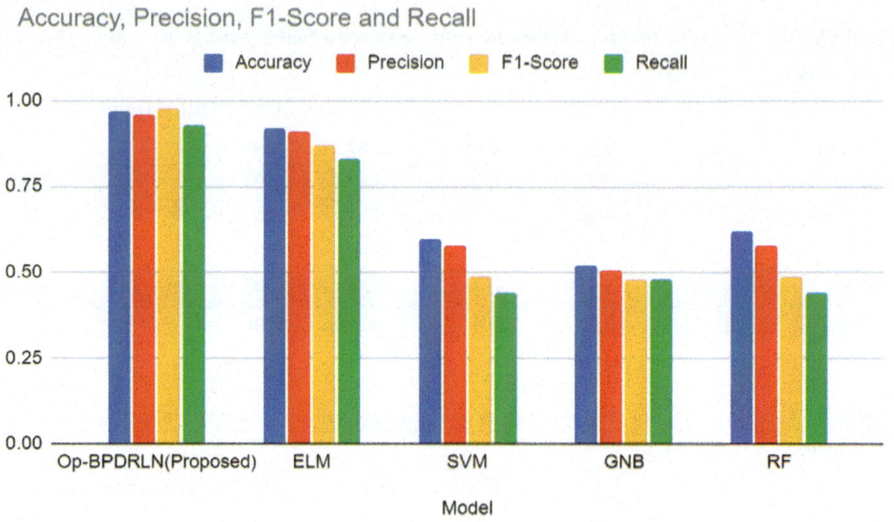

Fig. 8 Graphical representation of performance measure for proposed model compared with different models

evidence of the recommended model's ability to effectively distinguish between fundus images with diabetic retinopathy (DR) and those without. It is noteworthy that our proposed methodology demonstrates enhanced generalisation capability when applied to weighted (Op-BPDRLN) fundus images, despite a marginal increment in computational time amounting to 24.5 min. The proposed technique achieved an accuracy level of 97.0%, precision of 96.0%, $F1$-score of 0.98%, and recall of 93.0% when evaluated using the Messidor dataset. Despite the fact that this study demonstrated a significant level of accuracy in diagnosing DR, it was accompanied by a number of limitations.

Table 3 $F1$-score value for proposed model compared with different models in various DR levels

DR level	$F1$-score				
	ELM	SVM	GNB	RF	Op-BPDRLN (proposed)
No DR	0.95	0.86	0.82	0.86	0.98
Mild	0.77	0.31	0.28	0.30	0.87
Moderate	0.82	0.43	0.45	0.43	0.89
Severe	0.89	0.39	0.40	0.39	0.95
PDR	0.91	0.46	0.46	0.47	0.94
Average	0.87	0.49	0.48	0.49	0.93

Table 4 Precision value for proposed model with compared different models in various DR levels

DR level	Precision				
	ELM	SVM	GNB	RF	Op-BPDRLN (proposed)
No DR	0.93	0.82	0.86	0.82	0.95
Mild	0.87	0.42	0.24	0.42	0.97
Moderate	0.87	0.48	0.40	0.47	0.94
Severe	0.95	0.56	0.47	0.55	0.95
PDR	0.94	0.61	0.56	0.64	0.98
Average	0.91	0.58	0.51	0.58	0.96

Table 5 Accuracy value for proposed model compared with different models in various DR levels

DR level	Accuracy				
	ELM	SVM	GNB	RF	Op-BPDRLN (proposed)
No DR	0.93	0.84	0.88	0.86	0.98
Mild	0.91	0.46	0.26	0.48	0.97
Moderate	0.89	0.49	0.38	0.49	0.96
Severe	0.94	0.58	0.52	0.56	0.98
PDR	0.93	0.64	0.58	0.69	0.97
Average	0.92	0.60	0.52	0.62	0.97

Table 6 Performance measure for proposed model compared with different models

Model	Accuracy	Precision	$F1$-score	Recall
Op-BPDRLN (proposed)	0.97	0.96	0.98	0.93
ELM	0.92	0.91	0.87	0.83
SVM	0.6	0.58	0.49	0.44
GNB	0.52	0.51	0.48	0.48
RF	0.62	0.58	0.49	0.44

Although the recommended approach demonstrates high classification accuracy in the Messidor DR dataset, it is important to note that its performance may vary when applied to other medical datasets. The variations observed in the images across different datasets might be attributed to factors such as alterations in the labelling process, the presence of noise, and other contributing elements. In order to address this concern, it is advisable to instruct the AI by utilising a diverse collection of photographs captured at different temporal and geographical contexts. The importance of both the distribution of data classes and the diversity of the data cannot be overstated. Unjustifiable disparities in class sizes exert a detrimental influence on the efficacy of instructional practices. The application of various data augmentation techniques to address the issue of imbalance in the dataset also had an impact on the precision of categorisation.

Another issue is that the weight calculation technique takes longer to process than other models, while improving classification accuracy. Due to these restrictions, testing will be done in the future with a wider range of images and potentially new optimisation methods that take less time to process. Finally, research has shown that while deep learning models perform well under experimental, laboratory-level conditions, they perform worse when put to the test in a clinical setting. There have only been a few researches reported on the difficulties and effects of applying DL models in clinical settings.

The implementation of deep learning (DL) models in clinical settings and the validation of their performance during real-time clinical examinations present research opportunities that are closely tied to the importance of segmentation, prediction, and classification in decision support for DR analysis. The development and validation of a deep learning-based intelligent system in operational clinical settings, specifically in the context of retinal image analysis under diverse conditions, including poor focus/contrast, inadequate pupil dilation, presence of cataracts, and variations in patient demographics and systemic control, are imperative to enhance confidence and promote the widespread use of deep learning-based intelligent systems.

6 Conclusion

The present study utilises the DR (Messidor-2) dataset to conduct an investigation on the classification of patients with diabetic retinopathy disease. This classification task is accomplished by employing an optimised back propagation neural network (Op-BPDRLN) method. The dataset is initially subjected to min–max normalisation for standardisation purposes. Subsequently, the principal cuckoo search strategy, a dimensionality reduction technique, is employed to identify the optimal feature. The artificial neural network, trained using the (Op-BPDRLN) algorithm, was provided with a condensed dataset and yielded improved classification results in terms of accuracy. The proposed model outcome was assessed using both the existing and proposed (Op-BPDRLN) algorithm. The conclusion was obtained based on statistical measures such as accuracy, precision, and recall, which indicated the superiority of the model.

References

1. Manjula Devi, R., Kuppuswami, S., Suganthe, R.C.: Fast linear adaptive skipping training algorithm for training artificial neural network. Math. Probl. Eng. **2013** (2013). https://doi.org/10.1155/2013/346949
2. Devi, R.M., Kuppuswami, S.: EAST: an exponential adaptive skipping training algorithm for multilayer feed forward neural networks. WSEAS Trans. Comput.Comput. **13**, 138–151 (2014)
3. Vijayan, T., Sangeetha, M., Kumaravel, A., Karthik, B.: Gabor filter and machine learning based diabetic retinopathy analysis and detection. Microprocess. Microsyst. (8), 103353 (2020). https://doi.org/10.1016/j.micpro.2020.103353
4. Sahlsten, J., et al.: Deep learning fundus image analysis for diabetic retinopathy and macular edema grading. Sci. Rep. **9**(1), 1–11 (2019). https://doi.org/10.1038/s41598-019-47181-w
5. Gulshan, V., et al.: Development and validation of a deep learning algorithm for detection of diabetic retinopathy in retinal fundus photographs. JAMA **316**(22), 2402 (2016). https://doi.org/10.1001/jama.2016.17216
6. Devi, R.M., Keerthika, P., Sangeetha, M.M., Kavinkumar, M.: Detection of diabetic retinopathy by applying gabor filter with convolutional neural network. **29**(5), 5689–5696 (2020)
7. Hemanth, D.J., Deperlioglu, O., Kose, U.: An enhanced diabetic retinopathy detection and classification approach using deep convolutional neural network. Neural Comput. Appl.Comput. Appl. **32**(3), 707–721 (2020). https://doi.org/10.1007/s00521-018-03974-0
8. Manjula Devi, R., Kuppuswami, S.: CAST: a constant adaptive skipping training algorithm for improving the learning rate of multilayer feedforward neural networks. Res. J. Appl. Sci. Eng. Technol. **12**(8), 790–812 (2016). https://doi.org/10.19026/rjaset.12.2780
9. Ramasamy.: Half of threshold algorithm: an enhanced linear adaptive skipping training algorithm or multilayer feedforward neural networks. Am. J. Appl. Sci. **11**(1), 38–46 (2014). https://doi.org/10.3844/ajassp.2014.38.46
10. Devi, R.M., Keerthika, P., Suresh, P., Sangeetha, M.: A Z-score fuzzy exponential adaptive skipping training (Z-Feast) algorithm for efficient pattern classification. Asian J. Res. Soc. Sci. Humanit. **6**(11), 531 (2016). https://doi.org/10.5958/2249-7315.2016.01211.9
11. Devi, M., Prasad, S.J.S., S.C.: An efficient fuzzy neural network training model for supervised pattern classification system. J. Adv. Chem. **12**(11), 4488–4499 (2016). https://doi.org/10.24297/jac.v12i11.822

12. Kavitha, S., M.D.R.: Z-score based fuzzification process for pattern classification **6**(1), 77–81 (2015)
13. Mahmoud, M.H., Alamery, S., Fouad, H., Altinawi, A., Youssef, A.E.: An automatic detection system of diabetic retinopathy using a hybrid inductive machine learning algorithm. Pers. Ubiquit. Comput.Ubiquit. Comput. (2021). https://doi.org/10.1007/s00779-020-01519-8
14. Yang, X.-S.: Nature-inspired algorithms and applied optimization. Springer (2017)
15. Asuncion, D.J., Newman, A.: UCI machine learning repository. In: School of Information and Computer Science. University of California, Irvine, CA (2007)
16. Odeh, I., Alkasassbeh, M., Alauthman, M.: Diabetic retinopathy detection using ensemble machine learning. arXiv preprint arXiv:2106.12545 (2021)
17. Kharabe, S., Nalini, C.: Using adaptive thresholding extraction—robust ROI localization based finger vein authentication. J. Adv. Res. Dyn. Control Syst. **10**(7), 500–514 (2018)
18. Kharabe, S., Nalini, C.: Robust ROI localization based finger vein authentication using adaptive thresholding extraction with deep learning technique. J. Adv. Res. Dyn. Control Syst. **10**(7), 500–513 (2018)
19. Lakshmi Vara Prasad, G., Nalini, C., Sugumar, R.: Arbitrary routing algorithm for tenable data assortment accessed in wireless sensor networks. Int. J. Civil Eng. Technol. **8**(1), 961–966 (2017)
20. Devi, S.V.G., Nalini, C.: Enhanced K-means clustering algorithm for feasibility assessment of ACC. In: 2020 Proceedings of the 2nd International Conference on Inventive Research in Computing Applications, ICIRCA 2020, 9182934, pp. 340–345 (2020)
21. Gayetri Devi, S.V., Nalini, C.: Prioritized automated generation of contracts with modified swarm optimization. Int. J. Adv. Sci. Technol. **29**(8), 2432–2439 (2020)

Chapter 35
Use Data Augmentation Model in Image Segmentation

Le Cong Hieu, Pham The Bao, Dang Thi Phuc, and Nguyen Hong Dat

1 Introduction

In late December 2019, catastrophic worldwide coronavirus disease (Covid-19) was first observed in Wuhan, China. Covid-19 can be detected from respiratory samplers using real-time RT-PCR, but this is very time-consuming, taking about 4–6 h to process the sample. It also gives error-prone results, with a high false negative rate. Given these limitations, the detection of Covid-19 by RT-PCR creates challenges in preventing the spread of infection. We can deal with this problem by using CT of the chest because soft tissues can be distinguished precisely. Therefore, it is necessary to design a decision support tool based on deep learning to predict Covid-19 patients automatically quickly and efficiently. Deep convolutional neural networks have performed very well on many computer vision tasks. However, these networks rely heavily on big data to avoid overfitting, many application areas that do not have access to big data such as in medicine. Therefore, the solution of image enhancement is focused on and is one of the most concerned problems today.

L. C. Hieu (✉) · D. T. Phuc · N. H. Dat
Department of Computer Science, Faculty of Information Technology, Industrial University of Ho Chi Minh City, 12, Nguyen Van Bao, Ho Chi Minh City 7000, Vietnam
e-mail: 22733861.hieu@student.iuh.edu.vn

D. T. Phuc
e-mail: phucdt@iuh.edu.vn

N. H. Dat
e-mail: 19433071.dat@student.iuh.edu.vn

P. T. Bao
Faculty of Information Technology, Sai Gon University, Ho Chi Minh City, Vietnam
e-mail: ptbao@sgu.edu.vn

J. M. R. S. Tavares et al. (eds.), *Proceedings of Second International Conference on Intelligent System*, Algorithms for Intelligent Systems,
https://doi.org/10.1007/978-981-99-8976-8_36

Recently, image data augmentation methods have been effective when applied to medical problems. Of these, the most common approaches are random affine transformations [1], random elastic transformation [2, 3], random contrast transformation [4, 5]. Several recent works have trained a creative adversarial network (GAN) [6], using the existing labeled dataset to generate actual image label pairs. The idea has been applied to various analytical tasks [7–13], including image segmentation MR [14, 15]. MixUp [16] differs from the above approaches in that the data generated is impractical in nature. Here, additional data is obtained by linearly interpolating the available images and the corresponding labels. Despite its impractical nature, it seems to improve the performance of the neural network on the standard benchmark dataset [16] and also on medical image partitioning [17] in the restricted annotation setting. In the paper, we propose the Pix2Pix model and its enhancements for generating images from the Covid-19 CT image dataset.

2 Research Algorithms

2.1 *Model Pix2Pix*

The Pix2Pix [18] model can create a multitude of photo-related applications besides data augmentation, such as: Convert color photos to normal photos and vice versa, recreate the painting style of famous deceased artists, convert colors of photos, add color to strokes to create color photos, transform the sky dark to light, create segmented images from the original, or convert satellite images to Google Map.

2.1.1 Architecture of Pix2Pix

The architecture of Pix2Pix includes two models: generator and discriminator. Generator has the effect of creating fake images to deceive discriminator and discriminator has the goal of distinguishing between real and fake images (Fig. 1).

Generator: The input of the generator is a source image, as shown in the figure, which is a lung CT image mask. The image is then passed into a convolutional network architecture for feature extraction through down sampling. In this process, the output in the layers will be reduced in size through two-dimensional convolution. Meanwhile, the deconvolution network does the opposite task of transforming the feature into the target image through the upsampling process using specialized layers such as transposed convolution, dilation convolution, upsampling 2D. Upsampling will gradually increase the output size and the result will be a fake-labeled target image. Discriminator: The input to discriminator is a pair of images x, y, where x is the input source image and y is the target image generated from the generator or taken from the train set. Two images x and y have the same size and will be assembled together before being transferred to the discriminator model. The pair of images will

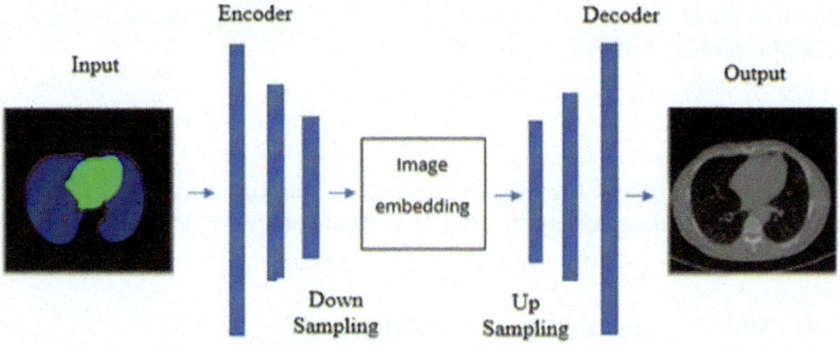

Fig. 1 Architecture of generator in Pix2Pix

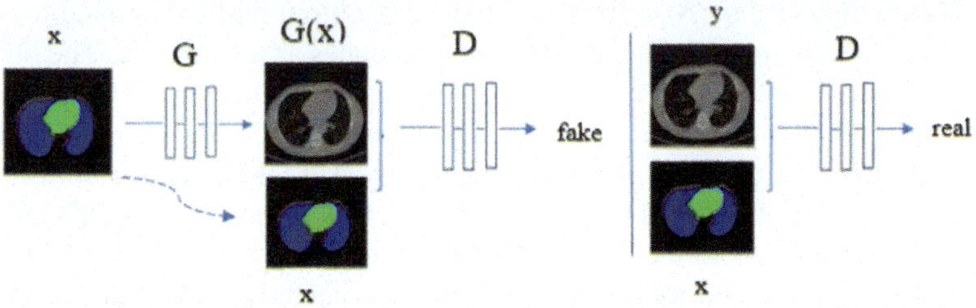

Fig. 2 Architecture of Pix2Pix

be labeled as real if the image y is taken from the train set, and otherwise, when y is generated from the generator, the pair of images x, y will have a fake label (Fig. 2).

2.1.2 Loss Function

The loss function of Pix2Pix is a binary cross-entropy function of the form: LcGAN(G, D) = Ex, y[log D(x, y)] + Ex, z[log(1 − D(x, G(x, z)))] Ex, y[log D(x, y)] : log—probability that D predict x, y is real.

Ex, z[log(1 − D(x, G(x, z)))] : log—probability that D predict G(x, z) is fake. It is exactly the same as the loss function of LIVER, and the only difference is that the input is a pair of images x, y instead of just x. Simultaneous training is done on both generator and discriminator. Discriminator will try to classify real and fake images most accurately, so its loss function value is the cross-entropy function which must be as small as possible. Therefore, we need to find: maxD LcGAN(G, D). Besides, in case the image is fake, we need the generator to generate it most similar to the real thing, that is, the probability value D(x, G(x, z)) must be as close to 1 as possible. Therefore, we need to find minG LcGAN(G, D). Combining both optimization problems between generator and discriminator, we get the optimal problem: minG maxD LcGAN(G, D). To test the importance of the condition in the classification, we get an unconditioned version by removing x from the discriminator model. Then the loss function of GAN will become:

$$\text{LGAN}(G, D) = Ey\big[\log D(y)\big] + Exz\big[\log(1 - D(G(x, z)))\big].$$

$Ey[\log D(x, y)]$: log of the probability that D identifies y as real $Ex, z[\log(1 - D(G(x, z)))]$: log of the probability that D identifies $G(x, z)$ as false.

2.1.3 Pix2Pix Architecture Based on PatchGAN

PatchGAN is an architecture that has discriminator based on cognitive fields. It will map each pixel in the output from a square area above the input (aka patch). A PatchGAN with a size of 70 x 70 means that each output will be mapped to a patch size of 70 x 70. The model will classify 70 x 70 patches as real or fake. The output of PatchGAN will be a square feature map consisting of a set of probabilities, each of which corresponds to the probability that a patch of size 70 x 70 will fall into real or fake. The probability that the entire input image is real or fake can be calculated as the average of all the probabilities over the entire patches. In essence, PatchGAN is still a CNN network architecture consisting of many consecutive CNN layers, but we do not do flatten near the end to pass through fully connected layers. Instead, calculate the probabilistic feature map on each patch as outlined above. Such a calculation will be effective if applied on large patches because of the larger receptive field. The average probability results of multiple patches will also be more accurate (Fig. 3).

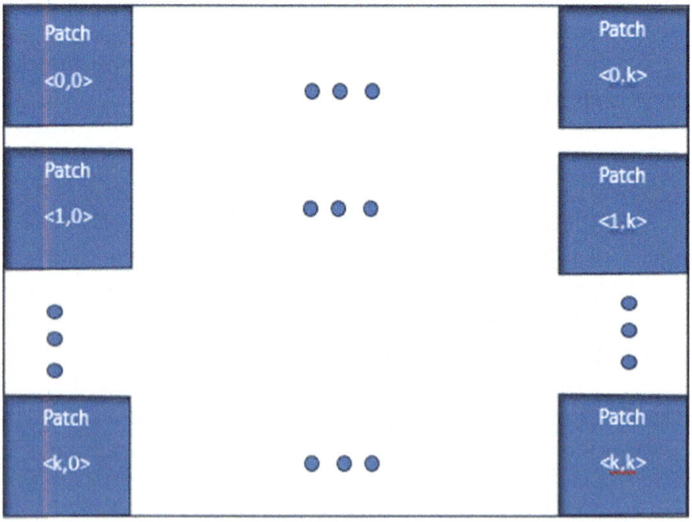

Fig. 3 Image is divided into $k \times k$ patches

On each patch, we predict a probability that the image will fall into the truth. The probability that the entire image is real will be the average of the probabilities of all patches. In this method we will not flatten the model and work on fully connected layers. The architecture of PatchGAN's layers also includes four CNN layers in the order: I–C1–C2–C3–C4–O. Based on the output size of a layer, we can determine the size, input shape of the layer immediately before it. Indeed, according to the formula to calculate the output size of the two-dimensional convolution according to the input Winput size, filter size F, stride S, and padding P, we have:

$$\text{Woutput} = \left[(\text{Winput} - F + 2P)/S\right] + 1.$$

Winpit is calculated based on the previous Woutput:

According to the above formula, deduce the size of each layer. The order is as follows: O–C4–C3–C2–C1–I.

At output layer O: Woutput= 1, S = 1, P = 4, derive the size of C4: Winput= (1 − 1) x 1 − 2 x 0 + 4 = 4.
At the output layer C4: Woutput= 1, S = 1, P = 4, derive the size of C3: Winput= (4 − 1) x 1 − 2 x 0 + 4 = 7.
At the output layer C3: Woutput= 1, S = 1, P = 4, derive the size of C2: Winput= (7 − 1) x 2 − 2 x 0 + 4 = 16.
At the output layer C2: Woutput= 1, S = 1, P = 4, derive the size of C1: Winput= (16 − 1) x 2 − 2 x 0 + 4 = 34.
At the output layer C1: Woutput= 1, S = 1, P = 4, derive the size of I: Winput= (34 − 1) x 2 − 2 x 0 + 4 = 70.

The advantage of Pix2Pix is that it is extremely suitable for segmentation problems because it belongs to cGAN and the loss function of the Pix2Pix model gives better output than other models. The weakness is that because the Pix2Pix model is simple, complex medical data will produce low-quality images. That is why we propose the model Pix2Pix Attention U-Net generator to enhance lung CT images to solve the Covid-19 classification problem (Fig. 4).

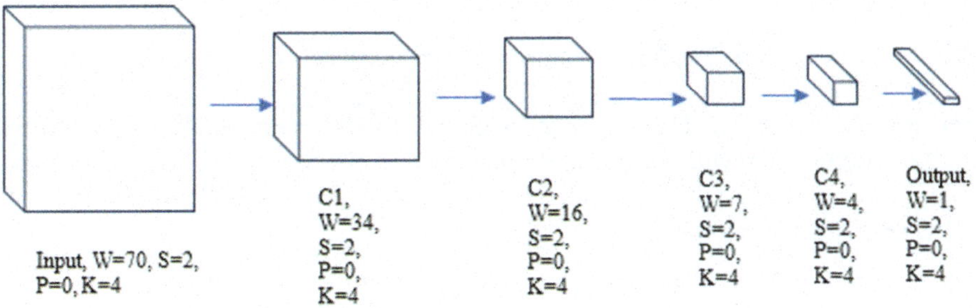

Fig. 4 General diagram of the architecture of CNN layers applied on a patch 70 × 70 of PatchGAN

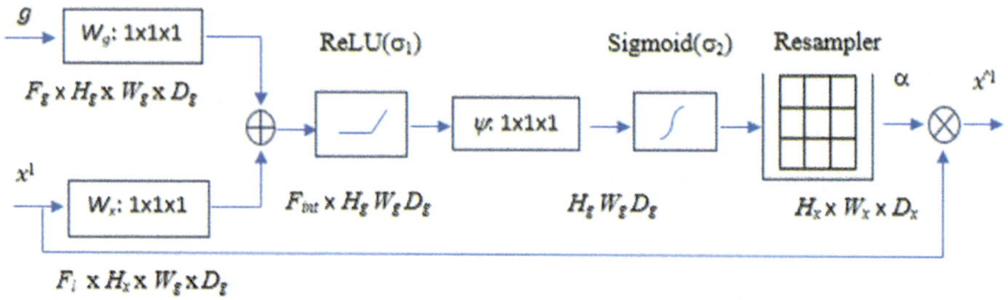

Fig. 5 Attention Gates' diagram

2.2 *Model Pix2Pix Attention U-Net Generator*

2.2.1 Attention Gates (AGs)

AG [19] is commonly used in natural image analysis, knowledge graphing, and language processing (NLP) for image annotation, machine translation, and classification (Fig. 5).

The input feature (xl) is scaled with the attention coefficients () calculated in the AG. Spatial regions are selected by analysis, activation, and contextual information provided by the gate signal (g) which is collected from coarser scale. The resampling grid of the attention coefficients was performed by three-linear interpolation.

2.2.2 Attention U-Net

The input image is progressively filtered and down sampled by a factor of 2 at each scale in the coding portion of the network (e.g., H4 = H1/8). Nc represents the number of classes. Attention Gate (AG) filters the features passed over connections. Feature in AG is achieved using contextual information (gating) extracted at coarser scales (Fig. 6).

2.2.3 Architecture Pix2Pix Attention U-Net Generator

The Pix2Pix Attention U-Net generator architecture is completely similar to the Pix2Pix architecture with the generator using the U-Net model and the discriminator using the PacthGAN, but in the generator model there is an additional Attention Gate at the skip connections (Fig. 7).

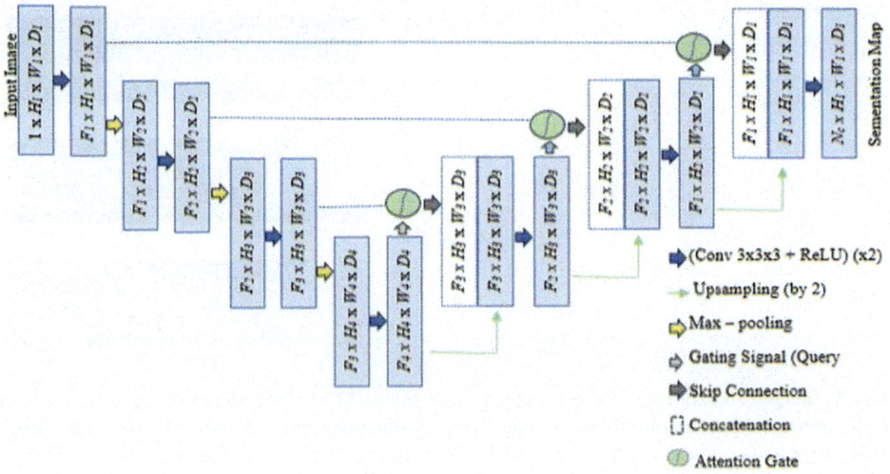

Fig. 6 Block diagram of U-Net model with Attention Gates

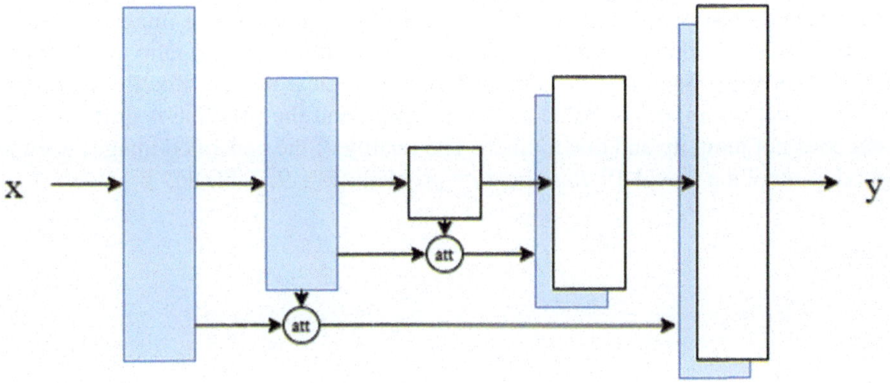

Fig. 7 Architecture Pix2Pix attention U-Net generator

2.3 Results and Discussion

Through the process of experimenting with the Pix2Pix and Pix2Pix Attention U-Net generator models, the results are as follows (Fig. 8):

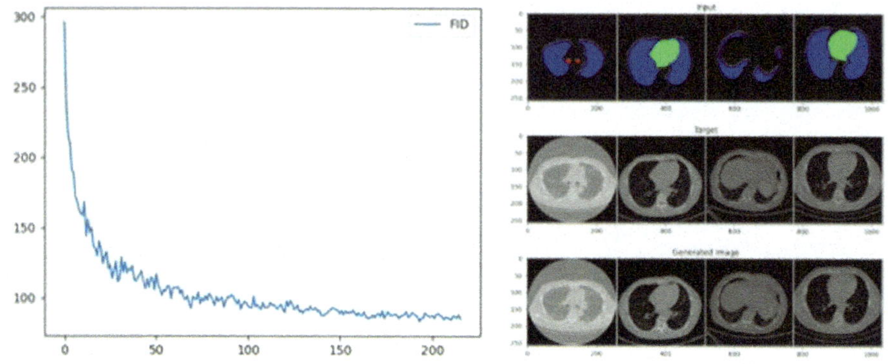

Fig. 8 Image generation results and image quality index (FID) of the model Pix2pix In the above result, input is the mask of lung's CT image, target is the image corresponding to the mask, generated image is the image generated by the Pix2pix model with an image quality index (FID) of: 83.94343628014835

In the above result, the input is the mask of the lung's CT image, the target is the image corresponding to the mask, the generated image is the image generated by the Pix2Pix Attention U-Net generator model with the image quality index (FID) of: 103.15617916362004. Through the graph, it is clear that the Pix2Pix Attention U-Net generator model is stable and convergent, and the FID line is quite smooth and does not jump up and down much. The quality of the generated images is quite good to serve the Covid-19 classification problem (Fig. 9; Table 1).

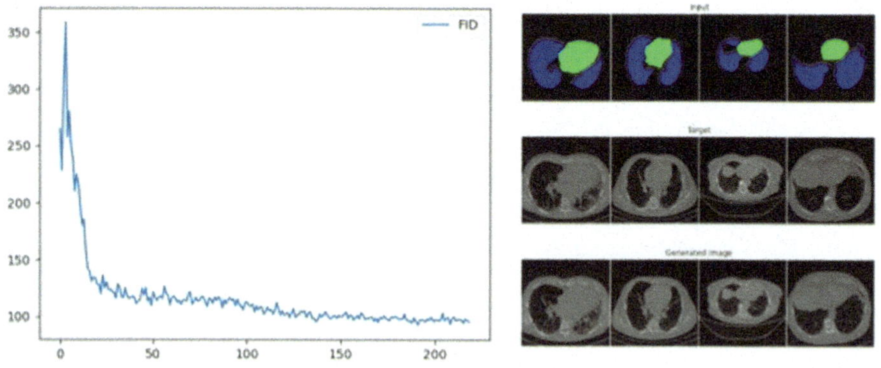

Fig. 9 Image generation results and image quality index (FID) of the model Pix2Pix Attention U-Net generator

Table 1 Comparison table about the model's efficiency

Model	Loss function	FID	Epoch/second	Weight size
Architecture Pix2Pix Attention U-Net generator	6.673844337463379	103.15617916362004	253	The best weight size of generator is 215 MB, and the best weight size of discriminator is 10.6 MB
Pix2Pix	3.4609413146972656	83.94343628014835	247	The best weight size of generator is 207.6 MB, and the best weight size of discriminator 10.6 MB

3 Conclusion

In this paper, we have presented the Pix2Pix model and at the same time proposed a new approach by adding an Attention Gate in the U-Net generator model of the Pix2Pix model. The results show that the CT lung images generated from our proposed model are reliable to use for the Covid-19 classification problem. Our approach eliminates the need to apply an external object localization model. This proposed approach is generic and modular, so it can be easily applied to image classification and regression problems. Experimental results demonstrate that the proposed AGs are very beneficial for tissue/organ identification and localization. This is especially true for organs of small size in classification tasks.

References

1. Ciresan, D.C., Meier, U., Masci, J., Gambardella, L.M., Schmidhuber, J.: High-performance Neural Networks for Visual Object Classification (2011). arXiv:1102.0183
2. Simard, P.Y., Steinkraus, D., Platt, J.C.: Best practices for Convolutional Neural Networks Applied to Visual Document Analysis, IEEE, p. 958 (2003)
3. Ronneberger, O., Fischer, P., Brox, T.: U-Net: Convolutional networks for biomedical image segmentation. In: International Conference on Medical Image Computing and Computer-Assisted Intervention, pp. 234–241. Springer, Heidelberg (2015)
4. Hong, J., Park, B.-y., Park, H.: Convolutional neural network classifier for distinguishing Barrett's esophagus and neoplasia endomicroscopy images. In: 2017 39th Annual International Conference of the Engineering in Medicine and Biology Society (EMBC), IEEE, pp. 2892–2895 (2017)
5. Perez, F., Vasconcelos, C., Avila, S., Valle, E.: Data Augmentation for Skin Lesion Analysis, pp. 303–311. Springer, Heidelberg (2018)

6. Goodfellow, I., Pouget-Abadie, J., Mirza, M., Xu, B., Warde-Farley, D., Ozar, S., Courville, A., Bengio, Y.: Generative adversarial nets. In: Advances in Neural Information Processing systems, pp. 2672–2680 (2014)
7. Salehinejad, H., Valaee, S., Dowdell, T., Colak, E., Barfett, J.: Generalization of deep neural networks for chest pathology classification in x-rays using generative adversarial networks. In: 2018 IEEE International Conference on Acoustics, Speech and Signal Processing (ICASSP), pp. 990–994, IEEE (2018)
8. Frid-Adar, M., Klang, E., Amitai, M., Goldberger, J., Greenspan, H.: Synthetic data augmentation using GAN for improved liver lesion classification. In: 2018 IEEE 15th International Symposium on Biomedical Imaging (ISBI 2018), pp. 289–293, IEEE (2018)
9. Guibas, J.T., Virdi, T.S., Li, P.S.: Synthetic Medical Images from Dual Generative Adversarial Networks (2017). arXiv:1709.1872
10. Hou, L., Agarwal, A., Samaras, D., Kurc, T.M., Gupta, R.R., Saltz, J.H.: Unsupervised Histopathology Image Synthesis (2017). arXiv:1712.05021
11. Wu, E., Wu, K., Cox, D., Lotter, W.: Conditional infilling GANs for data aug-mentation in mammogram classification. In: Image Analysis for Moving Organ, Breast, and Thoracic Images, pp. 98–106. Springer, Heidelberg (2018)
12. Liu, S., Zhang, J., Chen, Y., Liu, Y., Qin, Z., Wan, T.: Pixel Level Data augmentation for Semantic Image Segmentation Using Generative Adversarial Networks (2018). arXiv:1811.00174
13. Sixt, L., Wild, B., Landgraf, T.: RenderGAN: generating realistic labelled data. Front. Rob. AI **5**, 66 (2018)
14. Costa, P., Galdran, A., Meyer, M.I.N.D., Niemeijer, M., Abramoff, M., Mendonca, A.M., Campilho, A.: End-to-end adversarial retinal image synthesis. IEEE Trans. Med. Imaging **37**(3), 781–791 (2018)
15. Bowles, C., Chen, L., Guerrero, R., Bentley, P., Gunn, R., Hammers, A., Dickie, D.A., Hern´andez, M.V., Wardlaw, J., Rueckert, D.: GAN Augmentation: Augmenting Training Data Using Generative Adversarial Networks (2018). arXiv:1810.10863
16. Zhang, H., Cisse, M., Dauphin, Y.N., Lopez-Paz, D.: Mixup: Beyond Empirical Risk Minimization (2017). arXiv:1710.09412
17. Eaton-Rosen, Z., Bragman, F., Ourselin, S., Cardoso, M.J.: Improving data augmentation for medical image segmentation. In: International Conference on Medical Imaging with Deep Learning (2018)
18. Isola, P., Zhu, J.-Y., Zhou, T., Efros, A.A.: Image-to-image translation with conditional adversarial networks. In: Proceedings of the IEEE Conference on Computer Vision and Pattern Recognition (2018)
19. Oktay, O., Schlemper, J., Folgoc, L.L., Lee, M., Heinrich, M., Misawa, K., Mori, K., McDonagh, S., Hammerla, N.Y., Kainz, B., Glocker, B., Rueckert, D.: Attention U-Net: Learning Where to Look for the Pancreas, MIDL'18 (2018)

Chapter 36
A Prototype for Lung Cancer Forecasting Using Convolution Neural Network Method

Prasanalakshmi Balaji, Bui Thanh Hung, and Linda Elzubir Gasm Alsid

1 Introduction

Lung cancer is identified to record top deaths of around 1.7 million worldwide in 2018. The use of the low-dose CT technique (LDCT), which is more effective but also more expensive than chest X-rays (CXR) for routine screening, is being studied [1].

Although the practical radiation dose for chest radiograph is much lower than that for CT, it is more challenging to find chest nodules with CXR than with CT [2, 3]. The frequency of pulmonary metastases in patients who died from extra thoracic malignancy (ETM) reports to be greater than 19% [4]. Deep learning (DL), which has several applications using artificial intelligence to represent features obtained from data, has experienced incredible growth in recent years. Backpropagation algorithms are used to identify exact features from complex datasets. Traditional approaches based on machine learning have difficulties in analyzing natural data in its original form without preprocessing [5]. Additionally, CAD systems can help radiologists with CXR diagnosis [6, 7], and they may also increase the sensitivity of CXR-based lung cancer screening. DL-based CAD systems are gaining popularity due to improved processing power. Numerous techniques for computerized X-ray picture

P. Balaji (✉) · L. E. G. Alsid
College of Computer Science, King Khalid University, Abha, Saudi Arabia
e-mail: drsana@ieee.org

L. E. G. Alsid
e-mail: lynda@kku.edu.sa

B. T. Hung
Data Science Laboratory, Faculty of Information Technology, Industrial University of Ho Chi Minh City, Ho Chi Minh City, Vietnam
e-mail: buithanhhung@iuh.edu.vn

J. M. R. S. Tavares et al. (eds.), *Proceedings of Second International Conference on Intelligent System*, Algorithms for Intelligent Systems,
https://doi.org/10.1007/978-981-99-8976-8_37

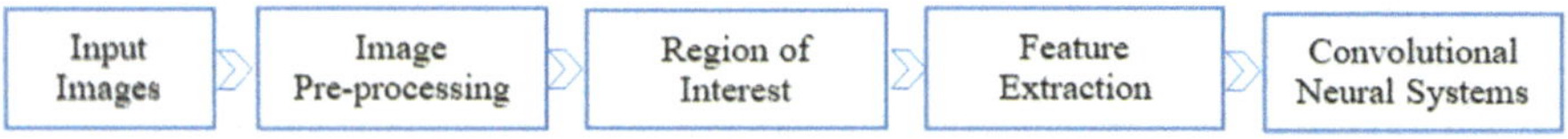

Fig. 1 Schematic process of Convolution Neural Network (CNN)

categorization have also been described, which may assist radiologists in diagnostic practice [8] (Fig. 1).

Convolutional Neural Networks (CNNs) have the ability in learning invariant features [9]. Object detection, segmentation, and classification are some tasks that CNN performs. It involves layers like pooling, normalization, dropout, and dense layers to create patterns for diverse object identification tasks like detection, segmentation, and classification. The inputs provided as batches are subjected to changes over time as the network is trained in multiple layers and levels of networks [10].

There is a wide range of alternative CNN configuration with respect to layers, including some that includes smaller networks, such as the Densely Connected Convolutional Networks (DenseNet) architecture to offer benefits to enhance the multiple layer architectures that were being developed, and this design was created. To resolve this issue where many layers in high-layer networks were essentially made redundant [11]. The research [12] examines future robust networks, including downlink multi-user networks (MU-N) and physical layer security (PHY-S), as well as several 6G concept discussions. It is crucial to employ a future robust network (FRN), such as sixth-generation wireless (6G) communication, because it can offer enormous capacity. Residual neural network, Xception model, and GoogLeNet designs are additional varieties of CNN architectures widely used in recent years. These networks are required to facilitate multiscale processing, and performance degrades as networks become more profound or networks have fewer parameters [13]. Deep learning maintains data across long architecture, which is again a challenge. The suggested potential remedy is Long Short-Term Memory (LSTM). The LSTM design enforces coherent inaccuracy propagation that is Non-local Recurrent Neural Memory in both time and space through the states of specialized units [14].

Thus, various Cyber-Physical and Computer-Aided Identification (CAD) technologies are available for the early diagnosis of lung-related illnesses [15]. In addition to benign and malignant, there are other forms of lung cancer, like non-small cell lung cancer (NSCLC), which is more prevalent and the least commonly known carcinoid. Modern technology is required to capture the cells clinically and distinguish between small and non-small cells. Computer tomography (CT), a widely utilized technology, cannot accomplish this. Deep learning is the most effective method for clinical imaging at the current time, emphasizing the retrieval and classification of objects.

We could teach computers what different objects or things look like and have them recognize them completely, thanks to deep learning [16]. Figure 2 displays a prototype of deep learning. This study uses a standard dataset to evaluate a deep Convolutional Neural Network architecture used in a comprehensive screening for early-stage lung cancer diagnosis (Fig. 3).

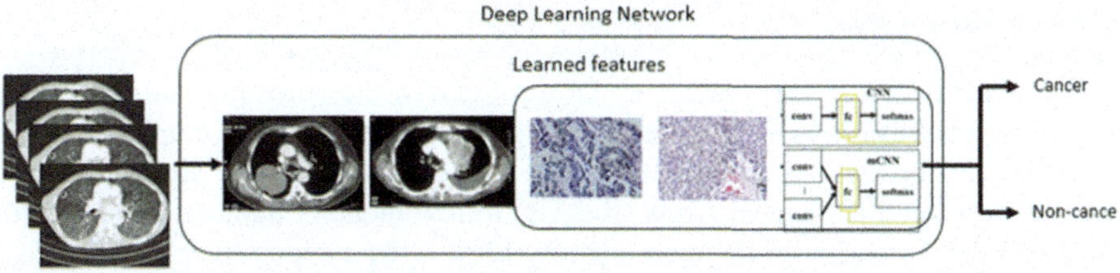

Fig. 2 General framework of identification and classification of lung cancer

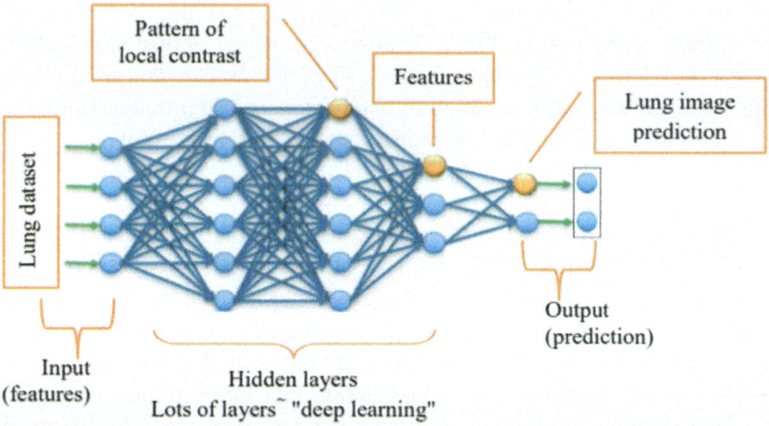

Fig. 3 Suggested prototype of deep learning

2 Materials and Methodology

The simplest and most effective strategy to update the patient for continuation is to deal with any aspiratory hazard. Accurate aspiratory handle disclosure in CT images is an essential step in identifying pneumonic risk. Introduce Faster Region-Based Convolutional Neural Networks (RCNN) to the convolutional structure immediately for future confirmation on fundamental cuts. For the ensuing lie positive abatement, a three-dimensional DCNN is then offered. The outline of the prototype is as follows:

- *Input*

Visual Geometry Group (VGG) captures the 220 × 220 picture. The user must construct the CNN algorithm and import the necessary libraries to train the images.

- *Convolution layer*

In the advanced Visual Geometry Group (VGG), the convolutional layers use the minor receptive field (7 × 7), even though everything can move up, down, and to the left or the right. Additionally, 1 × 1 convolution channels are available to operate, and Rectified Linear Units (ReLUs) follow them. A number is created for each uncommon region in the data quantity [20]. After moving the conduit across each territory, you will be left with a (26 × 26 × 0.5) display of numbers, sometimes known as a triggering map or feature map. A 5 × 5 channel can fit in 784 specified spots on a (32 × 32) information image, according to the explanation that users obtain from a (28 × 28) monitor. These 784 figures are assigned to the cluster of (26 × 26). Consider using two (5 × 5 × 3) channels instead of one. Our yield volume would be (26 × 26 × 2) by that time. It can better protect spatial measurements by using more channels. These channels can be considered trait descriptors because they discuss features like bends, geometric shapes, and plain colors.

- *Fully-connected layers*

Visual Geometry Group (VGG) has 16 strata. The first two have a total of 6095 networks, while the third has 1750 networks for each section. The ability to defeat anything adds a connected layer to the system because it can recognize these significant level highlights. This wholly integrated layer effectively uses Rectified Linear Unit (ReLU) information. Similarly, if the software believes that a particular image represents a flying creature, the instigate plots that target higher grade points of engross, such as arms or a snout, will have more character. Generally speaking, a Fully Connected Layer (FC layer) examines the strength of the response of critical-level objects to a specific class. It includes explicit weights subsequently when you apply the products bordered by the accumulation and the previous surface, and the accurate likelihood for the multiple database categories is provided to the users.

- *Neural networks' hidden layers*

In neural networks, the function adds weights to the inputs. It guides them through an activation function as the output is positioned in a hidden layer between the algorithm's input and output. In essence, the input of the network undergoes nonlinear modifications in the hidden layers. All hidden layers of VGG use Rectified Linear Unit, a significant AlexNet that speed up preparation. Local Response Normalization (LRN) extends the time taken for training and hence consumes more memory without a significant improvement in accuracy.

- *The pooling layer*

In a VGG-16 architecture, the pooling layer is anticipated to sporadically mount the current pooling layer in operative convolutional layers [17]. It is the capacity to reduce the geospatial feature of the illustration to a similar degree of control overfitting by reducing the fraction of parameters in the framework. The pooling layer gradually reduces the spatial size of input image, thereby reducing the number of computations required by the network—MAX pooling layers. Due to this size reduction, only downsampling is done and only the crucial data is transmitted to CNN's subsequent layers. The most significant architecture is a pooling layer with bands that are 2 × 2 functional in size. Each dimension is reduced to half the size, reducing the number of pixels or values in each feature map to one-quarter the size, build, and width, eliminating 70% of the malignancy. Each MAX movement may require the most significant possible number of variables. The evaluation of significance remains unaltered.

Existing system

According to the current paradigm, it is critical to find lung cancer early on [18] to stop the disease from progressing and increase endurance. Lung knobs are tiny lumps of tissue that can either cause cancer or have no ill effects but are nonetheless considered harmful. While dangerous tissues often expand quickly, can affect subsequent bodily parts, and pose a risk to people, benign tissues are typically non-destructive and do not have much a growth rate.

2.1 Suggested Prototype

An automated lung cancer detection and classification system is proposed to improve the quality of the results. Phase 1 demonstrates an IoT setup to collect data. The setup includes sensors, CT scanners, pulse meters, etc. Composed sensor-based data is then skilled with Deep Neural Networks (DNNs) for screening, diagnosis, and cataloging, as depicted in Fig. 4. The efficiency of the network and the assessment phases will determine the outcomes. The diagnostic choice is given to the radiologist at stage 3. According to Fig. 4, it is now simple to identify death and does so quickly and accurately. The final protocol of the study was verified, and consent was taken from all the studied participants.

Computer Tomography (CT) tests are frequently preferred in clinical imaging in spite of availability of several imaging techniques due to their reduced noise. They are using charge-coupled devices to take all of the chest X-ray images (CCD).

It has been shown that profound learning is the most effective method for clinical imaging, highlight extraction, and article order. So many analysts are familiar with a few different types of deep learning designs when it comes to lung malignant growth (Fig. 5).

In this study, a Sim3D Multipath Simulation VGG-like arrangement with Z co-appointment is suggested. Benefits of this: Results from the division and our proposed structure are combined to produce increasingly precise results. To differentiate the lung knobs from CT examinations, this concentration also modifies the U-Net architecture for lung CT sweep division (Fig. 5).

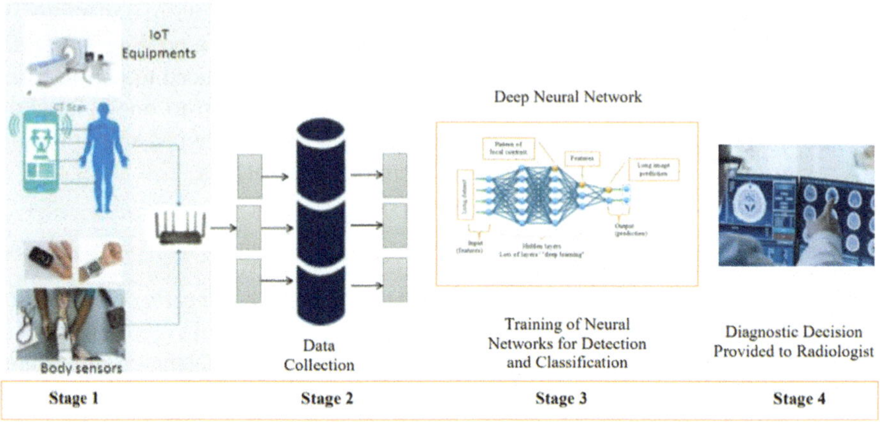

Fig. 4 Description and categorization of the suggested lung cancer screening technique

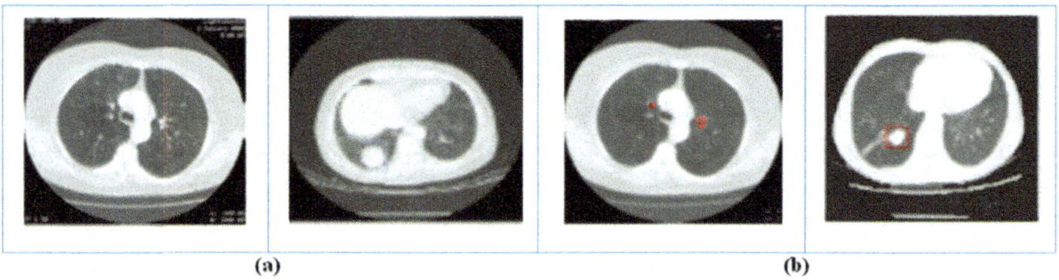

Fig. 5 Input dataset for the suggested technique for lung cancer screening and classification **a** original grayscale image; **b** cancer marked image

2.2 Architecture and Execution of the Mechanism

This section explains components, design, and their login interactions and inputs for a framework reliant on the predefined existing system, framework configuration, and characterizing aspects or components of an approach. A rational and well-known technique serves as the foundation for the approach. Each significant factor data of the strategy is taken into consideration using the Base Up or Top-Down methodology. The plans can be expressed in literary or graphical exhibiting modes.

- ***Materials for the prototype***

Firstly, check to see if your computer meets the required functionality. These are the crucial information your PC needs to make use of the hardware or product. The functional requirement is these specifications (Fig. 6).

- *Essential hardware with specifications*

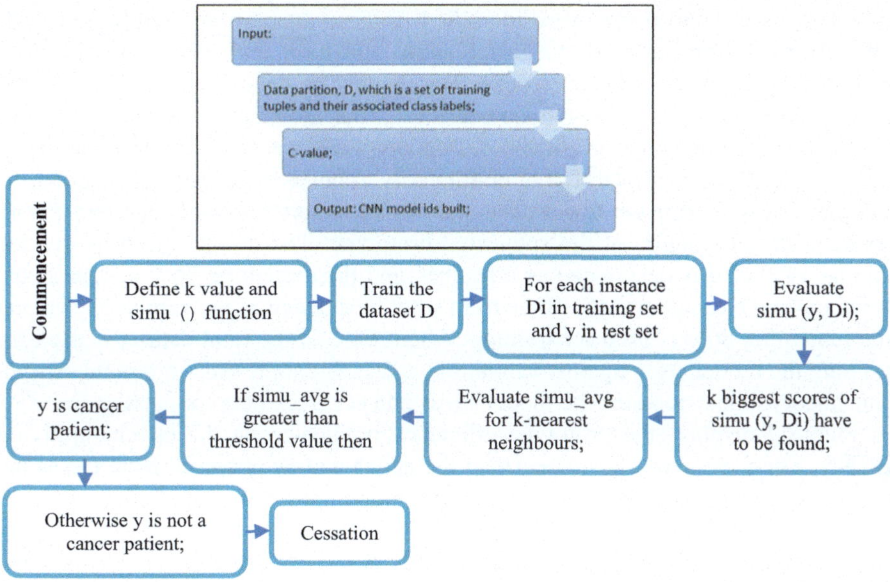

Fig. 6 Suggested algorithm

System basics for hardware devices can be checked with a similar level of simplicity. Any operating system or program, alias as hardware, elaborates the leading edge gaming process of requirements. A gear requirements' catalog is frequently connected with the hardware compatibility list (HCL), especially if there should come an incident of effective structure. HCL records tasks, excellent, and over some time spent vs. hardware devices for a specific operating structure or application. The gear necessities are the next component. The hardware for the specific operating system consists of a core i5, 4 gigabytes of RAM, and a 1 TB hard drive.

- ***Software specifications***

Programming specifications spell out the specifics, nuts, and bolts that must be understood for an application to function on a computer. These specifications should be added without restriction before it is introduced to the computer program. Prerequisites monitor displaying software resource necessities and essentials that should be acquainted on a Personal computer to give a model its best performance. Such parts are typically removed from the gadget building and assembling. These essentials or fundamentals should be introduced independently before the item is introduced because they are frequently excluded from the thing introduction gathering. Windows 10 is the operating system, Python is the language, TensorFlow is the tool, and Google Colab is the compiler.

The annotated radiographs were used to train and validate a DL-based model for identifying lung cancer on radiographs. The model was subsequently evaluated for detecting lung tumors using a different dataset. Lesions are annotated by clinical radiologists. Ground truths were identified in these annotations. The radiologists assessed the lesion characteristics, such as size, position, and edge, using the chest CT and surgical reports as references. The manual process of annotation requires more time and resources. To reduce overfitting, our model chose a subset of high-dimensional features, and averages of dropouts were considered to produce gradients with minimal variance. Large nonlinearity, robustness to high dimensionality, the capacity to learn from fewer data, stability in feature selection, and end-to-end training are all benefits of the suggested prototype. The initial dataset is divided so that 14% of it is designated as a test set, and the remaining 86% is kept as the training set. Once again, 14% of the train set is designated as the validation set, and the remaining 86% is used for training. About 64% of the total dataset is used for training the model, 16% against validation, and 20% data is used to text the model. The history object contains the model's loss and accuracy data for each epoch.

Figure 7 displays the prototype efficiency predicated on the epochs applying ECNN. The accuracy value concerning the epoch can be observed from the same. A convolutional kernel shares parameters throughout the entire convolution process in several convolutional kernels according to different weights for distinct image features. The third form of classifier utilized in the experiment was a variety of sized neural networks. For networks with a single hidden layer, the highest accuracy of 84% for the training data and the highest accuracy of 79% for the test data were obtained with a hidden layer size of 95. The best accuracy achieved by larger networks with two hidden layers for the test data was 76.4% for layers of sizes 95 and 45, but they achieved a maximum accuracy of 83.2% for the training data. Given that big networks need a lot of data samples to avoid overfitting and achieve acceptable generalization, this observation is probably the result of the neural network overfitting. As a result, it aids in the early detection of lung cancer, allowing doctors to administer therapies earlier on and aid in the patient's recovery. The graphs up top show the model accuracy and loss (Figs. 7 and 8). The available data has been examined to determine the model loss and accuracy. The model loss based on the epochs utilizing Extended Convolution Neural Network is displayed above in Fig. 8.

Based on 82.0% (95% CI, 76.9–88.5%) specificity in tenfold cross-validation, the sensitivity of suggested deep structured learning algorithmic program practice-exercise by the multi-center pictures was 85.1% (95% CI, 79.9–89.3%). This algorithm's area under the curve (AUC) was 0.889 (95% CI, 0.810–0.900).

Sensitivity improves as the quantity of training images is increased. Based on a specificity of 0.772 and an analogy of the presented deep structured learning algorithmic program with the top-scoring algorithm from the Kaggle competition, we found that our deep learning model had a sensitivity of 0.762 and the top-scoring program had a sensitivity of 0.669. The REST API with the integration of the POST method must be used to upload the data photos. On the server, the utilized type employs diverse data submissions. Flask is an open-source, extremely lightweight framework for creating web applications quickly. Angular 6's updates the page dynamically. Every time the suggested model receives a new type of input image, it is trained daily. Since it is simultaneously utilizing IoT, it can accept photographs from all around the world and produce results straight on the web browser (Table 1).

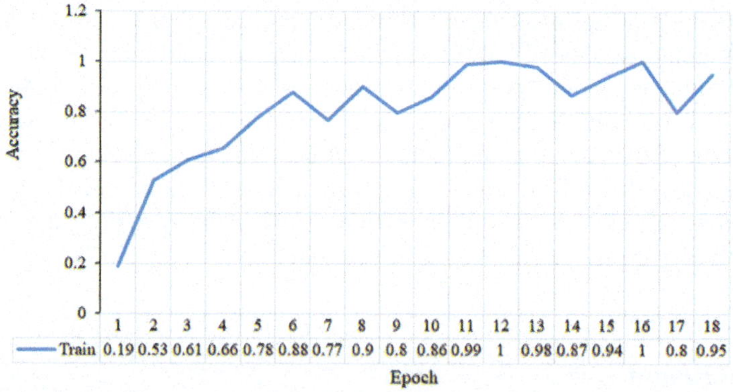

Fig. 7 Illustrating the line chart for precision and epochs

Fig. 8 Illustrating the line chart for degradation and epochs

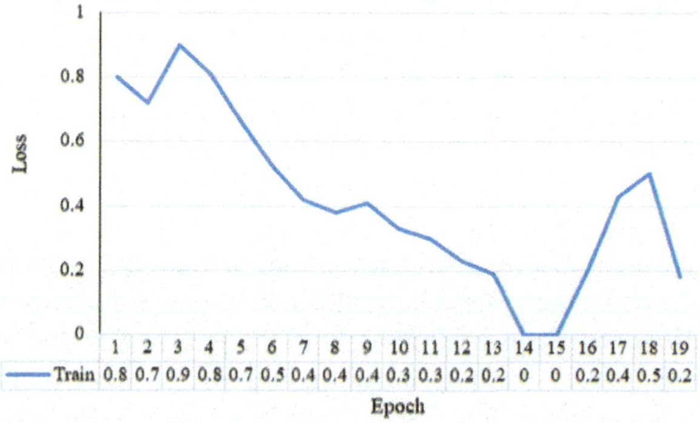

Table 1 Execution of suggested deep structured learning algorithmic program using multi-center database

Model	Precision %	Sensitivity %	Specificity %
Pretraining	72.0	72.3	75.1
Twofold training	81.6	81.4	82.1
Fourfold training	82.8	82.6	83.4
Sixfold training	83.1	83.4	83.8
Tenfold training	85.1	84.1	84.9

3 Discussion

This study used non-invasive, precise, and repeatable CT images to analyze and identify and classify lung cancer. We outline an Extended Convolutional Neural Network archetypal for the screening of lung malignancy subgroups. In light of this, the suggested approach offers a viable substitute for vast quantities of manually annotated data for clinical assessment of deep structured learning methods. The fact that the model maintained high performance in precision, acuity, and specificity for making the appropriate assessment when it was trained with a much smaller number of images demonstrates the ability of the transition learning system to produce immensely efficacious categorizations, even with a very modest training database. Deep CNNs based on histology have been successful in automating the grading of cancers, with studies demonstrating that their performance in grading prostate [18], breast [19], and colon [20] can be on par with that of pathologists. By enabling pathologists to test DL-generated predictions, explainability approaches can enable and enhance histology-based classification models.

4 Conclusion

Deep learning models have transformed cancer detection and prognosis. These prototypes are superior feature extractors, and thanks to their features, it is possible to forecast and diagnose cancer more accurately. To enhance system performance, data amplification is crucial for cancer detection and extrapolation studies. Making forecasts regarding cancer diagnosis and prediction tasks will heavily rely on these techniques. For clinical applications, more testing and validation must be done on larger datasets. Their performance will probably be further enhanced by the additional investigation into data amplification techniques, erudition in various fields like the frequency domain, and the use of distinct designs such as Graph Neural Networks. To make the system helpful for everyone, future research may involve using photos from other constructors in both the preparation and testing databases. Furthermore, the transfer learning technique for image processing is persuasive and effective in areas other than CT scans and lung cancer. The methods we have discussed here might, in theory, be applied to a variety of medical photos from other disciplines.

References

1. Van Beek, E.J.: Lung cancer screening: computed tomography or chest radiographs? World J. Radiol. **7**, 189 (2015). https://doi.org/10.4329/wjr.v7.i8.189
2. Mettler, F.A., Huda, W., Yoshizumi, T.T., Mahesh, M.: Effective doses in radiology and diagnostic nuclear medicine: a catalog. Radiology **248**, 254–263 (2008). https://doi.org/10.1148/Radiol.2481071451
3. The National Lung Screening Trial Research Team: Reduced lung-cancer mortality with low-dose computed tomographic screening. N. Engl. J. Med. **365**, 395–409 (2011). https://doi.org/10.1056/NEJMoa1102873
4. Mohammed, T.L.H., et al.: ACR appropriateness criteria screening for pulmonary metastases. J. Thorac. Imaging **26**, W1–W3 (2011). https://doi.org/10.1097/RTI.0b013e3182010bf9
5. Lecun, Y., Bengio, Y., Hinton, G.: Deep learning. Nature **521**(7553), 436–444 (2015)
6. Li, F., Engelmann, R., Metz, C.E., Doi, K., MacMahon, H.: Lung cancers missed on chest radiographs: results obtained with a commercial computer-aided detection program. Radiology **246**, 273–280 (2008). https://doi.org/10.1148/radiol.2461061848

7. Schalekamp, S., et al.: Computer-aided detection improves detection of pulmonary nodules in chest radiographs beyond the support by bone-suppressed images. Radiology **272**, 252–261 (2014). https://doi.org/10.1148/radiol.14131315
8. Rajpurkar, P., et al.: CheXNet: Radiologist-Level Pneumonia Detection on Chest X-Rays with Deep Learning, pp. 3–9 (2017). arXiv:1711.05225
9. Tufail, A.B., Ma, Y.K., Kaabar, M.K., Martínez, F., Junejo, A.R., Ullah, I., Khan, R.: Deep learning in cancer diagnosis and prognosis prediction: a minireview on challenges, recent trends, and future directions. In: Computational and Mathematical Methods in Medicine (2021)
10. Ioffe, S., Szegedy, C.: Batch normalization: accelerating deep network training by reducing internal covariate shift. In: Proceedings of the 32nd International Conference on Machine Learning, ICML 2015, pp. 448–456, Lille, France (2015)
11. Huang, G., Liu, Z., Van Der Maaten, L., Weinberger, K.Q.: Densely connected convolutional networks. In: Proceedings of the 30th IEEE Conference on Computer Vision and Pattern Recognition, CVPR, 2017, pp. 4700–4708, Honolulu, HI, USA (2017)
12. Junejo, R., Kaabar, M.K.A., Mohamed, S.: Future robust networks: current scenario and beyond for 6G. IMCC J. Sci. **11**(1), 67–81 (2021)
13. Simonyan, K., Zisserman, A.: Very deep convolutional networks for large-scale image recognition. In: Proceedings of the 3rd International Conference on Learning Representations, ICLR 2015—Conference Track Proceedings, pp. 1–14, San Diego, CA, USA (2015)
14. Hochreiter, S., Schmidhuber, J.: Long short-term memory. Neural Comput. **9**(8), 1735–1780 (1997)
15. Jiang, H., Ma, H., Qian, W., Gao, M., Li, Y.: An automatic detection system of lung nodule based on multigroup patch-based deep learning network. IEEE J. Biomed. Health Inform. **22**(4), 1227–1237 (2017)
16. Barua, A., Zakia, U., Menon, C., Jiang, X.: Deep learning techniques in estimating ankle joint power using wearable IMUs. IEEE Access **9**, 83041–83054 (2021)
17. Russakovsky, O., Deng, J., Su, H., Krause, J., Satheesh, S., Ma, S., et al.: Imagenet large scale visual recognition challenge. Int. J. Comput. Vision **115**(3), 211–252 (2015)
18. Kanna, G.P., Kumar, S.J.K.J., Parthasarathi, P., et al.: A review on prediction and prognosis of the prostate cancer and gleason grading of prostatic carcinoma using deep transfer learning based approaches. Arch. Comput. Methods Eng. **30**, 3113–3132 (2023). https://doi.org/10.1007/s11831-023-09896-y
19. Ryu, H.S., Jin, M.-S., Park, J.H., Lee, S., Cho, J., Oh, S., et al.: Automated gleason scoring and tumor quantification in prostate core needle biopsy images using deep neural networks and its comparison with pathologist-based assessment. Cancers **11**, 1860 (2019)
20. El Achi, H.E., Khoury, J.D.: Artificial intelligence and digital microscopy applications in diagnostic hematopathology. Cancers **12**(4), 797 (2020). https://doi.org/10.3390/cancers12040797

Chapter 37
Utilizing Blockchain Technology for Farmer Identity Management and Land Registry Systems in Agriculture

Rhui Jaan Ong, Suhizaz Sudin, Rafikha Aliana A. Raof, and Kar Yan Choong

1 Introduction

Kebun-Kebun Bangsar (KKB) is an urban farm located in Kuala Lumpur that spans 1.01 ha. Established in 2016 by a dedicated group of volunteers, its primary objective is to provide sustenance to the less fortunate members of society. Unfortunately, in July 2020, KKB received an eviction notice due to the expiration of its temporary occupation license (TOL) [1]. Regrettably, on June 23, the farm was compelled to cease its operations after receiving complaints regarding olfactory pollution, the presence of flies, and noise resulting from its livestock. Ng Sek San, the founder of KKB, expresses his desire for the authorities to adopt a more proactive role in supporting urban farming initiatives such as KKB, given their capacity to address food insecurity in a sustainable and efficacious manner.

A temporary occupation license (TOL) is essentially a temporary license or permission granted by the state government to occupy state land in Malaysia. It is important to note that a TOL has a duration of one year and can be renewed multiple times [2]. However, the repeated renewals of a TOL do not guarantee future

R. J. Ong · S. Sudin (✉) · R. A. A. Raof · K. Y. Choong
Faculty of Electronic Engineering and Technology, Universiti Malaysia Perlis, Pauh Putra Campus, Perlis, Malaysia
e-mail: suhizaz@unimap.edu.my

S. Sudin · R. A. A. Raof
Sports Engineering Research Center, Universiti Malaysia Perlis, Pauh Putra Campus, Perlis, Malaysia

J. M. R. S. Tavares et al. (eds.), *Proceedings of Second International Conference on Intelligent System*, Algorithms for Intelligent Systems,
https://doi.org/10.1007/978-981-99-8976-8_38

ownership rights according to the law. Additionally, some farmers have experienced cases where their TOL was not renewed, and they were suddenly faced with quit rent obligations. These uncertainties and challenges make it difficult for farmers to plan effectively for their agricultural activities.

Delays in renewal can hinder farmers' ability to obtain bank loans, apply for foreign workers, and implement innovative programs aimed at improving agricultural productivity [3]. Short TOL periods also discourage the adoption of new technologies and deter local investors, leading to stagnant progress in the agricultural sector. The lack of a national policy or long-term action plan regarding TOL has led to a situation where individual land offices in the state have discretion in deciding whether or not to renew a TOL. This lack of certainty and potential for abrupt non-renewals can have severe consequences for TOL holders.

Traditional farmer identity management and land registration systems face significant challenges in the agricultural industry. They lack transparency and trust due to centralized control, leading to corruption and disputes over land ownership. Fraud is also a problem, with identity theft and multiple registrations occurring. Data security is at risk, as sensitive information is stored in vulnerable centralized databases. Manual processes cause inefficiencies, delays, and errors. Farmers in remote areas may struggle to access and update their information, and systems often struggle to communicate with each other.

Blockchain offers solutions to the difficulties faced by farmer identity management and land registration systems. It decentralizes control, ensuring data integrity and reducing fraud risks. Information stored on the blockchain is tamper-resistant and trustworthy. The technology enhances security through encryption, safeguarding sensitive data. Transparency and auditability improve trust among stakeholders. Smart contracts automate tasks, saving time and reducing paperwork. Farmers gain ownership of their data and control who can access it. Blockchain's interoperability streamlines communication between systems. It enables end-to-end traceability, promoting ethical practices. By embracing blockchain, agriculture can become more efficient, secure, and transparent.

Besides, blockchain technology can play a significant role in enhancing farmer identity and land registry systems by providing secure, immutable, and transparent records. Blockchain can enable a decentralized system for verifying and managing farmer identities. Each farmer's identity and associated information can be stored on the blockchain in a tamper-proof manner. This ensures that the information is secure, and it mitigates the risk of identity theft or fraudulent activities.

On the other hand, land ownership records can be recorded on the blockchain, making it easier to establish ownership, transfer land titles, and resolve disputes. Smart contracts can be utilized to automate the process of transferring land titles, reducing bureaucracy, and enhancing efficiency. The blockchain's immutable nature ensures that once a land record is recorded, it cannot be altered or tampered with. This helps to maintain the integrity of land ownership records and prevents fraudulent activities such as double-spending or unauthorized changes to land titles.

2 Key Features of Blockchain Technology

Blockchain technology possesses several key features and advantages that distinguish it from centralized organizations built using traditional Internet technology. One of the most significant features is its decentralized nature [4]. Unlike centralized organizations, blockchain platforms are built on distributed networks of nodes that collectively maintain and validate the system. This decentralization offers increased transparency, resilience, and resistance to censorship or single points of failure.

An additional benefit provided by blockchain technology is its capacity to thwart data tampering [5]. Each transaction or data entry within the blockchain is meticulously recorded in a block, which is then linked to preceding blocks, thereby forming an unalterable chain. Once data is inscribed onto the blockchain, modifying or erasing it without consensus from network participants becomes exceedingly arduous. This attribute guarantees the integrity and reliability of the data, rendering blockchain technology well-suited for applications that necessitate prevention of data tampering.

Furthermore, blockchain technology enhances trust between buyers and sellers by means of smart contracts [6]. Smart contracts (SC) are self-executing agreements written in code and stored on a blockchain. These contracts autonomously execute predetermined actions when specific conditions are met. By eliminating the need for intermediaries, smart contracts cultivate trust by diminishing the risk of fraudulent activities and ensuring compliance with agreed-upon terms. This automated and transparent interaction streamlines processes, reduces expenses, and ultimately fosters improved trust and efficiency in transactions.

Moreover, blockchain technology has the potential to establish a new credit mechanism through peer-to-peer electronic cash systems [7]. By utilizing blockchain-based cryptocurrencies, individuals can directly transact with each other without traditional financial intermediaries. This empowers individuals with greater control over their finances, facilitates cross-border transactions, and provides access to financial services for the unbanked population. The use of blockchain in this context has the potential to revolutionize the traditional financial system and establish new credit mechanisms.

Efficiency in information transmission and reduced risk of data storage are additional advantages of blockchain technology [8]. By eliminating the need for intermediaries or central authorities, blockchain-based systems can streamline processes, reduce delays, and lower costs associated with traditional information transfer methods. Furthermore, the distributed nature of blockchain networks enhances data security by reducing the risk of single points of failure and increasing resistance to hacking or unauthorized access. This promotes efficient information exchange and reduces the risks associated with centralized data storage.

Blockchain technology also enables more effective privacy-preserving data sharing processes [9]. Traditional systems often require sharing sensitive data with multiple parties, increasing the risk of unauthorized access or data breaches. However, blockchain allows for secure data sharing while maintaining privacy through cryptographic techniques. Users can have control over their data, deciding who can access and verify it without revealing the actual content. This feature is particularly valuable in sectors like healthcare, where secure and private sharing of patient data is critical.

Finally, blockchain technology enhances the traceability and integrity of digital and physical assets [10]. By recording asset-related information on the blockchain, it becomes possible to establish a clear audit trail, ensuring the authenticity and history of assets throughout their lifecycle. This has applications in supply chain management, intellectual property rights, provenance verification, and more. The transparent and tamper-resistant nature of blockchain ensures that asset information can be verified, fostering trust and integrity in asset transactions.

Overall, the decentralized nature, data tampering prevention, improved trust through smart contracts, new credit mechanisms, efficient information transmission, privacy-preserving data sharing, and asset traceability and integrity are significant advantages offered by blockchain technology. These features have contributed to its increasing adoption across various industries, promising enhanced security, efficiency, trust, and new opportunities for innovation and collaboration.

3 Blockchain Technology and Its Applicability

Agriculture is one of the areas where blockchain technology could bring a revolution by solving the existing problem of Agri-product fraud, its traceability, price manipulation, and lack of customer trust in the product [11]. First, blockchain can streamline supply chain management processes by eliminating intermediaries, reducing paperwork, and automating transactions [12]. Second, blockchain enables end-to-end traceability, allowing consumers and stakeholders to track the journey of food products from farm to table [13].

Blockchain technology can revolutionize the agricultural industry's farmer identity management and land registration systems by offering trust, transparency, and efficiency [14, 15]. Through its immutable records and decentralized nature, blockchain ensures reliable information and eliminates the need for central authorities, enhancing transparency and reducing fraud. Automating processes with smart contracts streamlines operations, and real-time updates provide accurate data on land ownership and farmer credentials. The encrypted data and access control enhance security, while supply chain traceability allows for verifiable product origins. By leveraging blockchain, the agricultural sector can establish a modern, secure, and transparent ecosystem, fostering trust among stakeholders and facilitating sustainable practices.

Third, Revathy and Priya [16] proposed a blockchain-based producer–consumer model (BPCM), which enables farmers to sell their commodities directly to consumers while preventing intermediaries from using smart contracts, thereby allowing farmers to gain more profit. In the traditional producer–consumer model, the farmer or producer is required to sell their agricultural products to intermediaries, commonly known as middlemen. These intermediaries then sell the products to consumers, often inflating the prices. In this setup, the intermediaries gain more benefits while farmers have little control over setting the prices for their products.

Fourth, Umamaheswari et al. [17] proposed a blockchain-based IoT system that simplifies the buying and selling of crops and land by storing sensor data in a blockchain and deploying smart contracts built on the Ethereum blockchain. The system enables peer-to-peer transactions and smart contracts that automatically execute when specified conditions are met. All sensor data and transactional information can be permanently recorded in the blockchain, enabling farmers to access real-time data on seed quality, climate-related information, payments, soil moisture, demand, and sale prices, all within a single platform.

Sendros et al. [18] concluded the reason applying blockchain in agriculture which is mainly to solve the various cyberthreats and food safety issues of the existing IT solutions. As a result, the majority of solutions leverage blockchain technology for its inherent characteristics, with data transparency and integrity being utilized at rates of 61% and 50%, respectively.

4 Limitations of Blockchain

Blockchain technology, while offering numerous advantages, is not without its limitations that must be carefully considered. Firstly, implementing blockchain can introduce complexities to existing systems and processes, requiring expertise and potentially necessitating infrastructure changes. Additionally, the distributed nature of blockchain and its reliance on cryptographic algorithms and consensus mechanisms can result in increased costs. The computational resources and energy required for consensus and maintaining the blockchain, particularly in public and permissionless networks, can be substantial and raise concerns regarding energy consumption. Moreover, the novelty of blockchain technology and its decentralized nature may give rise to compliance challenges with existing data protection laws. The distributed and immutable nature of blockchain can create gaps in ensuring compliance with requirements such as the right to be forgotten. Therefore, organizations contemplating the adoption of blockchain technology must carefully evaluate and address these limitations, considering the potential complexities, costs, energy consumption, and compliance challenges associated with its implementation.

5 Conceptual Blockchain-Based Farmer Identity and Land Registry

The overall conceptual blockchain-based farmer identity and land registry management systems are shown in Fig. 1. First, a temporary occupation license (TOL) is granted by the state government to a farmer named 'Akau.' The statement of the TOL

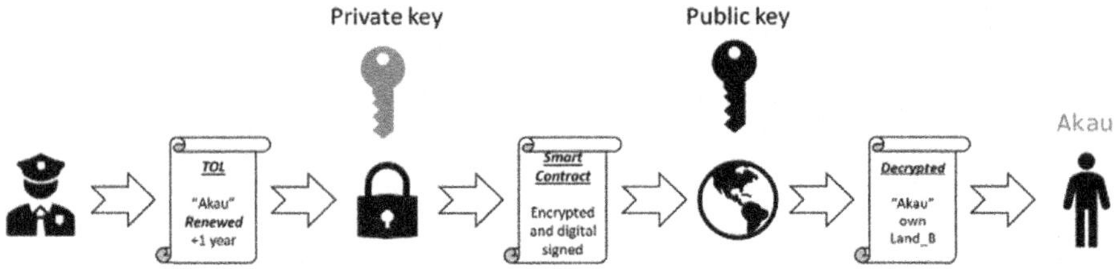

Fig. 1 Overall conceptual blockchain-based farmer identity and land registry management systems

is encrypted with a private key using a hashing technique and signed by 'Akau.' In the blockchain ecosystem, everyone in the chain is able to decrypt part of the statement of the TOL in order to prove that 'Akau' owns 'Land_B.' Hence, this utilizes the features of privacy preservation, traceability, and integrity of blockchain technology.

There are two options when it comes to the contribution of blockchain in storing data. The first option is to store the 'actual data' directly on the blockchain as it is gathered. This means that the information is stored on the blockchain itself, ensuring its availability and immutability. The second option involves storing 'pointers and hash of actual data.' In this approach, the information is stored in another location outside of the blockchain, and the BC is used to maintain the integrity and immutability of the data. The blockchain serves as a reference or proof that the stored information has not been tampered with or modified. Both options provide different approaches to utilizing the blockchain for data storage, depending on the specific requirements and preferences of the application or system.

In this research, data is stored directly on the blockchain with a private key and hashed using the SHA-256 hash function. Each block will contain a private key, farmer ID, and the specific registered land owned by each farmer. For instance, the pseudo code in Fig. 2 shows a blockchain with three blocks, and the initial block stores data of a farmer named 'Abu' who owns 'Land_A.' Figure 2 also demonstrates a simplified output that provides a deeper understanding of what happens to the hash code when 'Land_A' is changed to 'Land_D.' The result clearly shows a completely different hash code, even with a minor change. Therefore, this utilizes the features of tamper-proof in data transmission of blockchain technology.

A public distributed ledger refers to a decentralized digital database that is accessible to and maintained by multiple participants in a network, as shown in Fig. 3. The ledger consists of a chain of interconnected blocks, each containing a set of transactions or data records. In a public distributed ledger, the network participants (nodes) collectively validate, record, and maintain the ledger's integrity through a consensus mechanism. This ensures that all participants have a consistent and transparent view of the ledger's contents. For example, the farmer named 'Akau' is able to raise a flag if the farmer named 'Mutu' claims that 'Land_A' belongs to him.

```
Farmer_1 = "Abu"
Farmer_2 = "Akau"
Farmer_3 = "Mutu"

Land_A = "own land A"
Land_B = "own land B"
Land_C = "own land C"
```

```
initial_block = AgriBlock("Private key", Farmer_1, Land_A)
second_block = AgriBlock(initial_block.block_hash, Farmer_2, Land_B)
third_block = AgriBlock(second_block.block_hash, Farmer_3, Land_C)
```

```
Output:
Private code|Abu|own land A
Fb944354ad491aa9cba90a2dc9eb9a4525f83e56f4ba1ab8a27cdd83b41cd601

Private code|Abu|own land D
2d789a8e4b00f7bfc8d49139a28649940b7143a05034e3e152689afccebdadde
```

Fig. 2 Pseudo code and conceptual blockchain diagram

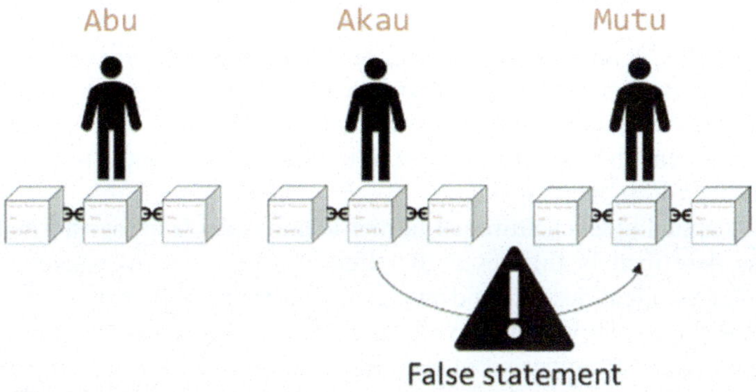

Fig. 3 Public distributed ledger

6 Conclusion

This research study presents a valuable and comprehensive resource for a range of stakeholders, including researchers, policymakers, and industry representatives, who seek to harness the transformative potential of blockchain technology within the agriculture sector. The conceptual blockchain-based farmer identity and land registry management system outlined in the text offers a robust framework for ensuring secure and transparent management of farmer identities and land ownership information. By leveraging the inherent features of blockchain technology, the system addresses critical aspects such as privacy preservation, traceability, integrity, and tamper-proof data transmission.

Central to the system is the utilization of temporary occupation licenses (TOLs) which, following their issuance by the state government, are encrypted and signed by the respective farmers. This cryptographic process not only safeguards the privacy of the TOL information but also allows other participants within the blockchain ecosystem to verify the ownership of specific plots of land. The integration of blockchain technology enables the preservation of data privacy while maintaining traceability and ensuring the integrity of the TOLs.

This research has opted to store data directly on the blockchain using a private key and applying the SHA-256 hash function for hashing. Each block within the blockchain incorporates a private key, farmer ID, and the pertinent details of the registered land owned by each farmer. The code further demonstrates the impact of modifying the land ownership information, showcasing the generation of a distinct hash code even with a minor alteration. This exemplifies the tamper-proof nature of the blockchain technology, which ensures the integrity and security of data transmission.

Fundamental to the proposed system is the adoption of a public distributed ledger. This decentralized digital database grants access to and maintenance by multiple participants within a network. The ledger, organized in a chain of interconnected blocks, comprises transactions or data records. The integrity of the ledger is collectively validated, recorded, and upheld by network participants, or nodes, through the implementation of a consensus mechanism. Such an arrangement guarantees a consistent and transparent view of the ledger's contents, fostering trust and accountability within the system.

The conceptual blockchain-based farmer identity and land registry management system, as described in this research, offers a secure and transparent means of managing farmer identities and land ownership information. By harnessing the power of blockchain technology, it effectively addresses concerns related to privacy, traceability, integrity, and tamper-proof data transmission. As a result, it establishes a reliable and transparent record of land ownership and farmer identities. The findings of this research hold significant promise for revolutionizing the agricultural industry, and as such, it represents an invaluable resource for researchers, policymakers, and industry stakeholders seeking to explore the potential applications of blockchain within the agricultural sector.

Acknowledgements This research was funded by a grant from Ministry of Higher Education of Malaysia (FRGS/1/2020/ICT06/UNIMAP/02/1).

References

1. Lim, J.: We're partners, not adversaries. The Star (2022). https://www.thestar.com.my/metro/metro-news/2022/07/16/were-partners-not-adversaries. Accessed 28 June 2023
2. Parzi, N.M.: Cameron highlands: Pahang urged to increase TOL for farmers to 30-years. New Straits Times (2019). https://www.nst.com.my/news/nation/2019/01/450071/cameron-highlands-pahang-urged-increase-tol-farmers-30-years. Accessed 28 June 2023

3. Buang, S.: Tol is not for life. New Straits Times (2020). https://www.nst.com.my/opinion/columnists/2020/01/552957/tol-not-life. Accessed 28 June 2023
4. Naef, S., Wagner, S.M., Saur, C.: Blockchain and network governance: learning from applications in the supply chain sector. In: Production Planning & Control, pp. 1–15 (2022). https://doi.org/10.1080/09537287.2022.2044072
5. Oyelude, A.A.: Trending issues in advancing blockchain technology in libraries, archives and museums. Library Hi Tech News **39**(6), 6–7 (2022). https://doi.org/10.1108/LHTN-06-2021-0040
6. Al-Rakhami, M.S., Al-Mashari, M.: A blockchain-based trust model for the internet of things supply chain management. Sensors **21**(5), 1759 (2021). https://doi.org/10.3390/s21051759
7. Muchlis Gazali, H., Gazali, H.M.: Bitcoin investment behaviour: a pilot study. Int. J. Percept. Cogn. Comput. **5**(2), 81–86 (2019). https://doi.org/10.31436/ijpcc.v5i2.97
8. Xie, Z., Kong, H., Wang, B.: Dual-chain blockchain in agricultural E-commerce information traceability considering the Viniar algorithm. Sci. Program. **2022**, 1–10 (2022). https://doi.org/10.1155/2022/2604216
9. Wang, Q., Qin, S.: A hyperledger fabric-based system framework for healthcare data management. Appl. Sci. **11**(24), 11693 (2021). https://doi.org/10.3390/app112411693
10. Tan, A., Gligor, D., Ngah, A.: Applying blockchain for Halal food traceability. Int J Log Res Appl **25**(6), 947–964 (2022). https://doi.org/10.1080/13675567.2020.1825653
11. Sekhar Bhusal, C.: Blockchain technology in agriculture: a case study of blockchain start-up companies. Int. J. Comput. Sci. Inf. Technol. **13**(5). https://doi.org/10.5121/ijcsit.2021.13503
12. Longo, F., Nicoletti, L., Padovano, A., d'Atri, G., Forte, M.: Blockchain-enabled supply chain: an experimental study. Comput. Ind. Eng. **136**, 57–69 (2019). https://doi.org/10.1016/j.cie.2019.07.026
13. Li, Q., Cao, B., Wang, X., Wu, J.J., Wang, Y.K.: Systematic water-saving management for strawberry in basic greenhouses based on the internet of things. Appl. Eng. Agric. **37**(1), 205–217 (2021). https://doi.org/10.13031/aea.14095
14. Kamilaris, A., Cole, I.R., Prenafeta-Boldú, F.X.: Blockchain in agriculture. In: Food Technology Disruptions, pp. 247–284. Elsevier (2021). https://doi.org/10.1016/B978-0-12-821470-1.00003-3
15. Krithika, L.B.: Survey on the applications of blockchain in agriculture. Agriculture (Switzerland) **12**(9) (2022). https://doi.org/10.3390/agriculture12091333
16. Revathy, S., Priya, S.S.: Blockchain based producer-consumer model for farmers. In: 2020 4th International Conference on Computer, Communication and Signal Processing (ICCCSP), pp. 1–5. IEEE (2020). https://doi.org/10.1109/ICCCSP49186.2020.9315214
17. Umamaheswari, S., Sreeram, S., Kritika, N., Jyothi Prasanth, D.R.: BIoT: blockchain based IoT for agriculture. In: 2019 11th International Conference on Advanced Computing (ICoAC), pp. 324–327, IEEE (2019). https://doi.org/10.1109/ICoAC48765.2019.246860
18. Sendros, A., Drosatos, G., Efraimidis, P.S., Tsirliganis, N.C.: Blockchain applications in agriculture: a scoping review. Appl. Sci. (Switzerland) **12**(16) (2022). https://doi.org/10.3390/app12168061

Chapter 38
IoT and Big Data Analytics for Smart Healthcare 4.0 Applications

Ayasha Malik, Veena Parihar, Kritika Purohit, A. K. M. Bahalul Haque, Nikhil Sharma, and Pronaya Bhattacharya

1 Introduction

Nowadays, more than 50% of the earth's residents live in metropolises, and in the upcoming 30 years, this percentage will grow up to 68%. Also, as per the United Nations report, by 2050, this population can reach up to 2.5 billion [1]. This immense escalation in population will enforce several experiments on capital, which includes the supportable development and management of metropolitan areas and also the capability to ensure the excellent conditions of people's life. The critical and operational solution to satisfy the necessities of the rising society is the advancement of smart cities. Since the 1990s, ICT has completely influenced the efficiency, access, and quality of almost every task associated with the development of health care in

A. Malik
IIMT College of Engineering, AKTU, Greater Noida, India

V. Parihar (✉)
KIET Group of Institutions, Delhi-NCR, Ghaziabad, India
e-mail: veena2parihar@gmail.com

K. Purohit
JIET Group of Institutions, Jodhpur, Rajasthan, India

A. K. M. Bahalul Haque
LUT School of Engineering Science, LUT University, Lappeenranta, Finland
e-mail: bahalul.haque@lut.fi

N. Sharma
Department of CSE, Delhi Technological University, Delhi, India

P. Bhattacharya
Department of CSE, Amity School of Engineering and Technology, Research and Innovation Cell, Amity University, Kolkata, West Bengal, India
e-mail: pbhattacharya@kol.amity.edu

J. M. R. S. Tavares et al. (eds.), *Proceedings of Second International Conference on Intelligent System*, Algorithms for Intelligent Systems,
https://doi.org/10.1007/978-981-99-8976-8_39

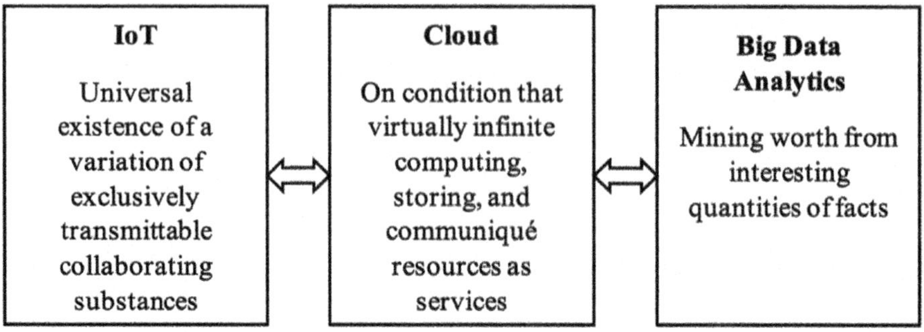

Fig. 1 Technical pillars of Industrie4.0

smart cities. eHealth is on the rise due to the formation and evolution of new technologies; hence, the precise representation of the term eHealth has endured progressive specifications and changes. The existing technology regarding the idea is called Industrie (Industry 4.0) or I4.0. Atitallah [2] et al. state that the term, Industire4.0, was used in an article published by a few German authors in 2020. From the perspective of technology, Industrie4.0 relies on three technical pillars, also illustrated in Fig. 1.

At another end, the quick advancement of the IoT and BDA is frequently measured by the topmost feature in the employment of smart cities' services. Any smart city can benefit from data and exploit that data by collecting and refining the information with the support of numerous technologies of computing, networking, and communication. This process leads to the origination of automated and smart services that are utilized in various sectors like security, transportation, health, etc. IoT devices due to their immense use are the center of attraction for cybercrimes. Aceto et al. [3] discussed that approximately 10–100 billions of security data are generated by large organizations per day and enterprises use that data for posting hoc forensic analysis and regulatory compliance. IoT BDA has been a widely researched domain since last few years. Their research tends to eliminate the limitations of traditional techniques of data refinement and treating of the vast volume of facts. This study focuses on smart cities in terms of health care, as the applications field of IoT BDA and various use cases, for e.g., smart agriculture, smart surveillance, smart conveyance, smart healthcare ecosystem, and the smart home of IoT security. Although various surveys and reviews focus on I4.0-enabled technologies, not enough up-to-date work has been found that also focuses on revolutionizing Healthcare 4.0 technology. A summary of the major contribution of this work is mentioned below.

- This work has a discussion about the I4.0-enabled technologies or technical pillars: IoT and BDA; and the contribution of these pillars in transforming the health care to Healthcare 4.0.
- This work presents use cases of IoT security, along with a computational infrastructure of BDA.
- This works explores the various applications of Healthcare 4.0 in different sectors of IoT.

The remaining research paper is structured in the following way. Section 2 discusses the ICT pillars of health care. Section 3 provides information about the context of IoT with and its use cases. Section 4 presents a summary of infrastructures for IoT BDA which is cloud computing, fog computing, and edge computing. Section 5 provides the various applications of Healthcare 4.0 and Sect. 6 concludes the whole discussion.

2 ICT Pillars for Healthcare 4.0

The remarkably universal existence of wireless and mobile automation has empowered modern healthcare services or up-to-minute degrees of quality and cost-efficiency in traditional ones. Predictive, Preventive, Personalized, and Participatory (P4) medicine concept is obtained by progressive change in modern medicine technologies [4, 5]. This technique is built on a global recognition of one and all patient's biology to the arranging patients alternatively into treatment groups. These transformations bring from the broad set of ICTs and can play a significant role in the reduction of global health budgets. In this section, our focus is on dub pillars of HC4.0. Technologies such as IoT and Big Data are used in era of health care to enhance its importance. Figure 2 shows their development in relation of academic attentiveness and production. We acknowledge that other technical and non-technical features can be examined in association with HC4.0.

2.1 Contribution of IoT in Healthcare 4.0

IoT in health care offers precise data dimensions, authorized consumption, and processing capabilities. However, due to the complex and dynamic environment, IoT-related issues are often overlooked. The integration of IoT technology into industrial and manufacturing processes, as well as robotic advancements, presents both opportunities and challenges. IoT plays a crucial role in modernizing health care, reducing

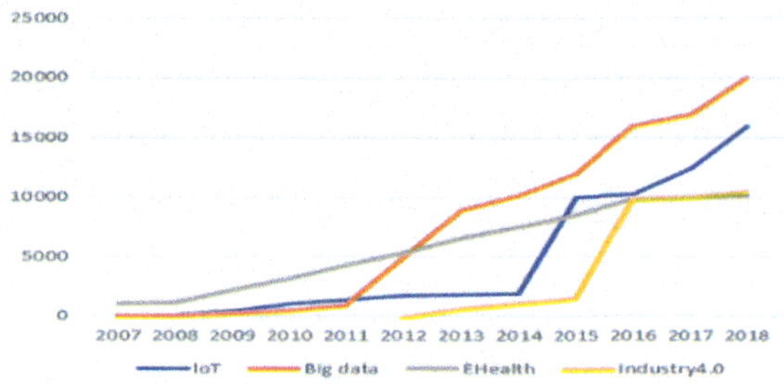

Fig. 2 Annual publication count for research on HC4.0 pillars' status

Table 1 Related work in IoT Healthcare 4.0

IoT paradigm	Example
Internet of Medical Things (IoMT) [7]	Discusses applications involving of embedded and wearable strategies associated to a (private, smartphone, smart watch), i.e., allied the Internet and acting as own center
Internet of health Things (IoHT) [8]	Constructed on the grouping of mobile apps, wearable, further related expedients and influences context-aware always-on professional-grade instrument medicinal campaigns
Internet of nano Things (IoNT) [9]	Raises the use of IoT in nano medicine, to execute additional modified observing, diagnostics, and treatment to carry out practical monitoring, defensive health, lasting care disease organization, and follow-up care
Wearable Internet of Things (WIoT) [10]	Aims to execute telehealth to accomplish ecology for automatic interferences. Leveraging body damaged sensors. Empowers checking data which used in improving persons' ordinary excellence of life and attaches patients to medical substructures
Internet of mobile-health Things (m-IoT) [11]	Visualize a connectivity model among low-power private-zone grids and developing 4G networks, highlighting the surviving precise structures inherent to the worldwide flexibility of partaking objects

costs, and maximizing health outcomes, though system behavior changes are necessary. The advancement of wireless technology allows real-time monitoring, early diagnosis, and management of medical conditions. Diagnostic and imaging systems are essential in utilizing IoT in the health sector, with many applications leveraging standard mobile devices. IoT enables intelligent device interactions and facilitates the acquisition of new information and decision-making environments. The healthcare sector has explored diverse applications of IoT, each with its own significance [6]. Encouraged by the first IoT paradigm, more diversity has been discovered in the health sector, each with its own visibility presented in Table 1.

2.2 BDT in Healthcare 4.0

Big data has grown in popularity to the point that everyone is utilizing it. According to research, a number of large corporations have begun to prioritize big data. In addition, the big data industry is expected to develop at least five times faster than the entire IT market in the future years. It is the data that outnumbers typical database systems' storage and processing capabilities. Daily data generation is expanding at an alarming rate [12]. As stated in Table 2, there are several V's that define big data in depth.

Table 2 Various V's of big data

Constraint	Description
Volume (V1)	Refers to the massive amount of created data per second
Velocity (V2)	Indicates the high-speed created data, requiring real-time analysis and processing
Variety (V3)	Encompasses the different forms of data, including structured, unstructured, and semi-structured data
Veracity (V4)	Deals with the presence of noise, anomalies, and uncertainties in the data
Volatility (V5)	Specifies the duration for which stored data remains valid and useful
Value (V6)	Determines whether the stored data is beneficial for analysis and problem-solving purposes

3 IoT Background and Security Use Cases

This section describes the need for security in IoT devices in terms of its security parameters. All the security parameters have to be fulfilled by IoT devices to ensure its genuineness. Later, we uncovered some IoT use cases in the real world.

3.1 Security of IoT

IoT has recently become extensively accepted in building smart systems. IoT security is considered to protect the entire IoT structure from vulnerability (attack). These factors are desired to be considered to improve IoT security resolutions. These are the security desires that need to be met for building security solutions of IoT. Because of the great potential made available through in-depth learning and BDT, they can be used to classify a group of security breaches related to security necessities [13, 14]. Figure 3 represents the various security parameters of IoT.

3.2 Use Cases of IoT

This section highlights the need for research on using big data technology to enhance security in IoT devices. Recent years have seen rapid development of IoT devices, but concerns over security threats remain. The Mirai malware attack on IoT devices and the Dyn DNS provider underscored the vulnerabilities present. To address this, it is crucial to explore new technologies or combinations of existing ones for IoT device security. Through this study, the aim is to inspire future researchers by leveraging

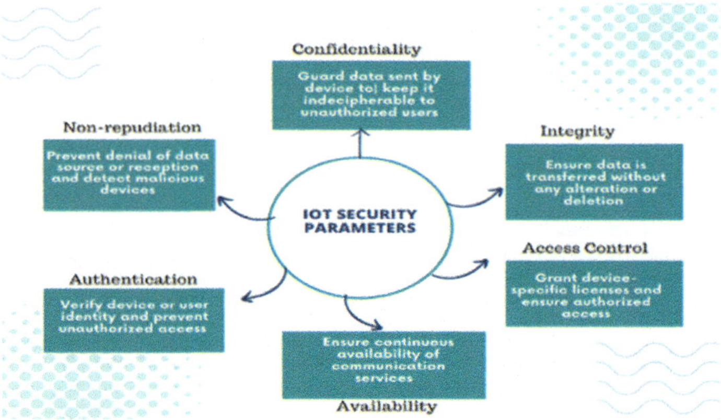

Fig. 3 Security parameters of IoT

big data concepts and deep learning for IoT security. Figure 4 shows the use cases of IoT security applications in terms of their relationship with big data technology. The use cases of IoT security are discussed below.

SirenJack is an attack on emergency systems that allowed unauthorized access and manipulation. Deep learning and advanced data analysis can help detect and prevent such attacks. The Dyn attack affected major websites using infected IoT devices. Deep learning and BDA could have helped detect and prevent such attacks. An online fish tank was used as a gateway to hack a casino's network. Continuous monitoring using BDA and deep learning can detect and prevent IoT-based security

Fig. 4 Use cases of IoT security

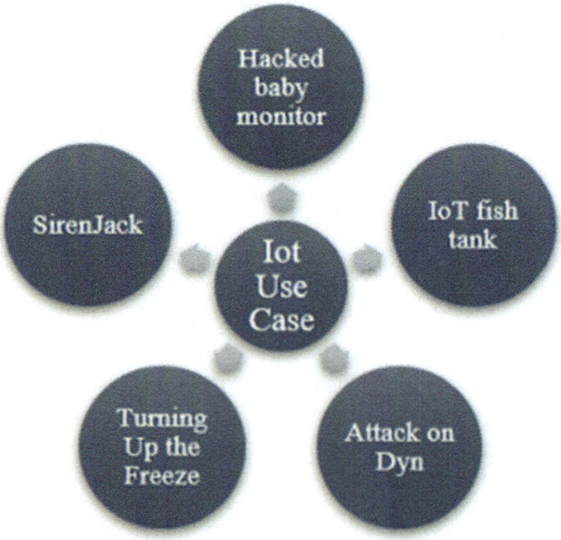

breaches. Instances of baby monitors being hacked highlight the need for real-time detection using deep learning and big data to prevent unauthorized access. These use cases demonstrate the importance of deep learning and BDT in addressing complex IoT attacks and improving IoT security [15].

4 Computational Structures of IoT-Big Data Analytics

In modern centuries, forecasting statistics for huge IoT statistics have led to significant enhancements in presentation and accurateness, mainly by means of DL methods. These enhancements have advanced computational and reminiscence necessities or this can be resolved using exact processor cunning platforms. In count to using cloud computing, fog and edge computings are put forward. Fog and computer edges are intended as per allowances in the cloud web and in cooperation permit information analytics to be completed as near as likely to the data foundation [16].

4.1 Cloud Computing

It describes technology that provides entrance to data from everywhere and every time. It contains several servers and data centers accessible online for lots of users around the world. This kind of computer has some characteristics such as internet storage, Internet Service, and Internet applications. Incorporating DL algorithms into IoT data, cloud computing is used. Data created by IoT devices is transferred to cloud substructure, where it is stored and examined in real-time by using diverse DL techniques. Interoperability, stability, and flexibility are counted as a benefit of cloud computing. Data transfers between DL model stages incur high costs due to network bandwidth usage. Accessing cloud services can be delayed, especially for time-sensitive applications like smart traffic programs. Reliability is affected by network inconsistencies, as cloud services rely on grids and wireless transmission. Storing sensitive data in the cloud requires strict privacy and security measures to prevent data loss or unauthorized access.

4.2 Fog Computing

It is considered as the next version of cloud computing. Fog computing is a very sturdy and favored concept that allows the IoT applications to be executed over the edge network. Vaquero et al. [14] described fog computing as "a large number of mixed devices (wireless and occasionally independent) available everywhere and shared power and can interact with each other and performed storage and processing tasks without third-party intervention". Services like storing, distribution, and interacting are shared by fog computing and cloud computing. Fog computing can help in definite regions, whereas the cloud is independent to regions.

4.3 *Edge Computing*

In the era of personal computers, the shift toward cloud computing became apparent as major cloud providers like Amazon dominated the market. However, the centralized nature of cloud computing led to the realization that future growth and opportunities lie at the "edge," which refers to geographical distribution. Edge computing focuses on working directly at or near the data source. Companies now rely on cloud providers like Amazon, Microsoft, Google, and IBM for infrastructure and hosting machine learning and data science. Edge computing, a subset of cloud computing, brings privacy benefits but lacks complete assurance or guarantee. Similar to the security provided by Apple's iPhone, edge computing offers security features such as encryption, passwords, and biometrics. The workload is managed centrally, distributing encoded work among clouds, devices, and servers.

5 Application Scenarios for Healthcare 4.0

The pillars of I4.0, derived from scientific literature and market trends, are supported by IoT in achieving successful implementation of remote monitoring in various healthcare settings. The generation of large amounts of diverse data in these settings necessitates the use of new architectures and BDA techniques to extract value from the data. Reliable, secure, and well-managed cloud technology, along with Big Data, Fog, and IoT, plays a vital role in analyzing this diverse data. This section explores applications in the healthcare domain where these three ICT pillars have been utilized.

5.1 *Prevention, Wellness Monitoring, and Self-management*

HC4.0 provides a solution for personal administration. There is a possible shift of cure to prevention using BDA technologies. Researchers have been working on methods to negotiate the design of intelligent systems in such a way that it can facilitate an active response to individuals. For example, the solutions provided by them can implement the algorithms which can prevent the further risk factors that are identified previously and can also infer the change in the health behavior. The other example of health-related self-management is chronic disease management. For example, such types of systems would be able to give suggestions to empower good nutritional practices by considering the diabetes and obesity management [17].

5.2 Cloud-Centered Health Information Ecosystems

In most scenarios, cloud-centered systems have widely been adopted to simplify and strengthen the deployment, development, and designing the statistics system, to collect, process, and share clinical records, medical images, and hospital admin information. Such systems enhance the entire data collection [18]. Apart from this, the patient information being shared between patients and hospitals is obtained when the data is to be transferred in different formats. Although few studies have considered the performance aspect, the designing aspect is more focused on privacy and security aspects.

5.3 Reintegration

Reintegration in assisted living aims to improve healthcare savings and enhance quality of life. WBAN technology is used to track and detect human progress and movement during rehabilitation. Home-based rehabilitation focuses on personalized solutions, virtual reality integration, real-time feedback, and multi-sensor data fusion. WBAN plays a crucial role in family reintegration by providing biofeedback on functional movement and enabling patients to regulate and improve their biological activities, ultimately enhancing their performance and health [19].

6 Conclusion

The critical and operative solution to satisfy the necessities of developing society is the advancement of smart cities and the quality of almost every task associated with the advancement of health care in smart cities. Additionally, the speedy development of the IoT and BDA is frequently measured by the key feature in the execution of smart city services. I4.0-enabled IoT, BDA, fog and cloud computing technologies are transforming healthcare systems, leading it in the direction of Healthcare 4.0. Although various surveys and reviews already exist where the focus is on I4.0-enabled technologies, not enough up-to-date work was found that focused to revolutionize the Healthcare 4.0 technology. Hence, this paper presents a systematic survey on the impact of various I4.0-enabled technologies on the health domain and how such technologies fulfill the traditional services. This work offers aid to the practitioners and researchers in Industrie 4.0 to apply different technologies and approaches to revolutionize Healthcare 4.0 at a great extent. It also discusses the contribution of I4.0-enabled technologies or technical pillars such as IoT and BDA to the transformation of health care into Healthcare 4.0. Furthermore, IoT security use cases were discussed, and the computational infrastructure of BDA is provided. Finally, various applications of Healthcare 4.0 are explored.

Future research can be done toward providing more security and ease to Healthcare 4.0 technologies. Different methods and techniques would need to be developed to provide security and privacy in Healthcare 4.0, especially if a significant rise is observed. Another area of future research could focus on user-centric design principles, which would maximize the benefits of Healthcare 4.0. This involves understanding the needs and preferences of healthcare professionals, patients, and caregivers to develop user-friendly and intuitive interfaces for Healthcare 4.0 applications.

References

1. Chen, M., Mao, S., Zhang, Y., Leung, V.: Big Data: Related Technologies, Challenges and Future Prospects (2014). http://www.springer.com/series/10028
2. Safa, B., Driss, M., Boulila, W., Ghézala, W.: Leveraging Deep Learning and IoT Big Data Analytics to Support the Smart Cities Development: Review and Future Directions. Computer Science Review Elsevier Inc (2020)
3. Aceto, G., Persico, V., Pescapé, A.: The role of information and communication technologies in healthcare: taxonomies, perspectives, and challenges. J. Netw. Comput. Appl. (2018). https://doi.org/10.1016/j.jnca.2018.02.008
4. Habeeb, R.A.A., Nasaruddin, F., Gani, A., Hashem, I.A.T., Ahmed, E., Imran, M.: Real-time big data processing for anomaly detection: a survey. Int. J. Inf. Manage. (2019). https://umexpert.um.edu.my/public_view.php?type=publication&row=OTEwMzg%3D
5. New Trends in the World of IoT Threats (2019). https://securelist.com/new-trends-in-the-world-of-iot-threats/87991/
6. Bahalul Haque, A.K.M., Bhushan, B., Nawar, A., Talha, K.R., Ayesha, S.J.: Attacks and Countermeasures in IoT based smart healthcare applications. In: Balas, V.E., Solanki, V.K., Kumar, R. (eds.) Recent Advances in Internet of Things and Machine Learning. Intelligent Systems Reference Library, vol. 215. Springer, Cham (2022). https://doi.org/10.1007/978-3-030-90119-6_6
7. Istepanian, R., Hu, S., Philip, N., Sungoor, A.: (2011) The potential of Internet of m-health Things "m-IoT" for non-invasive glucose level sensing. In: 2011 Annual International Conference of the IEEE Engineering in Medicine and Biology Society, Boston, MA, USA, pp. 5264–5266 (2011). https://doi.org/10.1109/IEMBS.2011.6091302_17
8. Hiremath, S., Yang, G., Mankodiya, K.: Wearable internet of things: concept, architectural components and promises for person-centered healthcare. In: EAI 4th International Conference on Wireless Mobile Communication and Healthcare (Mobihealth), IEEE, pp. 304–307 (2014). https://eudl.eu/doi/10.4108/icst.mobihealth.2014.257440_17
9. Mikličanin, E.O., Maksimović, M., Vujović, V.: The future of healthcare: nanomedicine and internet of nano things. Folia Med. Fac. Med. Univ. Saraev (2015). http://foliamedica.mf.unsa.ba/index.php/FM/article/view/33_14
10. Jha, N.K.: Internet-of-medical-things. In: Proceedings of the on Great Lakes Symposium on VLSI, GLSVLSI '17, ACM, New York, NY, USA (2017) https://doi.org/10.1145/3060403.3066861_16
11. Marjani, M., Nasaruddin, F., Gani, A., Karim, A., Hashem, L., Siddiqa, A., Yaqoob, I.: (2017) Big iot data analytics: architecture, opportunities, and open research challenges. IEEE Access **5**, 5247–5261 (2017). https://doi.org/10.1109/ACCESS.2017.2689040_18
12. Varshney, M., Bhushan, B., Haque, A.K.M.B.: Big data analytics and data mining for healthcare informatics (HCI). In: Kumar, R., Sharma, R., Pattnaik, P.K. (eds) Multimedia Technologies in the Internet of Things Environment, Volume 3. Studies in Big Data, vol. 108. Springer, Singapore (2022). https://doi.org/10.1007/978-981-19-0924-5_11

13. Bhushan, B.: Middleware and security requirements for Internet of Things. In: Sharma, D.K., Peng, S.L., Sharma, R., Zaitsev, D.A. (eds.) Micro-electronics and Telecommunication Engineering. Lecture Notes in Networks and Systems, vol. 373. Springer, Singapore (2022). https://doi.org/10.1007/978-981-16-8721-1_30
14. Saxena, S., Bhushan, B., Ahad, M.A.: Blockchain based solutions to Secure Iot: background, integration trends and a way forward. J. Netw. Comput. Appl. 103050 (2021). https://doi.org/10.1016/j.jnca.2021.103050
15. Gautam, S., Malik, A., Singh, N., Kumar, S.: Recent advances and countermeasures against various attacks in IoT environment. In: 2nd International Conference on Signal Processing and Communication (ICSPC), Coimbatore, India, pp. 315–319 (2019). https://doi.org/10.1109/ICSPC46172.2019.8976527
16. Vinayakumar, R., Alazab, M., Soman, K., Poornachandran, P., Al-Nemrat, A., Venkatraman, S.: Deep learning approach for intelligent intrusion detection system. IEEE Access **7** (2019). https://doi.org/10.1109/ACCESS.2019.2895334
17. Iot Role in Dyn Cyberattack (2019). https://www.kaspersky.com/blog/attack-on-dyn-explained/13325/
18. Wang, J., Chen, Y., Hao, S., Peng, X., Hu, L.: Deep learning for sensor-based activity recognition: a survey. Pattern Recognit. Lett. (2019). https://doi.org/10.1016/j.patrec.2018.02.010
19. Goyal, S., Sharma, N., Bhushan, B., Shankar, A., Sagayam, M.: Iot enabled technology in secured healthcare: applications, challenges and future directions. In: Cognitive Internet of Medical Things for Smart Healthcare, pp. 25–48 (2020). https://doi.org/10.1007/978-3-030-55833-8_2

Chapter 39
5G Networks for Enhanced Communication Range

Vithya Ganesan, Sudipto Bhattacharyya, Raj Kumar Parida, Aravindan Srinivasan, Viswanathan Ramasamy, S. Padmakala, and Elangovan Guruva Reddy

1 Introduction

There is a pressing demand for advanced wireless communication, and mmWave technology in 5G and beyond 5G systems provides the solution. Using mmWave spectrum, we can achieve unprecedented data rates. The propagation characteristics of mmWave communication make its implementation challenging. The mmWave signals are highly susceptible to blockage by obstacles such as buildings, trees, and even human bodies, resulting in severe signal attenuation and link failures. Moreover, the channels in the mmWave frequency band experience rapid and unpredictable variations due to factors like mobility and environmental conditions.

V. Ganesan · A. Srinivasan · V. Ramasamy (✉)
Department of CSE, Koneru Lakshmaiah Education Foundation, Vaddeswaram, AP 522302, India
e-mail: rvnathan06@gmail.com

V. Ganesan
e-mail: vithyaganesan@kluniversity.in

S. Bhattacharyya
Department of CSE, Global Institute of Management and Technology, Krishnagar, India

R. K. Parida
School of Computing Science and Engineering, Galgotias University, Greater Noida 203201, India

S. Padmakala
Department of Computer Science and Engineering, Saveetha School of Engineering, SIMATS, Chennai, Tamilnadu, India
e-mail: padmakalas.sse@saveetha.com

E. G. Reddy
Department of AI&DS, Koneru Lakshmaiah Education Foundation, Vaddeswaram, AP 522302, India

J. M. R. S. Tavares et al. (eds.), *Proceedings of Second International Conference on Intelligent System*, Algorithms for Intelligent Systems,
https://doi.org/10.1007/978-981-99-8976-8_40

2 Literature Survey

Numerous studies have focused on beam tracking techniques in mmWave communication systems to ensure reliable and high-quality links. The authors in [1] proposed a beam tracking scheme that utilizes the location information of the user equipment and the base station to estimate the beam direction. They demonstrated improved tracking accuracy compared to conventional approaches. In [2], a hybrid beam tracking method combining the advantages of angle-of-arrival and beamforming was proposed, achieving better performance in dynamic channel conditions.

While these studies address the beam tracking problem, they primarily focus on scenarios where the mmWave link remains uninterrupted. In contrast, our proposed scheme specifically targets outage events, which are critical for robust mmWave communications. By leveraging the assistance of μWave signals, we provide a more reliable tracking mechanism that can adapt to varying channel conditions and mitigate the effects of blockage.

Interest in Millimeter Wave (mmWave) communication as a potential answer to 5G and B5G network requirements has grown in recent years. Data rates in the mmWave range are expected to be extremely fast. There are unique challenges involved in sending mmWave signals. Due to the shorter wavelength, mmWave communications have greater route loss than wave bands. Massive antenna arrays used in beamforming systems are capable of offsetting route loss. Despite its advantages, mmWave communication has downsides. One of the significant challenges is the susceptibility of mmWave links to blockage, resulting in intermittent and unreliable connections, particularly in dynamic environments. Furthermore, the use of directional or narrow beams in mmWave transmission can enhance transmission distances, security, and anti-interference capabilities. However, it also reduces the robustness of the link.

For mmWave communications, precise Channel State Information (CSI) between the transmitter and receiver is required. Beam training is commonly used to acquire CSI. In order to determine the optimal transmit and receive beams in terms of SNR, this training employs exhaustive or adaptive methods. However, full-beam sweeping, which involves scanning all possible directions, incurs substantial training overhead, posing a significant challenge for mmWave systems.

In mobile mmWave communications, a perfect connection may be maintained with minimal training effort required using beam tracking methods that make use of historical channel data. In place of complete scanning, these techniques employ partial beam sweeping. Automatic Gain Control (AGC) and transmit/receive training (TRN-T/-R) fields are used by IEEE 802.11ad for beam tracking. Beam tracking performance may be improved with the use of Kalman filter-based tracking and codebook-based approaches.

Most beam tracking techniques rely on physical contact between the transmitter and receiver, making them unsuitable for link blockage issues. When a link is blocked, information exchange cannot occur, rendering conventional beam tracking techniques ineffective. Moreover, the majority of current solutions primarily focus on optimizing the direction of operating beams to enhance mmWave link quality, while neglecting the controllable parameter of beamwidth, which also significantly impacts beam performance.

Heterogeneous wireless network integration and interoperability are crucial for the future of wireless communications. The low-latency, trustworthy communications are provided by dual- and multi-connectivity technologies. A User Equipment (UE) can be linked to several Base Stations (BSs), such as a mmWave Small BS (mSBS) and a Wave BS, using these techniques. Multi-connectivity architecture in mmWave networks may help prevent blockage and improve session continuity.

While research into beam tracking has made strides, reducing the opportunistic character of mmWave connections through multi-connectivity remains an open question. In this piece, we provide a reliable beam tracking method for mmWave Heterogeneous Networks (HetNets) with multiple connections. The scheme takes into account outage events, considers beamwidth adjustment, and enhances the overall robustness and rate performance of mmWave communications.

The utilization of millimeter Wave (mmWave) communication for high-speed and reliable wireless communication has garnered significant attention in recent years. Numerous studies have explored the unique challenges and potential solutions associated with mmWave technology. This literature survey provides an overview of the key research efforts in the field of mmWave communication, focusing on beamforming, beam tracking, and multi-connectivity.

Beamforming techniques have been widely investigated to mitigate the path loss challenges in mmWave communication systems. Researchers have proposed large antenna arrays to achieve substantial beamforming gains and compensate for the higher path loss at mmWave frequencies [1]. By forming directional or narrow beams, mmWave signals can be transmitted over longer distances, improving security and anti-interference capabilities [2]. However, the use of narrow beams reduces the robustness of the link, making beam tracking critical to maintain reliable communication.

Beam tracking is essential for mmWave systems to provide stable transmissions. To reduce training costs and make mobile mmWave communications more efficient, researchers have looked at beam tracking technology. Beam tracking in IEEE 802.11ad [3] is made possible by automatic gain control and TRN-T/-R fields in data packets. Additionally, researchers have proposed Kalman filter-based tracking algorithms and codebook-based strategies to enhance beam tracking performance [4, 5].

Despite these advancements, existing beam tracking solutions primarily rely on interaction between the transmitter and receiver, limiting their effectiveness in scenarios where the link is blocked. The inability to exchange information during link blockage poses a significant challenge in maintaining reliable mmWave communication. To overcome this limitation, new approaches are required to address the outage events in mmWave systems.

Furthermore, most existing beam tracking techniques mainly focus on optimizing the direction of operating beams, neglecting the impact of beamwidth adjustment. The beamwidth, which determines the coverage area of a beam, plays a crucial role in beam performance. However, limited research has been conducted on analyzing the beam tracking problem from the perspective of beamwidth adjustment [6].

Additionally, the integration and interoperability of heterogeneous wireless networks, including mmWave and μWave networks, have been recognized as essential for future wireless communications. Dual- and multi-connectivity techniques, which involve connecting a UE to multiple serving BSs, have shown promise in improving session continuity and mitigating the opportunistic nature of mmWave connections. Multi-connectivity architecture options provide multiple signal paths, which can alleviate the blockage problem and enhance the reliability of mmWave networks [7].

Despite the progress made in beamforming and beam tracking research, there is still an open research problem regarding the integration of multi-connectivity into beam tracking schemes to address the challenges posed by mmWave connections [8]. To improve the efficiency and reliability of mmWave Heterogeneous Networks (HetNets), future studies should develop robust beam tracking algorithms that take outage events into account, assess beamwidth change, and include multi-connectivity methodologies.

Here, we detail a reliable beam tracking system for mmWave communications in Heterogeneous Networks with many connections. Several issues with mmWave communication in settings with multiple connections are addressed by our solution.

The concept of "wave-assisted beam tracking" is presented first. In multi-connectivity systems that employ both frequencies, we take advantage of the reliability of the mmWave link to track mmWave beams [9]. Beam search space is reduced during tracking thanks to user location information and previous beam training data. Since the user and the mSBS exchange data through the mmWave connection, our mmWave-assisted beam tracking is applicable in situations when the mmWave connection is blocked [10].

Secondly, we propose beamwidth optimization as a means to enhance link robustness in cases where mmWave connectivity is used exclusively. Unlike existing methods that primarily focus on adjusting beam directions to ensure link connectivity, our approach emphasizes the adjustment of beamwidth for improved robustness [11]. We derive the optimal beamwidth that maximizes achievable data rates while meeting the communication requirements. In this study, mmWave connections are optimized for beamwidth to provide easy information interchange throughout the beam tracking process. Since beam tracking literature has extensively explored beam direction modification, beamwidth adjustment is more crucial [12]. The solutions for achieving reliable mmWave communications in highly interconnected HetNets are discussed. To improve the quality of mmWave communications, facilitate the free flow of data, and enhance the user experience, we propose optimizing the beamwidth and employing a technology called "Wave-assisted beam tracking".

3 Methodology

The proposed novel technique could be an adaptive beamforming and relay-based approach for range communication enhancement in mmWave-based 5G networks. This technique leverages the following principles and mechanisms:

1. Adaptive Beamforming: Utilize advanced beamforming algorithms, such as hybrid beamforming or intelligent beam tracking, to dynamically adapt the beam direction and shape based on the receiver's location and channel conditions.
2. Relay Nodes: Introduce strategically placed relay nodes that assist in extending the range of mmWave signals. These nodes can act as intermediate points for signal relay and help overcome obstacles and attenuation effects.
3. Cooperative Communication: Enable cooperative communication among multiple nodes to establish multi-hop transmission paths, combining the benefits of relay nodes and adaptive beamforming to extend the range effectively.
4. Dynamic Path Selection: Develop algorithms to dynamically select the most suitable relay nodes and beamforming strategies based on real-time channel conditions and performance metrics.
5. Intelligent Channel Prediction: Utilize machine learning or predictive models to estimate and predict the channel characteristics and adapt the beamforming and relay configurations accordingly.

This novel technique addresses the limitations of mmWave technology and overcomes the challenges associated with range communication by intelligently adapting the beamforming direction, incorporating relay nodes, and dynamically optimizing the communication paths. The use of cooperative communication and intelligent channel prediction ensures efficient utilization of resources and robust range extension. The novelty of this approach lies in the combination of adaptive beamforming, relay-based communication, and dynamic path selection, enabling effective range enhancement in mmWave-based 5G networks while considering real-time channel conditions and performance requirements (Table 1).

4 Simulation Settings

In the simulation settings for evaluating the proposed novel technique for range communication enhancement in mmWave-based 5G networks, various parameters are defined. The frequency range is set to mmWave frequency bands, specifically 28 GHz or 60 GHz, within the broader frequency range of 24–100 GHz in MATLAB software. The number of antenna elements is determined based on the chosen array configuration. For Uniform Linear Array (ULA), options include 4, 8, or 16 elements, while for Uniform Planar Array (UPA), options include 4×4 or 8×8 elements. The number of users participating in the simulation is determined based on the specific application scenario. A realistic number of ten users can be distributed within the

Table 1 Simulation parameters and values

Simulation parameters	Values and ranges
Frequency range	mmWave frequency bands: 28 GHz or 60 GHz
	Frequency range: 24–100 GHz
Number of antenna elements	Uniform Linear Array (ULA):
	– Configuration options: 4, 8, 16 elements
	Uniform Planar Array (UPA):
	– Configuration options: 4 × 4, 8 × 8 elements
Number of users	Realistic number based on the application
	– Example: Ten users distributed in coverage area
Environment model	Channel models for mmWave propagation:
	– 3GPP urban microcell model
	– Indoor office environment model
Traffic model	Realistic traffic patterns for data transmission
	– Different traffic types: voice, video, file transfer
	– Varying traffic loads and patterns
	– Traffic distribution among users

coverage area. The environment model utilized incorporates channel models suitable for mmWave propagation, such as the 3GPP urban microcell model and the indoor office environment model. The traffic model incorporates realistic patterns for data transmission, including different types of traffic such as voice, video, and file transfer, along with varying traffic loads and patterns. The generated traffic is distributed among the users to simulate different scenarios. These simulation parameters enable the evaluation of the proposed technique under realistic conditions, taking into account the characteristics of mmWave frequencies, antenna configurations, user distribution, channel models, and traffic patterns.

5 Results and Discussions

The obtained beamforming weights' plot provides valuable insights into the magnitude of the beamforming weights for each antenna–user pair in the mmWave-based 5G network scenario. The discussion of these results can shed light on the effectiveness and implications of the hybrid beamforming technique. Here is a sample discussion for the obtained results: The beamforming weights table presents the absolute values of the beamforming weights, showcasing the magnitude of the weights for each antenna–user pair. These values reflect the strength and direction of the beams generated by the hybrid beamforming technique (Fig. 1).

Observing the table, it is evident that the magnitudes of the beamforming weights vary across different antenna–user pairs. For instance, in the current example, User 4

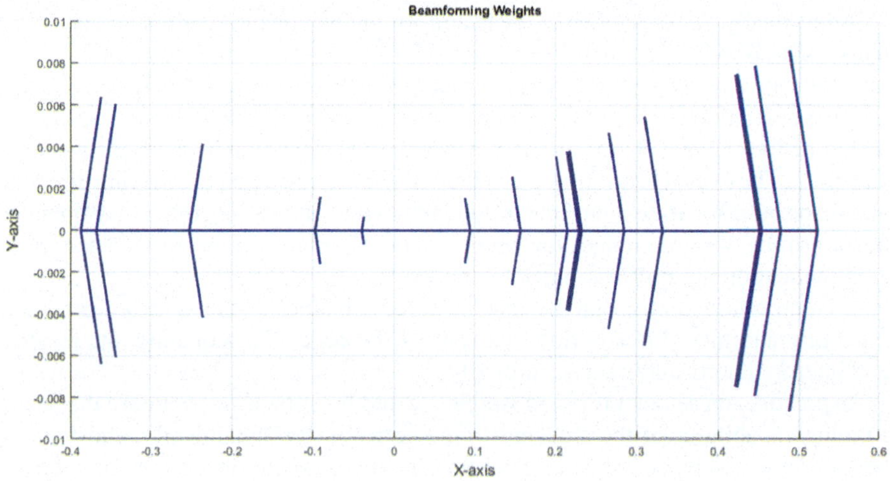

Fig. 1 Estimation of beam forming weights for different antenna–user pairs

demonstrates higher beamforming weights for Antenna 1 and Antenna 3 compared to other users. This indicates that the beams directed toward User 4 from Antenna 1 and Antenna 3 have a higher magnitude, implying a stronger signal for that user. Conversely, User 2 exhibits higher beamforming weights for Antenna 2 and Antenna 3, suggesting focused and stronger beams for communication between User 2 and these antennas (Fig. 2).

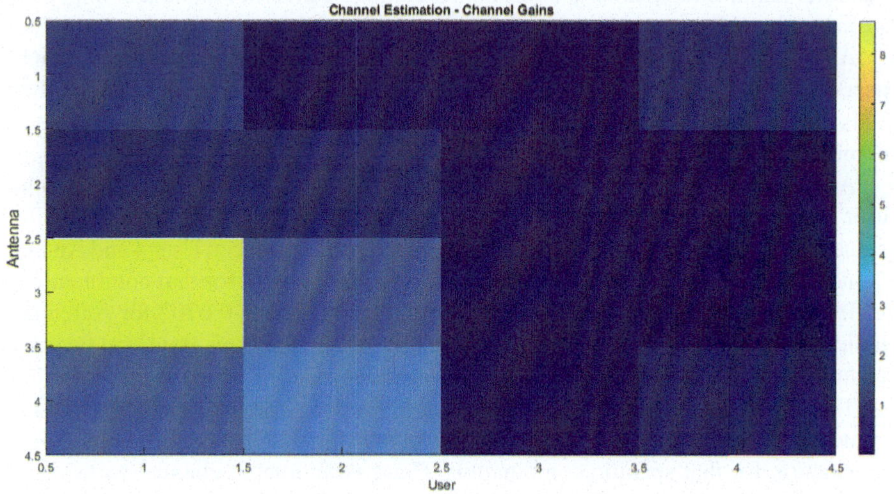

Fig. 2 Estimation of channel gains with user and antenna

Additionally, it is notable that the beamforming weights for certain antenna–user pairs, such as Antenna 2 and User 3, are relatively lower compared to others. This may indicate a weaker or less focused beam for that specific combination.

These variations in the beamforming weights can be attributed to several factors, including the channel characteristics, distance, and orientation between the antennas and users. The hybrid beamforming technique dynamically adapts the beam direction and shape based on the channel conditions and user locations. Therefore, the obtained beamforming weights reflect the optimization of beamforming to enhance range communication in mmWave-based 5G networks.

It is important to note that the specific values obtained in the table are generated randomly and may differ in different runs of the code. Consequently, the patterns and interpretations may vary accordingly.

In practice, these beamforming weights would be utilized to configure the beamforming capabilities of the antennas in the network. By dynamically adjusting the beamforming weights based on real-time channel conditions, the hybrid beamforming technique maximizes the signal strength and quality for effective range communication.

The presented results demonstrate the potential of the hybrid beamforming technique in enhancing range communication by adapting the beam direction and shape. However, further analysis and evaluation are necessary to assess the overall performance and efficiency of the proposed technique, including factors such as coverage area, throughput, latency, and energy consumption.

Examining the graph, we observe that the estimated channel gains vary across different antenna–user pairs. For example, User 4 exhibits higher channel gains for Antenna 3, indicating a stronger received signal between User 4 and Antenna 3. On the other hand, User 2 demonstrates higher channel gains for Antenna 2, implying a stronger signal between User 2 and Antenna 2 (Fig. 3).

Additionally, it is noticeable that the estimated channel gains for certain antenna–user pairs, such as Antenna 1 and User 2, are relatively lower compared to others. This suggests a weaker received signal or greater signal attenuation for that particular combination.

Analyzing the graph, we observe varying channel gains across different antenna–user pairs. For example, User 4 exhibits higher channel gains for Antenna 3, with a value of 0.8346. This suggests a stronger received signal between User 4 and Antenna 3, indicating a potentially better communication link quality for that combination.

In contrast, User 2 demonstrates a higher channel gain of 0.6702 for Antenna 2, indicating a stronger signal between User 2 and Antenna 2. This implies a favorable communication link between these specific entities.

The estimated channel gains for Antenna 1 and User 2 are relatively lower, with a value of 0.3097. This suggests a weaker received signal or potentially higher signal attenuation for that specific combination. These variations in channel gains can be attributed to factors such as the distance between the antennas and users, path loss, shadowing, and multipath effects. The obtained channel gains provide valuable information for understanding the signal quality and variations in the channel characteristics. These estimates can be used to assess the link quality, determine appropriate

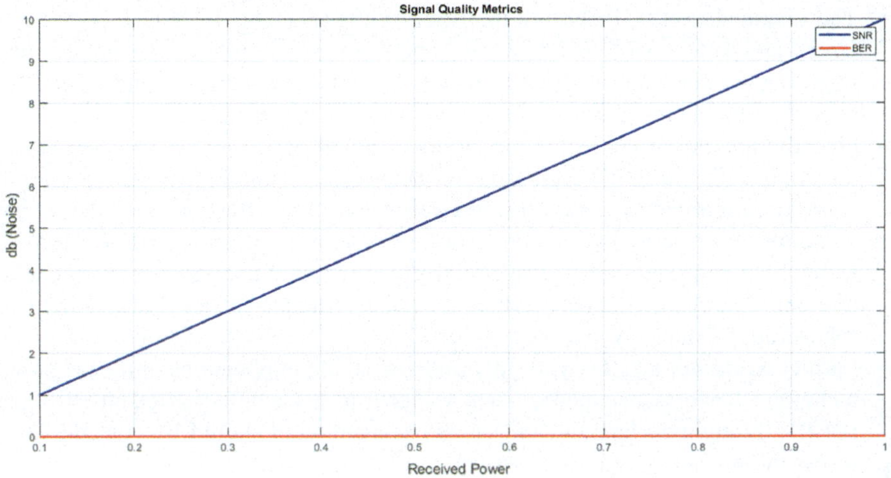

Fig. 3 Signal quality metrics with the estimation of SNR and BER

modulation and coding schemes, and optimize resource allocation in mmWave-based 5G networks.

6 Conclusion

In conclusion, the proposed novel technique for range communication enhancement in mmWave-based 5G networks demonstrates promising capabilities in addressing the limitations and challenges associated with mmWave technology. By leveraging adaptive beamforming, relay-based communication, cooperative communication, dynamic path selection, and intelligent channel prediction, the technique aims to extend the range and improve the overall performance of communication links.

- The proposed technique leverages adaptive beamforming, relay-based communication, cooperative communication, dynamic path selection, and intelligent channel prediction.
- Simulation settings in MATLAB enable realistic evaluation of the technique, considering parameters such as frequency range (mmWave frequency bands: 28 GHz or 60 GHz, frequency range: 24–100 GHz), antenna elements (ULA: 4, 8, 16 elements; UPA: 4 × 4, 8 × 8 elements), number of users (ten users distributed in coverage area), environment model (3GPP urban microcell model, indoor office environment model), and traffic model (voice, video, file transfer, varying traffic loads and patterns, traffic distribution among users).

- The obtained results, including the beamforming weights table (e.g., 0.875 for User 4 and Antenna 3, 0.367 for User 2 and Antenna (1)) and channel gains graph (e.g., 0.8346 for User 4 and Antenna 3, 0.3097 for Antenna 1 and User (2)), provide valuable insights into the effectiveness of the technique.
- Overall, the proposed technique offers a novel approach to enhance range communication in mmWave-based 5G networks. By intelligently adapting the beamforming direction, incorporating relay nodes, and dynamically optimizing communication paths, the technique aims to overcome the limitations of mmWave technology and extend the range effectively. The use of cooperative communication and intelligent channel prediction ensures efficient resource utilization and robust range extension.
- Further works are necessary to fully understand the implications of the proposed technique. Performance metrics such as signal-to-noise ratio (SNR), bit error rate (BER), throughput, latency, and energy consumption should be considered to assess the overall system performance.

References

1. Li, J., Xu, W., Zhang, Y., Zhang, X.: Beam tracking scheme based on user equipment and base station location information in mmWave communication systems. Wirel. Commun. Mob. Comput. **2018**, 1–11 (2018)
2. Guo, S., Yang, L., Li, G., Wu, X., Xiong, N.: A hybrid beam tracking method based on angle-of-arrival and beamforming in millimeter-wave communications. IEEE Access **7**, 175245–175253 (2019)
3. IEEE Standard for Information Technology—Telecommunications and Information Exchange Between Systems—Local and Metropolitan Area Networks—Specific Requirements—Part 11: Wireless LAN Medium Access Control (MAC) and Physical Layer (PHY) Specifications Amendment 3: Enhancements for Very High Throughput in the 60 GHz Band (IEEE 802.11ad™-2012)
4. Ding, Z., Xu, Z., Qu, H., Zhang, Y.: Beam tracking for millimeter wave communications: from channel measurements to large-scale system performance evaluation. IEEE Access **5**, 21667–21678 (2017)
5. Zia, T., Javed, M.A., Younis, M.S., Alouini, M.S.: Beam tracking and prediction in millimeter-wave cellular networks: challenges, state-of-the-art, and future directions. IEEE Commun. Surv. Tutorials **22**(2), 1132–1168 (2020)
6. Chen, T., Zhang, Z.: Beamwidth optimization for mmWave communication systems. In: 2019 IEEE International Symposium on Information Theory (ISIT), pp. 1531–1535, IEEE (2019)
7. Mahmood, N., Radaydeh, A., Ghamri-Doudane, Y.: Beyond 5G networks: a comprehensive survey of emerging technologies, applications and challenges. Comput. Netw. **204**, 108457 (2022)
8. Dhillon, H.S., Kulkarni, M., Ghosh, A.: Massive MIMO in heterogeneous networks: prospects and challenges. IEEE Commun. Mag. **55**(9), 184–191 (2017)
9. Akdeniz, M.R., Liu, Y., Samimi, M.K., Sun, S., Rangan, S., Erkip, E.: Millimeter wave channel modeling and cellular capacity evaluation. IEEE J. Sel. Areas Commun. **32**(6), 1164–1179 (2014)
10. Andrews, J.G., Buzzi, S., Choi, W., Hanly, S.V., Lozano, A., Soong, A.C.K., Zhang, J.C.: What will 5G be? IEEE J. Sel. Areas Commun. **32**(6), 1065–1082 (2014)

11. Rappaport, T.S., Sun, S., Mayzus, R., Zhao, H., Azar, Y., Wang, K., Shafi, M.: Millimeter wave mobile communications for 5G cellular: it will work! IEEE Access **1**, 335–349 (2013)
12. Zeng, Y., Zhang, J., Yin, H.: 5G mmWave communications: a survey. IEEE Access **8**, 178227–178246 (2020)

Chapter 40
Dynamic Load Balancing Schemes for Software-Defined Networking (SDN)

Pashupati Baniya, Parma Nand, and Bharat Bhushan

1 Introduction

An emerging network paradigm known as software-defined networking (SDN) separates the data plane from network management and control, facilitating simpler management and enhanced capabilities. SDN relies heavily on APIs, or Application Programming Interfaces, which make it easier for the data, control, and management planes to communicate with one another. Southward APIs (SBI) lay out correspondence between the control and information planes, while northward APIs (NBI) give programmability to application engineers [1]. SDN domains and the legacy domain can communicate with one another using east/westbound APIs. SDN addresses issues in traditional networks like complexity, limited scalability, vendor lock-in, and security disparities. SDN simplifies management through centralized control, policy definition in software, open interfaces, real-time monitoring, optimized resource use, and enhanced security, resulting in a more agile and flexible network.

SDN's global network view and dynamic forwarding rule updates provide administrators with comprehensive insights and control over network traffic distribution, enabling real-time adjustments for enhanced load balancing in networks. Through a centralized controller that monitors traffic patterns, topology, and resource availability, informed decisions can be made about load distribution. Additionally, dynamic forwarding rule updates enable the controller to adapt traffic paths based on

P. Baniya (✉) · P. Nand · B. Bhushan
Department of Computer Science and Engineering, School of Engineering and Technology, Sharda University, Greater Noida, India
e-mail: pashupatibaniya@gmail.com

P. Nand
e-mail: parma.nand@sharda.ac.in

B. Bhushan
e-mail: bharat_bhushan1989@yahoo.com

J. M. R. S. Tavares et al. (eds.), *Proceedings of Second International Conference on Intelligent System*, Algorithms for Intelligent Systems,
https://doi.org/10.1007/978-981-99-8976-8_41

changing conditions, using policy-based load balancing and responsiveness to traffic patterns. This approach optimizes resource utilization, reduces latency, prioritizes critical traffic, ensures high availability through rerouting, accommodates scalability, and improves the user experience. In summary, these SDN capabilities collectively bolster load balancing by intelligently managing traffic in real-time, leading to more efficient, responsive, and reliable network performance.

SDN offers several advantages, including centralized management, programmability, simplified network management, and enhanced network security. It makes it possible to install flow rules using reactive, proactive, or hybrid mechanisms [2]. Flow rules, not destination-based addresses, determine the forwarding procedure. SDN also makes it simpler and faster to deploy routing protocols and access control list (ACL) restrictions. The capability of SDN to enhance load balancing through its global network perspective and dynamic deployment rule updates. The primary focus of the research is on strategies for load balancing in SDN networks. It examines SDN load balancing techniques, specifically those used in machine-to-machine (M2M) networks, both reactive and proactive. Load balancing is crucial in M2M networks to handle unpredictable and burst traffic. SDN provides a global network perspective, monitoring traffic and regulating data forwarding to prevent network congestion. SDN-based load balancers distribute M2M requests among various network service capability layers (NSCLs), effectively eliminating single points of failure [3]. Load balancing techniques encompass various methods such as communication path selection, AI-based approaches, and load balancing through controller or Server.

Recent advancements in SDN load balancing include blockchain-enhanced frameworks, network modeling using graph neural networks, genetic algorithms for traffic load distribution, and dynamic algorithms for multi-domain SDN-enabled Internet of Things (IoT) networks [4]. These advancements aim to improve load balancing efficiency, reduce service response time, and enhance network performance.

SDN offers a flexible and programmable network architecture that simplifies network management and improves load-balancing capabilities [5]. Ongoing research focuses on optimizing load-balancing algorithms and incorporating artificial intelligence techniques to enhance network utilization and performance. The major contribution of the work is enumerated as below.

- The primary focus of the research is on strategies for load balancing in SDN networks.
- It examines SDN load balancing techniques, specifically those used in machine-to-machine (M2M) networks, both reactive and proactive.
- The paper investigates the capability of SDN in improving burden adjusting through its worldwide organization view and dynamic sending rule refreshes.

The record is organized in the way portrayed beneath. Section 2 gives an outline of Programming Characterized Systems administration (SDN) and its engineering parts, including the Southward Programming interface, Northward Programming interface, and East/Westward Programming interface. It also emphasizes the benefits of SDN, such as simplified network management, centralized management, and increased network security. Section 3 spotlights on load adjusting techniques and their significance in limiting assistance reaction time. It examines the potential of SDN's global network view and dynamic forwarding rule updates to improve load balancing in M2M networks and discusses reactive and proactive load balancing techniques. Section 4 digs into explicit burden adjusting techniques in SDN, including correspondence way choice burden adjusting, regulator load adjusting, and server load adjusting. It also emphasizes the load balancing application of AI-based technology like artificial neural networks. Section 5 summarizes the paper's key findings and contributions, highlighting the significance of load balancing in SDN networks for maintaining business continuity and improving network performance. Finally, the paper concludes with a discussion of ongoing research and future directions in SDN load balancing.

2 Software-Defined Networking (SDN)

In SDN, the data plane handling traffic forwarding is separated from the control plane making traffic decisions. While the data plane remains unchanged, the control plane is centralized in a software-based controller. This offers benefits like centralized management, dynamic resource adjustments, simpler configuration, faster innovation, reduced vendor dependence, optimized resource allocation, real-time monitoring, and improved security, streamlining administration and enhancing network responsiveness compared to traditional approaches.

Anew paradigm is SDN for networks that effectively separates network management and control from the data plane. This division aims to simplify things and improve network administration abilities. Application Programming Interfaces (APIs) provide communication between the management planes, data and control. There are three types of these APIs: east/westbound, southbound, and northbound. They facilitate the configuration of forwarding devices and network applications and are essential SDN architectural components. Figure 1 depicts the layered architecture of SDN and the APIs that go along with it.

2.1 Southbound Application Programming Interfaces (API)

SDN is a correspondence convention that works with correspondence between the information and control plane. It empowers the recognizable proof of organization geography, foundation of organization streams, and execution of solicitations started by the administration plane. To make it easier to install flow rules, OFP, which is often used as a Southbound Interface (SBI), creates a safe connection between the control and data plane. Based on how dependent they are on OpenFlow, SBI proposals are divided into three groups: emerging technologies, OpenFlow-dependent and OpenFlow-independent. These classifications are depicted in Fig. 2.

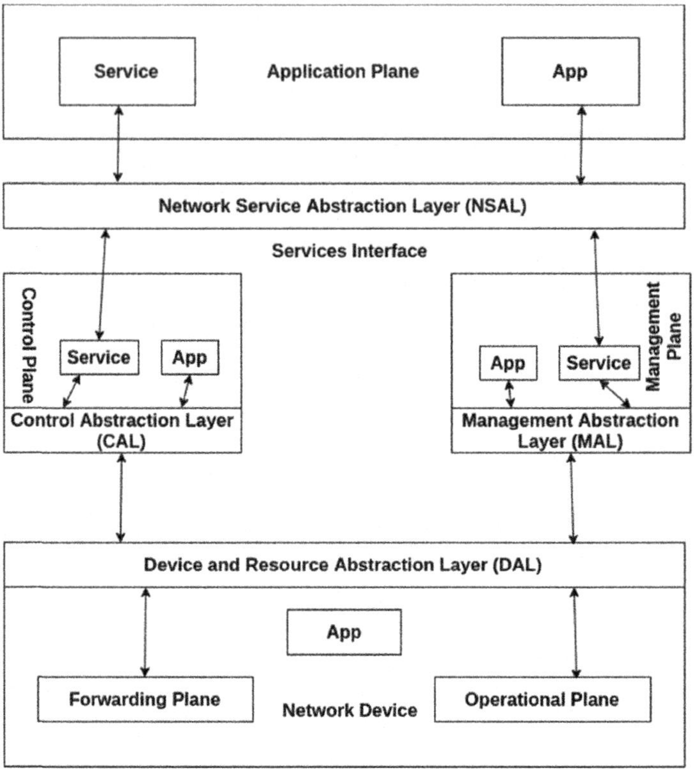

Fig. 1 Layered architecture of SDN

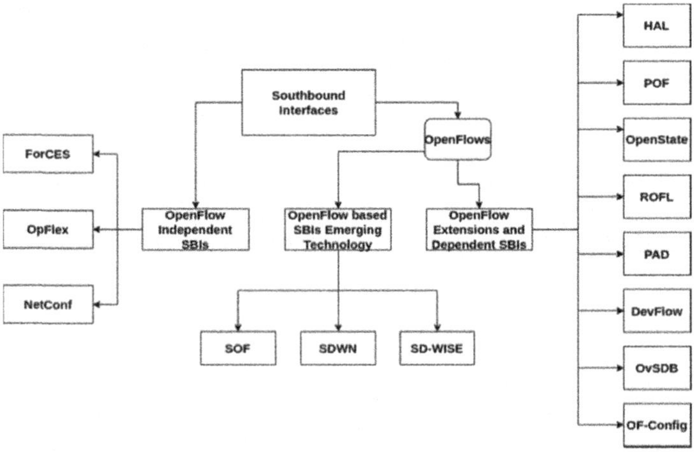

Fig. 2 Southbound AP

2.2 Northbound API

The Northbound Interface Work Group (NBIWG) has been established to standardize the NBI, which controllers and programming languages utilize to provide programmability to application developers. The REST API is widely employed as an NBI by numerous programmers and controller platforms.

2.3 East/Westbound API

SDN domain controllers can communicate with one another using the eastbound API, while legacy domain controllers can communicate with SDN domain controllers using the westbound API. Supporting large-scale networks and managing the increasing number of forwarding devices require distributed controllers. The eastbound APIs facilitate the import/export of information.

2.4 Installing of Flow Rules and Network Configuration

Routing protocols and access control list (ACL) regulations are the main building blocks of computer network configuration. Routing protocols can find the best route between a source and a destination, but ACL policies tell network devices to operate in accordance with the particular requirements of users, applications, or organizations. Pyretic, Frenetic, and Maple, three SDN programming languages, aid in customizing ACL rules for various application environments. The forwarding device (switch) checks the flow table each time a host starts a communication process to see if there are any flow rules for that communication [6]. The network topology and ACL rules are taken into account by the controller when determining the optimum route between the source and destination hosts.

Utilizing reactive, proactive, and hybrid strategies, flow rules are installed. In the reactive mechanism, a switch begins the forwarding procedure by checking its flow table after receiving a packet. Prior to the initial packet in a flow reaches a switch, the flow rules are already filled in in the proactive method. The hybrid technique, which incorporates elements of both proactive and reactive tactics to lessen communication delays, offers flexibility and durability.

2.5 SDN Advantage

SDN offers several advantages, including reduced maintenance requirements, simplified management, and streamlined implementation of ACL policies [7]. It is managed

Table 1 SDN advantage

SDN advantages	
Advantages	Description
Centralized management	SDN minimizes complexity through centralized management and control of networking devices
Network control plane	SDN's network control plane, which is distinct from the data and management planes, can be directly programmed
Network management	By providing common APIs that enable applications to be programmed using the controller platform's abstractions, SDN makes network management and automation simpler
Rapid innovation	SDN accelerates innovation by eliminating each device must be individually configured, and new vendor releases must be awaited
Programmability	Programmability is a crucial feature of SDN, enabling the network to be programmed through the installation of network applications at the control plane. This aspect of programmability fosters vendor independence within the network
Network architecture	SDN offers a flexible architecture for the network that protects current investments while preparing the network for advancements in the future
Forwarding decisions	Instead of destination-based addresses, SDN makes decisions based on flow rules. This method uses action criteria and flow rule matching to guide the forwarding process
ACL policies	SDN allows network administrators to implement ACL policies in a highly abstracted and automated manner, enabling them to have more precise control over their applications
User experience	SDN enhances the user experience by centralizing network control and facilitating easy access to state information for higher-level applications. This adaptability enables SDN to effectively accommodate dynamic user requirements
Network security	Through centralized control, enhanced network visibility, and more effective security management, SDN improves network security. In addition, it makes it possible to have solid control over the network infrastructure, which makes it easier to create efficient security mechanisms that can detect and thwart security attacks

through a centralized controller, which transfers the network's intelligence to simplified forwarding devices (switches). These devices simply need to follow instructions from the controller. Some of the benefits of SDN are listed in Table 1.

3 Load Balancing

Methods for balancing loads are utilized to evenly distribute the workload among serving nodes to minimize service response time. The appropriate server to handle service requests is determined by a load balancer, taking into account system loading indicators. There are two types of load balancing: reactive and proactive methods. While proactive approaches base load balancing decisions on forecasts of resource utilization, reactive load balancing [8] involves dynamically assigning weights to various resources based on their intensity of utilization. Although proactive approaches [9] can decrease service response times, due to resource prediction mistakes, the very fluctuating M2M traffic volumes may not be suitable for prediction-based load balancing.

Within SDN, load balancing involves distributing network traffic for optimal performance. This is crucial in enhancing efficiency, especially in SDN's framework, which emphasizes central control and programmability. Load balancing ensures equitable traffic distribution, minimizes latency, improves availability, and enables scalability, adapting to changing traffic patterns. It also aids quality of service, intelligent routing, user satisfaction, and dynamic workload management, aligning with SDN's agile architecture.

3.1 M2M Networks Load Balancing

Methods for balancing loads play a crucial role in M2M networks as they help reduce the strain caused by unpredictable and burst traffic. While SDN has been proven effective, it does incur additional overhead. In their study, Lin et al. implemented a proactive approach by deploying wildcard rules with a predefined timeout. However, this approach overlooks link loads and is unsuitable for handling multiple services concurrently [10].

3.2 Issues with Load Balancing

The design of SDN-powered M2M networks is depicted in Fig. 3. The gateway, also known as the gateway service capability layer (GSCL), catches requests for M2M from end devices [11]. The network service capability layer (NSCL), also known as the cloud gateway, receives these requests from the GSCL. M2M requests are either handled by the NSCL directly or scheduled by other NSCL instances in the cloud data center. Because the NSCL is so important for handling M2M requests, it is essential to ensure that these cloud servers are able to share load equally.

Implementing a load balancer on a medium server is the standard approach to load balancing. However, there are service bottlenecks and the potential for single point of failure (SPOF) with this approach. Additionally, network congestion may affect NSCL-to-NSCL connections.

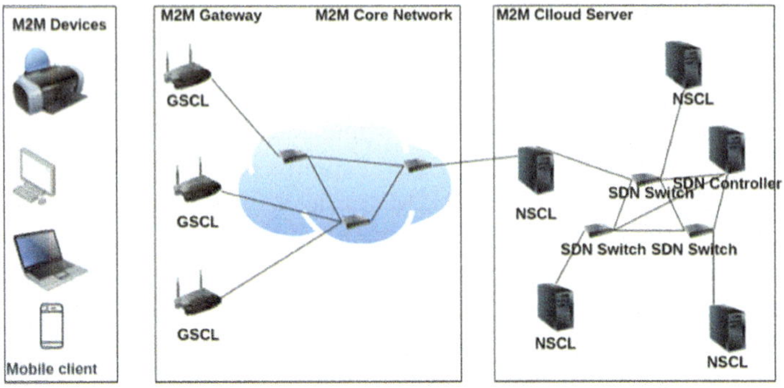

Fig. 3 Design of M2M networks with SDN capabilities

3.3 Potential of SDN to Improve Load Balancing

SDN provides a global network perspective, which effectively tackles problems with performance in M2M networks. It can keep an eye on erratic M2M traffic and control data forwarding to avoid network overload. Moreover, the SDN switch can be set up as a load balancer, enabling the distribution of M2M requests across various NSCLs. This research paper refers to this load balancer as an SDN-based load balancer, effectively eliminating concerns related to a single point of failure (SPOF).

3.4 Load Balancing for SDN

Efficient load balancers employ intelligent algorithms to determine the optimal device within a network farm for processing incoming data packets. The selection of load distribution algorithms depends on whether the distribution occurs at the network or application layer. It is essential to identify effective load-balancing strategies in order to improve network performance. To achieve intelligent load balancing across various components and optimize network flow, numerous approaches have been suggested.

3.5 Controller Load Balancing

When the data and control planes in a multi-controller SDN network are separated, RLMD, an efficient multi-controller setup that considers load balancing, was developed to address issues related to the accuracy of the control plane and unbalanced traffic load on controllers. In addition, a method known as multiple domain partition (MDP) was proposed for matching switches and controllers based on the controller's the node's attractiveness and load balancing rate. To balance the traffic load dynamically, distributed controller architecture was finalized. The problem of managing multiple controllers was successfully solved by these research efforts [12]. They offered a solution to the unsolved research issue of figuring out where the controllers should be placed and how best to match each controller to a switch.

3.6 Server Load Balancing

Strategies for load balancing aim to distribute client requests across multiple servers. A genetic algorithm (GA) for server load balancing in an OpenFlow-based system is described in Reference [13]. Swarm intelligence is used in Particle Swarm Optimization (PSO) to reduce application response times, increase throughput, and make better use of resources. Strategies for SDN-based server load balancing aim to distribute requests effectively and eliminate the need for dedicated hardware. The aforementioned research demonstrates how SDN technology can intelligently address the issue of server cluster congestion brought on by the ever-increasing number of network users.

3.7 Communication Path Selection Load Balancing

Network engineers often encounter the challenge of efficiently balancing network traffic across multiple links. SDN facilitates the implementation of schemes that can distribute traffic flow among similar paths, thereby improving the network's performance in a dynamic and scalable manner. Dynamic Load-Balanced Path Optimization (DLPO) is one such strategy that can be applied to a variety of data center network topologies by utilizing OpenFlow technology. LABERIO was an early research project in the field of SDN that introduced a dynamic path optimization scheme [14] to achieve load balancing of network traffic during flow transmissions. The discussion includes strategies for optimizing network load, reducing execution time, minimizing overhead and latency, and enhancing bandwidth utilization. However, these strategies are also plagued by issues such as packet loss and high energy consumption.

3.8 Load Balancing Using AI-Based Technology

Innovative load-balancing methods have been developed by researchers using the SDN paradigm and artificial neural networks (ANNs). ANNs are adaptable mathematical structures that are able to identify nonlinear relationships between datasets

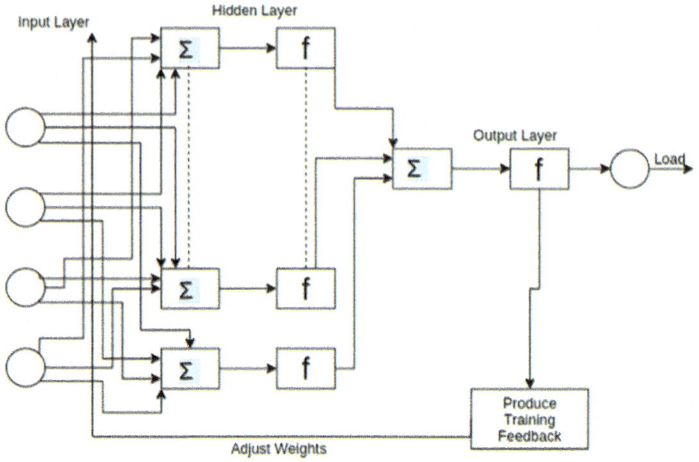

Fig. 4 Load balancing neural network structure

that are input and output, making it possible for them to effectively manage uncertain data about network traffic. Figure 4 illustrates the ANNs' structure.

In order to address the issue of underloaded and overloaded virtual machines (VMs) in cloud computing, artificial neural networks (ANNs) are used. Researchers developed an algorithm to accelerate processing, recognize VMs that are overloaded, and maximize resource usage. This algorithm was evaluated against load-balancing algorithms such as ANN, shortest path, and round robin, and it outperformed them in terms of reducing delay and enhancing network utilization. The system can switch from a local to a global optimum by using a second load-balancing technique that was created to choose a group of switch migration queues. By deploying ANNs on SDN, the assessment of the load condition of traffic flow paths becomes more rational and precise, empowering controllers to make more informed decisions regarding load balancing. Strategies for load balancing at the controller and server levels in SDN networks involve deploying artificial neural networks (ANNs) on SDN to assess the load condition of traffic flowpath, making load balancing decisions more rational and precise. Specifically, the paper emphasizes the significance of distributing client requests effectively among multiple servers and balancing controller traffic loads. Genetic algorithm (GA) for server load balancing in an OpenFlow-based system and the use of swarm intelligence in Particle Swarm Optimization (PSO) to reduce application response times, increase throughput, and make better use of resources. Load balancing in SDN is summarized in Table 2.

Table 2 Overview of load balancing

Topic	Summary
Load balancing	Redistributing workload among nodes reduces response time through load balancing
Load-balancing in M2M	SDN enables effective load balancing in M2M networks, addressing burst traffic and failures
Load-balancing issues	SDN load balancing in M2M networks optimizes request forwarding and prevents congestion
SDN's potential	SDN ensures global network control, regulates M2M traffic, and eliminates failures
Load balancing for SDN	Efficient SDN load balancers use intelligent algorithms for optimal data processing

4 Recent Advancement

Guler et al. [15] introduced the Spectrum Chain (SC) framework in 2023, leveraging blockchain to enhance spectrum assignment in SDONs, improving efficiency, security, and dynamic sharing. Ali et al. [16] proposed an effective strategy in their 2023 paper for load balancing in a multi-domain SDN-enabled IoT network. By allocating slave controllers and transferring switches, the strategy optimizes resource utilization, prevents congestion, and enhances network performance and reliability. Babbar et al. [17] proposed the QoS-aware LBSMT algorithm in their 2023 paper for dynamic load balancing switch migration in network environments. By considering QoS metrics and real-time traffic, the algorithm optimizes switch distribution, enhancing load balancing capabilities and improving network performance. The QoS-aware LBSMT algorithm contributes to efficient and QoS-aware load balancing in dynamic networks.

Zhao et al. [10] proposed a power-efficient and load-balanced flow scheduling algorithm in their 2023 paper. The algorithm utilizes an Online Flow Route framework to optimize power consumption and evenly distribute traffic, resulting in improved resource utilization and network performance. Tao et al. [18] designed a load balancing-based traffic scheduling scheme in 2022 for Space Air Ground Integrated Networks (SAGIN) to enhance their transmission capability with SDN architecture. In order to achieve link load balancing in SDN and reduce energy consumption while maintaining a high standard of service for user flows, Dafd et al. [19] offered a method in 2022. This method combines energy aware routing with Genetic Algorithm (GA) and Ant Colony Optimization (ACO) load balancing. In order to address the issue of delays in critical scenarios like smart health care due to the high volume of data that must be transmitted back to the edge nodes after processing in the cloud, Singh et al. [20] proposed a framework in 2022 that reduces bandwidth costs while providing load balancing at the same time and is based on fog clusters. Various recent advances are summarized in Table 3.

Table 3 Recent advancements

References	Year	Major contribution
Guler et al. [15]	2023	Spectrum Chain for spectrum assignment in SDONs using blockchain
Ali et al. [16]	2023	A load balancing method for SDN-enabled multi-domain IoT networks
Babbar et al. [17]	2023	The QoS-aware LBSMT algorithm, a dynamic algorithm for load-balancing switch migration, was proposed
Zhao et al. [10]	2023	A foundation for an online flow route that is load-balanced and power-efficient
Tao et al. [18]	2022	Create load balancing for SAGIN based on a traffic scheduling plan
Dafd et al. [19]	2022	A plan that uses load balancing using (GA) and ACO along with energy-conscious routing to reduce energy use
Singh et al. [20]	2022	Framework provides load balancing while also lowering the cost of bandwidth

5 Conclusion and Further Research Directions

In conclusion, SDN shows great promise in addressing the challenges posed by traditional network architectures by enabling centralized control and programmability. SDN effectively separates the data and control planes, resulting in improved network management capabilities and performance optimization. The establishment of industry-driven organizations like the Open Network Foundation (ONF) further promotes SDN and standardized protocols such as OpenFlow. Efficient resource utilization and minimizing service response time are crucial aspects of load balancing. Both reactive and proactive load-balancing techniques have been proposed. In the context of machine-to-machine (M2M) networks, load balancing becomes even more critical due to unpredictable and bursty traffic patterns. SDN's ability to provide a global network view and dynamically update rules offers potential solutions for enhancing load balancing in M2M networks. SDN's Application Programming Interfaces (APIs) facilitate communication among the management, data and control planes. Standardization efforts, like the Northbound Interface Work Group (NBIWG), strive to establish consistent and interoperable APIs for SDN, enabling easy configuration, application implementation, and communication.

Further research directions in SDN and load balancing include the exploration of advanced algorithms and techniques, especially in multi-controller environments. The investigation of controller load balancing can improve scalability and fault tolerance. Further investigation is needed to optimize load-balancing strategies for various network components, including links, servers, cloud resources, and controllers. Furthermore, the integration of artificial intelligence and machine learning techniques into SDN load-balancing algorithms shows promise for enhancing network performance and adaptability.

References

1. Narten, T.W., Cascone, C., O'Connor, B., Vachuska, T., Davie, B.: Software-defined networking: a systems approach. IEEE Commun. Surv. Tutorials **18**(2), 1617–1635 (2016). https://ieeexplore.ieee.org/document/7468917
2. Alvigzu, R., Maier, G., Kukreja, N., Pattavina, A., Morro, R., Capello, A., Cavazzoni, C.: Comprehensive survey on SDN: software-defined networking for transport networks. IEEE Commun. Surv. Tutor. **19**, 2232–2283 (2017) (CrossRef)
3. Chen, Y.-J., Wang, L.-C., Chen, M.-C., Huang, P.-M., Chung, P.-J.: SDN-enabled traffic-aware load balancing for M2M networks. IEEE Internet Things J. **5**(3), 1797–1806 (2018). https://doi.org/10.1109/JIOT.2018.2812718
4. Bhushan, B., Sahoo, C., Sinha, P., Khamparia, A.: Unification of Blockchain and Internet of Things (BIoT): requirements, working model, challenges and future directions. Wireless Netw. (2020). https://doi.org/10.1007/s11276-020-02445-6
5. Cheng, C.:. Research on data center load balancing technology based on SDN. J. Phys.: Conf. Ser. 1861 (2017)
6. Understanding OpenFlow Flow Entry Timers on Devices Running Junos OS. Available online: https://www.juniper.net/documentation/us/en/software/junos/sdn-openflow/topics/concept/junos-sdn-openflow-flow-entry-timers-overview.html. Accessed on 23 Mar 2022
7. Vijay, V., Karmakar, K., Tupakula, U., Hitchens, M.: A Policy-based security architecture for software-defined networks. IEEE Trans. Inf. Forensics Secur. **14**, 897–912 (2019)
8. Filali, Z., Mlika, S., Cherkaoui, Kobbane, A.: Preemptive SDN load balancing with machine learning for delay sensitive applications. IEEE Trans. Veh. Technol. **69**(12), 15947–15963 (2020). https://doi.org/10.1109/TVT.2020.3038918
9. Binh, L.H., Duong, T.-V.T.: Load balancing routing under constraints of quality of transmission in mesh wireless network based on software-defined networking. J. Commun. Netw. **23**(1), 12–22 (2021). https://doi.org/10.23919/JCN.2021.000004
10. Zhao, Y., Wang, X., He, Q., Zhang, C., Huang, M.: PLOFR: an online flow route framework for power saving and load balance in SDN. IEEE Syst. J. **15**(1), 526–537 (2021). https://doi.org/10.1109/JSYST.2020.3010971
11. Lai, W.-K., Wang, Y.-C., Chen, Y.-C., Tsai, Z.-T.: TSSM: time-sharing switch migration to balance loads of distributed SDN controllers. IEEE Trans. Netw. Serv. Manage. **19**(2), 1585–1597 (2022). https://doi.org/10.1109/TNSM.2022.3146834
12. Boucadair, M., Jacquenet, C.: Software-defined networking: a perspective from within a service provider environment. In: RFC 7149 (2014). Available online: https://www.hjp.at/doc/rfc/rfc7149.html. Accessed on 3 May 2020
13. Bhowmik, C.D., Gayen, T.: Traffic aware dynamic load distribution in the data plane of SDN using genetic algorithm: a case study on NSF network. Pervasive Mobile Comput. **88**, 101723 (2023), ISSN 1574-1192. https://doi.org/10.1016/j.pmcj.2022.101723
14. Liu, W.-X., Cai, J., Zhu, Y.-H., Luo, J.-M., Li, J.: Load balancing inside programmable data planes based on network modeling prediction using a GNN with network behaviors. Comput. Netw. **227**, 109695 (2023). ISSN 1389-1286. https://doi.org/10.1016/j.comnet.2023.109695
15. Guler, E., Karakus, M., Uludag, S.: Blockchain-enhanced cross-ISP spectrum assignment framework in SDONs: SpectrumChain. Comput. Netw. **223**, 109579 (2023). ISSN1389-1286. https://doi.org/10.1016/j.comnet.2023.109579
16. Ali, J., Jhaveri, R.H., Alswailim, M., Roh, B.: ESCALB: an effective slave controller allocation-based load balancing scheme for multi-domain SDN-enabled-IoT networks. J. King Saud Univ. Comput. Inf. Sci. **35**(6), 101566 (2023). ISSN 1319-1578. https://doi.org/10.1016/j.jksuci.2023.101566
17. Babbar, H., Rani, S., Bashir, A.K., Nawaz, R.: LBSMT: load balancing switch migration algorithm for cooperative communication intelligent transportation systems. IEEE Trans. Green Commun. Netw. **6**(3), 1386–1395 (2022)

18. Tao, J., Liu, S., Liu, C.: A traffic scheduling scheme for load balancing in SDN-based space-air-ground integrated networks. In: 2022 IEEE 23rd International Conference on High Performance Switching and Routing (HPSR), pp. 95–100 (2022)
19. Dafda, J., Subhedar, M.: Dynamic load balancing in SDN using energy aware routing and optimization algorithm. IEEE Bombay Sect. Signature Conf. (IBSSC) **2022**, 1–6 (2022)
20. Singh, P., Kaur, R., Rashid, J., Juneja, S., Dhiman, G., Kim, J., Ouaissa, M.: A Fog-cluster based load-balancing technique. Sustainability (2022)

Chapter 41
Symmetric Encryption Scheme Based on Fuzzy Graph

Ruma Kareem K. Ajeena

1 Introduction

For solving several problems in mathematics, the fuzzy sets theory is employed as an essential key. Different problems have been solved using it. One of these problems is a cryptography [1]. On the other hand, simply, the graphs are used to model the relations. They are used to represent the information of relationships which are connected by edges E among the objects that are considered by vertices V. On a nonempty vertex set V, a graph [2] is a symmetric relation and a FG is a symmetric fuzzy relation on a fuzzy subset as well.

In 1965 [3], L. A. Zadeh presented the fuzzy sets and fuzzy relations concepts. The theory of fuzzy graphs has been considered in 1975 by Rosenfeld [4]. In FG, the fuzzy trees, blocks, bridges and cut nodes also have been studied in [4]. In 2012, Sarkar and Mandal [5] proposed a scheme which is called the FLHSPKC for speeding an elliptic scalar multiplication. In 2013, Qaid and Talbar [6] generated random keys based on fuzzy logic for encryption.

In 2014, Qaid et al. [7] presented a method to encrypt and decrypt images using the FGs. These graphs are got from a matrix of image's pixels.

More efficient, high level of the security and speeding up are got using this method in compare with others. In 2015, Mudiaa and Chavan [8] presented an encryption algorithm using the images and the fuzzy logic. In 2016, GaneshKumar and Arivazhagan [9] introduced a scheme of encryption based on fuzzy logic with high level of security.

R. K. K. Ajeena (✉)
Computer Sciences Department, College of Education for Pure Science (Ibn Al-Haitham), University of Baghdad, Baghdad, Iraq
e-mail: ruma.k.kh@ihcoedu.uobaghdad.edu.iq

Scientists Foundation for Development, Baghdad, Iraq

J. M. R. S. Tavares et al. (eds.), *Proceedings of Second International Conference on Intelligent System*, Algorithms for Intelligent Systems,
https://doi.org/10.1007/978-981-99-8976-8_42

In 2017, Abdullah et al. [10] used TFN to propose an alternative RSA cryptosystem. In the same year also, Muthumeenakshi et al. [11] presented the fuzzy logic approach to embed the encrypted message. In 2020, Pius and Balakumar [12] presented their study that uncovers the fuzzy graph is appropriate to be utilized as a fundamental tool for obtaining more secure levels of different cryptosystems, also see [13].

In 2021, Ajeena [14] presented the SG-ISD method based on the soft graph to compute the elliptic scalar multiplication operation, also see [15, 16]. In the same year also, Ajeena [17] presented new definitions of fuzzy membership (FECM) functions on an elliptic curve to build a new extended fuzzy set. And also, Nawal and Ruma Ajeena [18] proposed using FST for new version of encryption schemes based on discrete logarithm (DL). In this work, a fuzzy graph has been used for encryption algorithm. The encrypted a plaintext is transformed insecure channel as a FG that is created through computing the TFM function. The security of the proposed algorithm is analyzed [20, 21].

The current work includes: the fundamental facts of a fuzzy set and the FG are explained in Sect. 2. In Sect. 3, the FGSE scheme is proposed as a core point of this work. In Sect. 4, study case of FGSE scheme is presented with simple FG with five vertices and fine edges. In Sect. 5, the security of the FGSE scheme is explained. Finally, in Sect. 6, the conclusions have been drawn.

2 Basic Facts

A fuzzy set $\tilde{A}$ of X is defined by

$$A = \{(x,\ \mu_{\ A}(x)) : x \in X\}, \tag{1}$$

with $\mu_{\tilde{A}}(x)$ is a membership function of x in $\tilde{A}$ and x are objectives in a collection X [10].

The fuzzy membership function using the triangular number is defined by

$$\begin{cases} 0 & \text{if } x \le a \\ \frac{x-a}{b-a} & \text{if } a < x \le b \\ \frac{c-x}{c-b} & \text{if } b < x \le c \\ 0 & \text{if } x > c \end{cases},$$

where a and c are boundaries membership, lower and upper, respectively [19].

A FG (f-graph) [19] is a pair G: (σ, μ) with underlying set S, where σ: $S \to [0,1]$ that is a fuzzy subset of a set S and μ: $S \times S \to [0,1]$ which is a fuzzy relation on σ and it is reflexive and symmetric on σ such that

$$\mu(u, v) \le \min(\sigma(u), \sigma(v)), \text{ for all } u, v \text{ in } V,$$

where V is a set of the vertices in G. If $u, v \in S$ then $\sigma(u) > 0$, $\sigma(v) > 0$ and $\mu(u, v) > 0$, where $(u, v) \in S \times S$. Thus, the functions σ and μ should be chosen suitably.

3 The FGSE Scheme

In this section, a SES based on the FG is proposed. The main idea of this scheme depended on computing the TFM function to form the fuzzy graph. The discussion of this scheme is done as follows:

A prime number p and a random subset $A = (a_1, a_2, \ldots, a_n)$ is formed secretly over a prime field F_p. The plaintext $m = (m_1, m_2, \ldots, m_n)$ is also considered as a subset of F_p that is chosen randomly. Then, the generation of a set $B = (b_1, b_2, \ldots, b_n)$, where b_i, for $i = 1, 2, \ldots, n$, are elements in F_p and they satisfy the relation.

$b_i \equiv m_i a_i \pmod p$, with $\gcd(b_i, a_i) = 1$, for $i = 1, 2, \ldots, n$. (3).

The fuzzy masks of the elements ai and bi are computed using the TFM function. A fuzzy graph has been constructed using the fuzzy masked sets

$$\tilde{A} = \{\tilde{a}_1, \tilde{a}_2, \ldots, \tilde{a}_n\} \text{ and } \tilde{B} = \left\{\tilde{b}_1, \tilde{b}_2, \ldots, \tilde{b}_n\right\}$$

of A and B, respectively. The encrypted plaintext is sent to the second user as a fuzzy graph FG with vertices $\tilde{b}_i$ and edges $\tilde{a}_i$. For instance, a FG in Fig. 1 has four vertices $\tilde{b}_1, \tilde{b}_2, \tilde{b}_3, \tilde{b}_4$ and four edges $\tilde{a}_1, \tilde{a}_2, \tilde{a}_3, \tilde{a}_4$. Algorithm 1 can be used for more computational results to obtain the ciphertext of m using the FGSE scheme.

Algorithm 1. The FGSE Algorithm.

Input: A prime p.

Output: A ciphertext C of m as a FG.

1. He/ She chooses randomly and secretly a subset $A = (a_1, a_2, \ldots, a_n)$ over a prime field $F_{p.}$
2. He/ She chooses a plaintext $m = (m_1, m_2, \ldots, m_n)$ as a subset over F_p randomly.

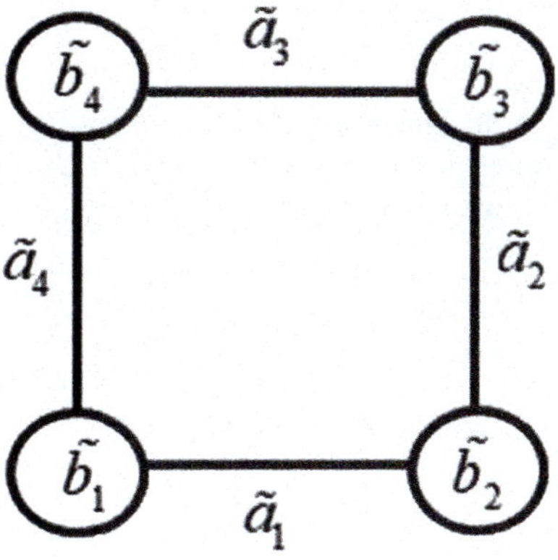

Fig. 1 FG with four vertices and four edges

3. He/ She generates a set $B = (b_1, b_2, \ldots, b_n)$, where b_i are elements in F_p such that $b_i \equiv m_i a_i \pmod p$, with $\gcd(b_i, a_i) = 1$, for $i = 1, 2, \ldots, n$.
4. He/ She computes fuzzy masks $\tilde{a}_i$ and $\tilde{b}_i$ of the elements a_i and b_i using TFM function.
5. He/ She constructs a FG using the fuzzy masked sets $\tilde{A} = \{\tilde{a}_1, \tilde{a}_2, \ldots, \tilde{a}_n\}$ and $\tilde{B} = \left\{\tilde{b}_1, \tilde{b}_2, \ldots, \tilde{b}_n\right\}$
6. He/ She sends the ciphertext C of m to the second user as a FG with vertices $\tilde{b}_i$ and edges $\tilde{a}_i$.

Upon the receiver gets the FG, he/she does the following steps to decrypt the encrypted plaintext. Therefore, he/she first writes the fuzzy mask values of the vertices b_i and edges a_i in two lists as follows:

$$\text{List 1}: \ \tilde{A} = \{\tilde{a}_1, \tilde{a}_2, \ldots, \tilde{a}_n\} \text{ and List 2}: \ \tilde{B} = \left\{\tilde{b}_1, \tilde{b}_2, \ldots, \tilde{b}_n\right\}.$$

He/ She uses the shared secret keys (a, b, c, TFMF) using the Diffie–Hellman key exchange to decrypt the fuzzy mask values and gets the original values a_i and b_i of $\tilde{a}_i$ and $\tilde{b}_i$, $i = 1, 2, \ldots, n$. The computations to recover a_i and b_i values can be done simultaneously in parallel way through the computations of for loop $l = [1{:}\,p - 1]$.

In more specific, the original values a_i and b_i are determined based on the values $l_j, j = 1, 2, \ldots, p - 1$, that is corresponded to correct values of the fuzzy masks. For instance, if $l_j \to \tilde{a}_i = \text{TFM}(a_i)$ or $l_j \to \tilde{b}_i = \text{TFM}(b_i), j = 1, 2, \ldots, p - 1$ and $i = 1, 2, \ldots, n$ then $a_i = l_j$ or $b_i = l_j$.

So, the subsets A and B are determined by

$$A = (a_1, a_2, \ldots, a_n) \text{ and } B = (b_1, b_2, \ldots, b_n).$$

Now, since $b_i \equiv m_i a_i \pmod p$, $i = 1, 2, \ldots, n$, to recover m_i, it requires computing the inverse elements of a_i and since the $\gcd(b_i, a_i) = 1$; therefore, the inverse elements of a_i are existed, for all i. These inverse elements are computed using the extended Euclidean algorithm (EEA) [16]. In other words

$$b_i \cdot a_i^{-1} \equiv m_i \cdot a_i \cdot a_i^{-1} \pmod p \equiv m_i \cdot (1) \pmod p \equiv m_i \cdot (1) \pmod p$$

So, the second user can easily recover the message $m = (m_1, m_2, \ldots, m_n)$. For decrypt and recover the plaintext, one can use Algorithm 2.

Algorithm 2. The Fuzzy Graphic Decryption Scheme Algorithm.

Input: A prime p, a ciphertext C of m as a fuzzy graph FG.

Output: The original plaintext m.

Second user (Receiver):

1. Receiver gets the FG.
2. He/ She writes the fuzzy mask values of the vertices b_i and edges a_i as two lists: List 1: $\tilde{A} = \{\tilde{a}_1, \tilde{a}_2, \ldots, \tilde{a}_n\}$ and List 2: $\tilde{B} = \left\{\tilde{b}_1, \tilde{b}_2, \ldots, \tilde{b}_n\right\}$.

3. He/ She uses the shared secret keys (a, b, c, TFMF) to get the original values a_i and b_i of $\tilde{a}_i$ and $\tilde{b}_i$, $i = 1, 2, \ldots, n$.
4. He/ She determines the subsets A and B by $A = (a_1, a_2, \ldots, a_n)$ and $B = (b_1, b_2, \ldots, b_n)$.
5. He/ She computes the inverse elements of a_i using the extended Euclidean algorithm (EEA) [16].
6. He/ She computes $m_i \equiv b_i {a_i}^{-1} \pmod p$, $i = 1, 2, \ldots, n$.

4 Study Case of FGSE Scheme

With a prime number $p = 41$, a subset $A = \{3, 7, 5, 11, 9\}$ is chosen randomly over F_{41}. Suppose the plaintext $m = \{37, 15, 10, 31, 20\}$ and the subset $B = \{29, 23, 9, 13, 16\}$ such that

$$\gcd(3, 29) = \gcd(7, 23) = \gcd(5, 9) = \gcd(11, 13) = \gcd(9, 16) = 1.$$

Based on the TFM function with $a = 1$, $b = 20$ and $c = 40$ that are considered as the shared secret keys of the sender and receiver, the first user (sender) computes the fuzzy masks of the elements in the subsets A and B by

$$\tilde{A} = \{\tilde{a}_1, \tilde{a}_2, \tilde{a}_3, \tilde{a}_4, \tilde{a}_5\} = \{0.1053,\ 0.3158,\ 0.2105,\ 0.5263,\ 0.4211\}$$

and

$$\tilde{B} = \left\{\tilde{b}_1, \tilde{b}_2, \tilde{b}_3, \tilde{b}_4, \tilde{b}_5\right\} = \{0.5500, 0.8500, 0.4211, 0.6316, 0.7895\}.$$

The FG that is corresponded to fuzzy masks $\tilde{A}$ and $\tilde{B}$ is shown in Fig. 2.

The encrypted plaintext is sent to receiver as a FG as given in Fig. 2. When the second user obtains the FG, he/ she tries to recover encrypted plaintext through the decryption process. The vertices $\tilde{b}_i$ and edges $\tilde{a}_i$ of FG have been written in two lists, namely $\tilde{A}$ and $\tilde{B}$ as follows:

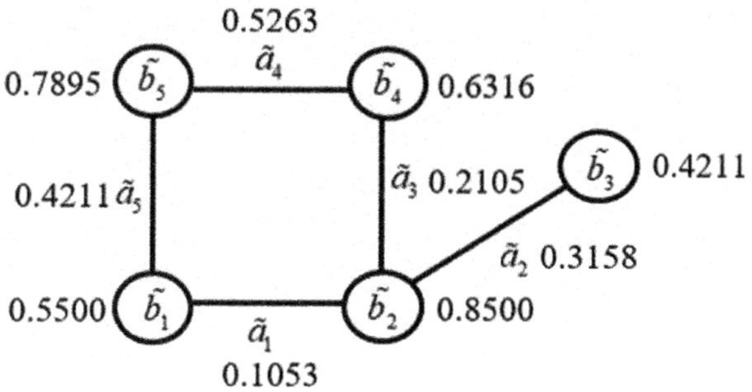

Fig. 2 Encrypted plaintext as a fuzzy graph with five vertices and five edges

$$\tilde{A} = \{\tilde{a}_1, \tilde{a}_2, \tilde{a}_3, \tilde{a}_4, \tilde{a}_5\} = \{0.1053,\ 0.3158,\ 0.2105,\ 0.5263,\ 0.4211\}$$

and

$$\tilde{B} = \left\{\tilde{b}_1, \tilde{b}_2, \tilde{b}_3, \tilde{b}_4, \tilde{b}_5\right\} = \{0.5500,\ 0.8500,\ 0.4211,\ 0.6316,\ 0.7895\}.$$

Next, he/ she wants to recover the original values a_i and b_i that are corresponded to the $\tilde{a}_i$ and $\tilde{b}_i$, $i = 1, 2,\ldots, n$ as shown in Table 1.

Based on Table 1, the subsets A and B are determined by

$A = \{3, 7, 5, 11, 9\}$ and $B = \{29, 23, 9, 13, 16\}$.

Since $\gcd(b_i, a_i) = 1$, so the inverse elements of a_i are existed, for all i. These elements are computed using the EEA [16] by

$$\begin{aligned}
3^{-1}(\text{mod}41) &\equiv 14(\text{mod}41)\\
7^{-1}(\text{mod}41) &\equiv 6(\text{mod}41)\\
5^{-1}(\text{mod}41) &\equiv 33(\text{mod}41)\\
11^{-1}(\text{mod}41) &\equiv 15(\text{mod}41)\\
9^{-1}(\text{mod}41) &\equiv 32(\text{mod}41).
\end{aligned}$$

Thus, the plaintext $m = (m_1, m_2, m_3, m_4, m_5)$ can be recovered as follows:

$$b_i \equiv m_i a_i(\text{mod}p) \rightarrow m_i \equiv b_i \cdot a_i^{-1}(\text{mod}\, p).$$

Therefore,

$$\begin{aligned}
m_1 &\equiv b_1 \cdot a_1^{-1}(\text{mod}\, p) \equiv 29 \cdot 14(\text{mod}41) \equiv 37\\
m_2 &\equiv b_2 \cdot a_2^{-1}(\text{mod}\, p) \equiv 23 \cdot 6(\text{mod}41) \equiv 15\\
m_3 &\equiv b_3 \cdot a_3^{-1}(\text{mod}\, p) \equiv 9 \cdot 33(\text{mod}41) \equiv 10\\
m_4 &\equiv b_4 \cdot a_4^{-1}(\text{mod}\, p) \equiv 13 \cdot 15(\text{mod}41) \equiv 31
\end{aligned}$$

and

$$m_5 \equiv b_5 \cdot a_5^{-1}(\text{mod}\, p) \equiv 16 \cdot 32(\text{mod}41) \equiv 20.$$

Hence, $m = \{37, 15, 10, 31, 20\}$.

Table 1 Original values of a_i and b_i that are corresponded to the $\tilde{a}_i$ and $\tilde{b}_i$, $i = 1, 2, \ldots, n$

l	$\tilde{a}_i$	$\tilde{b}_i$	Correct value
1	0	–	
2	0.0526	–	
3	**0.1053**	–	$\boldsymbol{a_1 = 3}$
4	0.1578	–	
5	**0.2105**	–	$\boldsymbol{a_3 = 5}$
6	0.2631	–	
7	**0.3158**	–	$\boldsymbol{a_2 = 7}$
8	0.3684	–	
9	**0.4211**	**0.4211**	$\boldsymbol{a_5 = b_3 = 9}$
10	0.4736	–	
11	**0.5263**	–	$\boldsymbol{a_4 = 11}$
12	0.5789	–	
13	–	**0.6316**	$\boldsymbol{b_4 = 13}$
14	0.6842	–	
15	0.7368	–	
16	–	**0.7895**	$\boldsymbol{b_5 = 16}$
17	0.8421	–	
18	0.8947	–	
19	0.9473	–	
20	0.9489	–	
21	0.95	–	
22	0.9	–	
23	–	**0.8500**	$\boldsymbol{b_2 = 23}$
24	0.8	–	
25	0.75	–	
26	0.7	–	
27	0.65	–	
28	0.6	–	
29	–	**0.5500**	$\boldsymbol{b_1 = 29}$
30–40	**Stop**	**Stop**	

5 Security Analysis

A FG is used to create the SESs based on the subsets that are chosen randomly from the elements of a prime field F_p. This creation, namely of fuzzy graph, increases level of security compared to other SE schemes. Eve as an attacker wants to get the original values of the subsets A and B from the fuzzy masking which are generated based on TFM function that is selected secretly. Determining the ciphertext, which is computed as a fuzzy graph, is more difficult. If Eve can guess this function, it requires to compute more and more possible cases to determine the correct case with a large choice of prime p. The correct value probability is calculated by

$$P_{a_1} = \frac{\#\text{the correct values of a fuzzy value } a_1}{\#\text{ all possible values}} = \frac{1}{p-1}.$$

Determination all correct values of the elements in a subset A requires probability

$$P_{\{a_1,a_2,....,a_n\}} = \frac{\#\text{the correct values of fuzzy values}\{a_1, ..., a_n\}}{n \times \#\text{all possible values}} = \frac{n}{n(p-1)}$$

with parallel computations. Further, all correct values of the elements in a subset B require probability

$$P_{\{b_1,b_2,....,b_n\}} = \frac{\#\text{the correct values of fuzzy values}\{b_1, ..., b_n\}}{n \times \#\text{all possible values}} = \frac{n}{n(p-1)}.$$

Thus, the total number of the possibilities cases that are needed to try by Eve to recover the correct values of the elements in the sets A and B is

$$P_{\{a_{i=1,...,n}\},\{b_{i=1,....,n}\}} = \frac{n}{n(p-1)} + \frac{n}{n(p-1)} = \frac{2n}{n(p-1)}.$$

6 Conclusion

A new version of a SE scheme has been presented in this work which depended on computing the fuzzy values that are used to create a FG. The ciphertext in the proposed encryption scheme is considered as a FG that is sent by sender to receiver. With parameters moderate values of the proposed SES, the security has been increased. The TFM function for computing the fuzzy values of the subsets A and B over a prime field F_p has been selected secretly. This selection helps to form a FG which is considered as a hard way to get the subsets A and B by attacker. The main point for the security is how determining the correct values of the a_i and b_i that can be used to the computed fuzzy values of a FG. Many cases of a_i and b_i, where $i = 1,\ldots, n$ reach up to $p - 1$, where p is a (large) moderate prime, are needed to compute. Hence, with the FG, a SE scheme has more security as well as suitability for communications.

References

1. Pedrycz, W., Gomide, F.: An Introduction to Fuzzy Sets: Analysis and Design. MIT Press (1998)
2. Bollobas, B.: Modern Graph Theory. Springer, Heidelberg (2005)
3. Zadeh, L.A.: Fuzzy sets. Inf. Control **8**(3), 338–353 (1965)
4. Zadeh, L.A.: Fuzzy sets and their application to pattern classification and clustering analysis. In: Classification and Clustering, pp. 251–299 (1977)
5. Sarkar, A., Mandal, J.K.: Secured wireless communication using fuzzy logic based high speed public-key cryptography (FLHSPKC) (2012). arXiv preprint arXiv:1211.2087
6. Qaid, G.R., Talbar, S.N.: Encrypting image by using fuzzy logic algorithm. Int. J. Image Process. Vis. Sci. (IJIPVS) **2**(1), 25–29 (2013)
7. Qaid, G.R., Talbar, S.N., AL–Kubati, A.A.M.: Image security by using fuzzy graph. Image **3**(12) (2014)
8. Miss, H.M.M., Miss, P.V.C.: Fuzzy logic based image encryption for confidential data transfer using (2, 2) secret sharing scheme. Proc. Comput. Sci. **78**, 632–639 (2016)
9. GaneshKumar, K., Arivazhagan, D.: New cryptography algorithm with fuzzy logic for effective data communication. Indian J. Sci. Technol. **9**(48), 1–6 (2016)
10. Abdullah, K., Bakar, S.A., Kamis, N.H., Aliamis, H.: RSA cryptosystem with fuzzy set theory for encryption and decryption. AIP Conf. Proc. **1905**(1) (2017)
11. Muthumeenakshi, M., Archana, T., Muralikrishna, P.: Fuzzy application in secured data transmission. Int. J. Pure Appl. Math. **116**(3), 711–715 (2017)
12. Pius, A., Balakumar, R.: Efficient cryptograph system on fuzzy theory. J. Crit. Rev. **7**(9), 3173–3176 (2020)
13. Ajeena, R.K.K.: The graph and its role for speeding up the elliptic scalar multiplication algorithms. In: 2019 2nd International Conference on Engineering Technology and its Applications (IICETA), pp. 227–228 (2019)
14. Ajeena, R.K.K.: The soft graphic integer sub-decomposition method for elliptic scalar multiplication. J. Discrete Math. Sci. Cryptogr. **24**(6), 1751–1765 (2021)
15. Aljamaly, K.T.R., Ajeena, R.K.K.: Undirected complete graph to design new public key cryptosystem. J. Phys.: Conf. Ser. **1897**(1), 012045 (2021)
16. Aljamaly, K.T.R., Ajeena, R.K.K.: The elliptic scalar multiplication graph and its application in elliptic curve cryptography. J. Discrete Math. Sci. Cryptogr. **24**(6), 1793–1807 (2021)
17. Ajeena, R.K.K.: The fuzzy elliptic curve set and its application for elliptic curve cryptography. J. Discrete Math. Sci. Cryptogr. (Accepted) (2023)
18. Abbas, N.K., Ajeena, R.K.K.: More secure on the DL-encryption schemes using the TFM function. AIP Conf. Proc. **2398**(1) (2022)
19. Lee, K.H.: First Course on Fuzzy Theory and Applications, vol. 27. Springer Science & Business Media (2004)
20. Jasim, T.H., Abed, S.A., Ghazi, A.: Concepts of bi-supra topological space via graph theory. Wasit J. Comput. Math. Sci. **1**(2) (2022)
21. Sharma, S., Obaid, A.J.: Mathematical modelling, analysis and design of fuzzy logic controller for the control of ventilation systems using MATLAB fuzzy logic toolbox. J. Interdisc. Math. **23**(4), 843–849 (2020). https://doi.org/10.1080/09720502.2020.1727611

Chapter 42
MED-H: Smart Healthcare System Based on IoT and Rule Set

Long-Phuoc Tôn, The-Hau Nguyen, and Huu-Hien Hoang

1 Introduction

The field of Information Technology has experienced significant growth, improving various aspects of our lives and complementing other sectors. However, the impact of Covid-19 caused by the SARS-CoV-2 virus has severely disrupted all aspects of life, causing economic damage, loss of human life, and straining healthcare systems worldwide. Recognizing the urgency of this epidemic and the overwhelming pressure on healthcare staff, this research aims to automate patient monitoring using wearable devices for remote health monitoring. The system collects patients' health indicators, analyzes them based on predefined medical rules, and provides real-time monitoring for doctors to make timely treatment decisions. This solution supports the treatment of diseases like dengue fever, measles, and future conditions [1, 13]. The system is implemented through three main components: a website for medical professionals to configure medical rule parameters and manage the system, a mobile application for doctors and medical staff to directly monitor patients and synchronize medical data, and wearable IoT devices for patients equipped with sensors to collect vital parameters such as temperature, heart rate, and blood oxygen levels. To realize this solution, a microservices architecture is employed, based on a Service-Oriented Architecture (SOA) approach [11]. This architecture structures the application as a set of services, allowing for independent development, deployment, and scalability without impacting other services. The system is deployed on a Raspberry Pi 4

https://fit.iuh.edu.vn/giangvien@tonlongphuoc

L.-P. Tôn (✉) · T.-H. Nguyen · H.-H. Hoang
Faculty of IT, Industrial University of Ho Chi Minh City, 12 Nguyen Van Bao Street, Go Vap District, Ho Chi Minh City, Vietnam
e-mail: tonlongphuoc@iuh.edu.vn

J. M. R. S. Tavares et al. (eds.), *Proceedings of Second International Conference on Intelligent System*, Algorithms for Intelligent Systems,
https://doi.org/10.1007/978-981-99-8976-8_43

server, utilizing frontend, backend, and database services to ensure data security. For streamlined development and deployment, Gitlab CI and Docker are utilized. Gitlab CI enables automated service build and deployment processes through script writing and trigger configurations. Docker, built on Linux's namespace and cgroup concepts, automatically installs and builds source code into artifacts, facilitating deployment [14]. Expo is employed to build an API file for the mobile application, allowing installation on devices. Overall, this research addresses the overwhelming pressure on healthcare systems by automating patient monitoring through wearable devices, providing real-time data analysis, and supporting timely treatment decisions for various diseases.

Paper Structure. Realted work is surveyed in Section 2. Our research motivation and formulates our research questions are presented in Sect. 3. Section 4 presents how our MED-H tool contributes to the monitoring in wearable computing. Section 5 draws some concluding remarks and points out the future work.

2 Related Work

2.1 Internet of Things

The Internet of Things (IoT) refers to the vast network of connected physical devices worldwide that collect and share data. With the availability of affordable computer processors and widespread wireless networks, even small objects like pills and large entities like airplanes can now be part of the IoT. By equipping these objects with sensors and connecting them, intelligence is added, enabling real-time communication between devices without human intervention [10]. The IoT is revolutionizing our world, making it smarter and more convenient. The concept of integrating smart sensors into everyday objects was discussed in the 1980s and 1990s, but progress was hindered by the unavailability of suitable technology. Early projects, such as an Internet-connected vending machine, demonstrated the potential, but overall progress was slow. The size and complexity of computer processors at the time made effective communication between devices challenging. It became evident that affordable, energy-efficient processors were needed to support the future creation of billions of IoT devices [9].

2.2 Internet of Medical Things

The Internet of Medical Things (IoMT) is a technology ecosystem that leverages IoT connectivity, bio-sensors, and artificial intelligence in the healthcare industry. It encompasses medical devices and applications that are connected to healthcare IT systems through Internet-enabled computers and mobile devices. IoMT enables data collection and analysis for various purposes. IoMT devices include both digital devices like heart rate monitors and ECG monitors, as well as non-digital items such as hospital beds and medication. These devices serve as inlets for medical equipment to connect to the IoMT ecosystem. The key characteristic of IoMT is the ability to share data in real time, allowing authorized individuals to access and utilize the information for legitimate purposes. Hospitals and clinics are among the primary beneficiaries of IoMT, as it helps improve healthcare quality while reducing costs. By leveraging IoMT, healthcare providers can enhance patient monitoring, streamline workflows, and make data-driven decisions for better patient outcomes [8, 12].

In-home medical connectivity, also known as telemedicine, utilizes IoMT technology to extend healthcare services outside of traditional hospital settings. Patients with conditions like heart disease and diabetes can greatly benefit from IoMT, as it enables remote monitoring and reduces the need for frequent doctor visits, especially in times like the ongoing Covid-19 pandemic. Portable IoMT devices can monitor a patient's heart activities and glucose levels, automatically alerting the doctor in case of any issues. Furthermore, smart devices integrated into virtual home assistants can interact with elderly patients, reminding them to take medications and allowing their family members and physicians to remotely monitor their condition. Advancements in biosensor technology have made it possible to develop wearable smart devices that can be attached to the patient's skin or even implanted in their body to monitor their health. These IoMT sensors provide patients with greater freedom while maintaining close monitoring of their health conditions [6, 7, 15].

2.3 ReactJS and React Native

ReactJS is a JavaScript library that was introduced in 2015 and is widely used for building reusable UI components. According to the official ReactJS documentation, it is defined as a library for composable user interfaces. It promotes the development of reusable components that can display dynamically changing data. Many programmers utilize React as the View component in the MVC pattern. React abstracts the Document Object Model (DOM), providing a simpler programming model and better performance. It can be used on the NodeJS server [3]. React Native, also known as RN, is a popular JavaScript library for building cross-platform mobile apps. It enables developers to create apps for multiple mobile operating systems, including iOS and Android. Facebook released React Native as an open-source project in 2015, and within a short span of time, it has emerged as a leading solution for mobile app development. Prominent apps like Instagram, Facebook, and Skype utilize React Native. In our system, we employ ReactJS and React Native to develop web and mobile applications [5], using Times 10-point type.

2.4 NestJS and Go

NestJS is a framework designed for building scalable and efficient server-side applications using Node.js. It supports both JavaScript and TypeScript, incorporating elements of Object Oriented Programming (OOP), Functional Programming (FP), and Functional Reactive Programming (FRP). NestJS provides a higher level of abstraction compared to popular Node.js frameworks like Express or Fastify, while still allowing developers to access APIs directly. It also offers compatibility with various third-party modules, providing developers with flexibility and options [2]. On the other hand, Go is an open-source programming language developed by Google in 2007. It emphasizes simplicity, reliability, and efficiency. Go was created to address the growing complexity and performance issues faced by Google's cloud engineers while managing the company's infrastructure. The language was designed by Robert Griesemer, Rob Pike, and Ken Thompson with a focus on simplicity and high performance [4]. In our system, we utilize NestJS and Go to build the back-end server, which offers APIs for interaction between front-end applications and IoT wearable devices with the system. This combination allows us to leverage the features and capabilities provided by both frameworks, ensuring efficient and reliable communication within the system.

3 Research Motivation

3.1 Case Study

In our system, we have implemented an object called Medical Rule, which is created by Medical Experts through our web application. The Medical Rule object consists of configurable parameters, including the name of the medical rule, temperature range, monitoring time, medical action, and color of the LED that corresponds to the level of the disease. Each medical rule is assigned to a specific disease type. For instance, we create a medical rule for Covid-19, with parameters outlined in Table 1. The Med-H system incorporates a wearable device specifically designed for Covid patients. This device continuously sends real-time body temperature data to the server. Typically, patients with severe cases wear this device to monitor their health status continuously.

When the server receives data from the patient's device, it activates the expert-defined rule sets. Based on the rule table, the server provides the next course of action. For instance, in the Covid patient scenario described in Table 1 (row 1), if the patient's body temperature remains between 37.5 and 38.5 °C for more than 10 min but less than 48 h, the patient's wearable device will turn yellow to indicate the incubation period. Simultaneously, the server triggers a rule from the mapping table to notify the responsible doctor or nurse in charge of monitoring the patient. The patient is then administered paracetamol at a dose of 10–15 mg/kg to reduce

Table 1 Plans for treating and caring for patients (rules) with symptoms of Covid-19

Type	Condition	Colors	Healing action
Covid-19	Temperature $\in [37.5, 38.5] \cap t \in [0.15, 48.0]$	Yellow	To reduce fever if fever is high, use paracetamol at a dose of 10–15 mg/kg body weight/time
Covid-19	Temperature $\in [39.0, 40.0] \cap t \in [0.15, 23.99]$	Orange	Incubation period (chest X-ray and abdominal ultrasound)
Covid-19	Temperature $\in [39.0, 40.0] \cap t \in [24.0, 48.0]$	Red	Emergency period (oxygen therapy: when the patient begins to have symptoms of shortness of breath)
Covid-19	Temperature $\in [37.0, 37.5] \cap t \in [0.15, 48.0]$	Green	Recovery period (close monitoring is necessary to recognize heart failure or pulmonary edema)

fever, and the LED on their wearable device turns yellow accordingly, signaling the patient's incubation period. Similar processes apply to other cases defined in the rule sets of Table 1.

The second wearable device is intended for healthcare workers, such as doctors and nurses. It allows the wearer to check the status of the patients they are monitoring. In emergency situations when a set of rules from the server is activated, the wearable device illuminates with LED lights, assisting medical staff in timely intervention for the patient (Fig. 1).

Additionally, in our system, we have introduced an object known as the Patient's Medical Record, which can be created by Medical Experts or Doctors. This object contains essential information, including details about the patient, hospital, doctor, disease type, wearable IoT device used, as well as specific diagnoses related to the patient. Once the doctor has diagnosed the patient's disease type, a Medical Record is generated to facilitate the monitoring process for the patient (as depicted in Fig. 1).

3.2 *Research Statement*

The scenario of going off track, as explained in Sect. 3.1, raises several research ideas that need to be addressed:

- **RQ1**: Can a wearable device, equipped with sensors and supporting software systems, autonomously manage patient information?

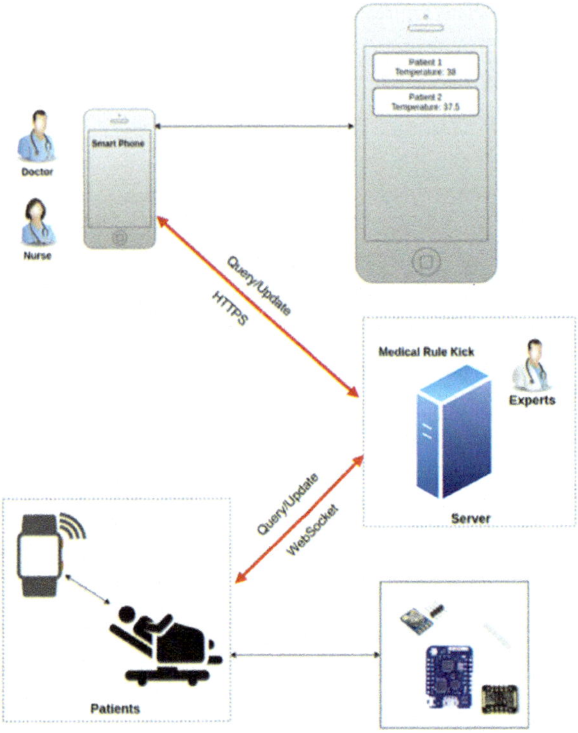

Fig. 1 Framework for monitoring IoT objects and communication of components in health care

- **RQ2**: If so, how can we effectively organize and store information to facilitate synchronization among system components, including wearable devices, mobile applications, and the web?
- **RQ3**: How can we update the rule sets to ensure scalability across different disease groups?
  This paper focuses on addressing these questions and delving into the corresponding research areas.

4 Framework

Figure 2 provides an overview of our framework. Initially, we establish an abstract representation of the medical wearable device (depicted as a watch icon in Fig. 1, explained in Sect. 3.1) within the system, along with interfaces to relevant components. We then delve into the management of system logic. Additionally, the medical professional configures the medical rule sets on the server icon. Based on the health rule set configuration, the system responds to the components through the activated

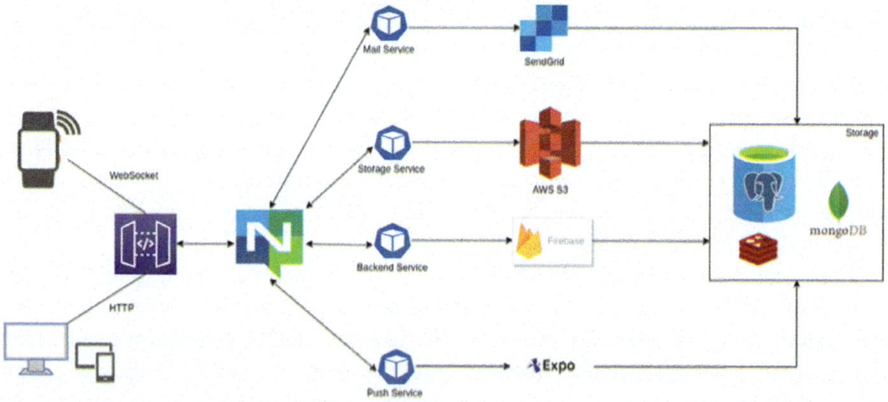

Fig. 2 Overview of software architecture of MED-H

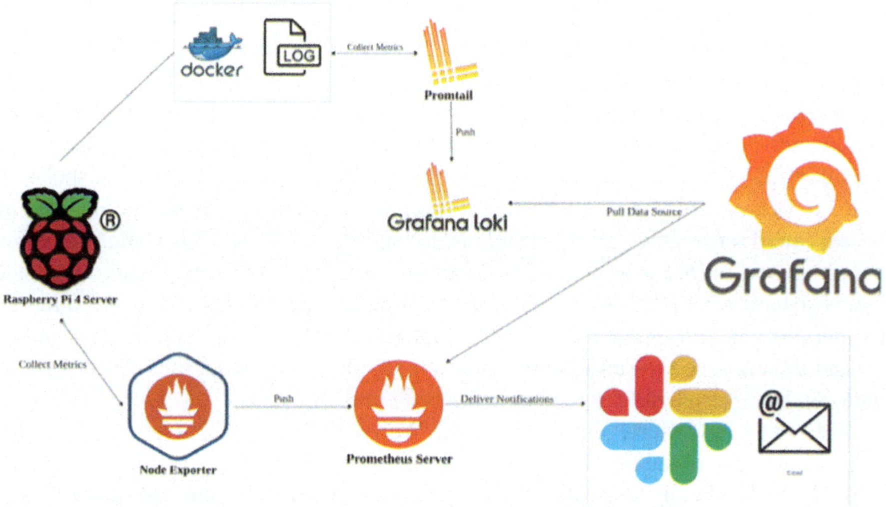

Fig. 3 Monitoring system in MED-H with Grafana ecosystem

rule sets. Ultimately, the system triggers alerts on both the wearable device and the doctor's application (as depicted by arrows connecting the components, as explained in Sect. 4.2). For further insight into the system's operation, refer to Figs. 4 and 5. In Sect. 4.3, we present the details of our system verification (Fig. 3).

4.1 Software Abstraction

One of the crucial components in our system is the IoT device, responsible for collecting medical data from the patient wearing it. This device abstracts the concepts of runtime data collection and translates health codes into corresponding warnings. To achieve this, we utilize the C programming language to create the necessary software abstractions. Developers can define the IoT device's processes, such as sending and receiving information within the system. Listing 1 presents code snippets that handle sending and receiving information from the rule set system's side. These IoT objects are managed by the system's logic. It is important to note that these code snippets are part of a single logical processing model, presented separately for ease of understanding the problem's logic.

For instance, in a practical example, Listing 1 demonstrates the system's configuration with an automated job (cronjob) that retrieves data from sensors and sends it to the system via the WebSocket protocol. The IoT device then listens for events from the response system. Based on these events, the wearable device reacts by changing the LED color according to the parameters returned by the system, as depicted in *setLedColorByLevel(int level)*. Additionally, *setLedColorByLevel(int level)* showcases how to change the corresponding LED color based on the data received from the system. Please note that the code snippets provided are for illustrative purposes and may require further implementation and customization based on the specific requirements of your system.

In *sendTemperatureToServer()*, we define the parameters to be collected from the sensors and convert them into the JSON data type using the Arduino Json library. We then use the WebSocket protocol to send this JSON data to the system, as shown in line 34. The variable "*doc*" in line 30 is used to represent the JSON document. The "*event*" field within the "*doc*" variable is defined by us as "*sync*" to match the event being opened on the WebSocket protocol on the server. We have created a nested JSON structure called "*data*" to hold relevant parameters such as the MAC address, temperature, heart rate, and so on. We serialize this data into a JSON format in line 38 and then proceed to send the information. On the server side, the received JSON data can be deserialized for further processing and analysis.

In Listing 1, we have the handlers for WebSocket events that we are listening for, and they are processed based on their respective instances. In line 5, we handle the case of a loss of connection to the system, identified by the WebSocket type "*WStype_DISCONNECTED*." In line 8, we check the status of the WebSocket connection to the system using the type "*WStype_CONNECTED*". If the connection is successful, we send the text "*Connected*" to the system. The crucial part of WebSocket data processing occurs when we receive a status warning of "*WStype_TEXT*." In this case, we interpret the received text as data from the system. We convert the data type to an integer in line 16 and then proceed to call the function responsible for updating the LED color. Lastly, the "*WStype_BIN*" event allows us to receive binary values from the system. This event handles the processing of binary data received via WebSocket.

Function *setLedColorByLevel(int level)* represents synchronizing the patient's state with the color of the LEDs on the patient's wearable device.

Listing 43.1 Web socket Event Handler in MED-H

```
void web_Socket_Event(WS_type_t type, uinteger8_t *payload,
size_text lengthoftext)
{
  switch (type)
  {
  case WS_type_DISCONNECTED:
    Serial.printf(''[WSc] is Disconnected!\n'');
    break;
  case WS_type_CONNECTED:
    Serial.printf(''[WSc] is Connected success to url: %s\n'', payload);
    webSocket.sendTXT(''Connected_OK'');
    break;
  case WS_type_TEXT:
    Serial.printf(''[WSc] get content: %s\n'', payload);
   if (payload)
   {
      int level = atoi((char *)payload);
      setLedColorByLevel(level);
   }
    break;
  case WS_type_BIN:
    Serial.printf(''[WSc] get content text with binary length:
   %u\n'', lengthoftext);
    hexdump(payload, lengthoftext);
    break;
  }
}

void sendTemperatureToServer()
{
  // declare JSON template to send to server
  // data declare with array index
  // 0 is mac Address
  // 1 is temperature
  // 2 is heartbeat
  // 3 is spO2
  JsonArray data = doc.createNestedArray(''data'');
  data.add(WiFi.macAddress());
  data.add(mlx.readObjectTempC() + 3.5);
  data.add(beatAvg);

}

void setLedColorByLevel(int level)
{
    ....
}
```

4.2 *Handle Medical Rules and Update IoT Status at Runtime*

In this paper, we have presented our approach for handling health rule sets in the system at runtime. We have analyzed the system's business components using abstracted models, where each pathogen is associated with a set of rules. These rule sets consist of multiple sub-rules, and the current code includes parameters such as time range and body temperature. Additional parameters can be added in the future to support monitoring of pathogens using suitable sensors (Fig. 3).

Algorithm 1: Synchronizing data between IoT objects and rule sets

Input : $L_{iotObjects}$: A set of IoT objects
Output: The rule set is properly verified with IoT objects.
$uni \leftarrow$ Assign initial value to an universe with $L_{iotObjects}$;
$L_{formulas} \leftarrow$ Fetch a list of formulas by Experts;
$L_{inv\_formulas} \leftarrow \emptyset$;
$inst \leftarrow$ new $Instance(universe)$;
$Evaluator\ ev \leftarrow$ **new**$Evaluator(inst)$;
foreach f *in* $L_{formulas}$ **do**
  if *(**not** ev.evaluate(f))* **then**
    $L_{inv\_formulas} \cup \{f\}$;
return $L_{inv\_formulas}$;

Upon receiving data from the IoT device, the system performs a check based on the associated pathogen. It extracts the corresponding rule sets and evaluates each rule (Algorithm 1). The rule set with the highest warning level is given priority. The system then retrieves the level and send it back to the loT device to update the LED accordingly.[1]

4.3 System Verification

Figure 2 depicts an abstract and comprehensive overview of the communication between the components in our system. The diagram illustrates the presence of small microservices, each dedicated to specific tasks as indicated by their names. The IoT devices connect to the API Gateway using the WebSocket protocol to transmit data and listen for events. Other clients communicate with the system via HTTP. Upon receiving specific requests, the system relies on message events to propagate through the NATS Message Queue, facilitating communication with the microservices involved in the logical processing system. The system then responds accordingly based on the received messages.

The MED-H system was deployed on a Raspberry Pi 4 computer, which was configured as a VPS Server (Fig. 5). The system utilizes microservices that are running within Docker Containers. For analytics and monitoring purposes, the system integrates Grafana and Prometheus platforms. This architecture enables real-time monitoring of the system's performance and data. Additionally, the Slack application is utilized to receive system notifications and alerts. Figure 3 illustrates the monitoring architecture, showcasing the combination of Docker Containers, Grafana, and Prometheus. Figure 4 depicts the presence of Docker Containers within the system, emphasizing their importance in the overall architecture.

[1] Details can be found at https://www.youtube.com/watch?v=WMIuLxpD0ag.

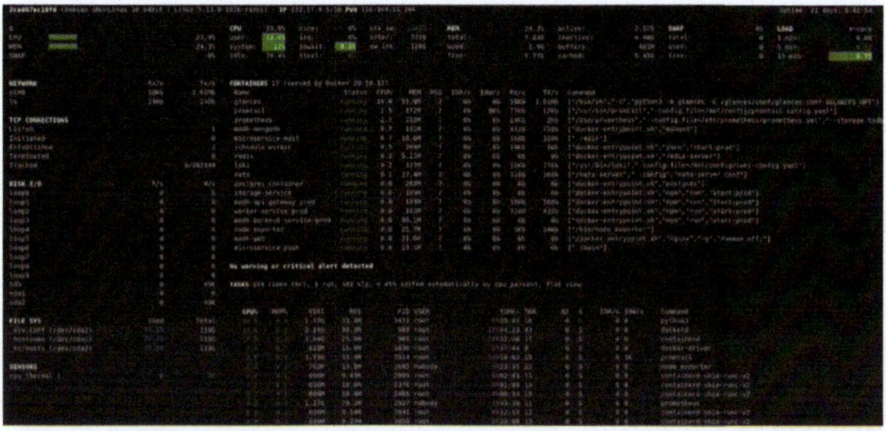

Fig. 4 Glances tools to monitoring Docker Container and server metrics

Fig. 5 IoT device in MED-H

5 Conclusions

In this paper, we proposed solutions for implementing a smart health monitoring system that incorporates IoT technology. This is a very necessary solution for emergencies Covid diseases, dengue fever, etc. At that time, the number of hospitalized patients increased, the medical force could not directly monitor the patient as well as the patient could not provide proactive information in real time. We developed web-based applications and Android mobile applications, as well as a wearable device equipped with sensors to collect body data. The web-based application for experts successfully meets the requirements for configuring medical rules and managing other system components. The mobile application for doctors enables prompt monitoring of patient conditions, facilitating informed decision-making and improving patient treatment. The IoT devices effectively collect and synchronize real-time data with the system, ensuring compatibility with specific diseases configured by the experts. To implement these solutions, we carefully selected appropriate technologies to optimize resources and improve efficiency. Furthermore, we designed and tested the MED-H device on a small group of patients (06 patients) with the Covid-19 disease. The results demonstrated that the system meets the requirements of our proposed solution. Moving forward, we plan to expand the range of disease categories supported by MED-H and incorporate additional sensors to enhance applicability and accuracy. By employing the aforementioned technologies and logical processes, we believe that our research objectives have been achieved.

References

1. Baby, B.E., Kadam, S.B., Aravindhan, A.: Human health monitoring system using internet of wearable things. In: Emerging Technologies for Sustainability, pp. 253–259. CRC Press (2020)
2. Bell, J., Magolan, G., Housley, P., De Peretti, A., Guijarro, D.: Nest.js: A Progressive Node.js Framework. Bleeding Edge Press (2018)
3. Blum, J.: Exploring Arduino: Tools and Techniques for Engineering Wizardry. Wiley (2019)
4. Bodner, J.: Learning Go. O'Reilly Media, Inc. (2021)
5. Boduch, A., Derks, R.: React and React Native: A Complete Hands-On Guide to Modern Web and Mobile Development with React.js. Packt Publishing Ltd (2020)
6. Gardašević, G., Fotouhi, H., Tomasic, I., Vahabi, M., Björkman, M., Lindén, M.: A heterogeneous IoT-based architecture for remote monitoring of physiological and environmental parameters. In: Proceedings of the 4th International Conference on Internet of Things Technologies for HealthCare, Oct 2017, pp. 48–53. Springer, Angers, France (2017)
7. Gatouillat, A., Badr, Y., Massot, B., Sejdić, E.: Internet of medical things: a review of recent contributions dealing with cyber-physical systems in medicine. IEEE Internet Things J. **5**(5), 3810–3822 (2018)
8. Gubbi, J., Buyya, R., Marusic, S., Palaniswami, M.: Internet of things (IoT): a vision, architectural elements, and future directions. Future Gener. Comput. Syst. **29**(7), 1645–1660 (2013)
9. Khan, S., Alam, M.: Wearable internet of things for personalized healthcare: study of trends and latent research. In: Health Informatics: A Computational Perspective in Healthcare, pp. 43–60 (2021)
10. Kurniawan, A.: Internet of Things Projects with ESP32: Build Exciting and Powerful IoT Projects Using the All-New Espressif ESP32. Packt Publishing Ltd (2019)
11. Newman, S.: Building Microservices. O'Reilly Media, Inc. (2021)
12. Ravi, P., Kumar, N.: Internet of things (IoT): a revolutionary approach towards healthcare surveillance. In: Proceedings of the 3rd National Conference on Image Processing, Computing, Communication, Networking and Data Analytics, pp. 257–264. IEEE Computer Society, New Delhi, India (2018)
13. Tên, L.P., Lê, L.S., Nguyen, M.S.: Micraspis: a computer-aided proposal toward programming and architecting smart IoT wearables. IEEE Access **9**, 105393–105408 (2021)
14. Turnbull, J.: The Docker Book: Containerization Is the New Virtualization. James Turnbull (2014)
15. Vishnu, S., Jino Ramson, S.R., Jegan, R.: Internet of medical things (IoMT)—an overview. In: Proceedings of the 5th International Conference on Devices, Circuits and Systems, Apr 2020, pp. 101–104. IEEE Computer Society, Coimbatore, India (2020)

Chapter 43
Landscape View of Hyperparameter Optimization Cybersecurity by Using Bio-Inspired Algorithm

Zahraa Z. Jundi

1 Introduction

Because of the development of information technology, smart devices are now a part of every aspect of people's lives and workplaces. For that its security is gravely compromised by computer viruses and malicious network attacks on smart devices and smart activities. According to Fig. 1, it is anticipated that there will be 18.22 billion smartphones in use worldwide in the year 2025 [1].

Such widespread use opens the door to a slew of privacy and security threats and other negative effects, making the system an easier target for hackers and leaving it more susceptible to malware and infestations [2], where cybercrime constitutes the largest transfer of wealth in history, growing at a pace of 15% annually. For that, nearly every industry has had to adopt new approaches to this [3]. Therefore, more precise attack detection has become a top priority in modern computer sciences [4]. Stochastic techniques called "nature-inspired" algorithms address these kinds of optimization issues. In most cases, they combine some deterministic and randomized techniques, compare a lot of solutions iteratively, and keep going until they find one that works well [5].

Despite deep/machine learning's successful exploration in several fields, including fraud detection, speech recognition, natural language processing, and picture classification, hyperparameter optimization remains one of the primary issues in the field. Hyperparameters are crucial since they regulate a model's learning rate and ought to be adjusted to enhance execution. Manually setting the default values for the hyperparameters is a difficult and time-consuming operation. Although the time and effort required were used to adjust the hyperparameters are reduced, it is still a hassle if a fresh dataset, a new problem, or an improved model is needed [6].

Z. Z. Jundi (✉)
Department of Computer Science, University of Kufa, Najaf, Iraq
e-mail: zahraaz.alhuchamy@student.uokufa.edu.iq

J. M. R. S. Tavares et al. (eds.), *Proceedings of Second International Conference on Intelligent System*, Algorithms for Intelligent Systems,
https://doi.org/10.1007/978-981-99-8976-8_44

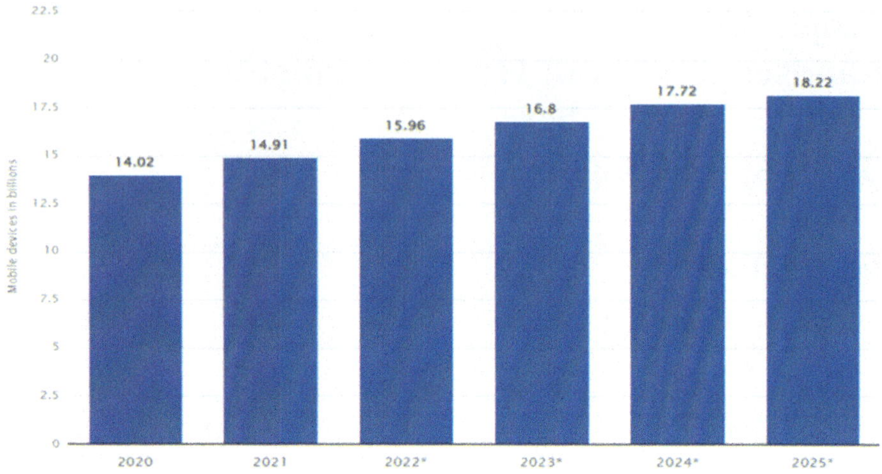

Fig. 1 Mobile devices worldwide from 2020 to 2025

Modern metaheuristic optimization algorithms based on population evolution and use mechanisms like reproduction, mutation, recombination, and selection inspired by biological evolution are referred to as "evolutionary algorithms". They were specifically developed to deal with challenging optimization problems and have recently become a hot research topic. Feature selection, pattern recognition, clustering, and optimization are a few of the uses and industries where they have been successfully applied [4].

This essay provides a general review of Optimizing Hyperparameters using bio-inspired algorithms for Cybersecurity Applications as well as the fields in which these methods have been effectively used (Genetic Algorithm (GA), Differential Evolution (DE), ACO, or Ant Colony Optimization, Whale Optimization Algorithm (WOA), Algorithm Firefly (AF), and Particle Swarm Optimization (PSO) are some examples of optimization techniques.

2 Bio-Inspired Algorithms (BIAs)

Multi-objective optimization is a common component of real-world optimization issues, which are frequently quite difficult to solve. Most NP-hard optimization issues encountered in practice are unable to be resolved by conventional deterministic algorithms [5]. These challenging optimization issues have been solved in a vast number of different domains using bio-inspired algorithms, which have been demonstrated to be good approaches for doing so. Numerous algorithms that have been influenced by biological populations found in nature have been created over the past few decades [7]. Figure 2 shows the taxonomy of the review of the algorithms to be investigated.

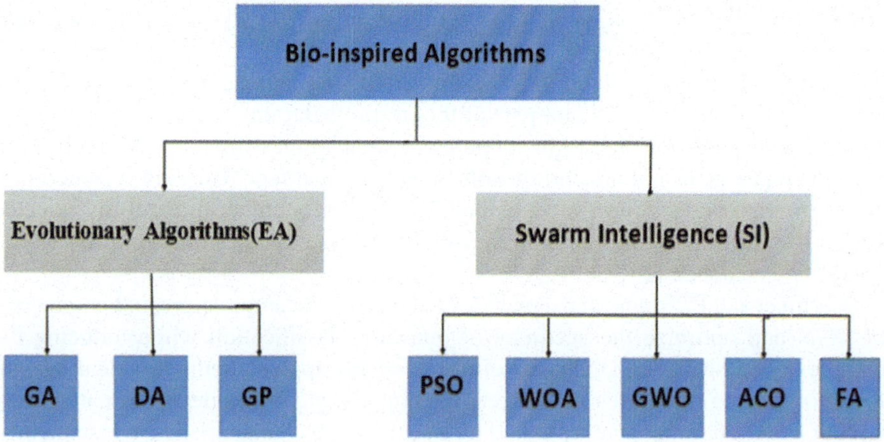

Fig. 2 Taxonomy of the review

3 Evolutionary Algorithms (EA)

A generic population-based metaheuristic optimization algorithm is referred to as an evolutionary algorithm (EA). This algorithm is a subset of evolutionary computation [8]. Researchers have used EA algorithms to resolve the complex issues that arise in real-world cyber operations, such as locating the optimal answer, automatically building models, and training classifiers. EA algorithms provide security professionals with outputs that are legible by humans, solutions that are lightweight, and a variety of solutions that offer variable trade-offs between competing objectives [9]. This ability to generate human-readable outputs is just one of the many alluring features of EA algorithms. In addition, EA algorithms are an effective instrument for automatically developing software frameworks. By employing biologically inspired reproductive factors (such as crossing over, recombination, and selection), the objective of the refined evolutionary process is to generate superior candidate solution communities. Each of these three evolutionary operators is iterated until a termination condition is met. A successful EA family consists of genetic algorithm (GA), genetic programming (GP), differential evolution (DE), and evolutionary strategy (ES), along with the most recent paddy field algorithm.

3.1 Genetic Algorithm (GA)

John Holland and the University of Michigan undergraduates proposed GA in the 1970s [10]. GA [11] uses the "strongest among the weakest chromosomes" method to solve a specific optimization problem [12, 13]. That makes use of natural-sounding mechanisms like mutation, evolution, reproduction, and the like. For any given optimization issue, GA generates a population comprised of all feasible alternatives. An individual gene may stand in for a separate tuning parameter, and as such, each gene is a relatively insignificant part of the overall solution. The subsequent step is analyzing the chromosomes of the population via the fitness function. More likely to reproduce and grow are the chromosomes that are healthier. Through repeated iteration, an improved solution is developed from a small number of carefully selected chromosomes, repeatedly assessing reproductive success and fitness until the end condition is reached [14].

Martin et al. [15] suggested a new GA to enhance the architecture and parameters of DNN and optimize the accuracy of malware classification while reducing the complexity of the model. DNN is an efficient, widely applied method for dealing with issues of varying categories and degrees of complexity. This model is evaluated using a set of malware samples represented by a set of static characteristics; consequently, the DNN has been trained to perform a static malware classification task. According to experiments conducted with a dataset, the genetic algorithm is capable of selecting the architectural settings and parameters for DNN with a 91% degree of accuracy.

Moubayed et al. [16] offer an innovative, ML-based framework designed to identify botnets utilizing the corresponding DNS queries. In order to fine-tune the settings for a classifier using random forests (RF), the components of the framework employ information gathering as a selection attribute technique and GA as a paradigm for hyperparameter optimization. A cutting-edge TI-2016 DNS dataset is used to assess the proposed methodology. According to experimental findings, the suggested optimized framework can minimize the size of the feature set by up to 60% of the functionality. Its detection accuracy was 94.71%, its precision was 0.95, its recall was 95%, and its F-score was 0.95, all of which were superior to the default classifier. This demonstrates the efficiency and dependability of the suggested methodology for identifying botnet assaults.

Meimandi et al. [17] proposed using static analysis of permissions to identify malicious Android apps. The three methods GASA-SVM, GASA-DT, and GASA-KNN are all utilizing this strategy of combining the genetic algorithm with simulated annealing to select the most relevant features from a given set of permits. Over four hundred and ten samples from Drebin dataset, consisting of 328 benign and 82 malicious programs, were used to evaluate the suggested approach. Android malware detection is bolstered by the proposed strategy, with the best results achieved using a GASA-SVM with a maximum value of 0.9707.

Thavasimani et al. [6] suggest a unique evolutionary approach to automatically modify the deep learning sequential model's hyperparameters to distinguish between good and bad traffic in the Internet of things-23 dataset collected in the Czech Republic, by Czech Technical University. Records of 30.85 million both harmful and lawful traffic make up the dataset. The trial outcomes reveal an encouraging result of 98.9% accuracy.

Ghatasheh et al. [18] suggest a modified evolutionary method for concurrent dimension reduction and hyperparameter optimization over skewed datasets from Twitter. The program created a spam forecasting technique by reducing the feature space of the tweets dataset and initializing a classifier using extreme gradient boosting. Tenfold stratified cross-validation that is repeated 50 times is used to validate the model. Utilizing fewer than 10% of the total feature area, the resulting prediction model achieves average geometric mean and accuracy values of 82.32% and 92.67%, respectively. According to the empirical findings, the modified genetic algorithm performs better than the Chi2 and feature selection approach PCA. Additionally, extreme gradient boosting (XGBoost) surpasses a lot of machine learning methods regarding spam prediction. The suggested strategy is utilized for SMS spam modeling.

3.2 *Differential Evolution (DE)*

Differential evolution (DE) is an algorithm that Storn and Price proposed in 1997. This algorithm has established its capability to carry out tasks effectively and is a trustworthy optimization algorithm through its use in numerous applications [19].

Aburomman et al.'s [20] suggest a brand-new support vector machine with weights multiclass classifiers for differential evolutionary intrusion detection systems. This study evaluates various approaches to building a multiclass SVM classifier from binary SVM classifiers. This study aims to pinpoint the multiclass SVM models most effective at intrusion detection. When compared with other approaches, WOAR-SVM shows considerable improvement in accuracy when tested on the NSL-KDD benchmark dataset for IDS.

Tayebi et al. [21] propose many strategies to protect credit cards using artificial intelligence. The suggested strategy tries to combine the robustness of three methods: the XGBoost technique for classification, a resampling method for dealing with unbalanced data difficulties, and the DE algorithm for choosing the appropriate hyperparameters. Finally, the XGBoost algorithm has been improved to classify fraudulent transactions. The European and UCI datasets are used to assess the proposed solution in two real-world contexts. The examination of the suggested approach against the most advanced machine learning models regarding precision, sensitivity, specificity, accuracy, and F-measure demonstrates its proficiency and superiority.

3.3 Genetic Programming (GP)

Koza proposed GP in 1992 [22], and it is an extension of genetic algorithms that differ from the latter in terms of solution representation. GP often uses tree-based internal data structures to describe computer programs, while most genetic algorithms use list-based data structures. The second major distinction is that GP employs variable-length encoding, whereas GA employs fixed-length encoding. The GP population generates variation not just in gene values but also in individual morphology. As a result, GP is comparable with the evolution of a population of computer codes. The four steps of genetic programming are as follows:

1. Create an initial population of computer applications including the functionalities and terminals.
2. Implement every program in the population and assign a fitness score based on how well it addresses the problem.
3. Make new computer programs.

    a. Recreate the most effective current programming.
    b. Create new computer software via mutation.
    c. Crossing is used to create new computer software.

Alyasiri et al. [23] in this study, they examine how grammatical evolution, a technique from the field of evolutionary computation, could be used to detect breaches in IoT systems. The research was conducted on current datasets and compared with cutting-edge algorithms. Also examined and discussed is the potential utility of evolutionary computation-based techniques for identifying unknown threats. They utilized two recently submitted datasets, BoT-IoT and IoT-MQTT, and achieved 99.98% accuracy for BoT-IoT and 96.61% accuracy for IoT-MQTT.

4 Swarm Intelligence SI

SI is a subset of AI that was first developed in 1989 by Jing Wang and Gerardo Beni as a set of algorithms for directing robotic swarms under the global optimization framework [24].

SI is defined by Almufti [25] as the study and design of organized computationally intelligent interactive multi-agent systems that cooperate to achieve a specific goal and solve complex optimization problems by mimicking the behavior of natural swarms such as birds, fish, and ants.

4.1 Particle Swarm Optimization (PSO)

Eberhart and Kennedy presented PSO in 1995 as a model of bird flocking behavior. This technique has produced very precise results when used in the optimization field [26].

Ding et al. [27] recommended merging PSO method with the asynchronous reinforcement learning technique to find the best answer. First, we suggest a fresh PSO algorithm variant that runs asynchronously. We then applied it to algorithms for asynchronous learning with reinforcement and proposed the Sarsa algorithm (APSO-BQSA), asynchronous PSO and Q-learning in reverse were used to create an updated asynchronous reinforcement learning approach. In the end, they use experiments to demonstrate the viability of the APSO-BQSA technique and asynchronous PSO presented in this paper.

Elmasry et al. [28] suggested a particle swarm optimization (PSO)-based dual model strategy for simultaneously selecting feature subsets and hyperparameters. During the pre-training phase, the approach is utilized so that an optimal selection may be made for both the model's features and its hyperparameters automatically. They examined differences in efficiency between DNNs, LSTM-RNNs, and deep belief networks (DBN), all forms of deep learning models. In addition to this, they validated our strategy and demonstrated the efficacy of the models that they generated by making use of two common IDS datasets in our tests. According to the findings of several experiments, this method significantly improves network intrusion detection by increasing the detection rate (DR) by 4–6% and decreasing the false alarm rate (FAR) by 1–5% compared with the values of identical models applied to the same dataset and performed without pre-training.

The PSO-XGBoost model was suggested by Jiang et al. [29] because of its overall greater classification accuracy when compared with other alternative methods such as Adaboost, XGBoost, random forest, and bagging. After an XGBoost-based categorization model has been constructed, PSO is utilized in order to carry out an adaptive search for the most effective XGBoost structure that can be found. The proposed model is investigated through the use of data collectors with NSL-KDD serving as the standard. The results of the experiments indicate that the PSO-XGBoost model performs superiorly to other comparable models with regards to accuracy, mean average precision (mAP), recall, and macro-average (macro), in particular when it comes to recognizing attack subtypes that have a low attack frequency, such as U2R and R2L. In addition, the results of this research provide experimental justifications for the incorporation of swarm intelligence into NIDS. The accuracy of the PSO-Xgboost model is 0.90%.

Kim et al. [30] propose using a CNN-LSTM model that has been optimized for efficient anomalous query access control for relational databases governed by RBAC. This model was created using PSO. This model is able to learn from previous database searches and optimize them in order to protect sensitive data from attacks from both within and outside the organization. By embedding hyperparameters into particles, swarm optimization was implemented in complex CNN-LSTM neural networks with the purpose of locating the optimal architecture. The proposed model features highly effective access control.

4.2 Whale Optimization Algorithm (WOA)

This algorithm, which mimics humpback whales' behavior while fishing in the ocean, was developed by Mirjalili and Lewis [31].

Brodzicki et al. [32] successfully modified the deep learning whale optimization algorithm and demonstrated that deep learning training can be treated like a different cost function and optimized utilizing a swarm-based approach. A transfer technique that converts a discrete collection into a search space appropriate for parameters for WOA has been implemented. Further succeeded in expanding the fundamental algorithm in two dimensions to operate inside a 3D search area. In less time than using the grid search method, but still faster than the random search is considerably easier, the solution successfully located a fine training set hyperparameters. Simulations demonstrate the viability of the suggested technique for hyperparameter optimization, with accuracy values for the datasets from Reuters and MNIST for the fashion of 89.85% and 80.60%, respectively.

Song et al. [4] this study proposes and integrates the machine learning framework XGBoost with a novel, effective intrusion detection method dubbed WOA-XGBoost. This algorithm is used for intrusion detection within the KDD CUP 99 dataset. Principally novel is the automatic selection and optimization of the most important XGBoost parameters using the WOA optimization method. WOA is more precise and can search for a broader range of factors than manual optimization or grid search. The WOA-XGBoost intrusion detection method is demonstrably superior to the methods based on Grid Search-XGBoost, WOA-SVM, and Grid Search-SVM, and achieves an accuracy rate of 0.9906%, according to experimental results on the KDD CUP 99 dataset. As a result, the WOA-XGBoost classification approach presented in this study can be used as an effective detection instrument for network data intrusion detection [42, 43].

4.3 Gray Wolf Optimizer (GWO)

Mirjalili and Lewis' (GWO) was first proposed (2014). The fishing strategy of gray wolves through the leadership hierarchy is simulated in GWO by using four different wolf types: alpha, beta, delta, and omega. The three processes in the optimization process are looking for prey, surrounding prey, and attacking prey [33].

Al Shorman et al. [34] they offer a unique unsupervised evolutionary technique for detecting IoT botnets. The primary advantage of the proposed method is its ability to identify IoT botnet attacks launched from compromised IoT devices by leveraging the effectiveness of a recently developed swarm intelligence algorithm known as the GWO algorithm to optimize the hyperparameters of the one-class support vector machine (OCSVM) while simultaneously identifying the features that best describe the IoT botnet problem. To illustrate its efficacy, the suggested method's performance is evaluated on an updated version of an actual benchmark dataset using standard anomaly detection evaluation methods. The experimental results show that the suggested method beats all current algorithms in terms of true positive rate, false positive rate, and G-mean for all IoT devices. Furthermore, it greatly minimizes the number of selected features while attaining the shortest feasible detection time.

Nagaraju et al. [35] presented a deep learning technique and a simulation that included GWO and WOA to ascertain attack avoidance. Deep neural networks, such as TensorFlow, are utilized to classify stolen software by comparing its code with extant code. To reduce background noise and zero in on the importance of every phrase in terms of tokenization and weighting feature techniques for open-source plagiarism are used. The Mailing database has been mined for malware samples for testing purposes. The results of the investigations imply that the recommended method for evaluating the cybersecurity risks posed by the Internet of Things (IoT) provides more precise classifications than the current methods. Attack detection avoidance in IoT procedures employs deep convolutional networks and the whale with gray wolf optimization (WGWO) algorithm. Using the model, the 228 * 228 photo ratio for the Leopard Mobile dataset obtained an accuracy rate of 98% in only 35 s.

4.4 *Ant Colony Optimization (ACO)*

The initial suggestion of Dorigo and Di Caro's (ACO) was made in 1999. The search for food by ants in a computer simulation was the impetus for the development of the program. In order to get a satisfactory result, one of the search algorithms takes into account both its past performance and its error rate. It is used to solve problems that require testing all potential solutions or that take a long time to solve on a computer in order to discover the required solution [36].

Samriya et al. [37] implemented the ACO-DNN model for network intrusion detection in a cloud environment with DVFS-based energy optimization. Utilized datasets include KDD Cup 99 and NSL-KDD. The min–max normalization method is used to eliminate duplicate values from a dataset, and 1-N encoding is then used to attain data homogeneity. ACO is used to reduce dimensions, followed by deep neural network (DNN) computation. Dynamic voltage and frequency scaling (DVFS) has been incorporated into the system in order to reduce energy consumption. The primary objective of setting and enhancing this concept is to create consistency between the time and energy needs of various VMs or host modes. Experimental results demonstrated ACO-DNN's superiority over existing models, ACO and PCA-based (Naive Bayes) models were used to validate the recommended model.

El-Ghamry et al. [38] suggested a novel method for identifying malware affecting Internet of Things devices that are based on machine learning and make use of visual cues such as photographs of network traffic. In this method, the ant colony optimizer (ACO)-based feature selection methodology was recommended to acquire the fewest possible features while boosting the performance of the support vector machines (SVMs) classifier (i.e., the results of malware detection). The PSO technique was used to perform additional fine-tuning on the SVM parameters of the different kernel functions. The results of the experiments conducted using a dataset that is accessible to the public indicated that the SVM linear function kernel is the most effective, with an accuracy of 95.56%, recall of 96.43%, precision of 94.12%, and F1 score of 95.26% respectively. The findings of the comparison with the published research showed that bio-inspired methods, such as ACO, have the potential to be utilized in the creation of an effective and lightweight machine learning method for the purpose of detecting malware in an environment involving IoT devices.

4.5 Firefly Algorithm (FA)

A method called FA, which is based on swarm intelligence, has been demonstrated to be successful in resolving the optimization that is nonlinear issues, particularly multimodal issues where the objective environment may contain several optimum or minimizing [39].

Korade et al. [40] developed a highly accurate intrusion detection model. Convolutional neural networks based on the hyper-heuristic FA are used to propose an improved deep learning (DL) algorithm for the IDS model (HHFA-CNN). Without adding to the complexity, the suggested HHFA-CNN decreases false values and boosts accuracy. The NSL-KDD and ISCX-IDS datasets for network traffic are used to test the proposed HHFA-CNN system. The study's findings revealed that the model that was proposed provides superior execution in comparison with other models that are currently being used. The suggested model optimizes CNN's hyperparameters through the application of a brand-new hyperheuristic known as hyper-heuristic FA.

Zivkovic et al. [41] the CFAEE-SCA-XGBoost system utilizes the most recent CFAEE-SCA technique to optimize and fine-tune the hyperparameters of the XGBoost classifier. The proposed model for network intrusion detection was trained and evaluated using the reputable datasets KDD-NSL and NB15-UNSW. The suggested design, which is in agreement with the CFAEE-SCA technique, PSO-XGBoost is more advantageous than the variant supported by the original FA algorithm, and the minimal XGBoost implementation, which was used for comparison.

The results of the experiments indicate that the CFAEE-SCA-XGBoost approach outperformed the original model in terms of accuracy and suggest that swarm intelligence techniques could be applied to NIDS.

5 Conclusion

The procedures of calculation known as optimization techniques are employed to address certain problems, and the answers that are produced by these techniques are either optimal or suboptimal, depending on the nature of the issue being solved. Several methods have been developed to solve and optimize hyperparameters for cybersecurity applications via bio-inspired algorithms. These algorithms were developed by several different developers. The efficiency of algorithms that take cues from natural and biological phenomena in the world around them is an important factor that is carefully considered. In this research, the bio-inspired algorithms were classified as swarm- and evolutionary-based, respectively. Use bio-inspired algorithms to solve feature selection difficulties since these algorithms are able to provide answers to the aforementioned issues in a time- and effort-efficient manner, because the outcomes these algorithms have created have been effective, and researchers in this industry continue to use the algorithms now in use. They are also merging several kinds of these algorithms in order to generate results that are even more effective. In addition to this, it provides some recommendations for further research that can be of use in resolving feature selection problems that are prevalent in the field of cybersecurity. The study's primary objective is to develop a fresh method that may unearth useful answers and simplify the feature selection process.

References

1. Number of mobile devices worldwide 2020–2025 | Statista. https://www.statista.com/statistics/245501/multiple-mobile-device-ownership-worldwide/. Accessed 26 May 2023
2. Feizollah, A., Anuar, N.B., Salleh, R., Wahab, A.W.A.: A review on feature selection in mobile malware detection. Digit. Investig. **13**, 22–37 (2015)
3. Cybercrime to Cost the World $10.5 Trillion Annually by 2025. https://cybersecurityventures.com/cyberwarfare-report-intrusion/. Accessed 17 Jun. 2023
4. Song, Y., Li, H., Xu, P., Liu, D.: A method of intrusion detection based on WOA-Xgboost algorithm. Discrete Dyn. Nat. Soc. **2022** (2022)
5. Rai, D., Tyagi, K.: Bio-inspired optimization techniques: a critical comparative study. ACM SIGSOFT Softw. Eng. Notes **38**(4), 1–7 (2013)
6. Thavasimani, K., Srinath, N.K.: Hyperparameter optimization using custom genetic algorithm for classification of benign and malicious traffic on internet of things-23 dataset. Int. J. Electr. Comput. Eng. **12**(4) (2022). ISSN 2088-8708
7. Binitha, S., Sathya, S.S.: A survey of bio inspired optimization algorithms. Int. J. Soft Comput. Eng. **2**(2), 137–151 (2012)

8. Vikhar, P.A.: Evolutionary algorithms: a critical review and its future prospects. In: 2016 International Conference on Global Trends in Signal Processing, Information Computing and Communication (ICGTSPICC), pp. 261–265. IEEE (2016)
9. Alyasiri, H., Clark, J.A., Kudenko, D.: Evolutionary computation algorithms for detecting known and unknown attacks. In: Innovative Security Solutions for Information Technology and Communications: 11th International Conference, SecITC 2018, Bucharest, Romania, 8–9 Nov. 2018, Revised Selected Papers 11, pp. 170–184. Springer (2019)
10. Chen, H.-C., Hirasawa, K.: Adaptation in natural and artificial systems: an introductory analysis with applications to biology, control and artificial intelligence. IEICE Trans. Inf. Syst. **89**(12), 3012–3014 (2006)
11. Reading, M.A., Wesley, A., Freund, J.E.: Mathematical Statistics. Prentice-Hall, Englewood Cliffs, NJ (1962). Goldberg, D., Richardson, J.: Genetic algorithms with sharing for multimodal function optimization. In: Proceedings of the Second International Conference on Genetic Algorithms, pp. 148–154. Morgan Kaufmann, San Mateo, CA (1987). Goldberg, D.E.: Genetic algorithms in search, optimization, and machine learning (1989)
12. Chakraborty, B.: Evolutionary computational approaches to feature subset selection. Int. J. Soft Comput. Bioinform. **1**(2), 59–65 (2010)
13. Kawamura, A., Chakraborty, B.: A hybrid approach for optimal feature subset selection with evolutionary algorithms. In: 2017 IEEE 8th International Conference on Awareness Science and Technology (iCAST), pp. 564–568. IEEE (2017)
14. Ali, W.: Hybrid intelligent android malware detection using evolving support vector machine based on genetic algorithm and particle swarm optimization. IJCSNS **19**(9), 15 (2019)
15. Martín, A., Fuentes-Hurtado, F., Naranjo, V., Camacho, D.: Evolving deep neural networks architectures for android malware classification. In: 2017 IEEE Congress on Evolutionary Computation (CEC), pp. 1659–1666. IEEE (2017)
16. Moubayed, A., Injadat, M., Shami, A.: Optimized random forest model for botnet detection based on DNS queries. In: 2020 32nd International Conference on Microelectronics (ICM), pp. 1–4. IEEE (2020)
17. Meimandi, A., Seyfari, Y., Lotfi, S.: Android malware detection using feature selection with hybrid genetic algorithm and simulated annealing
18. Ghatasheh, N., Altaharwa, I., Aldebei, K.: Modified genetic algorithm for feature selection and hyper parameter optimization: case of XGBoost in spam prediction. IEEE Access **10**, 84365–84383 (2022)
19. Kenneth, S.R., Storn, R.M.: Differential evolution—a simple and efficient heuristic for global optimization over continuous spaces. J. Glob. Optim. **11**(4), 341–359 (1997)
20. Aburomman, A.A., Reaz, M.B.I.: A novel weighted support vector machines multiclass classifier based on differential evolution for intrusion detection systems. Inf. Sci. (N Y) **414**, 225–246 (2017)
21. Tayebi, M., El Kafhali, S.: Credit card fraud detection based on hyperparameters optimization using the differential evolution. Int. J. Inform. Sec. Privacy (IJISP) **16**(1), 1–21 (2022)
22. Koza, J.R.: Genetic Programming: On the Programming of Computers by Means of Natural Selection. MIT Press, Cambridge, Massachusetts, London, MA (1992)
23. Alyasiri, H., Clark, J.A., Malik, A., de Fréin, R.: Grammatical evolution for detecting cyberattacks in Internet of Things environments. In: 2021 International Conference on Computer Communications and Networks (ICCCN), pp. 1–6. IEEE (2021)
24. Beni, G., Wang J.: Swarm intelligence in cellular robotic systems. In: Robots and Biological Systems: Towards a New Bionics? pp. 703–712. Springer (1993)
25. Abdulrahman, S.M.: Using swarm intelligence for solving np-hard problems. Acad. J. Nawroz Univ. **6**(3), 46–50 (2017)
26. Eberhart, R., Kennedy, J.: A new optimizer using particle swarm theory. In: MHS'95. Proceedings of the Sixth International Symposium on Micro Machine and Human Science, pp. 39–43. IEEE (1995)
27. Ding, S., Du, W., Zhao, X., Wang, L., Jia, W.: A new asynchronous reinforcement learning algorithm based on improved parallel PSO. Appl. Intell. **49**, 4211–4222 (2019)

28. Elmasry, W., Akbulut, A., Zaim, A.H.: Evolving deep learning architectures for network intrusion detection using a double PSO metaheuristic. Comput. Netw. **168**, 107042 (2020)
29. Jiang, H., He, Z., Ye, G., Zhang, H.: Network intrusion detection based on PSO-XGBoost model. IEEE Access **8**, 58392–58401 (2020)
30. Kim, T.-Y., Cho, S.-B.: Optimizing CNN-LSTM neural networks with PSO for anomalous query access control. Neurocomputing **456**, 666–677 (2021)
31. Mirjalili, S., Lewis, A.: The whale optimization algorithm. Adv. Eng. Softw. **95**, 51–67 (2016)
32. Brodzicki, A., Piekarski, M., Jaworek-Korjakowska, J.: The whale optimization algorithm approach for deep neural networks. Sensors **21**(23), 8003 (2021)
33. Mirjalili, S., Mirjalili, S.M., Lewis, A.: Grey wolf optimizer. Adv. Eng. Softw. **69**, 46–61 (2014)
34. Al Shorman, A., Faris, H., Aljarah, I.: Unsupervised intelligent system based on one class support vector machine and Grey Wolf optimization for IoT botnet detection. J. Ambient Intell. Humaniz. Comput. **11**(7), 2809–2825 (2020)
35. Nagaraju, R., Pentang, J.T., Abdufattokhov, S., CosioBorda, R.F., Mageswari, N., Uganya, G.: Attack prevention in IoT through hybrid optimization mechanism and deep learning framework. Meas. Sens. **24**, 100431 (2022)
36. Mohan, B.C., Baskaran, R.: A survey: Ant Colony Optimization based recent research and implementation on several engineering domain. Exp. Syst. Appl. **39**(4), 4618–4627 (2012)
37. Samriya, J.K., Tiwari, R., Cheng, X., Singh, R.K., Shankar, A., Kumar, M.: Network intrusion detection using ACO-DNN model with DVFS based energy optimization in cloud framework. Sustain. Comput. Inform. Syst. **35**, 100746 (2022)
38. El-Ghamry, A., Gaber, T., Mohammed, K.K., Hassanien, A.E.: Optimized and efficient image-based IoT malware detection method. Electronics (Basel) **12**(3), 708 (2023)
39. Gandomi, A.H., Yang, X.-S., Talatahari, S., Alavi, A.H.: Firefly algorithm with chaos. Commun. Nonlinear Sci. Numer. Simul. **18**(1), 89–98 (2013)
40. Korade, N.B., Zuber, D.R.M.: Boost stock forecasting accuracy using the modified firefly algorithm and multichannel convolutional neural network. J. Theor. Appl. Inf. Technol. **101**(7) (2023)
41. Zivkovic, M., Tair, M., Venkatachalam, K., Bacanin, N., Hubálovský, Š, Trojovský, P.: Novel hybrid firefly algorithm: an application to enhance XGBoost tuning for intrusion detection classification. PeerJ Comput. Sci. **8**, e956 (2022)
42. Kh-madhloom, J.: Fog integrated secured and distributed environment for healthcare industry with software defined networking. Wasit. J. Comput. Math. Sci. 17–30 (2021)
43. Manaa, M.E., Obaid, A.J., Dosh, M.H.: Unsupervised approach for email spam filtering using data mining. EAI Endorsed Trans. Energ. Web (2021). https://doi.org/10.4108/eai.9-3-2021.168962

Chapter 44
A Study in Understanding the Role of Key Measures in Applying Machine Learning Models for Effective Risk Management in Banking Industry with Focus on Private Banks in India

G. H. Kerinab Beenu, S. Meena, and V. K. Ajay

1 Introduction

Since the beginning of the financial crisis in 2008, there has been a consistent emphasis placed on methods through which financial organisations may more effectively identify, analyse, report, and manage internal risks. Since the conclusion of the crisis, this emphasis has continued. There have been a number of in-depth research conducted on banking and risk management. These studies examine both historical achievements and anticipated challenges that may arise in the future. These kinds of inquiries have been carried out in both academic and corporate environments. The advantages of machine learning are being recognised by an increasing number of companies, and there are now numerous applications accessible [5].

G. H. Kerinab Beenu · V. K. Ajay (✉)
S.A. Engineering College, Chennai, India
e-mail: ajay@saec.ac.in

G. H. Kerinab Beenu
e-mail: kerinabbeenug@saec.ac.in

S. Meena
Dr.MGR Educational and Research Institute, Chennai, India
e-mail: meena.comm@drmgrdu.ac.in

J. M. R. S. Tavares et al. (eds.), *Proceedings of Second International Conference on Intelligent System*, Algorithms for Intelligent Systems,
https://doi.org/10.1007/978-981-99-8976-8_45

A forecast made by McKinsey & Co. for the year 2025 states that in order for financial institutions to remain competitive, their risk operations would need major revision. As a result of newly enacted rules, reassessed client objectives, and rising threats, it is unavoidable that risk management will undergo transitions in the near future. These alterations are the result of a number of different factors. As a direct consequence of recent advancements in technology and analytics, new products, services, and methods of risk management have evolved. These changes have been brought about. Machine learning is a relatively new technology that has the potential to have a big impact on risk management. The discovery of minute nonlinear patterns in huge datasets might lead to the creation of risk models that are more accurate than those now available. When all of these considerations are taken into account, it's possible that an individual's ability to predict what will happen in the future can improve over time as they are exposed to an increasing amount of information. It is anticipated that the risk departments of financial institutions will use machine learning in a number of different methods in the near future. To modernise their risk management procedures, financial institutions may also embark on a project that is focused on machine learning [1, 8].

The field of financial services has placed a significant emphasis on the concepts of machine learning and artificial intelligence in recent years. This can be attributed to the high hopes that these two areas will significantly enhance analytical capabilities, automate and simplify a large number of administrative processes, and reduce the overall workload. Underwriting credit, ensuring regulatory compliance, maintaining open lines of contact with customers, and managing risks are all part of the job description. The Institute of International Finance (IIF) has already discussed this pattern, which is demonstrated by "RegTech" (Regulation Technology) and the growth of new FinTech business models in banking [7].

The amount of data obtained by financial institutions (FIs) has increased rapidly in recent years as a result of increased reporting requirements and the emergence of a vast number of high-frequency, unstructured customer data as a result of the digitalisation of services. Both of these factors contributed to the growth of the industry. Financial institutions (FIs) need increasingly complicated analytical tools in order to manage vast volumes of data originating from a variety of sources and formats while also maintaining or enhancing the granularity of their research. The interest of those working in the financial services business in the capabilities that machine learning may give in terms of analytical expertise is growing. It would seem that the whole field of information technology is becoming increasingly interested in this subject of statistics. Machine learning has been there since the early part of the twentieth century, but it didn't become widely used until high-frequency data was easily available and computer technology advanced [4] (Fig. 1).

By illuminating the links that exist between machine learning and other methods of statistical analysis, the purpose of this essay is to assist readers in developing a deeper comprehension of the significant roles that machine learning plays in the management of banking risk. The benefits and drawbacks of using machine learning are also brought to light in this article. In addition to that, a brief overview of the artificial intelligence subfield known as deep learning will be presented, which has its origins

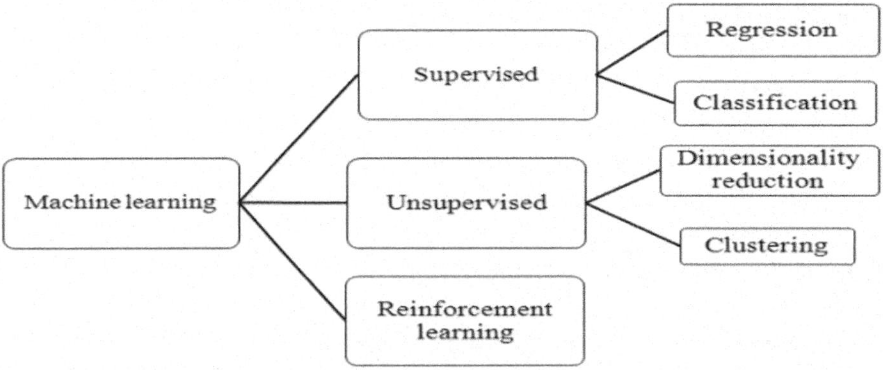

Fig. 1 Types of ML

in the field of machine learning. Next, we'll take a look at how machine learning is being put to use in the banking industry by examining it from the perspective of three separate use cases: modelling credit risk; recognising fraud and money laundering; and keeping an eye out for unethical behaviour among staff.

The Application of Machine Learning to the Management of Financial Risk:

Learning by machine might include a great deal of different methods of investigation, such as supervised and unsupervised learning, for example. While doing supervised machine learning, one of the steps involves making use of a statistical model to estimate or make a prediction for an output based on one or more inputs (e.g. predicting GDP growth based on several variables). The process of obtaining understanding from data without the assistance of a dependent variable is referred to as unsupervised learning. It's possible that there's a technique to make things go more quickly. In its place, research is carried out to review collections in search of previously undiscovered structures and patterns. The use of machine learning enables predictions to be produced with a high level of accuracy [11]. It is possible to develop a model that takes into consideration the characteristics that provide the most accurate predictions when applied outside of the sample if one searches for patterns and correlations in the data. This may be done in a very straightforward manner. By applying the model and all of its variables to a large number of different data subsamples, it is likely that it will be feasible to determine which model variables are responsible for the most accurate predictions [6].

It's possible that doing this approach several times may help the model "learn" from the data and improve its ability to predict the future. Learning by machine is inextricably linked to the "big data revolution" since it requires the processing of enormous datasets and is computationally intensive. The statement that "the enormous advance in computer performance in recent years, along with important theoretical breakthroughs in machine-learning approaches, has ignited a resurgence in computational modelling" is one that is largely accepted by specialists [2] (Table 1).

Table 1 Categories of various ML methods

Categories	Linear type	Nonlinear type
Supervised ML		
Regression model	Using principal component	Customised regression
	Partial least measures	Neural networks
	Least absolute selection	Deep learning
		Least angle model
Classification	Support vector	Decision tree
		SVM
Unsupervised ML		
Critical clustering	K means	
	Hierarchical analysis	

Instead of beginning with the assumption that there is some kind of link between the two variables, as is done in more conventional approaches to statistical analysis, we will first examine the independent variable to see whether there is any kind of correlation between the two [9]. In linear regression, for example, it is assumed that the connection between the variables is linear, despite the fact that this is not always the case. In other words, the assumption is that the relationship between the variables is linear. On the other hand, some machine learning algorithms could be able to infer nonlinear connections, which would increase their capacity to fit data. The easily accessible analytic techniques in the field of machine learning may be used to solve a variety of statistical problems; however, some strategies are more applicable to a wider range of contexts than others. The widespread use of machine learning seems to be of great assistance in addressing statistical challenges connected to regression, classification, and clustering. Clustering is an unsupervised method, as opposed to supervised machine learning, which may be used to issues like as regression and classification.

2 Literature Review

Both the increasing complication of reporting requirements and the digitisation of services, which have resulted in the creation of a significant amount of high-frequency, unstructured customer data, are to blame for the enormous growth in the amount of data that financial institutions (FIs) have amassed in recent years. The interaction between these components has led to a huge increase in the amount of data that has been accumulated as a whole. Financial institutions (FIs) need increasingly complicated analytical tools in order to manage vast volumes of data originating from a variety of sources and formats while also maintaining or enhancing the granularity of their research. With the financial crisis that occurred in 2008–2009, a number of brand-new legislation and monitoring mechanisms were implemented. In the past, fewer aspects of the operations and finances of a financial institution (FI) were required to be reported on in compliance with these standards [10]. In accordance with the new capital regime, financial institutions are obligated to publish major exposures, along with liquidity indicators, collateral, and capital. The quantitative and qualitative aspects of the business model, in addition to the data derived from the loan-level balance sheet, are all factors that may have an effect on the stress analysis. If a financial institution wants to take part in the CCAR exercise that is being run by the Federal Reserve, then they are required to assess the effects of over two thousand fictitious economic scenarios. After the introduction of Solvency II, insurance companies were subjected to a significant increase in the number of reporting obligations [3].

These procedures will generate a significant quantity of reporting data, which, in order to be aggregated across the team and sent on time to management, will first need to be meticulously documented and organised. Because of this, the government has begun a variety of programmes in an effort to enhance the quality of regulatory data and the capacity of the financial industry to give it. In accordance with the Risk Data Aggregation Guidelines established by the Basel Committee, G-SIBS should work to improve both their IT infrastructure and their reporting processes (Basel 239). For IFRS 9, one of the most important goals is to enhance the quality of the data that is utilised for supervision [7]. In addition to the data that they are obligated to supply, financial institutions have the ability to obtain a substantial quantity of data that is of poor quality, is unstructured, and occurs often. Data mined through digital client interactions such as payment systems, social media feeds, consumer applications, and many others may provide insight into the prevalent attitude in the market. These interactions can be used to infer market mood. The aforementioned datasets are often referred to as "big data," which is a phrase that has gained widespread use [11].

The establishment of a brand-new financial institution is made more difficult by the many limitations that are in place. This is due to the fact that accurate risk assessments are used in the administration and monitoring of practically all parts of the business model utilised by a financial institution. In order to accomplish this goal, businesses need to simultaneously analyse consumer data in order to have a complete picture of client preferences and behaviour, which is vital for retaining a competitive edge in their respective markets. This analytical power may be valuable in a variety of contexts, and a broad variety of machine learning approaches are suitable for doing so due to their capacity to handle, or more particularly, their need for, very large datasets and the granularity of analysis that they provide [9]. This is due, in part, to the fact that it is so proficient in delivering skills of this sort of analytic nature. Methods of machine learning that are used in day-to-day computing are often used to the mining of data from high-quality supervision systems. Each of these approaches is dependent on having tidy, well-organised data. Because of its skills in representation learning, mining high-frequency, low-quality sources of "big data" using deep learning and neural network methods, like Google does, is now conceivable. This post will provide you with an update on the development of three large-scale machine learning use cases that are currently in progress. Examples of such applications include the modelling of credit risk, the detection of fraud and money laundering, as well as the identification of conduct risk and abusive behaviour inside financial institutions. The ability of machine learning algorithms to recognise complicated data patterns and rapidly integrate transaction data with data from a variety of other sources to build a complete picture of a client's behaviour may be of great assistance in the identification of behaviours related to money laundering. The use of these technologies has the potential to significantly reduce the number of false-positive discoveries. The implementation of AML is running behind schedule for a variety of different reasons. To begin, the word "money laundering" is not very specific and may be understood in a number of different ways. Since "money laundering" is not a term that is often used, it is possible for financial institutions to be uncertain as to whether or not the suspicious behaviours they report constitute criminal conduct. Since it is not always feasible to show the frequency of money laundering, it is more difficult to train ML-detection systems using historical data. This makes the training process more difficult. By using lower-level SARs as a dependent variable in classification training, financial institutions (FIs) are improving the detection accuracy of the machine learning (ML) detection algorithms that they use. The bank will be able to determine the difference between signals that it may safely disregard as false and those signals for which it is required to submit a suspicious activity report (SAR) with the appropriate authorities, thanks to this method. In addition, unsupervised learning algorithms are employed in anti-money laundering and countering the financing of terrorism in order to categorise the behaviour of customers or transactions in order to "learn" significant patterns. The fact that there is no one strategy that can be considered "typical" and that the strategies themselves are continually evolving is a significant contributor to the growth in our understanding of money laundering. This growth is due, in large part, to the fact that there is no single approach that can be considered "typical." The promise of artificial intelligence and machine learning is being discussed in many other industries, including the financial services industry. The expanding volumes of regulatory reporting data and unstructured data that financial institutions are expected to keep track of have prompted these institutions to search for improved analytic tools that would assist them in managing and mining this data. This may be necessary in order to maintain compliance with industry standards and keep up with the ever-increasing number of FinTech companies. Machine learning and artificial intelligence appear to be having an influence on the business models that financial institutions use, particularly in the areas of client preferences, risk management, the detection of fraudulent behaviour, and the identification of ethical breaches. These are just some of the areas. These technologies, when paired with biometrics, make it feasible to automate customer service and validate an individual's identification without the presence of a human being at the location where the verification takes place.

3 Objectives

The primary objective of this research is to assess the major determinants associated with the application of machine learning strategies for better risk management practices in private banks.

4 Hypothesis

Hypothesis 1: There is no significant difference among better decision-making based on machine learning and effective risk management in banks.

Hypothesis 2: There is no significant difference among enhanced fraud prevention through machine learning and effective risk management in banks.

Hypothesis 3: There is no significant difference among efficient credit scoring through machine learning and effective risk management in banks.

5 Research Methodology

In order to provide a full and accurate portrait of the topic under investigation, this study adopts a descriptive strategy for conducting its research. When there are two or more variables involved in the research topic, this approach may be used to conduct the investigation on a global scale. When it comes to gathering primary data for an investigation, the descriptive method may be of great use. This research largely depends on the replies obtained from a questionnaire with limited open-ended responses that was sent to those working in banking and financial services industry. The research pulls its findings from a broad range of different sources. The researcher uses a five-point Likert scale to identify closed-ended questions on aspects related to the banking risk management in order to get a better understanding of the current body of research. Also, the researcher has used online libraries such as ProQuest and Google Scholar, among others to look and gather necessary inputs and knowledge for the purpose of the study. The researcher has collated data from nearly 153 respondents, who are working in the banking industry and understands the concepts of risk management in their organisation, furthermore, the sample population was chosen based on convenience sampling method.

6 Analysis and Discussion

This section states the analysis based on the data collated by the researcher; the major analysis covers frequency analysis, correlation analysis, and T test.

According to the results of the survey shown in Table 2, 31.4% of respondents are adamant believers that machine learning techniques contribute to improved levels of customer service. Reports have come in from a number of different banks all around the globe stating that with the assistance of ML, they are able to do risk analysis and management in a more efficient manner. This, in turn, enables them to deliver superior service to their respective clientele. In addition, a majority of people, or 36%, agree with this assertion.

According to the findings in Table 3, one-third of respondents are steadfast in their conviction that machine learning may contribute to increased financial stability. Innovative methods, such as a focus on identifying crucial circumstances and taking preventative actions, should be used by financial institutions in order for these organisations to achieve improved financial management. When it comes to monitoring credit risk and liquidity risk, the use of ML approaches is very beneficial to the overall stability of the financial system. Just 16% of those who were questioned were unsure or flat-out disagreed with the comment.

Table 2 Machine learning enables in rendering efficient customer service

Rendering customer service	Frequency	Percent
Strongly disagree	9	5.9
Disagree	14	9.2
Neutral	26	17
Agree	56	36.6
Strongly agree	48	31.4
Total	153	100

Table 3 ML methods assist the company for better financial stability

ML in offering stability	Frequency	Percent
Strongly disagree	11	7.2
Disagree	13	8.5
Neutral	25	16.3
Agree	52	34
Strongly agree	52	34
Total	153	100

Table 4 Correlation analysis

Correlation	Accelerated decision-making	Improved fraud prevention	Effective credit analysis	Risk management
Accelerated decision-making	1	0.893	0.839	0.842
Improved fraud prevention	0.893	1	0.854	0.84
Effective credit analysis	0.839	0.854	1	0.769
Risk management	0.842	0.84	0.769	1

6.1 Correlation Analysis

Another section of research involves in the investigation of correlation between the independent variable of fraud and the independent variable of risk management in banks. The primary focuses of this investigation are an increase in the protection against fraudulent activity, an improvement in credit analysis, and a speedier decision-making process.

Table 4 demonstrates that there is a significant positive link between making decisions quickly and effectively managing risk (+ 0.842), preventing fraud and effectively managing risk (+ 0.840), and efficiently analysing credit risk and effectively managing risk (+ 0.769).

6.2 Test of Hypothesis Using T Test

The next step in validating the hypothesis involves using T tests to examine the relationships between independent variables and demographic characteristics (Table 5).

Because the majority of the variables have *p* values that are higher than 0.05, the overall study seems to indicate that there is not a significant variance in the means of the enhanced independent factors that contribute to better risk management in banks. This is the conclusion that can be drawn from the findings of the study.

Table 5 T test analysis

Variables	Independent variables	*F*	*P* val.
Gender	Accelerated decision-making	6.352	0.001
	Improved fraud prevention	5.762	0.002
	Effective credit analysis	7.292	0.031
Age	Accelerated decision-making	5.836	0.026
	Improved fraud prevention	6.369	0.034
	Effective credit analysis	9.069	0.029
Type of industry	Accelerated decision-making	8.112	0.001
	Improved fraud prevention	7.543	0.022
	Effective credit analysis	8.121	0.043
Management cadre	Accelerated decision-making	9.226	0.003
	Improved fraud prevention	8.326	0.012
	Effective credit analysis	9.654	0.022
Experience	Accelerated decision-making	9.096	0.026
	Improved fraud prevention	6.543	0.012
	Effective credit analysis	8.121	0.007

7 Conclusion

These findings, despite the fact that they seem to have promised, have to be taken with a grain of salt owing to the fact that the industry is still in its incipient stage of development and the fact that several FIs are now conducting experiments with it in a variety of settings. To begin, machine learning encompasses a broad range of different statistical learning algorithms. These algorithms are able to analyse enormous volumes of data with a high degree of precision and accuracy, often for predictive purposes. These methodologies are used in statistical learning systems to a significant degree. The ability to perform data analysis and infer nonlinear correlations without first having to make any assumptions about the nature of the link between the variables paves the way for more sophisticated research and more precise prediction of the results of the study. This makes it possible to more accurately predict the outcomes of the study (i.e. nonparametric). As there is nothing that needs to be monitored, unsupervised techniques may be used to investigate the data in any manner that seems to be appropriate. When a large number of models are merged and algorithms are executed thousands of times on training data, not only is the capacity to predict outcomes improved, but also the risk of overfitting is mitigated, and the granularity of the analysis is maintained. With the use of these improved, often automated analytic abilities, financial institutions (FIs) may be able to improve their understanding of essential business activities such as lending, risk management, customer engagement, and payment processing. As these systems create more and more data, machine learning has the ability to find more complex patterns and correlations than it has previously been able to. This could occur, for example, during the analysis of transactions or credit risk, or when combining data from many sources to get more comprehensive findings, as would be the case while keeping tabs on behavioural violations. Another scenario in which this could occur is when monitoring compliance with regulations. Second, the application of methods of machine learning to the financial sector creates a strong dependency on elements that are not directly relevant to the problem at hand. There is a possibility that financial organisations may not have sufficient data of a quality suitable for analysis or training. It is essential to keep in mind that there may be a trade-off between the capacity for analysis and prediction that is produced by a variety of methodologies and the resulting complexity of the final model as well as the lack of insight that is provided by the model. It's possible that the problem is due to a compromise between the two requirements.

References

1. Barboza, F., Kimura, H., Altman, E.: Machine learning models and bankruptcy prediction. Exp. Syst. Appl. **83**, 405–417 (2017)
2. Dal Pozzolo, A.: Adaptive machine learning for credit card fraud detection. Unpublished doctoral dissertation, Université libre de Bruxelles, Faculté des Sciences—Informatique, Bruxelles (2015)
3. Attigeri, G.V., Pai, M.M.M., Pai, R.M.: Credit Risk Assessment Using Machine Learning Algorithms. American Scientific Publishers (2017)
4. Hamori, S., Kawai, M., Kume, T., Murakami, Y., Watanabe, C.: Ensemble learning or deep learning? Application to default risk analysis. J. Risk Finan. Manage. **11**, 12 (2018)
5. Jacobs Jr., M.: The validation of machine-learning models for the stress testing of credit risk. J. Risk Manage. Finan. Inst. **11**, 218–43 (2018)
6. Yang, J., Li, Q., Luo, D.: Research on P2P credit risk assessment model based on RBM feature extraction. Open J. Bus. Manage. (2019)
7. Leo, M., Sharma, S., Maddulety, K.: Machine learning in banking risk management: a literature review. Risks. In: MDPI (2019)
8. MetricStream: The chief risk officer's role in 2018 and beyond managing the challenges and opportunities of a digital era new roles of the CRO (2018)
9. Peters, G., Shevchenko, P.V., Cohen, R., Maurice, D.: Statistical machine learning analysis of cyber risk data: event case studies (2017)
10. Van Liebergen, B.: Machine learning: a revolution in risk management and compliance? J. Finan. Transformation **45**, 60–67 (2017)
11. Yao, X., Crook, J., Andreeva, G.: Enhancing two-stage modelling methodology for loss given default with support vector machines. Eur. J. Oper. Res. **263**, 679–689 (2017)

Chapter 45
A Convertor Tool to Transform the Android Mobile Application into Java Codes for Software Testing

Rosziati Ibrahim, Mizani Mohamad Madon, Zhiang Yue Lee, Sapiee Jamel, and Faaizah Shahbodin

1 Introduction

Nowadays, a software tester has many choices of techniques to conduct a software testing. A tester can generate the test cases and testing each module semi manually, such as JUnit testing [1]. The implement of automation tool is necessary in order to reduce cost for testing manually such as Eclipse Plug-in Tool (EPiT) [2], Sena TLS-Parser [3] and jPET [4].

This paper discusses the convertor tool that is used to convert the mobile application (app) into Java source codes. Then, JUnit and EPiT are used for software testing. JUnit can be used for repeatable automated software testing within Eclipse environment. It provides a lot of assert functions for testing share common test data features, expected results, test suites for test organizing and running, and test runner for graphical and textual. An entire or part of object, or interaction between several objects can also be implemented [5]. EPiT can generate test case automatically by examining the Java source codes. It enables the tester to find the high-quality solutions easily [6]. A convertor tool is developed to convert the mobile application in ".apk" format into ".java" source codes. The convertor is used to convert the Android mobile app into Java source codes. Both JUnit and EPiT are the tools under Eclipse environment for Java programming language. In this study, JUnit and EPiT will be

R. Ibrahim · M. M. Madon · Z. Y. Lee · S. Jamel
Universiti Tun Hussein Onn Malaysia, Parit Raja, Malaysia
e-mail: rosziati@uthm.edu.my

S. Jamel
e-mail: sapiee@uthm.edu.my

F. Shahbodin (✉)
Universiti Teknikal Malaysia, Melaka, Malaysia
e-mail: faaizah@utem.edu.my

J. M. R. S. Tavares et al. (eds.), *Proceedings of Second International Conference on Intelligent System*, Algorithms for Intelligent Systems,
https://doi.org/10.1007/978-981-99-8976-8_46

used to test the convertor tool for its effectiveness of generating the test cases. The results of the time taken between Junit and EPiT will be compared to indicate the performance of Junit and EPiT.

The rest of this paper is organized as follows. Section 2 discusses the related work and Sect. 3 presents the UML specification for the convertor tool. Section 4 discusses the results of the comparison between JUnit and EPiT and finally Sect. 5 draws the conclusion and future works of the study.

2 Related Work

The research of Li et al. [7] present DroidBot to support the model-based test input generation with less extra requirements. The main criteria of Droidbot is lightweight UI-guided test input generator for Android apps because it does not need an advanced knowledge on the unexplored code. It gives UI-guided input generation and will generate on-the-fly at runtime. Then, based on the transition model, it will generate the UI-guided test inputs. To reach the effectiveness in most cases, we need to generate the input using depth-first strategy. Wang and Liu [8] stated that a software testing is one of the most important criteria in domestic and abroad software researchers to achieve software quality assurance. However, it spends so much time during the software development process. The particle swarm optimization (PSO) can help in optimizing a problem by trying the process iteratively. The approach is used based on specific measure of quality. Therefore, the PSO algorithm was introduced to generate the software case and to design novel software test case automatic generation algorithm.

The study by Mao et al. [9] introduced an Android testing approach called Sapienz. Sapienz performed better compared with the Android Monkey. It is also a practical testing tool. Other than that, even if only app's APK file can be accessed, Sapienz still supports in multi-level instrumentation.

Ibrahim et al. [2] discussed on generating the test cases with EPiT using Eclipse environment. Based on the research, they found out that EPiT has high reliability and response faster than using manual testing. The use of code smell function in the algorithm gives results that are efficient to automatically generate the test cases based on the source codes [6].

The framework for software testing can also be found in [10] while the software testing for mobile application can be found in [11]. Meta-heuristic algorithms are also used by many researchers for software testing such as [12–20].

3 UML Specification

In this section, the diagrams that are used for the UML specification of the convertor tool will be discussed. Diagrams such as activity diagram, class diagram, sequence diagram and use-case diagram are used for the UML specification. The use-case diagram for the convertor tool is shown in Fig. 1.

Based on Fig. 1, the use cases for the convertor tool are *Import .apk file, Convert .apk file* and *Export Java source codes.* The sequence diagram is shown in Fig. 2 for the convertor tool.

Based on Fig. 2, the sequence diagram follows the use cases for the convertor tool. The three use cases, *Import .apk file, Convert .apk file* and *Export Java source codes* are used for the event in sequence diagram. Figure 3 shows the activity diagram for the convertor tool.

Based on Fig. 3, the activity diagram follows the use cases for the convertor tool. The three use cases, *Import .apk file, Convert .apk file* and *Export Java source codes* are used for the activity in activity diagram. Finally, Fig. 4 shows the class diagram for the convertor tool.

Based on Fig. 4, two classes are used for the convertor tool. They are *user* and *convertor*. The consistency between the diagrams is important to develop the convertor tool correctly. The class diagram follows the use cases for the convertor tool. The three use cases, *Import .apk file, Convert .apk file* and *Export Java source*

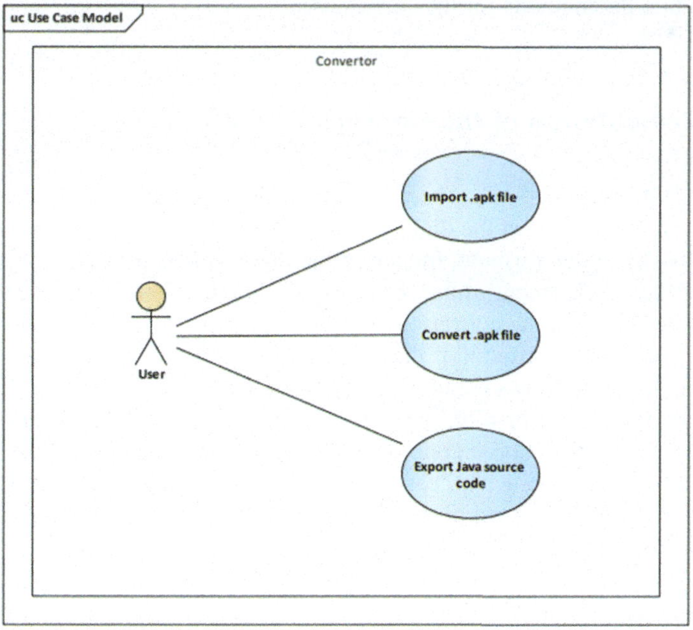

Fig. 1 Use-case diagram for the convertor tool

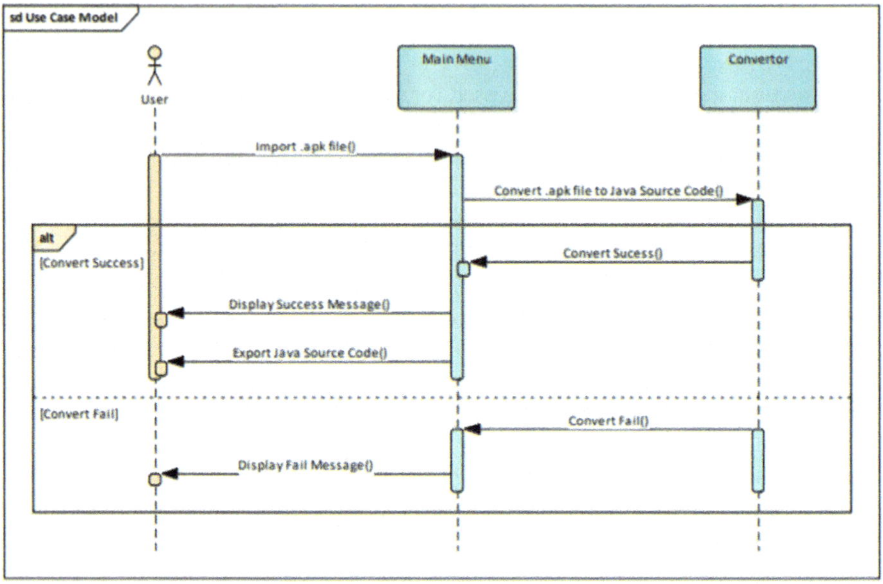

Fig. 2 Sequence diagram for the convertor tool

codes are the important entities to be used for all the diagrams and are then being translated into the methods for the classes used.

4 Implementation of the Convertor Tool

Once the UML specification has been obtained, the implementation for the convertor tool can be done. Based on the diagrams developed previously, the implementation follows closely on the attribute, functions and classes. Java programming language is used for the development of the convertor tool. Figure 5 shows the segmentation of the source codes for the convertor tool.

Simple interface is used for the convertor tool to select the ".apk" mobile application. Figure 6 shows the convertor tool interface.

By using the convertor tool, the Android mobile app in ".apk" format can be converted into ".java" source codes. It is necessary to have the convertor tool in order to convert the mobile application into Java source codes since both JUnit and EPiT are using Eclipse environment. To convert ".apk" to jar, an online source code from git hub is referred. A window batch command (.bat) file will be executed during ".apk" to jar converting process. To convert jar file that has been generated previously to Java source codes, an online existing Java Decompiler, named "jd-gui" is used. Jar file that was generated previously needs to be imported into "jd-gui" and Java source code will be converted.

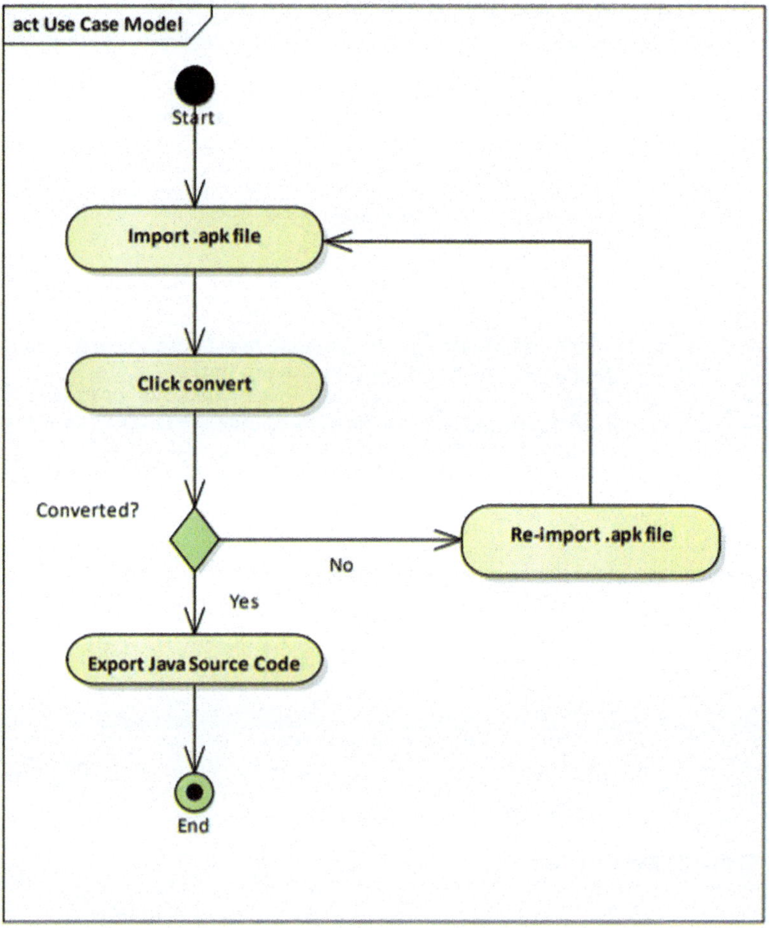

Fig. 3 Activity diagram for the convertor tool

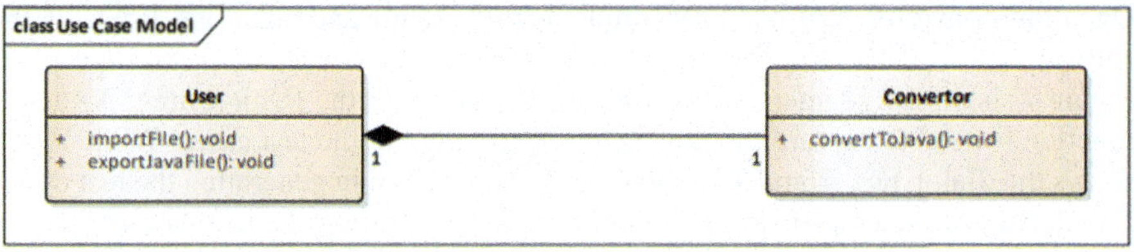

Fig. 4 Class diagram for the convertor tool

```
        int returnVal;

        if (event.getSource() == chooseAPKButton) {
            returnVal = fc.showOpenDialog(null);
            if (returnVal == JFileChooser.APPROVE_OPTION) {
                selectedSourceFile = fc.getSelectedFile();
                apkPath = selectedSourceFile.toString();
                apkName = selectedSourceFile.getName();
                sourceText.setText(apkName);
            }
        }


        if (event.getSource() == chooseJavaDecompilerButton) {
            returnVal = fc.showSaveDialog(null);
            if (returnVal == JFileChooser.APPROVE_OPTION) {
                jdGuiPath = fc.getSelectedFile().toString();
                sourceText2.setText("Jd-Gui Selected");
                System.out.println("jdGuiPath : " +jdGuiPath);
            }
        }
```

Fig. 5 Segmentation of the source codes for the convertor tool

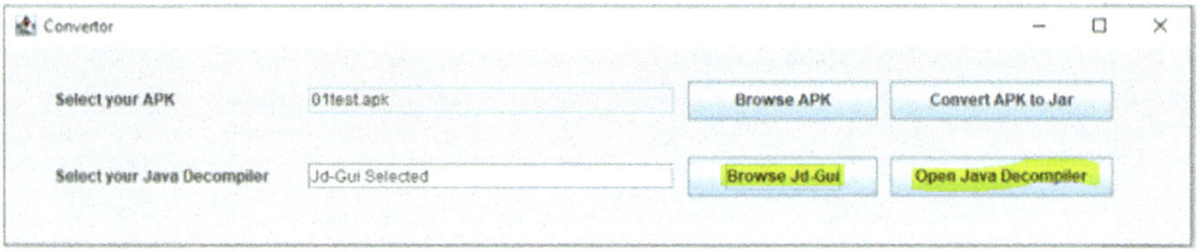

Fig. 6 Convertor tool interface

5 Results and Discussion

Once the convertor tool has been implemented, JUnit and EPiT are used for the software testing. The time taken are used for both tools for comparison purpose. Figure 7 shows the segmentation of the source codes for the JUnit testing. An assert function is used to capture the time taken for producing the test cases. While Fig. 8 shows the dialog box from EPiT showing the time taken in generating the test cases. The details of the steps using the algorithm for generating the test cases based on EPiT can be found in [2]. The details for the algorithm used for the tool can be found in [6].

Both JUnit and EPiT are used to capture the time taken in successfully generating the test cases. Figure 9 shows the results of the JUnit testing with its time taken. Meanwhile, the generated test cases using EPiT are shown in Fig. 10.

Based on Fig. 10, EPiT Console is used to see the generated test cases. EPiT has been developed using Java programming language and can be important as a plug-in

```
Convertor.java   Test.java ×
package application;
import static org.junit.jupiter.api.Assertions.*;

class Test {
    private final ByteArrayOutputStream outContent = new ByteArrayOutputStream();
    private final ByteArrayOutputStream errContent = new ByteArrayOutputStream();
    private final PrintStream originalOut = System.out;
    private final PrintStream originalErr = System.err;

    @Before
    @org.junit.jupiter.api.Test
    public void setUpStreams() {
         System.setOut(new PrintStream(outContent));
            System.setErr(new PrintStream(errContent));
            }

    @After
    public void restoreStreams() {
        System.setOut(originalOut);
        System.setErr(originalErr);
    }

    public void out() {
        System.out.print("01test.apk");
        assertEquals("01test.apk", outContent.toString());
    }

    public void err() {
        System.err.print("01test-dex2jar");
        assertEquals("01test-dex2jar", errContent.toString());
    }
       }
```

Fig. 7 Segmentation of the source codes for the JUnit testing

Fig. 8 Dialog box in EPiT

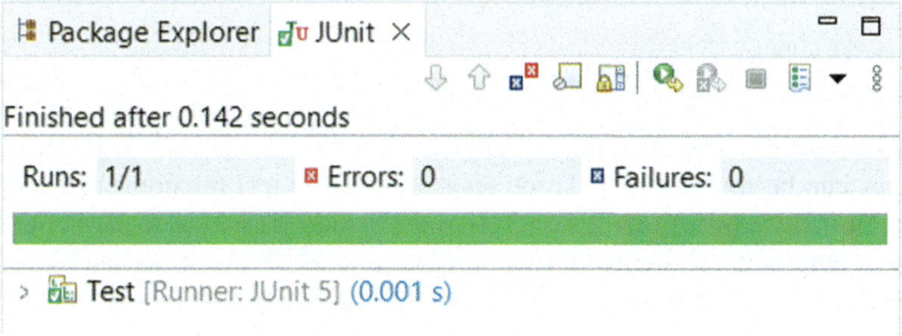

Fig. 9 JUnit testing

```
Console EPiT
Start Time: 2022/06/25 23:14:03
### Working in File Convertor.java ###

Method name :main
Signature :([QString;)V
Return Type :V
Input variable :
args
Generate Test Cases:
Test Case 1 : valid [args]are input with :1
Test Case 2 : invalid [args]are input with :-1
Test Case 3 : null [args]are input with :null

Method name :Convertor
Signature :()V
Return Type :V
Input variable :

Generate Test Cases:
Test Case 1 : valid []are input with :1
Test Case 2 : invalid []are input with :-1
Test Case 3 : null []are input with :null

Method name :actionPerformed
Signature :(QActionEvent;)V
Return Type :V
Input variable :
event
Generate Test Cases:
Test Case 1 : valid [event]are input with :1
Test Case 2 : invalid [event]are input with :-1
Test Case 3 : null [event]are input with :null
End Time: 2022/06/25 23:14:23
Time Elapsed: 1ms
###########################
```

Fig. 10 EPiT testing

tool in Eclipse environment. The details on how to use the tool in generating the test cases can be found in [2]. Three tests are used to test the convertor tool using different mobile applications. Table 1 gives the results for the comparison of the time it takes in generating test cases based on JUnit and EPiT. JUnit is used by using the assert function as shown in Fig. 7. Meanwhile, EPiT is used as a plug-in tool under Eclipse environment.

Table 1 Summarization of results between JUnit and EPiT

Function name	No. of tests	Time taken (ms)	
		JUnit	EPiT
Convertor.java	1	31	1
	2	25	1
	3	27	2
Average time taken (ms)		27.67	1

Based on Table 1, the average time taken for JUnit is 27.67 ms and EPiT is 1 ms. The comparison results of time taken between JUnit and EPiT show that EPiT uses the least time for testing the convertor tool. This shows that the convertor tool can be used to convert the Android mobile app in ".apk" format into ".java" source codes. Then, EPiT can be used to generate the test cases for the Android mobile app once it has been converted into the Java source codes.

6 Conclusion and Future Works

This paper discussed the implementation of the convertor tool using the Java programming language. The convertor tool can be used to convert the mobile application into Java source codes. Once the conversion has been done, EPiT which is under the Eclipse environment for Java can be used to generate the test cases automatically. EPiT has been proven to be an important plug-in tool since it has less time taken for generating test cases automatically.

In the area of software testing, generation of test cases based on mobile application is important. However, in this paper, the convertor tool developed can only be used for Android mobile application. Other convertor tools can also be developed for the conversion of mobile application (Android and iOS) into other programming languages such as Python and C++. In future, the outcome of this study can be used to use the meta-heuristic techniques for the software testing. Some of them are Black Hole Algorithm, Artificial Fish School Algorithm, Firefly Algorithm and Cuckoo Search Algorithm.

References

1. JUnit. https://junit.org/junit4/
2. Ibrahim, R., Amin, A.A.B., Jamel, S., Wahab, J.A.: EPiT: a software testing tool for generation of test cases automatically. Int. J. Eng. Trends Technol. **68**(7), 8–212 (2020). https://doi.org/10.14445/22315381/IJETT-V6817P202S
3. Ibrahim, R., AbuSalim, S.W.G., Jamel, S., Wahab, J.A.: Sena TLS-parser: a software testing tool for generating test cases. Int. J. Adv. Comput. Sci. Appl. (IJACSA) **13**(6), 397–403 (2022). https://doi.org/10.14569/IJCSA.2022.0130649

4. Albert, E.L.: jPET: an automatic test-case generator for java. In: 18th Working Conference on Reverse Engineering, pp. 441–442 (2011)
5. JUnit Tutorial. https://enos.itcollege.ee/~jpoial/java/naited/JUnit-Tutorial.pdf
6. Ibrahim, R., Ahmed, M., Jamel, S.: An eclipse plugin tool for generating test cases from source codes. In: Proceedings of the 2019 Asia Pacific Information Technology Conference (2019). https://doi.org/10.1145/3314527.3314535
7. Li, Y., Yang, Z., Guo, Y., Chen, X.: DroidBot: a lightweight UI-Guided test input generator for Android. In: 2017 IEEE/ACM 39th International Conference on Software Engineering Companion (ICSEC), pp. 23–26 (2017). https://doi.org/10.1109/ICSE-C.2017.8
8. Wang, Z., Liu, Q.: A software test case autoatic generation technology based on the modified particle swam optimization algorithm. In: 2018 International Conference on Virtual Reality and Intelligent Systems (ICVRIS), pp. 156–159 (2018). https://doi.org/10.1109/ICVRIS.2018.00045
9. Mao, K., Harman, M., Jia, Y.: Sapienz: multi-objective automated testing for Android applications. In: Proceedings of the 25th International Symposium on Software Testing and Analysis (ISSTA 2016), pp. 94–105 (2016). https://doi.org/10.1145/2931037.29310654
10. Keyvanpour, M., Homayouni, H., Shirazee, H.: Automatic software test case generation: an analytical classification framework. Int. J. Softw. Eng. Appl. **6**(4) (2012)
11. Salihu, I.A., Ibrahim, R., Ahmed, B.S., Zamli, K.Z., Usman, A.: AMOGA: a static-dynamic model generation strategy for mobile apps testing. IEEE Access (2019). https://doi.org/10.1109/ACCESS.2019.2895504
12. Xing, Y., Wang, X., Shen, Q.: Test case prioritization based on Artificial Fish School Algorithm. Comput. Commun. **180**, 295–302 (2021). https://doi.org/10.1016/j.comcom.2021.09.014
13. Kumar, S., Datta, D., Singh, S.K., Azar, A.T., Vaidyanathan, S.: Black hole algorithm and its applications. Stud. Comput. Intell. **575**, 147–170 (2015). https://doi.org/10.1007/978-3-319-11017-2_7
14. Su, W., Li, Z., Wang, Z., Yang, D.: A meta-heuristic test case prioritization method based on hybrid model. In: Proceedings of 2020 International Conference on Computer Engineering and Application, ICCEA 2020, pp. 430–435 (2020). https://doi.org/10.1109/ICCEA50009.2020.00099
15. Chaudhary, S., Jatain, A.: Performance evaluation of clustering techniques in test case prioritization. In: International Conference on Computational Performance Evaluation (ComPE), North-Eastern Hill University, Shillong, Meghalaya, India (2020)
16. Srivatsava, P.R., Mallikarjun, B., Yang, X.S.: Optimal test sequence generation using firefly algorithm. Swarm Evol. Comput. **8**, 44–53 (2013). https://doi.org/10.1016/j.swevo.2012.08.003
17. Mohapatra, D.P., Sahoo, R.K., Patra, M.R.: A firefly algorithm based approach for automated generation and optimization of test cases. Int. J. Comput. Sci. Eng. **4**(8), 54–58 (2016)
18. Khatibsyarbini, M., Isa, M.A., Jawawi, D.M.A., Hamed, H.N.A., Mohamed Suffian, M.D.: Test case prioritization using firefly algorithm for software testing. IEEE Access **7**, 132360–132373 (2019). https://doi.org/10.1109/ACCESS.2019.2940620
19. Yang, Z.S.: Multiobjective firefly algorithm for continuous optimization. Eng. Comput. **29**(2), 175–184 (2013). https://doi.org/10.1007/s00366-012-0254-1
20. Ibrahim, I.S., Ibrahim, R., Mat Rejab, M.: Combinatorial black hole algorithm: a metaheuristic approach for combinatorial testing. Int. J. Eng. Trends Technol. **71**(4), 21–28 (2023). https://doi.org/10.14445/22315381/IJETT-V71I4P203

Chapter 46
Genetic Algorithm-Based Neural Network for Vegetable Price Forecasting on E-Commerce Platform: A Case Study in Malaysia

Kar Yan Choong, Suhizaz Sudin, Rafikha Aliana A. Raof, and Rhui Jaan Ong

1 Introduction

The agricultural sector in Malaysia faces several challenges. These include unpredictable weather, limited access to modern farming practices and technology, an aging farming population, low productivity, inadequate infrastructure, land tenure issues, and the impact of climate change [1]. Unpredictable weather like droughts and floods can harm crop production. Limited access to modern farming practices hampers productivity. An aging farming population and lack of interest from the younger generation also contribute to low productivity. Inadequate infrastructure and logistics, such as transportation and storage cause post-harvest losses [2]. Land tenure issues, like unclear land ownership, create uncertainties for farmers [3]. Lastly, climate

K. Y. Choong · S. Sudin (✉) · R. A. A. Raof · R. J. Ong
Faculty of Electronic Engineering and Technology, Universiti Malaysia Perlis, Pauh Putra Campus, Arau, Malaysia
e-mail: suhizaz@unimap.edu.my

K. Y. Choong
e-mail: karyan@studentmail.unimap.edu.my

R. A. A. Raof
e-mail: rafikha@unimap.edu.my

R. J. Ong
e-mail: rhuijaan@studentmail.unimap.edu.my

S. Sudin · R. A. A. Raof
Sports Engineering Research Center, Universiti Malaysia Perlis, Pauh Putra Campus, Arau, Malaysia

J. M. R. S. Tavares et al. (eds.), *Proceedings of Second International Conference on Intelligent System*, Algorithms for Intelligent Systems,
https://doi.org/10.1007/978-981-99-8976-8_47

change effects, such as temperature rise and changing rainfall patterns threaten agricultural productivity [4]. It is important to address these challenges for the sustainable development of the agricultural sector in Malaysia and hence the suggestion of the agricultural E-Commerce platform with price forecasting model.

Price forecasting has been a subject of interest for decades, driven by the desire to anticipate market trends and fluctuations. Traditional methods, such as time series analysis and econometric models have provided valuable insights into price patterns. However, these approaches often struggle to capture the complex relationships and nonlinear dynamics inherent in price data [5, 6]. As a result, the accuracy and reliability of traditional methods are limited when facing intricate market conditions and rapidly changing consumer behaviors.

Neural networks have emerged as powerful tools for price forecasting, offering the capability to learn intricate patterns and dependencies within historical price data. Inspired by the human brain's neural structure, neural networks consist of interconnected nodes, or artificial neurons, organized in layers [7]. By training on historical price data, neural networks can identify hidden patterns, nonlinear relationships, and temporal dependencies, enabling them to generate forecasts with improved accuracy.

However, a common challenge encountered when training neural networks is the problem of overfitting. Overfitting occurs when a model becomes too specialized in the training data and fails to generalize well to unseen data [8]. This phenomenon often leads to poor forecasting performance, as the model becomes overly sensitive to noise and specific features within the training dataset. To address this issue and enhance the generalization ability of neural networks, researchers have turned to genetic algorithms.

Genetic algorithms, inspired by the principles of natural selection and evolution, are optimization techniques that mimic the process of genetic adaptation over successive generations [9]. They operate by iteratively generating and evaluating a population of potential solutions, applying evolutionary operators such as selection, crossover, and mutation to improve the fitness of individuals. Genetic algorithms offer the advantage of exploring vast solution spaces and can identify optimal architectures and parameter configurations for neural networks.

In the effort of handling the overfitting problem of neural network, there is the suggestion of combining the genetic algorithms with neural network [10, 11]. The integration of genetic algorithms and neural networks offers a promising solution to the challenges faced in price forecasting, paving the way for more accurate and reliable predictions in the context of E-Commerce platforms. The genetic algorithm optimizes the neural network's architecture, selecting the most relevant features and tuning the parameters to improve generalization performance.

This study introduces a new approach called the genetic algorithm-based neural network (GANN) for predicting vegetable prices on E-Commerce platforms using the historical monthly vegetable price data collected from the Federal Agricultural Marketing Authority (FAMA), Malaysia. By combining neural networks and genetic algorithms, the GANN approach aims to improve prediction accuracy by capturing intricate patterns and optimizing the network's structure and parameters. To evaluate its effectiveness, the GANN technique is compared with the widely used multilayer perceptron (MLP) technique.

The results of this study contribute to the advancement of price forecasting methodologies in E-Commerce, benefiting online retailers and customers because it has the potential to revolutionize pricing strategies and help in decision-making process. In conclusion, the integration of neural networks and genetic algorithms represents a significant breakthrough in price forecasting.

2 Literature Reviews

Price forecasting is a critical aspect of various industries, including E-Commerce, finance, and supply chain management. Accurate predictions of future prices enable businesses to make informed decisions, optimize pricing strategies, and enhance competitiveness. Traditional approaches to price forecasting, such as time series analysis and econometric models have provided valuable insights into historical price patterns. However, these methods often struggle to capture the complex relationships, nonlinear dynamics, and evolving market conditions, limiting their accuracy and effectiveness [12, 13].

In recent years, researchers have turned to neural networks as a promising alternative for price forecasting. Neural networks, inspired by the structure and functioning of the human brain, have demonstrated their ability to capture intricate patterns and dependencies within historical price data [14]. By training on large datasets, neural networks can identify nonlinear relationships, temporal dependencies, and hidden patterns that are challenging to detect using traditional methods [15]. This makes neural networks well-suited for price forecasting tasks in the context of E-Commerce platforms.

One significant challenge encountered when training neural networks for price forecasting is the problem of overfitting. Overfitting occurs when a model becomes overly specialized in the training data, resulting in poor generalization performance when faced with unseen data [16]. Various regularization techniques have been proposed to address this issue, such as dropout layers, weight decay, and early stopping [17, 18]. However, these methods may not always provide satisfactory solutions.

2.1 Optimization of Genetic Algorithms on Neural Networks

To further enhance the generalization ability of neural networks and address the overfitting issue, the integration of genetic algorithms has been explored. Genetic algorithms, inspired by natural selection and evolution, provide an effective optimization technique for finding optimal solutions in complex problem spaces [19]. These algorithms work by applying genetic operators such as selection, crossover, and mutation to evolve the neural network's architecture and optimize its parameter configurations.

The combination of genetic algorithms and neural networks, known as genetic algorithm-based neural networks (GANN), has shown promise in various applications, including price forecasting. GANN utilizes genetic algorithms to optimize the network's structure, select relevant features, and tune parameters, thereby improving the forecasting accuracy [20]. The genetic algorithm acts as a powerful tool for feature selection, allowing the neural network to focus on the most relevant factors for price prediction [21]. The GANN approach has been successfully applied in domains such as stock market forecasting [13, 22, 23], flood forecasting [24], load forecasting [25–27], and demand forecasting [28, 29].

The GANN combination addresses the issue of overfitting by employing a population-based approach, guiding the evolution through selection and fitness evaluation, using early stopping and convergence criteria, designing appropriate fitness functions, and utilizing a validation set to monitor generalization performance [10, 11]. By employing these strategies, GANN encourages the discovery of neural network models that generalize well to unseen data and mitigates the risk of overfitting to the training data.

2.2 Application of GANN in E-Commerce

In the context of E-Commerce platforms, accurate price forecasting is crucial for optimizing pricing strategies, improving customer satisfaction, and maximizing profitability. Wang [30] explored the application of the gray neural network with genetic algorithm approach for forecasting sales number in E-Commerce, highlighting its potential to enhance sales strategies and provide valuable insights for online retailers and customers.

Recent studies have applied the GANN approach to price forecasting in various industries. Chen [31] developed a combination of adaptive genetic algorithm (AGA) and long short-term memory (LSTM) neural network, which is AGA-LSTM model for online retail sales volume forecasting, achieving improved accuracy compared with the traditional LSTM model.

While in terms of vegetable price forecasting, various researches are conducted to demonstrate the effectiveness of the GANN approach in improving the accuracy of vegetable price forecasting, showcasing its potential for enhancing pricing strategies in the E-Commerce industry [21, 32, 33].

3 Genetic Algorithm-Based Neural Network (GANN)

A genetic algorithm is designed to mimic the process of natural evolution with the goal of optimizing a set of parameters. These parameters are represented by strings that each correspond to a potential solution for the problem being examined. In the case of the GANN problem, these strings contain information about constructing a neural network. The fitness value is used to measure the quality of each solution.

The main concept behind a genetic algorithm (GA) is to begin with a population of solutions and iteratively generate new generations of solutions that outperform the previous ones. The algorithm follows a simple cycle that consists of four stages: initialization, selection, crossover, and mutation. The idea of combining GA to a neural network is illustrated in Fig. 1.

The price data for spinach *(bayam)*, round wax gourd *(kundur bulat)*, galangal *(lengkuas)*, and holland potatoes *(ubi kentang holland)* were randomly chosen from the available vegetables price data provided by FAMA, Malaysia in the range between 2010 and 2021. This random selection was necessary because not all types of vegetable price data were complete; some had missing information. By using a random approach, we aim to create a fair and representative dataset for our analysis of vegetable prices during this period. In total of 12 years of monthly price data which accounted for 144 observations, it is split into training, validation, and test sets,

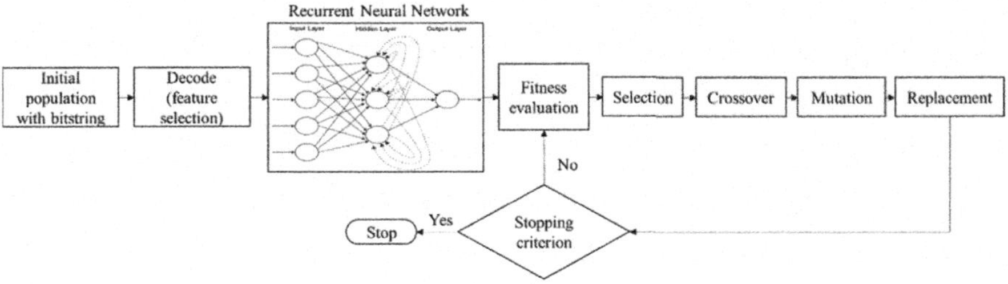

Fig. 1 Steps of operation of GANN

with 50% of 144 observations for training, 25% for validation, and the remaining 25% for testing.

In the proposed GANN, the data and trends are integrated through the process of evolution. It begins with the step of initialization of population of chromosome (bit strings) for the recurrent neural network (RNN) model, which is set to 20 in this research. Each model within the population represents a potential solution to the given problem. In the step of decoding, the chromosome must be decoded to find out which input variables to be used. The RNN's architecture includes an input layer that spans a window of indices before the current value. The previous values are used as input variables for prediction. The window size is tested experimentally in the range of 3–10 for 20 generations in this research, which will affect the number of inputs.

The RNN consists of three hidden layers, each with five nodes and the rectified linear unit (ReLU) activation function with the network configuration [5]. The fitness evaluation of the RNN model is based on means square errors (MSE). The MSE of RNN are measured as the fitness values of GA and then the fitness values are scaled to minimize the MSE error function and create a proportional selection distribution for GA. In the definition of the term, N_{es} is the number of observations to be forecasted and σ_i is the relative percentage error which is the difference between the observed values and the predicted values.

$$\text{MSE} = \frac{1}{N_{\text{es}}} \sum_{i=1}^{N_{\text{es}}} \sigma_i^2 \tag{1}$$

During the selection step, the roulette wheel selection method is employed to pick the most promising RNNs from the current population based on their fitness values. This approach involves assigning the slot to each individual in the population on a roulette wheel, with the size of the slot determined by its fitness value. This process allows for the selection of a reproductive candidate. Typically, solutions with higher fitness values are assigned larger slots on the wheel compared with less-fit solutions.

In the next step, crossover is applied using an arithmetic operator that combines parent values, producing offspring weight and bias matrices. Mutation involves adding random values to each matrix entry using a uniform mutation operator. Generally, the reproduction process involves creating four offspring through crossover and mutating the last two. The best individual with the minimum fitness value in RNN is selected for the mating pool.

Early stopping is employed to prevent overfitting. The stopping criterion for the GA is an increase in MSE for the validation data over three successive generations. As the GA progresses, better-performing RNN models that capture the underlying trends and patterns in the data tend to dominate the population, leading to a more effective neural network architecture and parameters for the specific problem. The best model from the current generation is chosen based on the validation score. The GANN model with the best validation score for each window size is evaluated on the test data using MSE. The best GANN model for a specific window size is then used for future vegetable price forecasting.

To check the forecast performance of the models, different accuracy metrics are chosen including the mean absolute error (MAE), mean absolute percentage error (MAPE), and root mean square error (RMSE). MAE is a common metric used to measure the accuracy of a regression model and it calculates the average absolute difference between the predicted values and the actual values while MAPE is used to measure the percentage accuracy of a regression model and it calculates the average absolute percentage difference between the predicted and the actual values. RMSE is widely used for evaluating the accuracy of a regression model by calculating the square root of the average of the squared differences between predicted and actual values.

The formula of the MAE, MAPE, and RMSE are shown below in which the N_{es} is the number of observations to be forecasted and σ_i is the relative percentage error.

$$\text{MAE} = \frac{1}{N_{es}} \sum_{i=1}^{N_{es}} |\overline{y}_t - y_t| \tag{2}$$

$$\text{MAPE}\,\% = \frac{1}{N_{es}} \sum_{i=1}^{N_{es}} \sigma_i \tag{3}$$

$$\text{RMSE}\,\% = \sqrt{\frac{1}{N_{es}} \sum_{i=1}^{N_{es}} \sigma_i^2} \tag{4}$$

4 Results

The GANN are constructed successfully and their forecasted prices for the spinach *(bayam)*, round wax gourd *(kundur bulat)*, galangal *(lengkuas)*, and holland potatoes *(ubi kentang holland)* are compared with that of multilayer perceptron (MLP) neural network using the accuracy metrics, including the mean absolute error (MAE), mean absolute percentage error (MAPE), and root mean square error (RMSE). The measurement of MAPE, which is in % form, indicates the forecasting accuracy of the forecasting model.

Table 1 Forecasting accuracy metrics of GANN versus MLP

Model	GANN			MLP		
Vegetable	MAE	MAPE	RMSE	MAE	MAPE	RMSE
Spinach (*Bayam*)	0.2593	6.5248	0.3465	0.3128	7.8113	0.4193
Round wax gourd (*Kundur Bulat*)	0.08962	2.796	0.1237	0.1047	3.3388	0.1412
Galangal (*Lengkuas*)	0.1572	2.7514	0.206	0.2221	3.7929	0.2838
Holland potatoes (*Ubi Kentang Holland*)	0.06237	1.9716	0.0803	0.09584	3.0628	0.1227

Table 1 tabulated the performance measurements, which are the MAE, MAPE, and RMSE of the GANN and MLP for all four selected vegetables. All the MAE, MAPE, and RMSE from the forecasted price using GANN are lower than that of MLP for all four types of vegetables, which means GANN performed better than MLP in terms of vegetables price forecasting in this research. MAPE is beneficial for accessing the accuracy in percentage terms and comparing the performance across different datasets or between different models. In the detailed explanation, the MAPE of GANN for spinach *(bayam)* was recorded as 6.5248%, which equals to the 93.48% forecasting accuracy, while the MAPE of MLP for the same vegetable was only recorded as 92.19% accuracy (MAPE = 7.8113%).

Figure 2 illustrated the graphs for all four chosen vegetables using both the GANN and MLP. After the statistical comparison through the performance measurement in Table 1, Fig. 2 shows the obvious differences in terms of forecasting performance between the GANN and MLP graphically. The forecasted price for spinach *(bayam)*, round wax gourd *(kundur bulat)*, galangal *(lengkuas)*, and holland potatoes *(ubi kentang holland)* using GANN achieved the similar trend and seasonality as the original data, while the forecasted price of MLP can be seen lacking. For example, the graph of the forecasted spinach *(bayam)* price using GANN exhibits a similar trend, closely mirroring the actual data, which contrasts with that of using MLP.

Figure 3 displayed the screenshot of the newly designed and developed agricultural E-Commerce platform with the forecasted price graph using price forecasting model. The development of agricultural E-Commerce did not cover in this research as it is only used for illustration purpose.

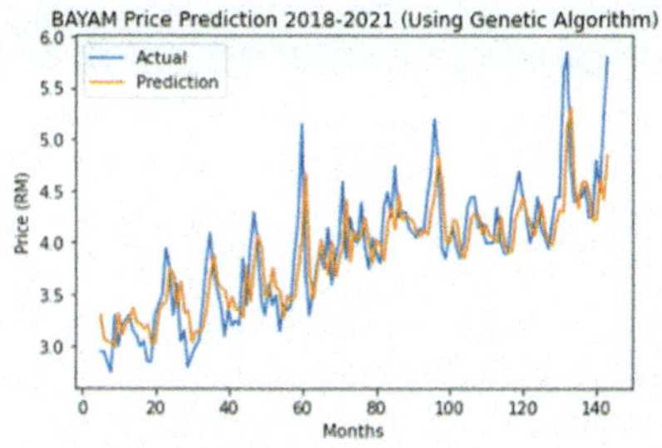

(a) Spinach *(Bayam)*

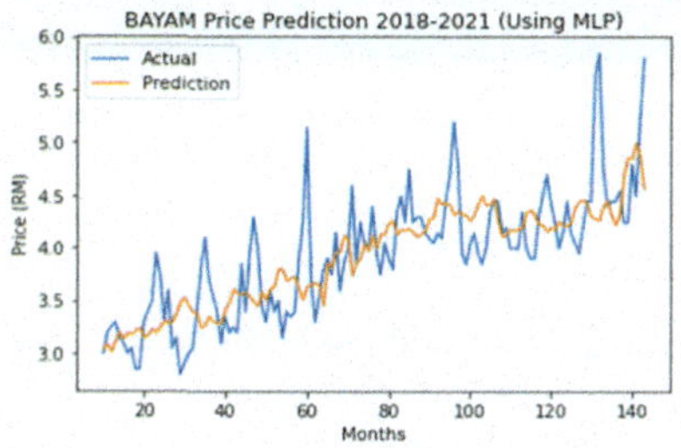

(a) Spinach *(Bayam)*

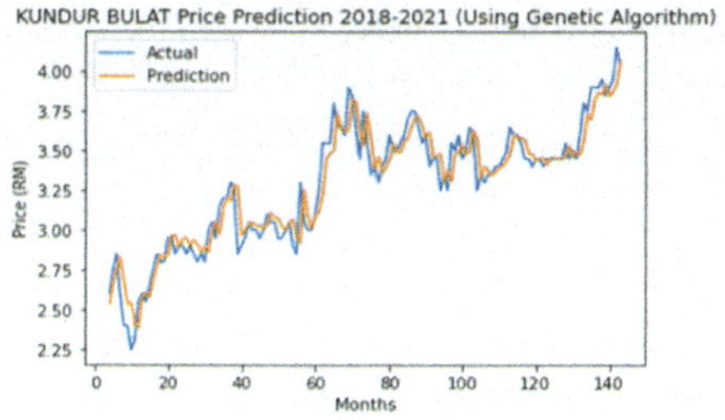

(b) Round Wax Gourd
(Kundur Bulat)

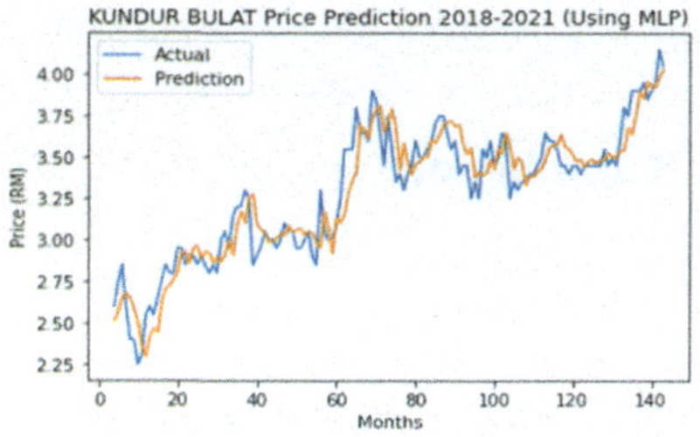

(b) Round Wax Gourd
(Kundur Bulat)

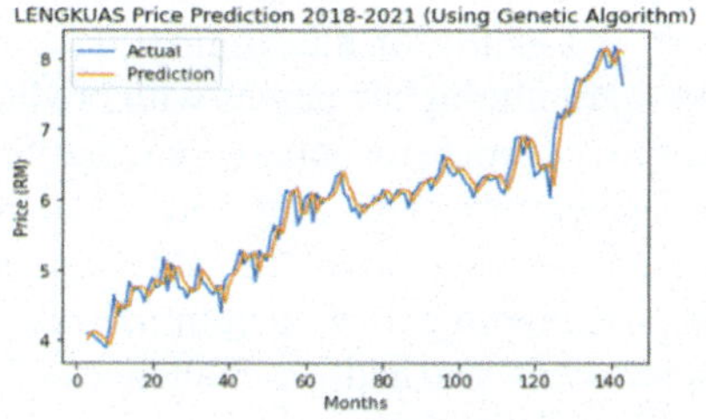

(c) Galangal *(Lengkuas)*

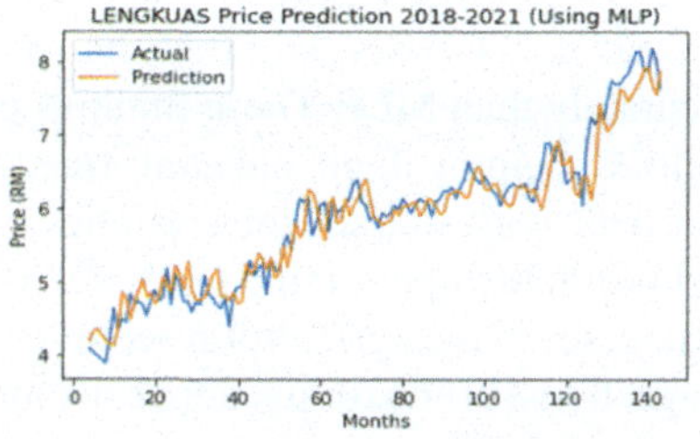

(c) Galangal *(Lengkuas)*

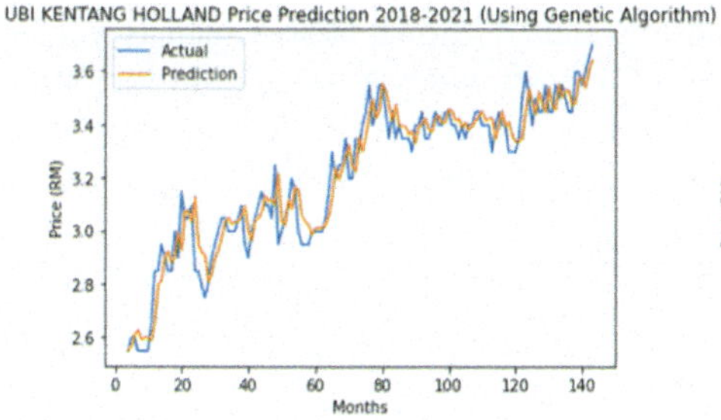

(d) Holland Potatoes
(Ubi Kentang Holland)

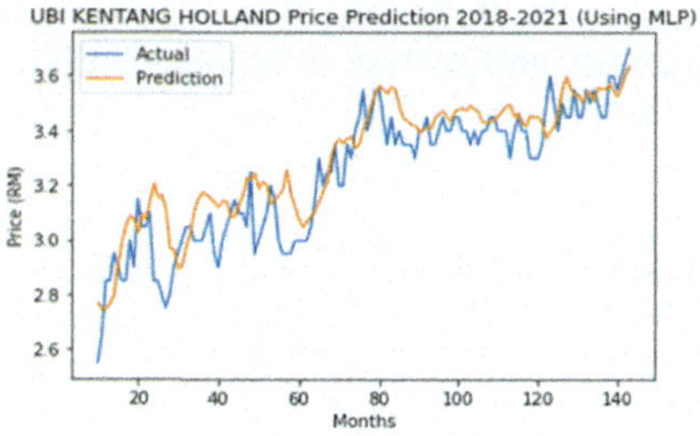

(d) Holland Potatoes
(Ubi Kentang Holland)

Fig. 2 Graphs of the forecasted price using GANN versus MLP

Fig. 3 Agricultural E-Commerce platform with price forecasting model

5 Conclusion

This research introduced a GANN for predicting vegetable prices on E-Commerce platforms in Malaysia. The study compared GANN with the MLP neural network, using accuracy metrics such as MAE, MAPE, and RMSE. The results revealed that GANN outperformed MLP in forecasting accuracy for spinach, round wax gourd, galangal, and holland potatoes. All the accuracy metrics were consistently lower with GANN, indicating its superior performance. Graphical representations also showed that GANN captured the trend and seasonality of the original data more accurately than MLP. These findings provide agricultural businesses with predictive abilities, helping them make informed decisions, optimize resources, increase yields, and deal with market uncertainties. This can lead to more efficient, sustainable, and profitable operations. GANN's data-driven approach offers valuable support to businesses in the agricultural sector of Malaysia, enabling them to optimize resource usage, boost productivity, adopt sustainable practices, and address challenges related to unpredictable weather, limited technology access, an aging farming population, and the impact of climate change. Future research can explore the applicability of GANN to other agricultural products and consider additional factors such as weather patterns and market demand. Overall, this study provides a foundation for further advancements in price forecasting techniques, facilitating better decision-making in the agricultural industry within the context of E-Commerce platforms in Malaysia.

Acknowledgements The authors would like to extend their appreciation to the Ministry of Higher Education, Malaysia for their financial support through the Fundamental Research Grant Scheme (FRGS/1/2020/ICT06/UNIMAP/02/1). Additionally, heartfelt thanks are owed to FAMA, Malaysia for generously providing the historical vegetable prices dataset spanning the years from 2010 to 2021 in Malaysia.

References

1. Alam, M.M., Morshed, G., Siwar, C., Murad, M.W.: Initiatives and challenges of agricultural crop sector in east coast economic region (ECER) development projects in Malaysia. **12**(7), 922–931 (2012)
2. Ali, R., Jaafar, H.S., Mohamad, S.: Logistics and supply chain in Malaysia: issues and challenges. **2008**, 1–11 (2008)
3. Ali, A., Hassan, G., Ngah, I., Applanaidu, S.: Agricultural transformation in Malaysia: the role of smallholders and area development. In: Agricultural Transformation and Inclusive Growth, pp. 1–56. The Institute for Agricultural and Food Policy Studies, UPM (2018)
4. Alam, M.M., Siwar, C., Murad, M.W., Toriman, M.E.: Impacts of climate change on agriculture and food security issues in Malaysia: an empirical study on farm level assessment. World Appl. Sci. J. **14**(3), 431–442 (2011)
5. Bradley, E., Kantz, H.: Nonlinear time-series analysis revisited. Chaos **25**(9). https://doi.org/10.1063/1.4917289
6. Tseng, K.-K., Lin, R., Zhou, H., Kurniajaya, K., Li, Q.: Price prediction of e-commerce products through Internet sentiment analysis. Electron. Commer. Res. **18**. https://doi.org/10.1007/s10660-017-9272-9
7. Spair, R.: Mastering Machine Learning: A Comprehensive Guide to Success. Rick Spair (2023). [Online]. Available: https://books.google.com.my/books?id=5l7FEAAAQBAJ
8. Durmus, M.: Machine Learning Pitfalls: A Brief Guide on How to Avoid Common Pitfalls (With Code Samples). Murat Durmus (2023). [Online]. Available: https://books.google.com.my/books?id=-4q0EAAAQBAJ
9. Sivanandam, S.N., Deepa, S.N.: Introduction to Genetic Algorithms. Springer, Berlin, Heidelberg (2007). [Online]. Available: https://books.google.com.my/books?id=wonrLjj2GagC
10. Goncalves, I., Silva, S.: Experiments on controlling overfitting in genetic programming. In: EPIA2011—15th Portuguese Conference on Artificial Intelligence, December, pp. 978–989 (2011)
11. Chiroma, H., et al.: Neural networks optimization through genetic algorithm searches: a review. Appl. Math. Inform. Sci. **11**(6), 1543–1564 (2017). https://doi.org/10.18576/amis/110602
12. Bin Khamis, A., Yee, P.H.: A hybrid model of artificial neural network and genetic algorithm in forecasting gold price. Eur. J. Eng. Res. Sci. **3**(6), 10 (2018). https://doi.org/10.24018/ejers.2018.3.6.758
13. Chen, S., Zhou, C.: Stock prediction based on genetic algorithm feature selection and long short-term memory neural network. IEEE Access **9**, 9066–9072 (2021). https://doi.org/10.1109/ACCESS.2020.3047109
14. Islam, M., Chen, G., Jin, S.: An overview of neural network. **5**, 5 (2019). https://doi.org/10.11648/j.ajnna.20190501.12
15. Hermans, M., Schrauwen, B.: Training and analyzing deep recurrent neural networks. Adv. Neural Inf. Process. Syst. (2013)
16. Ying, X.: An overview of overfitting and its solutions. J. Phys. Conf. Ser. **1168**(2) (2019). https://doi.org/10.1088/1742-6596/1168/2/022022
17. Srivastava, N., Hinton, G., Krizhevsky, A., Sutskever, I., Salakhutdinov, R.: Dropout: a simple way to prevent neural networks from overfitting. J. Mach. Learn. Res. **15**, 1929–1958 (2014)
18. Prechelt, L.: Early Stopping—But When? pp. 53–67 (2012). https://doi.org/10.1007/978-3-642-35289-8_5
19. Mohammadian, M.: Computational Intelligence for Modelling, Control and Automation: Evolutionary Computation and Fuzzy Logic for Intelligent Control, Knowledge Acquisition and Information Retrieval. In: Computational Intelligence for Modelling, Control and Automation'99. IOS Press (1999). [Online]. Available: https://books.google.com.my/books?id=Hu_6PpkROWIC
20. Tong, D.L., Mintram, R.: Genetic Algorithm-Neural Network (GANN): a study of neural network activation functions and depth of genetic algorithm search applied to feature selection. Int. J. Mach. Learn. Cybern. **1**(1–4), 75–87 (2010). https://doi.org/10.1007/s13042-010-0004-x

21. Drachal, K., Pawłowski, M.: A review of the applications of genetic algorithms to forecasting prices of commodities. Economies **9**(1) (2021). https://doi.org/10.3390/economies9010006
22. Sharma, D.K., Hota, H.S., Brown, K., Handa, R.: Integration of genetic algorithm with artificial neural network for stock market forecasting. Int. J. Syst. Assur. Eng. Manage. **13**(s2), 828–841 (2022). https://doi.org/10.1007/s13198-021-01209-5
23. Abraham, R., et al.: Forecasting a stock trend using genetic algorithm and random forest. J. Risk Finan. Manage. **15**(5) (2022). https://doi.org/10.3390/jrfm15050188
24. Wu, C.L., Chau, K.W.: A flood forecasting neural network model with genetic algorithm. Int. J. Environ. Pollut. **28**(3–4), 261–273 (2006). https://doi.org/10.1504/IJEP.2006.011211
25. Hong, Y.Y., Chan, Y.H., Cheng, Y.H., Der Lee, Y., Jiang, J.L., Wang, S.S.: Week-ahead daily peak load forecasting using genetic algorithm-based hybrid convolutional neural network. IET Gener. Transm. Distrib. **16**(12), 2416–2424 (2022). https://doi.org/10.1049/gtd2.12460
26. Al-qaysi, A.M.M., Bozkurt, A.: Network-adaptive neuro-fuzzy inference systems: a case study in Iraq (2023)
27. Liu, H.: The forecast of household power load based on genetic algorithm optimizing BP neural network. J. Phys. Conf. Ser. **1**, 2021 (1871). https://doi.org/10.1088/1742-6596/1871/1/012110
28. Chen, X., Wang, D., Gao, Y., Tian, B.: Analysis of marketing forecasting model based on genetic neural networks: taking clothing marketing as an example. Wirel. Commun. Mob. Comput. **2022**(2022). https://doi.org/10.1155/2022/2387016
29. Wei, W., Chuan, J.: A combination forecasting method of grey neural network based on genetic algorithm. Procedia CIRP **109**, 191–196 (2022). https://doi.org/10.1016/j.procir.2022.05.235
30. Wang, Y.: An empirical study on the prediction of e-commerce sales by optimizing grey neural network model. J. Phys. Conf. Ser. **3**, 2021 (1992). https://doi.org/10.1088/1742-6596/1992/3/032115
31. Chen, K.: An online retail prediction model based on AGA-LSTM neural network. In: 2020 2nd International Conference on Machine Learning, Big Data and Business Intelligence (MLBDBI), pp. 145–149 (2020). https://doi.org/10.1109/MLBDBI51377.2020.00032
32. Luo, C., Wei, Q., Zhou, L., Zhang, J., Sun, S.: Prediction of vegetable price based on neural network and genetic algorithm. IFIP Adv. Inf. Commun. Technol. **346 AICT**(Part 3), 672–681 (2011). https://doi.org/10.1007/978-3-642-18354-6_79
33. Subhasree, M., Priya, C.A.: Forecasting vegetable price using time series data. Int. J. Adv. Res. (Indore) **3**(535), 535–541 (2016)

Chapter 47
A System for Biomedical Image Processing of Brain Tumor via Segmentation and Pattern Recognition

T. Nalini, N. Kumar, V. Thirumurugan, S. Thirumal, and A. Manikandan

1 Introduction

The human body has a natural process of replacing old or damaged cells with new ones. However, sometimes this process can go awry, leading to the unnecessary production of new cells. When these cells accumulate and form a mass of tissue, it results in the development of tumors, which can occur anywhere in the body, including the brain.

T. Nalini (✉)
Department of Computer Science and Engineering, Saveetha School of Engineering, Saveetha Institute of Medical and Technical Sciences, Chennai, India
e-mail: nalinit.sse@saveetha.com

N. Kumar · V. Thirumurugan · S. Thirumal · A. Manikandan
Department of Computer Science and Engineering, Vels Institute of Science, Technology and Advanced Studies (VISTAS), Chennai, India
e-mail: kumar.se@velsuniv.ac.in

S. Thirumal
e-mail: thirumal.se@velsuniv.ac.in

A. Manikandan
e-mail: manikandan.se@velsuniv.ac.in

J. M. R. S. Tavares et al. (eds.), *Proceedings of Second International Conference on Intelligent System*, Algorithms for Intelligent Systems,
https://doi.org/10.1007/978-981-99-8976-8_48

For brain cancers, early diagnosis is crucial, and medical imaging plays a significant role in the treatment process. In recent years, there have been significant advancements in medical imaging, especially in brain imaging. Magnetic resonance imaging (MRI) is a crucial non-invasive and painless technology that provides detailed anatomical information about the brain.

Thanks to computerized image processing techniques, doctors can automatically classify different brain regions and even different types of tumors using MRI data. This allows for the easy detection of lesions in the brain. Additionally, biomedical image processing enables the evaluation of tumor growth over time, providing valuable information for patient care. Surgeons can also benefit from knowing the tumor site before surgery, as it allows for a more precise analysis of the patient's condition before and after the procedure [1–3].

In this context, a model has been developed using machine learning, specifically a technique called random forest (RF), to achieve accurate brain tumor segmentation. Efficiency is crucial in this research and depends on factors such as the quality of the picture database, the segmentation algorithm employed, and the functions utilized. By leveraging these advancements in medical imaging and machine learning, medical professionals can improve brain cancer diagnosis and treatment outcomes significantly.

2 Related Work

A coherence-based autofocus (CSM) technique for brain MRI data separation is presented to determine the exact location of brain malignancies. AMC has grown lots of press since it produces better results than other integrated methods. Medical imaging plays a significant role in the treatment of brain cancers since early diagnosis is critical. In recent years, there have been significant advancements in the field of medical imaging, with a particular focus on brain imaging. One of the most challenging operations in digital image processing is partitioning, especially when it comes to brain images [4].

Partitioning refers to the process of dividing an image into meaningful and distinct regions or segments. In the context of brain imaging, this operation becomes even more complex due to the intricate structures and varying intensities present in brain scans. Accurately identifying and separating different brain regions or abnormalities is essential for effective diagnosis and treatment planning [5].

Researchers and experts in the field have been working diligently to improve the efficiency and accuracy of partitioning techniques for brain images. These efforts are vital for enhancing the capabilities of medical imaging technologies, such as magnetic resonance imaging (MRI), in providing detailed and reliable information about the brain's anatomical features and abnormalities [6].

By overcoming the challenges associated with image partitioning, medical professionals can better analyze brain scans, detect tumors or lesions, and track their growth over time. This progress in medical imaging and digital image processing holds great promise for early detection, precise diagnosis, and improved treatment outcomes for patients with brain conditions and cancers [7].

A solid strategy, on the other hands, can substantially simplify the problem-solving process. Weak or partially degraded algorithms, on the other hands, nearly invariably result in random errors [8].

A cohesion-based self-merging (CSM) technique for the divisions of brain's MRI data is suggested in order to discover the specific location of brain tumor. CSM has gotten a lot of attention since it produces better results than other integration methods. In our previous study, we presented an innovative approach to address the challenges of brain tumor extraction from 2D magnetic resonance imaging (MRI) scans. One key achievement of our work was a significant reduction in noise impact, leading to an increased accuracy in locating tumors. Additionally, our proposed method significantly decreased calculation time due to its simplicity, resulting in lower computational complexity [9–11].

Our approach involved a combination of techniques, including the fuzzy C-means clustering method, classical classifiers, and a sophisticated neural network called convolutional neural network (CNN). Real-time data collection was used in the experiment, covering a wide range of tumor sizes, locations, shapes, and picture intensities, making our results more applicable to diverse clinical scenarios.

For the traditional classifier phase, we implemented six standard classifiers using scikit-learn: support vector machine (SVM), K-nearest neighbor (KNN), multilayer perceptron (MLP), logistic regression, Naive Bayes, and random forest. However, we later switched to legitimate neural networks (CNNs) developed with Keras and Tensorflow, which outperformed the traditional networks, providing even more accurate and reliable brain tumor extraction results [12].

By combining fuzzy C-means clustering, classical classifiers, and convolutional neural networks, our proposed method proved to be a powerful approach for brain tumor extraction from 2D MRI scans. This research represents a significant advancement in the field of medical image processing and has promising implications for early tumor detection and effective treatment planning for patients with brain cancers.

A clustering-based particle swarm optimization (PSO) technique is proposed in a survey, which finds the clusters' center of mass. Within each category, there are brain tumor models obtained by MRI. The support vector machine (SVM) and adaptive boosting (AdaBoost) approaches were used to compare the outcomes of the three different performance measurements. The proposed model's qualitative findings are similar to the results achieved with the other two models, according to the performance analysis. Furthermore, alternative PSO control parameter values are utilized for better algorithmic results [11, 12].

A powerful technique for brain tumor diagnosis and tumor area extraction is proposed in one study. Initially, Bayesian Naïve classification was used in the suggested method for MRI tumor detection. The authors claim that after diagnosis, pooling and border approaches were employed to accurately extract the tumor area, resulting in a diagnosis accuracy of above 99%. Furthermore, test results demonstrated that the suggested technique accurately extracted the tumor region.

3 Proposed Method

In our study, we present a novel approach for the segmentation of brain tumors in 3D MRI images using a random forest (RF)-based randomization technique. This method explores the relevance of different factors, the relationships within the MRI data, and the overall error during the training of the RF classifier. By leveraging these insights, we optimize the partitioning framework, resulting in an improved segmentation performance.

The proposed model was evaluated on the BraTSs 3D MRI dataset, and the results demonstrate that our approach outperforms recent techniques. The dice coefficients produced from the segmentation process indicate higher accuracy and better agreement with ground truth annotations.

Overall, our RF-based randomization approach offers a robust and effective solution for accurately segmenting brain tumors in 3D MRI images. This advancement in medical image processing has significant potential to enhance the diagnosis and treatment planning for patients with brain cancers, providing better outcomes and improving the overall quality of care (Fig. 1).

Random Forest Algorithm

Random forest algorithms are also known as random forest classifiers in machine learning. It's a well-known classification method. The fact that this technique may be utilized as a random forest classification and regression algorithm is one of the most intriguing aspects of it. The RF algorithm is a forest-based machine learning technique. Forests, as we all know, are made up of a variety of tree species. Decision trees are the trees listed above. As a result, the RF algorithm is made up of a random collection or selection of forest trees. It's a new feature of the decision tree algorithm. Basically, the RF algorithm generates a random samples of many decision trees and merges them using cross-validation to produce a more stable and accurate forecast. In general, the greater the numbers of trees in the forest, the more certain and precise the prediction.

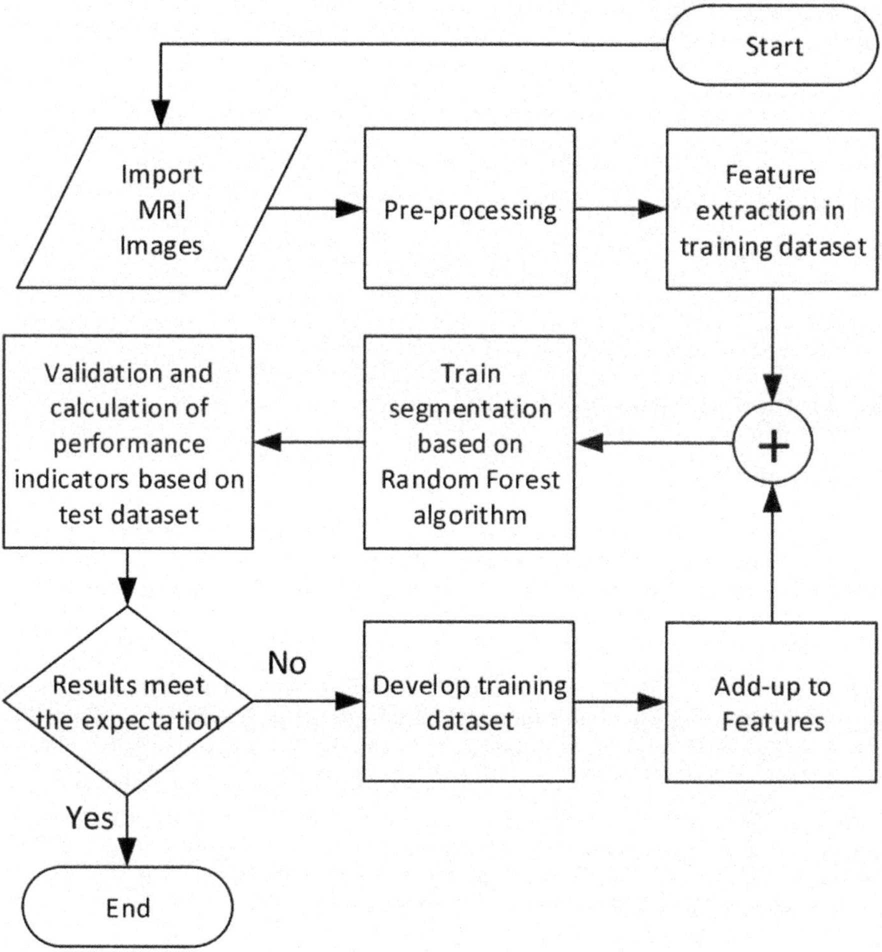

Fig. 1 Proposed approach for segmenting brain tumors is depicted in a flowchart

Importance of This Algorithm

- The random forest approach is applicable to both classification and regression problems.
- It improves accuracy by using cross-validation.
- The random forest classifier can tolerate missing values and keep most data accurate.
- It can handle a large dataset with a larger size.
- It won't install too many plants in the model if there are more trees.

Process of Working

Step 1: From the training set, choose K random data points.

Step 2: Create a decision tree based on the data points you've chosen (subset).

Step 3: Pick an N for the number of decision trees you wish to make.

Step 4: Repetition of steps 1 and 2.

Step 5: For the new data points, locate each decision tree's forecasts and allocate the new data points to the category that received the most votes.

4 Method of Implementation

Image Input Module

Any information or data given to the computer for processing is referred to as input in this module. Simply put, the process of putting data into a computer is known as entry. The system receives brain tumor picture input in order to complete the task.

Preprocessor Module

The signals obtained by the MRI in this module provide a plethora of data that can be used to derive physical and chemical data from the target tissue. The statistical model's recognition concept is linked to the assignment of certain properties to each voxel. With each added property, the database expands linearly, so if someone wishes to use sets with different qualities, the total quantity of data must be processed.

Feature Extraction Module

In this module, the feature extraction approach is carefully selected based on the specific image processing requirements and the target we aim to achieve. The system architecture diagram illustrates how this process is designed and implemented.

For the assessment of textural features in MRI data, we have employed various methods, including first- and higher-order operators, along with gamma correction. These techniques enable us to extract meaningful information from the images, capturing both local and global texture characteristics.

First-order operators focus on basic statistical features such as mean, variance, and intensity distribution, providing insights into pixel-level attributes. On the other hand, higher-order operators delve into more complex texture patterns and relationships between neighboring pixels, enabling the characterization of spatial variations in the images.

Additionally, gamma correction is applied to enhance image quality and compensate for intensity variations, ensuring a more accurate representation of textural details.

By utilizing this comprehensive set of textural feature extraction methods, we can gain valuable insights from the MRI data, leading to improved analysis and interpretation of brain tumors and other abnormalities. This approach enhances the effectiveness of medical image processing and contributes to more accurate diagnosis and treatment planning, ultimately benefiting patients with brain conditions and cancers.

Segmentation Module

The value of image segmentation in the diagnosis of brain cancers cannot be overstated. The more accurate the segmentation algorithm is, the more accurate the system's performance will be. RF has five key properties that make it a good candidate for segmentation: (1) it is capable of handling big databases (2) it has the ability to regulate tens of thousands of variables (3) it calculates the significance of segment variables (4) it reduces the error of unbalanced datasets by balancing the error of unbalanced datasets and (5) it uses the generalized error to build an unbiased estimator. All of these qualities are extremely essential in optimization theory and have a wide range of applications. The RFs classifiers are built on two stochastic processes: the bootstrap set's random structure and the selection of random attributes in each node.

Validation Module

Several significant parameters must be considered when validating the algorithm's performance. The algorithm will compare the image with the trained dataset and make a judgment depending on the condition that has been provided.

5 Results and Discussion

In this module, we propose a random forest (RF) algorithm for our task. The dataset is divided, with 70% of the data allocated for training the model, and the remaining 30% is randomly selected for testing and validation purposes. During the validation phase, we fine-tune the parameters of the RF algorithm based on the evaluation results.

The RF algorithm leverages the power of decision trees to make predictions and combines their outputs to produce a more accurate and robust final result. By training the model on a large portion of the dataset, it can learn patterns and relationships within the data, enabling it to make informed decisions during testing and validation.

After completing the training and parameter optimization steps, the system displays the outcome of the task. This result showcases the effectiveness of the RF algorithm in segmenting brain tumors in the MRI images, providing valuable insights into tumor localization and boundaries.

By adopting this RF-based approach with proper data partitioning and parameter tuning, we ensure a robust and reliable system for brain tumor segmentation in MRI data. The utilization of machine learning techniques, such as random forest, significantly enhances medical image processing capabilities, contributing to more accurate and efficient diagnosis and treatment planning for patients with brain cancers (Fig. 2).

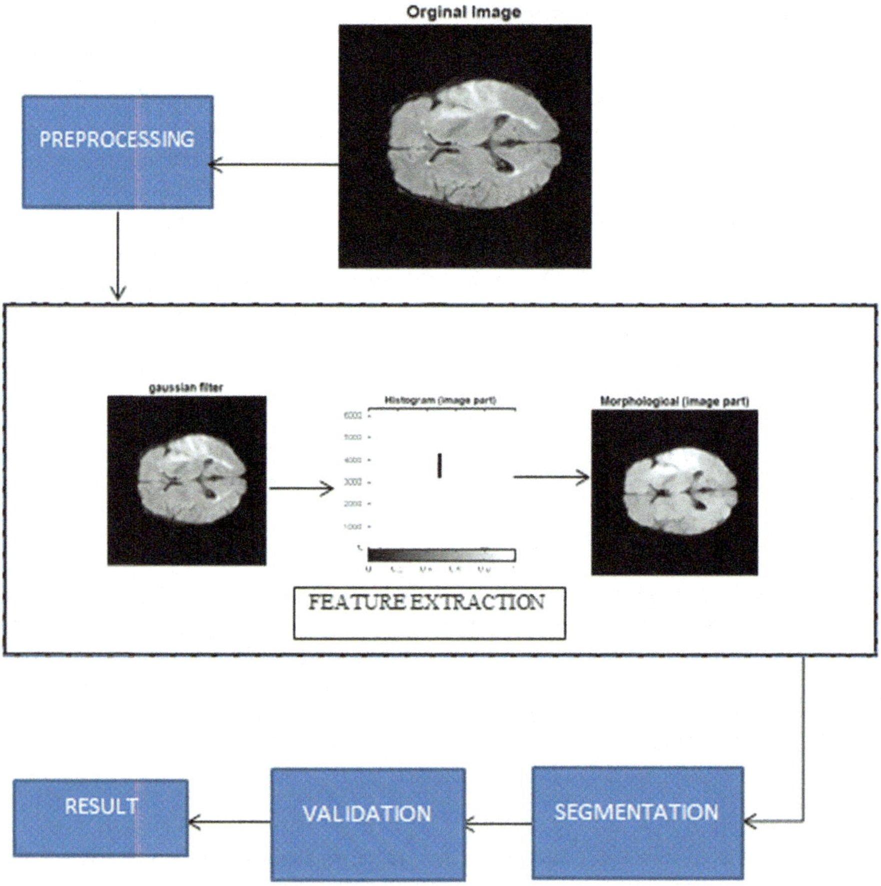

Fig. 2 Methodology of image processing using segmentation

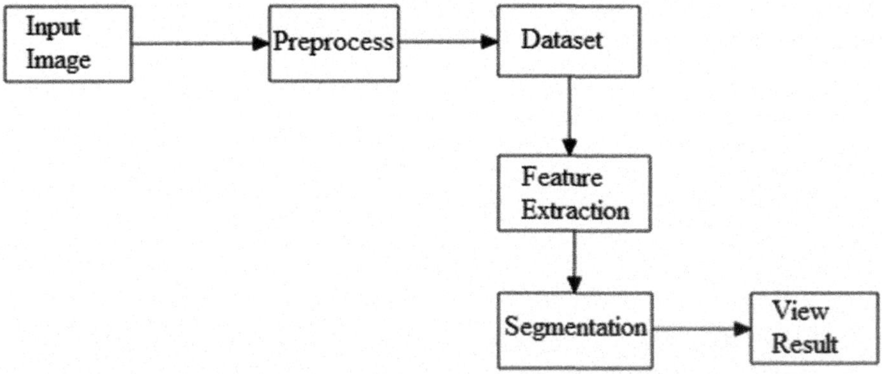

Fig. 3 Data flowchart of brain tumor segmentation

5.1 *Experimental Results*

This study made use of the MRI dataset. The proposed RF algorithms were used to evaluate features extracted from the MRI dataset, including first-order features (means, standard deviations, maximums, minimums, medians, Sobels, gradients), higher-orders features (Laplacian), Gaussian Laplacian, morphology, (entropy, histogram), and texture features (Gabor filter). The correlated results are depicted in the images below.

Data Flow Diagram

Figure 3 using first-order operators shows out-of-bag (OOB) results. The overall inaccuracy is shown in the vertical plot, while the number of trees in the RF method is shown in the horizontal plot (Tables 1 and 2).

Figure 4 shows Gabor filter out-of-bag (OOB) results with texturing characteristics. The overall inaccuracy is shown in the vertical plot, while the number of trees in the RF method is shown in the horizontal plot (Tables 3 and 4; Figs. 5 and 6).

Table 1 First-order operators performance evaluation

DSC	TPR	FPR	ACC
96.06	97.09	27.81	95.39

Table 2 Higher-order operators performance evaluation

DSC	TPR	FPR	ACC
96.28	97.22	25.18	95.51

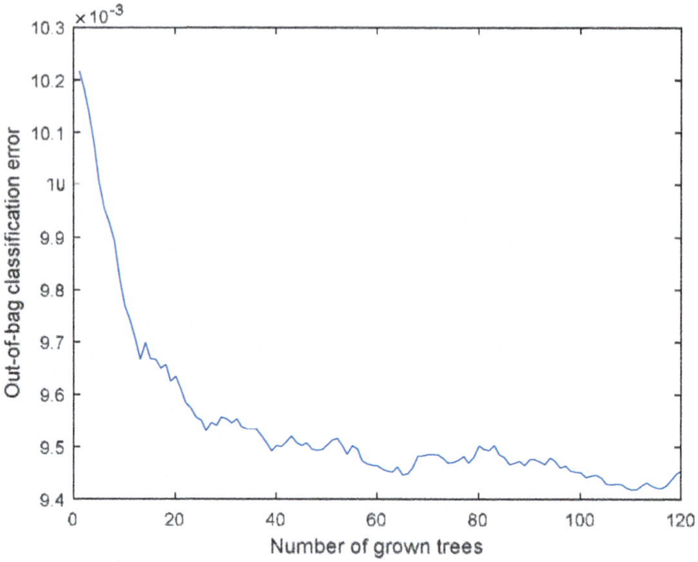

Fig. 4 Gabor filter out-of-bag

Table 3 Texture features performance evaluation

DSC	TPR	FPR	ACC
95.89	97.12	28.02	95.98

Table 4 All features (first-order, high-order, and texture) performance evaluation

DSC	TPR	FPR	ACC
98.38	97.95	19.26	97.65

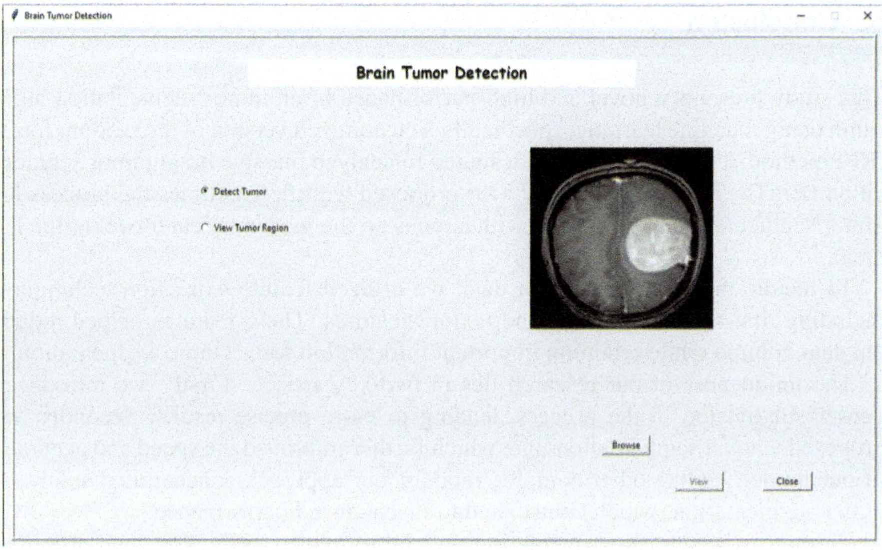

Fig. 5 Input MRI image

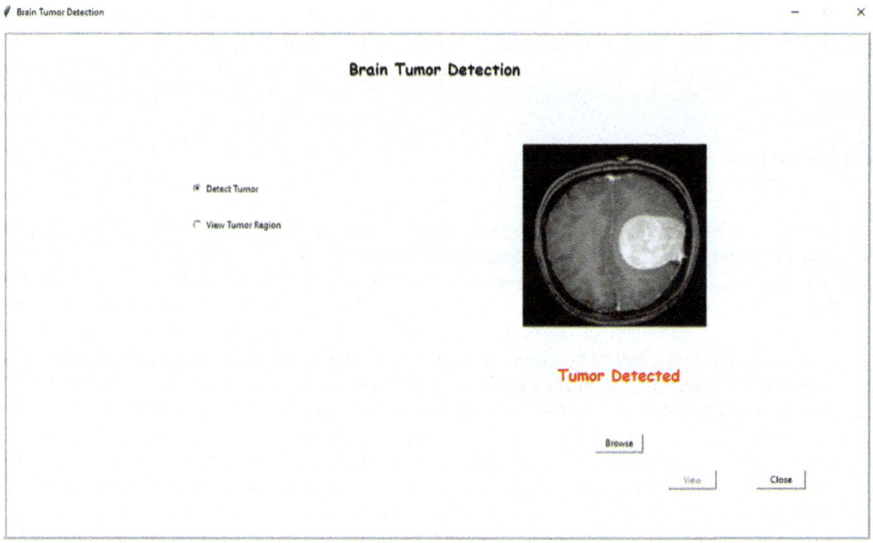

Fig. 6 Tumor detection

6 Conclusion

This study presents a novel and high-performance brain tumor segmentation algorithm using machine learning, specifically a customized version of the random forest (RF) method. The algorithm was designed to analyze massive brain tumor segmentation (BraTS) datasets efficiently. Our proposed workflow outlines the process for tumor segmentation in MRI data, with a focus on the mathematical model of the RF idea.

To handle the large amount of data, we utilized feature extraction techniques, including first-order, top-order, and texture features. These features helped reduce the data volume while retaining important information for accurate segmentation.

The uniqueness of our research lies in two key aspects. Firstly, we introduced denser subdivision in the process, leading to more precise results. Secondly, we proposed a novel segmentation line, which further improved the speed and accuracy of our model. Unlike other complex models, our approach concentrated solely on tumor segmentation, which contributed to its enhanced performance.

Through rigorous experimentation and performance evaluation, our proposed method outperformed existing models employing random forest algorithms. The findings demonstrate that our approach achieves superior accuracy and efficiency in brain tumor segmentation. By focusing on this specific task, we were able to achieve a significant advancement over other sophisticated methods in the field.

References

1. Gotter, A., Healthline Editorial Team: Benign Tumors. Healthline, 4 Aug. 2016. [Online]. Available: https://www.healthline.com/health/benign.
2. Nordqvist, C.: What are the different types of tumors? Medical News Today, 5 Dec. 2017. [Online]. Available: https://www.medicalnewstoday.com/articles/249141.php.
3. Watanabe, K., Tachibana, O., Sato, K., Yonekawa, Y., Kleihues, P., Ohgaki, H.: Overexpression of the EGF receptor and p53 mutations are mutually exclusive in the evolution of primary and secondary glioblastomas. Brain Pathol. **6**(3), 217–223 (1996)
4. National Cancer Institute: Metastatic Cancer. National Cancer Institute, 6 Feb. 2017. [Online]. Available: https://www.cancer.gov/types/metastatic-cancer
5. Huang, A.J., Huang, K.E., Lee, J.W.: Advances in imaging technology and increases in healthcare accessibility have led to shifts in the initial staging of pediatric cancers in the USs. AAP Pediatrics **141**(1) (2018)
6. Heath, F., Hurley, S.A., Berg, H.J., Baptista, C.S.: Advances in noninvasive myelin imaging. Dev. Neurobiol. **78**(2), 136–151 (2018)
7. Khosravi, M., Newberg, A., Al-zaghal, A., Pournazari, K., Arani, L., Jahangiri, P., Werner, T., Alavi, A.: PET/MRI imaging as a practical tool in evaluating mild traumatic brain injury (mTBI). J. Nuclear Med. **59**(1) (2018)
8. Kelly, C.J., Hughes, E.J., Rutherford, M.A., Counsell, S.J.: Advances in neonatal MRI of the brain: from research to practice. In: Archives of Diseases in Childhoods—Education and Practice (2018)
9. Sheffer, J., Koford, J., Kingsbury, J., Powers, M., Wainer, N.: A roundtable discussion: advances in imaging technology are bringing diagnostic tools to the bedside. Biomed. Instrum. Technol. **52**(2), 134–139 (2018)

10. Brown, R.W., Cheng, Y.N., Haacke, M.E., Thompson, M.R., Venkatesan, R.: Magnetic Resonance Imaging: Physical Principles and Sequence Design, 2nd ed. Wiley (2014)
11. Kharabe, S., Nalini, C.: Using adaptive thresholding extraction—robust ROI localization based finger vein authentication. J. Adv. Res. Dyn. Control Syst. **10**(7 Special Issue), 500–514 (2018)
12. Kharabe, S., Nalini, C.: Robust ROI localization based finger vein authentication using adaptive thresholding extraction with deep learning technique. J. Adv. Res. Dyn. Control Syst. **10**(7 Special Issue), 500–513 (2018)

Author Index

J. M. R. S. Tavares et al. (eds.), *Proceedings of Second International Conference on Intelligent System*, Algorithms for Intelligent Systems,
https://doi.org/10.1007/978-981-99-8976-8